LABORATORY EXPERIMENTS

John H. Nelson
University of Nevada, Reno

Kenneth C. Kemp
University of Nevada, Reno

Michael Lufaso
University of North Florida

CHEMISTRY

THE CENTRAL SCIENCE

14TH EDITION IN SI UNITS

BROWN | LEMAY
BURSTEN | MURPHY
WOODWARD | STOLTZFUS

Courseware Portfolio Manager: Terry Haugen
Acquisitions Editor, Global Edition: Sourabh Maheshwari
Managing Producer, Science: Kristen Flatham
Product Marketing Manager: Elizabeth Bell
Content Producer, Science: Beth Sweeten
Assistant Project Editor, Global Edition: Sulagna Dasgupta
Media Production Manager, Global Edition: Vikram Kumar
Full Service Vendor: Cenveo® Publisher Services
Main Text and Supplement Cover Designer: Lumina Datamatics Ltd.
Buyer: Stacey Weinberger
Senior Manufacturing Controller, Production, Global Edition: Caterina Pellegrino

Credits and acknowledgments borrowed from other sources and reproduced, with permission, in this textbook appear on the appropriate page within the text.

Pearson Education Limited
KAO Two
KAO Park
Harlow
CM17 9NA
United Kingdom

and Associated Companies throughout the world

Visit us on the World Wide Web at:
www.pearsonglobaleditions.com

ISBN 10: 1-292-22133-X
ISBN 13: 978-1-292-22133-5

British Library Cataloguing-in-Publication Data
A catalogue record for this book is available from the British Library

10 9 8 7 6 5 4

Typeset by Cenveo® Publisher Services
Printed and bound in Great Britain by Ashford Colour Press Ltd.

Contents

*Approximate time required to complete experiment.

**The numbers in brackets after experiment titles refer to chapter(s) in the fourteenth edition of *Chemistry: The Central Science* by Brown, LeMay, Bursten, Murphy, Woodward, and Stoltzfus that are relevant to the experiment.

alkanes

In Memoriam

Kenneth C. Kemp, a cherished colleague, friend, and coauthor of this laboratory manual, passed away on November 6, 2010. He was born in Chicago, Illinois on August 7, 1925. He obtained his Bachelor's degree in chemistry from Northwestern University and his PhD. from the Illinois Institute of Technology. The two degrees were separated by a three-year stint during World War II in the U.S. Navy as a radar technician in Alaska. He taught chemistry at the University of Nevada, Reno for 35 years where more than 10,000 Introductory Chemistry students benefitted from his bountiful teaching skills. He was known as a rigorous and compassionate instructor. He was named Outstanding Teacher at UNR in 1981, and earlier recognized as an Outstanding Educator in America. His former students will never forget his opening lecture that featured his amazing ability to write forwards and backwards simultaneously with both hands.

Preface

Most students who take chemistry in the first year of their undergraduate studies are not planning for a career in this discipline. As a result, the introductory or general chemistry course usually serves several functions at various levels. It begins the training process for those who seek to become chemists. It introduces nonscience students to chemistry as an important, useful, and, we hope, interesting and rewarding part of their general education. It also should stimulate those students who are seeking the intellectual challenges and sense of purpose they hope to obtain from a career.

This manual has been written with these objectives in mind and to accompany the fourteenth edition of the text *Chemistry: The Central Science* by Theodore L. Brown, H. Eugene LeMay, Jr., Bruce E. Bursten, Catherine J. Murphy, Patrick M. Woodward, and Matthew W. Stoltzfus. Each of the experiments is self-contained, with sufficient background material to conduct and understand the experiment. Each has a pedagogical objective to exemplify one or more specific principles. Because the experiments are self-contained, they may be undertaken in any order; however, we have found for our General Chemistry course that the sequence of Experiments 1 through 7 provides the firmest background and introduction.

To assist the student, we have included prelab questions to be answered before the experiments are begun. These are designed to help the student understand the experiment, to learn how to do the necessary calculations to treat their data, and as an incentive to read the experiment in advance. As a further incentive, answers to some of these questions are provided in Appendix K.

We have made an effort to minimize the cost of the experiments. We have at the same time striven for a broad representation of the essential principles while keeping in mind that many students gain no other exposure to analytical techniques. Consequently, balances, pH meters, and spectrophotometers are used in some of the experiments. A list of necessary equipment and chemicals is given at the beginning of each experiment.

Each of the experiments contains a report sheet that is easily graded, and most experiments contain unknowns. Very few of the experiments may be "dry-labbed."

In this fourteenth edition we have

- carefully edited all experiments for clarity, accuracy, safety, and cost.
- added new experiments concerning solutions, polymers, and hydrates.
- checked all answers to questions for accuracy.

We owe a sincere debt of gratitude to the tens of thousands of students who have tested these experiments and commented on the directions. We wish to thank all of you who have sent us corrections and suggestions for improving these experiments and hope that you continue to make recommendations for their improvement. Please e-mail them to michael.lufaso@unf.edu.

We are particularly appreciative of the many helpful suggestions and criticisms made by reviewers and accuracy checkers throughout these 14 editions:

Ted Clark	The Ohio State University
A. Dale Marcy	North Idaho College
David L. Cedeño	Illinois State University
Eric Goll	Brookdale Community College
Joseph Ledbetter	Contra Costa College
Juma Booker	Fayetteville State University
Kerri Scott	University of Mississippi
Sean Birke	Jefferson College
Thomas R. Webb	Auburn University

We gratefully acknowledge the help of the Pearson Education editorial and production staff. The quality of this manual is the result of their effective skills and expertise.

MICHAEL W. LUFASO
michael.lufaso@unf.edu
University of North Florida

Pearson would like to thank the following people for their contributions to the Global Edition.

Contributor

Burkhard Kirste	Institut für Chemie und Biochemie, Freie Universität Berlin

Reviewer

Burkhard Kirste	Institut für Chemie und Biochemie, Freie Universität Berlin

Mahendra Kumar Sharma

To the Student

You are about to engage in what, for most of you, will be a unique experience. You will be collecting experimental data on your own and using your reasoning powers to draw logical conclusions about the meaning of these data. Your laboratory periods are short, and in most instances, there will not be enough time to come to the laboratory unaware of what you are to do, collect your experimental data, make conclusions and/or calculations regarding them, clean up, and hand in your results. Thus, you should *read the experimental procedure in advance* so that you can work in the lab most efficiently.

After you've read through the experiment, try to answer the prelab questions we've included at the end of each experiment. These questions will help you to understand the experiment, learn how to do the calculations required to treat your data, and give you another reason to read over the experiment in advance. You can check most of your own answers against the answers we've included in Appendix K. Also try to answer the Give It Some Thought questions that we have added to aid you in understanding the principals exemplified by the experiment.

Some of your experiments will also contain an element of *danger*. For this and other reasons, there are laboratory instructors present to assist you. They are your friends. Treat them well, and above all, don't be afraid to ask them questions. Within reason, they will be glad to help you.

Chemistry is an experimental science. The knowledge that has been accumulated through previous experiments provides the basis for today's chemistry courses. The information now being gathered will form the basis of future courses. There are basically two types of experiments that chemists conduct:

1. Qualitative—noting observations such as color, color changes, hardness, whether heat is liberated or absorbed, and odor.
2. Quantitative—noting the amount of a measurable change in mass, volume, or temperature, for example, for which the qualitative data are already known.

This laboratory manual includes both qualitative and quantitative analyses. The former determines what substances are present, and the latter determines the amounts of the substances.

It is much easier to appreciate and comprehend the science of chemistry if you actually participate in experimentation. Although there are many descriptions of the scientific method, the reasoning process involved is difficult to appreciate without performing experiments. Invariably there are experimental difficulties encountered in the laboratory that require care and patience to overcome. There are four objectives for you, the student, in the laboratory:

1. To develop the skills necessary to obtain and evaluate a reliable original result.
2. To record your results for future use.
3. To be able to draw conclusions regarding your results (with the aid of some coaching and reading in the beginning).
4. To learn to communicate your results critically and knowledgeably.

By attentively reading over the experiments in advance, and by carefully following directions and working safely in the laboratory, you will be able to accomplish all these objectives. Good luck and best wishes for an error-free and accident-free term.

Laboratory Safety and Work Instructions

Attention Student! Read the following carefully because your instructor may give you a quiz on this material.

LABORATORY SAFETY

The laboratory can be—but is not necessarily—a dangerous place. When intelligent precautions and a proper understanding of techniques are employed, the laboratory is no more dangerous than any other classroom. Most of the precautions are just common-sense practices. These include the following:

1. Wear *approved* eye protection (including splash guards) at all times while in the laboratory. (*No one will be admitted without it.*) Your safety eye protection may be slightly different from that shown, but it must include shatterproof lenses and side shields to provide protection from splashes.

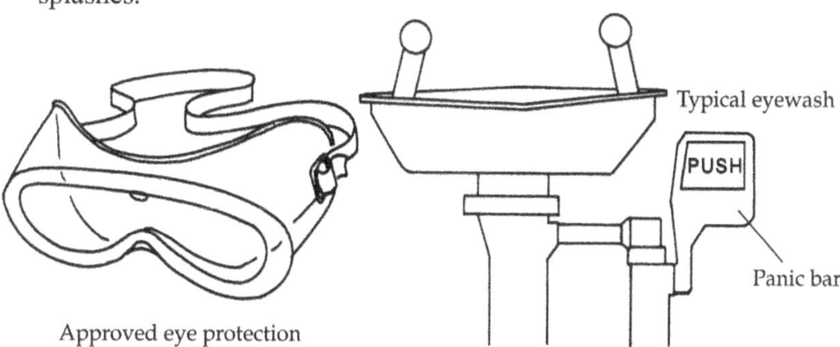

   The laboratory has an eyewash fountain available for your use. In the event that a chemical splashes near your eyes, you should use the fountain ***before the material runs behind your eyeglasses and into your eyes.*** The eyewash has a "panic bar," which enables its easy activation in an emergency.
2. Wear shoes at all times. (*No one will be admitted without them.*)
3. Eating, drinking, and smoking are strictly prohibited in the laboratory at all times.
4. Know where to find and how to use all safety and first-aid equipment (see the first page of this book).
5. Consider all chemicals to be hazardous unless you are instructed otherwise. ***Dispose of chemicals as directed by your instructor.*** Follow the explicit instructions given in the experiments.

6. If chemicals come into contact with your skin or eyes, wash immediately with copious amounts of water and then consult your laboratory instructor.
7. Never taste anything. Never directly smell the source of any vapor or gas. Instead, by means of your cupped hand, bring a small sample to your nose. Chemicals are not to be used to obtain a "high" or clear your sinuses.

8. Perform in the fume exhaust hood any reactions involving skin-irritating or dangerous chemicals, or unpleasant odors. This is a typical fume exhaust hood. Exhaust hoods have fans to exhaust fumes out of the hood

and away from the user. The hood should be used when you are studying noxious, hazardous, and flammable materials. It also has a shatterproof glass window, which may be used as a shield to protect you from minor explosions. Reagents that evolve toxic fumes are stored in the hood. Return these reagents to the hood after their use.

9. Never point a test tube that you are heating at yourself or your neighbor—it may erupt like a geyser.

10. Do not perform *any* unauthorized experiments.
11. Clean up all broken glassware *immediately*.
12. Always pour acids into water, not water into acid, because the heat of solution will cause the water to boil and the acid to spatter. "Do as you oughter, pour acid into water."
13. Avoid rubbing your eyes unless you *know* that your hands are clean.
14. When inserting glass tubing or thermometers into stoppers, *lubricate the tubing and the hole in the stopper with glycerol or water.* Wrap the rod in a towel and grasp it as close to the end being inserted as possible. Slide the glass into the rubber stopper with a twisting motion. Do not push. Finally, remove the excess lubricant by wiping with a towel. Keep your hands as close together as possible in order to reduce leverage.

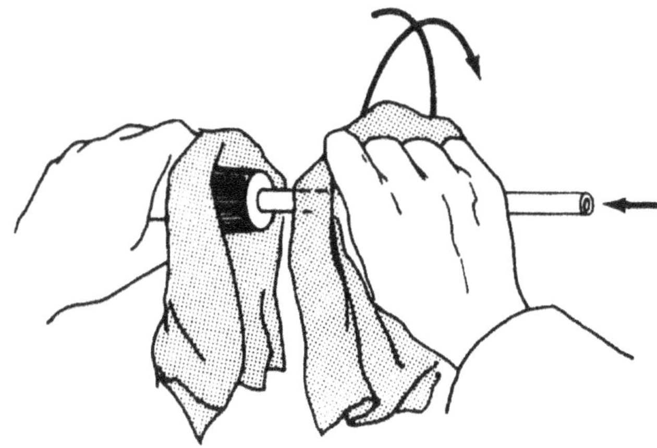

15. For safety purposes, always place the ring stand as far back on the laboratory bench as comfortable, with the long edges of the base perpendicular to the front of the bench.

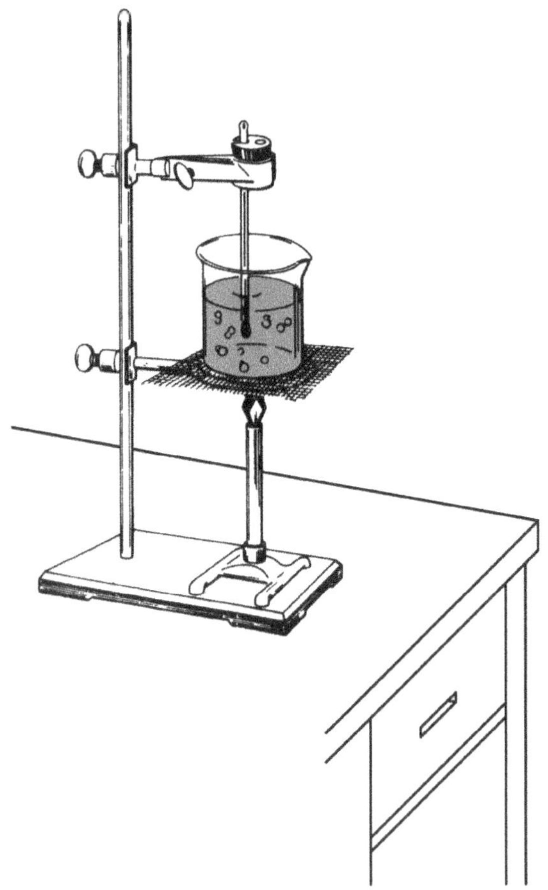

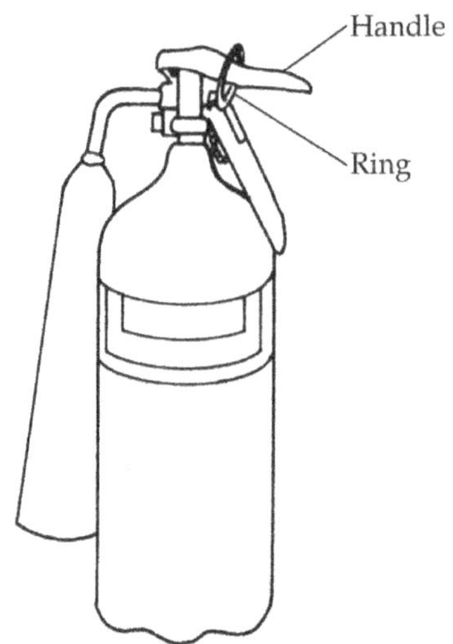

16. NOTIFY THE INSTRUCTOR IMMEDIATELY IN CASE OF AN ACCIDENT.
17. Many common reagents—for example, alcohols, acetone, and especially ether—are highly flammable. *Do not use them anywhere near open flames.*
18. Observe all special precautions mentioned in experiments.
19. Learn the location and operation of fire-protection devices.

In the unlikely event that a large chemical fire occurs, carbon dioxide fire extinguishers are available in the lab (usually mounted near one of the exits in the room). A typical carbon dioxide fire extinguisher is shown on the previous page.

In order to activate the extinguisher, you must pull the metal safety ring from the handle and then depress the handle. Direct the output from the extinguisher at the base of the flames. The carbon dioxide smothers the flames and cools the flammable material quickly. If you use the fire extinguisher, be sure to turn the extinguisher in at the stockroom so that it can be refilled immediately. If the carbon dioxide extinguisher does not extinguish the fire, evacuate the laboratory immediately and call the fire department.

One of the most frightening and potentially most serious accidents is the ignition of one's clothing. Certain types of clothing are hazardous in the laboratory and must *not* be worn. Since *sleeves* are most likely to come closest to flames, ANY CLOTHING THAT HAS BULKY OR LOOSE SLEEVES SHOULD NOT BE WORN IN THE LABORATORY. Ideally, students should wear laboratory coats with tightly fitting sleeves. Long hair also presents a hazard and must be tied back.

If a student's clothing or hair catches fire, his or her neighbors should take prompt action to prevent severe burns. Most laboratories have a water shower for such emergencies. A typical laboratory emergency water shower has the following appearance:

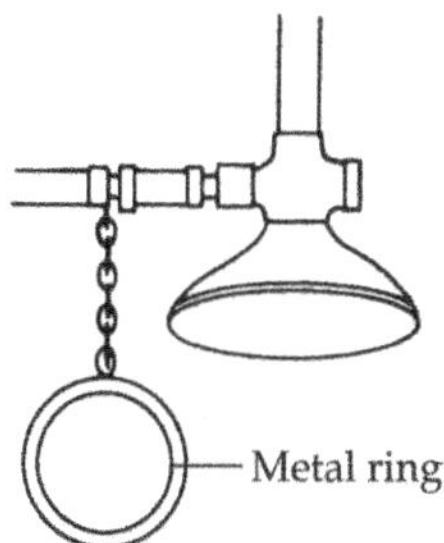

In case someone's clothing or hair is on fire, immediately lead the person to the shower and pull the metal ring. Safety showers generally dump 40 to 50 gallons of water, which should extinguish the flames. These showers generally cannot be shut off once the metal ring has been pulled. Therefore, the shower cannot be demonstrated. (Showers are checked for proper operation on a regular basis, however.)

20. Whenever possible, use hot plates in place of Bunsen burners.

BASIC INSTRUCTIONS FOR LABORATORY WORK

1. Read the assignment *before* coming to the laboratory.
2. Work independently unless instructed to do otherwise.
3. Record your results directly onto your report sheet or notebook. DO NOT RECOPY FROM ANOTHER PIECE OF PAPER.
4. Work conscientiously to avoid accidents.
5. Dispose of excess reagents as instructed by your instructor. NEVER RETURN REAGENTS TO THE REAGENT BOTTLE.
6. Do not place reagent-bottle stoppers on the desk; hold them in your hand. Your laboratory instructor will show you how to do this. Replace the stopper on the same bottle, never on a different one.
7. Leave reagent bottles on the shelf where you found them.
8. Use only the amount of reagent called for avoid excesses.
9. Whenever instructed to use water in these experiments, use distilled water unless instructed to do otherwise.
10. Keep your area clean.
11. Do not borrow apparatus from other desks. If you need extra equipment, obtain it from the stockroom.
12. When weighing, do not place chemicals directly on the balance.
13. Do not weigh hot or warm objects. Objects should be at room temperature.
14. Do not put hot objects on the desktop. Place them on a wire gauze or heat-resistant pad.

"I have read and understand these instructions as well as the laboratory safety and work instructions"

__________________________ ____________

Student Signature Date

COMMON LABORATORY APPARATUS

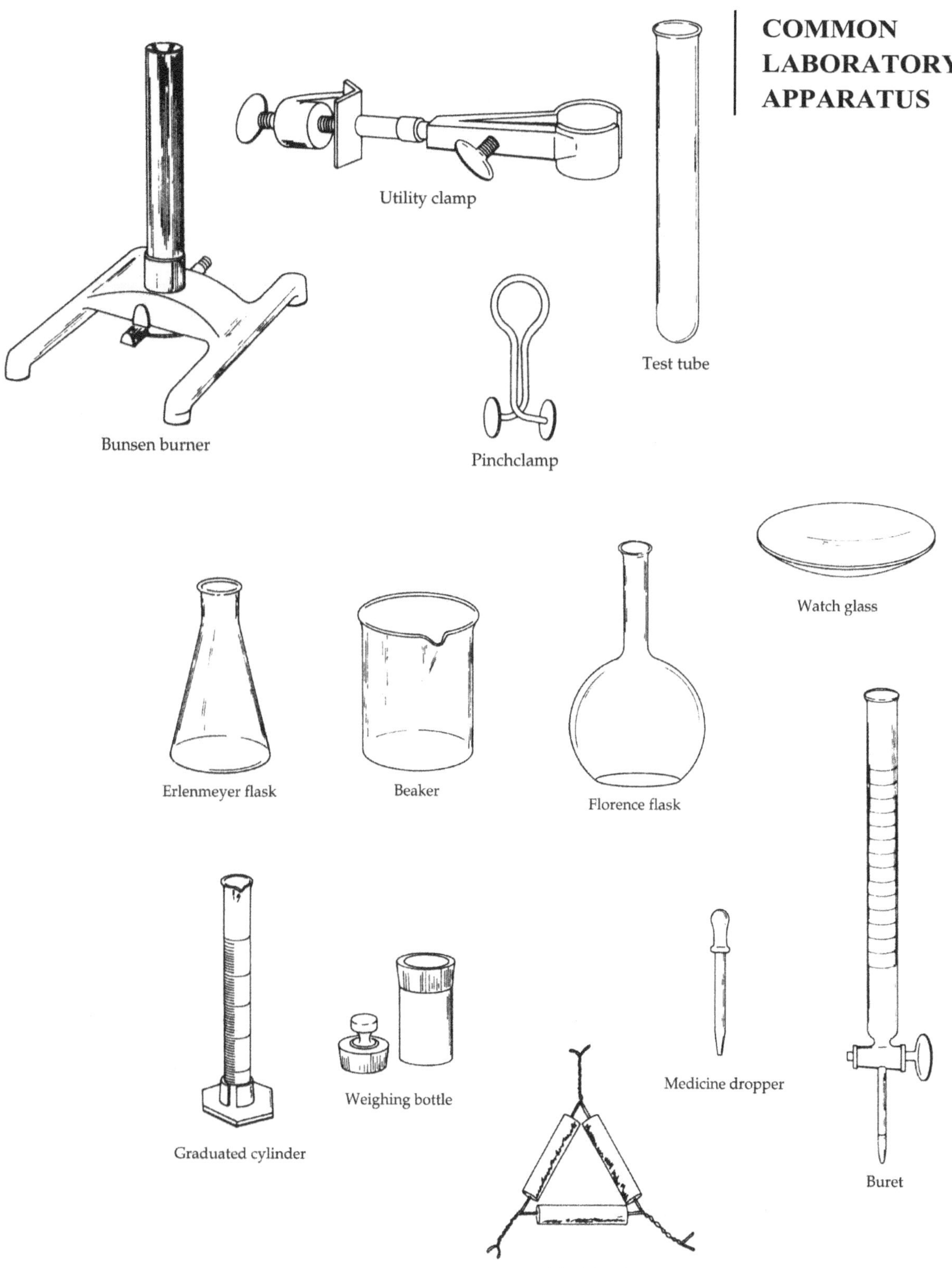

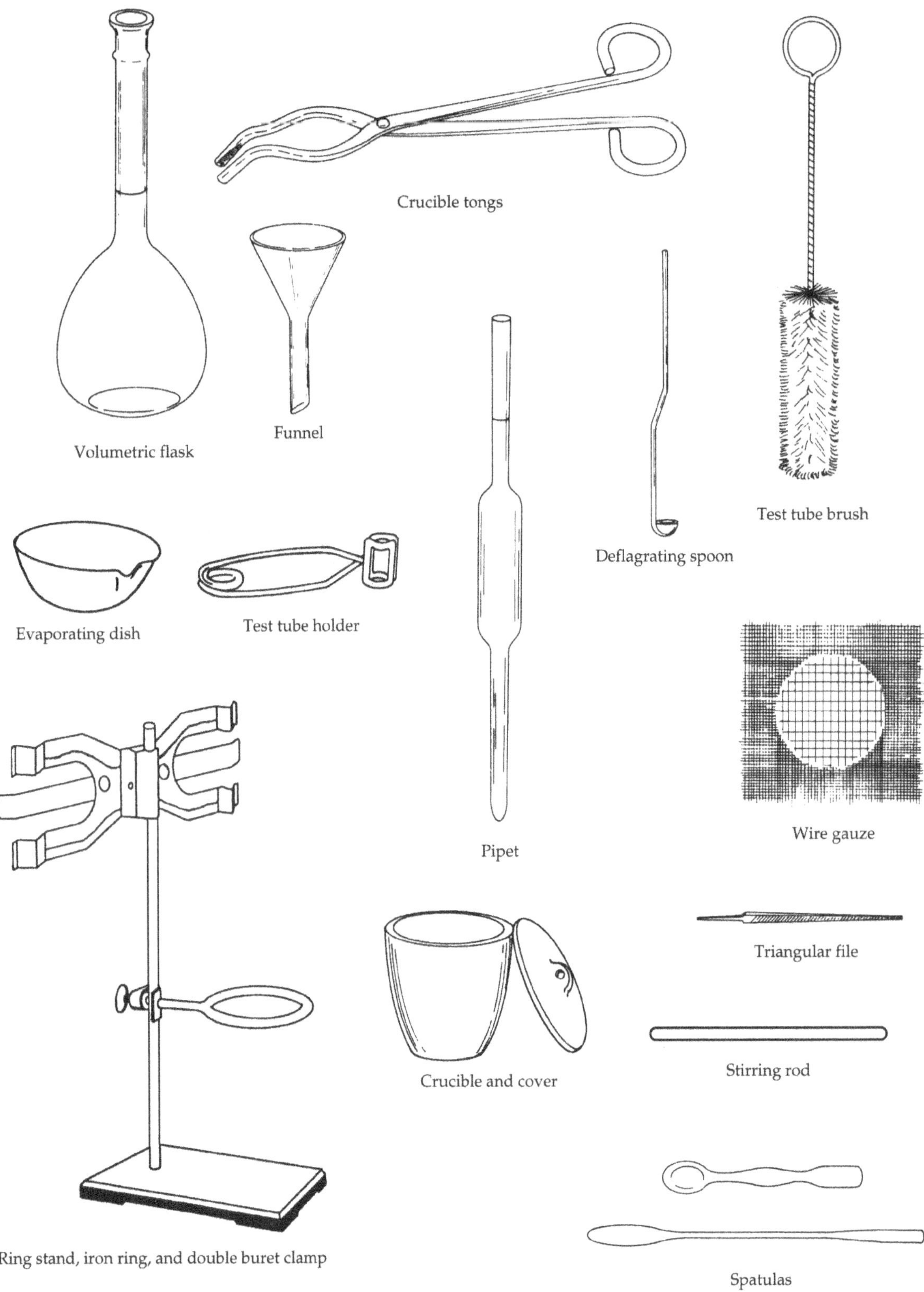
Volumetric flask
Crucible tongs
Funnel
Test tube brush
Deflagrating spoon
Evaporating dish
Test tube holder
Pipet
Wire gauze
Ring stand, iron ring, and double buret clamp
Crucible and cover
Triangular file
Stirring rod
Spatulas

Experiment

1

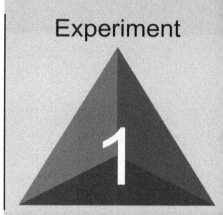

Basic Laboratory Techniques

OBJECTIVE

To learn the use of common, simple laboratory equipment.

APPARATUS AND CHEMICALS

Apparatus

balance
150 and 250 mL beakers
50 and 125 mL Erlenmeyer flasks
50 or 100 mL graduated cylinder
barometer
clamp
large test tube
Bunsen burner and hose
meterstick or ruler
10 mL pipet
rubber bulb for pipet
ring stand and iron ring
thermometer

Chemicals

ice
antifreeze (ethylene glycol)

DISCUSSION

Chemistry is an experimental science. It depends upon careful observation and the use of good laboratory techniques. In this experiment, you will become familiar with some basic operations that will help you throughout this course. Your success as well as your safety in future experiments will depend upon your mastering these fundamental operations.

Because every measurement made in the laboratory is really an approximation, it is important that the numbers you record reflect the accuracy and precision of the device you use to make the measurement. Appendix B of this manual contains a section on significant figures and measurements that you may find helpful in performing this experiment (Section 1.5). Our system of weights and measures, the metric system, was originally based mainly upon fundamental properties of one of the world's most abundant substances: water. The system is summarized in Table 1.1. Conversions within the metric system are quite simple once you have committed to memory the meaning of the prefixes given in Table 1.2 and you use dimensional analysis.

In 1960 an international agreement was reached, specifying a particular choice of metric units in which the basic units for length, mass, and time are the meter, the kilogram, and the second, respectively. This system of units, known as the International System of Units, is commonly referred to as the SI system and is preferred in scientific work (Section 1.5). A comparison of some common SI, metric, and English units is presented in Table 1.3.

In Table 1.2, the prefix *means* the power of 10. For example, 5.4 *centi*meters means 5.4×10^{-2} meters; *centi-* has the same meaning as $\times10^{-2}$.

TABLE 1.1 Units of Measurement in the Metric System

Measurement	Unit and definition
Mass*	Gram (g) = mass of 1 cubic centimeter (cm^3) of water at 4 °C and 760 mm Hg Mass = quantity of material
Weight*	Newton (N) = Force to accelerate 1 kg at a rate of 1 m/s^2 Weight = mass × gravitational force Weight = force exerted by gravity on a mass
Length	Meter (m) = 100 cm = 1000 millimeters (mm) = 39.37 in.
Volume	Liter (L) = volume of 1 kilogram (kg) of water at 4 °C
Temperature	°C, measures heat intensity: $°C = \frac{5}{9}(°F - 32)$ or $°F = \frac{9}{5}°C + 32$
Heat	1 calorie (cal), amount of heat required to raise 1 g of water 1 °C: 1 cal = 4.184 joule (J)
Density	*d*, usually g/mL for liquids, g/L for gases, and g/cm^3 for solids: $d = \frac{\text{mass}}{\text{unit volume}}$
Specific gravity	sp gr, dimensionless: $\text{sp gr} = \frac{\text{density of a substance}}{\text{density of a reference substance}}$

*Mass and weight are not the same. We will use the term "weigh an object" in this manual but you will actually be determining the mass of the object.

TABLE 1.2 The Meaning of Prefixes in the Metric System

Prefix	Meaning (power of 10)	Abbreviation
femto-	10^{-15}	f
pico-	10^{-12}	p
nano-	10^{-9}	n
micro-	10^{-6}	μ
milli-	10^{-3}	m
centi-	10^{-2}	c
deci-	10^{-1}	d
kilo-	10^{3}	k
mega-	10^{6}	M
giga-	10^{9}	G

TABLE 1.3 Comparison of SI, Metric, and English Units

Physical quantity	SI unit	Some common metric units	Conversion factors between metric and English units
Length	meter (m)	Meter (m) Centimeter (cm)	$1\text{ m} = 10^2\text{ cm}$ $1\text{ m} = 39.37\text{ in.}$ $1\text{ in.} = 2.54\text{ cm}$
Volume	cubic meter (m^3)	Liter (L) Milliliter (mL)*	$1\text{ L} = 10^3\text{ cm}^3$ $1\text{ L} = 10^{-3}\text{ m}^3$ $1\text{ L} = 1.06\text{ qt}$
Mass	kilogram (kg)	Gram (g) Milligram (mg)	$1\text{ kg} = 10^3\text{ g}$ $1\text{ kg} = 2.205\text{ lb}$ $1\text{ lb} = 453.6\text{ g}$
Energy	joule (J)	Calorie (cal)	$1\text{ cal} = 4.184\text{ J}$
Temperature	kelvin (K)	Degree celsius (°C)	$0\text{ K} = -273.15\text{ °C}$ $\text{°C} = \frac{5}{9}(\text{°F} - 32)$ $\text{°F} = \frac{9}{5}\text{°C} + 32$

*A mL is the same volume as a cubic centimeter: 1 mL = 1 cm^3.

Example 1.1

Convert 6.7 nanograms to milligrams.

SOLUTION:

$$(6.7\ \cancel{\text{ng}})\left(\frac{10^{-9}\ \cancel{\text{g}}}{1\ \cancel{\text{ng}}}\right)\left(\frac{1\text{ mg}}{10^{-3}\ \cancel{\text{g}}}\right) = 6.7\times10^{-6}\text{ mg}$$

Notice that the conversion factors have no effect on the magnitude (only the power of 10) of the mass measurement.

The quantities presented in Table 1.1 are measured with the aid of various pieces of apparatus. A brief description of some measuring devices follows.

Laboratory Balance

A laboratory balance is used to obtain the mass of various objects. There are several varieties of balances, with various limits on their accuracy. Two common kinds of balances are depicted in Figure 1.1. These single-pan balances are found in most modern laboratories. Generally, they are simple to use but are very *delicate* and *expensive.* The amount of material to be weighed and the accuracy required determine which balance you should use.

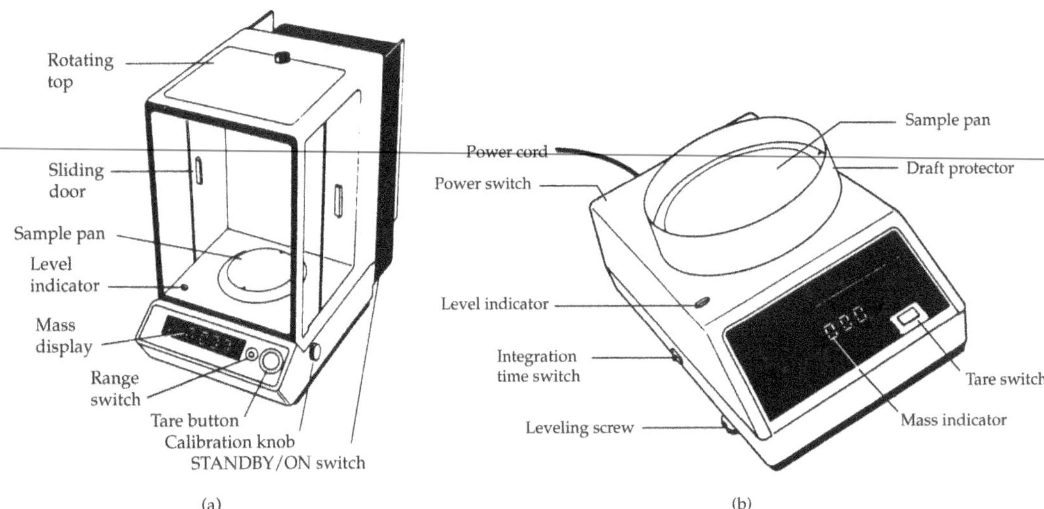

▲**FIGURE 1.1** Digital electronic balances. The balance gives the mass directly when an object to be weighed is placed on the pan. (a) Analytical balance. (b) Top loading balance.

Meter Rule

The standard unit of length is the meter (m), which is 39.37 in. in length. A metric rule, or meterstick, is divided into centimeters (1 cm = 0.01 m; 1 m = 100 cm) and millimeters $(1\,\text{mm} = 0.001\,\text{m}; 1\,\text{m} = 1000\,\text{mm})$. It follows that 1 in. is 2.54 cm (1 in. is currently defined as exactly 2.54 cm). (Convince yourself of this because it is a good exercise in dimensional analysis.)

Graduated Cylinders

Graduated cylinders are tall, cylindrical vessels with graduations scribed along the side of the cylinder. Because you measure volume in these cylinders by measuring the height of a column of liquid, the cylinder must have a uniform diameter along its height. Obviously, a tall cylinder with a small diameter will be more accurate than a short cylinder with a large diameter. A liter (L) is divided into milliliters (mL), such that 1 mL = 0.001 L and 1 L = 1000 mL.

Thermometers

Most thermometers are based upon the principle that liquids expand when they are heated. Most common thermometers use mercury or colored alcohol as the liquid. These thermometers are constructed so that a uniform diameter capillary tube surmounts a liquid reservoir. To calibrate a thermometer, you define two reference points—normally the freezing point of water (0 °C, 32 °F) and the boiling point of water (100 °C, 212 °F) at 1 atm of pressure (1 atm = 101.3 kPa).* Once these points are marked on the capillary, its length is then subdivided into uniform divisions called *degrees*. There are 100° between the freezing and boiling points on the Celsius (°C, or centigrade) scale and 180° between the points on the Fahrenheit (°F) scale.

*Old units: 1 atm = 760 mm Hg = 760 torr.

Pipets

Pipets are glass vessels that are constructed and calibrated to deliver a precisely known volume of liquid at a given temperature. The markings on the pipet illustrated in Figure 1.2 signify that this pipet was calibrated to deliver (TD) 10.00 mL of liquid at 25 °C. *Always* use a rubber bulb to fill a pipet. NEVER USE YOUR MOUTH! A TD pipet should not be blown empty.

Be aware that *every* measuring device has limitations in its accuracy. Moreover, to take full advantage of a given measuring instrument, you should be familiar with or evaluate its accuracy. Careful examination of the subdivisions on the device will indicate the maximum accuracy you can expect of the particular tool. In this experiment, you will determine the accuracy of your 10 mL pipet. The approximate accuracy of some of the equipment you will use in this course is given in Table 1.4.

GIVE IT SOME THOUGHT

a. Which is easier to use—a pipet or a graduated cylinder?

b. What does that tell you about the precision and accuracy of each instrument?

You should not only obtain a measurement to the highest degree of accuracy the device or instrument permits, but also record the reading or measurement in a manner that reflects the accuracy of the instrument (see the section on

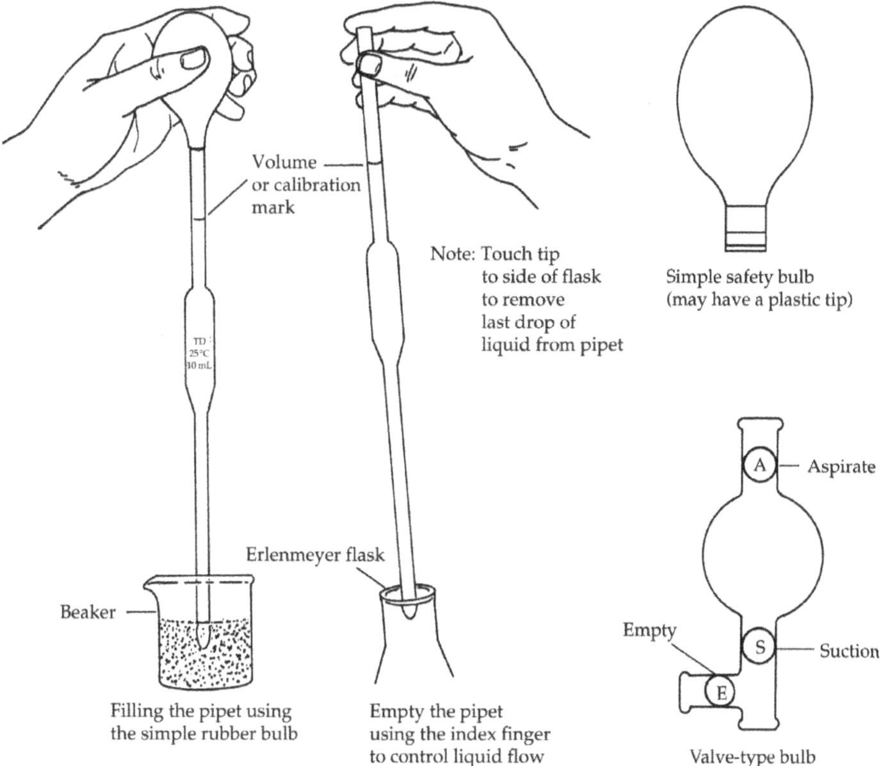

▲**FIGURE 1.2** A typical volumetric pipet, rubber bulbs, and the pipet-filling technique.

TABLE 1.4 Equipment Accuracy

Equipment	Accuracy
Analytical balance	±0.0001 g (±0.1 mg)
Top-loading balance	±0.001 g (±1 mg)
Meterstick	±0.1 cm (±1 mm)
Graduated cylinder	±0.1 mL
Pipet	±0.02 mL
Buret	±0.02 mL
Thermometer	±0.2 °C

significant figures in Appendix B). For example, a mass obtained from an analytical balance should be observed and recorded to the nearest 0.0001 g, or 0.1 mg. If the same object were weighed on a top-loading balance, its mass would be recorded to the nearest 0.001 g. This is illustrated in Table 1.5.

PROCEDURE

A. The Meterstick

Examine the meterstick. Observe that one side is ruled in inches and the other side is ruled in centimeters. Measure and record the length and width of your lab book in both units. Mathematically convert the two measurements to show that they are equivalent.

B. The Graduated Cylinder

Examine the 100 mL graduated cylinder and notice that it is scribed in milliliters. Fill the cylinder approximately half full with water. Notice that the *meniscus* (curved surface of the water) is concave (Figure 1.3).

The *lowest* point on the curve, never the upper level, is read as the volume. Avoid errors due to parallax; you will obtain different and erroneous readings if your eye is not perpendicular to the scale at the bottom of the meniscus. Read the volume of water to the nearest 0.1 mL. Record this volume. Measure the maximum amount of water your largest test tube will hold by transferring the water to the graduated cylinder. Record this volume.

C. The Thermometer and Its Calibration

You perform this part of the experiment to check the accuracy of your thermometer. These measurements will show how measured temperatures (read from the thermometer) compare with true temperatures (the boiling and freezing points of water). The freezing point of water is 0 °C; the boiling point depends

TABLE 1.5 Significant Figures Used in Recording Mass

Analytical balance	Top loader
85.9 g (incorrect)	85.9 g (incorrect)
85.93 g (incorrect)	85.93 g (incorrect)
85.932 g (incorrect)	85.932 g (correct)
85.9322 g (correct)	

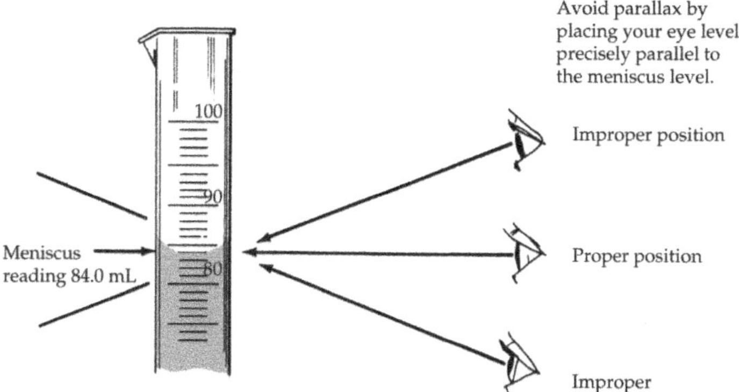

▲**FIGURE 1.3** Proper eye position for taking volume readings.

upon atmospheric pressure and is calculated as shown in Example 1.2. Place approximately 50 mL of ice in a 250 mL beaker and cover the ice with distilled water. Allow about 15 min for the mixture to come to equilibrium and then measure and record the temperature of the mixture. *Theoretically, this temperature is 0 °C.* Now set up a 250 mL beaker on a wire gauze and iron ring as shown in Figure 1.4. Fill the beaker about half full with distilled water. Adjust your burner to give maximum heating and begin heating the water. *(You can save time by heating the water while conducting other parts of the experiment.)* Periodically determine the temperature of the water with the thermometer, but be careful not to touch the walls of the beaker with the thermometer bulb. The boiling point (b.p.) is reached when the temperature becomes constant. Record the boiling point of the water. Using the data given in Example 1.2, determine the *true boiling point at the observed atmospheric pressure.* Obtain the atmospheric pressure from your laboratory instructor. Determine the temperature correction to apply to your thermometer readings.

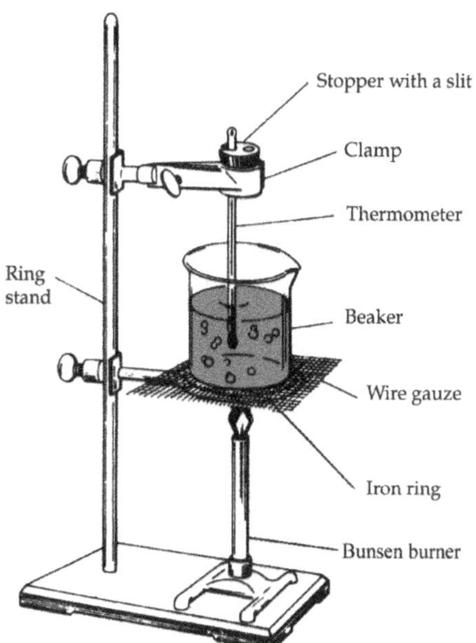

▲**FIGURE 1.4** Apparatus for thermometer calibration.

EXAMPLE 1.2

Determine the boiling point of water at 85.6 kPa.

SOLUTION: Temperature corrections to the boiling point of water are calculated using the following formula:

$$\text{b.p. correction} = (101.3 \text{ kPa} - \text{atmospheric pressure}) \times (0.28 \text{ K/kPa})$$

Therefore, the correction at 85.6 kPa is as follows:

$$\text{b.p. correction} = (101.3 \text{ kPa} - 85.6 \text{ kPa}) \times (0.28 \text{ K/kPa}) = 4.4 \text{ K}$$

Thus, the true boiling point is as follows:

$$100.0\ °\text{C} - 4.4\ °\text{C} = 95.6\ °\text{C}$$

Using the graph paper provided, construct a thermometer-calibration curve like the one shown in Figure 1.5 by plotting observed temperatures versus true temperatures for the boiling and freezing points of water.

D. Using the Balance to Calibrate Your 10 mL Pipet

Determining the mass of an object on a single-pan balance is simple to do. Because of the sensitivity and expense of the balance (some balances cost more than $2500), you must be careful when using it. Directions for operating single-pan balances vary with make and model. Your laboratory instructor will explain how to use the balance. Regardless of the balance you use, proper care of the balance requires that you observe the following:

1. Do not drop an object on the pan.
2. Center the object on the pan.
3. Do not place chemicals directly on the pan; use a beaker, a watch glass, a weighing bottle, or weighing paper.
4. Do not weigh hot or warm objects; objects must be at room temperature.
5. Return all readouts to zero after making the measurement.
6. Clean up any chemical spills in the balance area.
7. Inform your instructor if the balance is not operating correctly; do not attempt to repair it yourself.

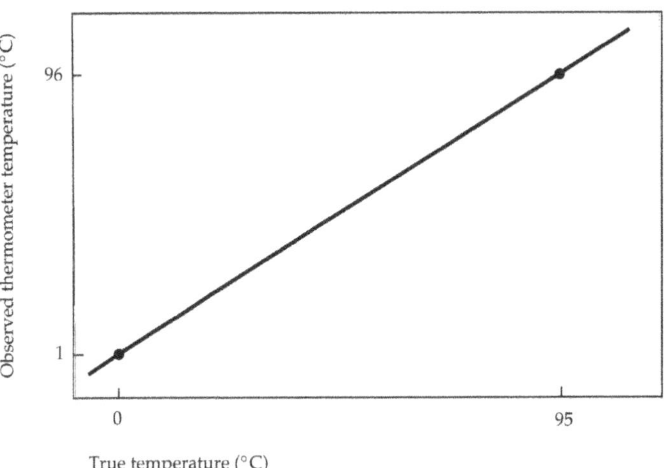

▲FIGURE 1.5 Typical thermometer-calibration curve.

The following method is used to calibrate a pipet or other volumetric glassware: Obtain about 40 mL of distilled water in a 150 mL beaker. Allow the water to sit on the desk while you determine and record the mass of an empty, dry 50 mL Erlenmeyer flask (tare) to the nearest 0.1 mg. Measure and record the temperature of the water. Using your pipet and rubber bulb, pipet exactly 10 mL of water into this flask and determine the mass of the flask with the water in it (gross) to the nearest 0.1 mg. Obtain the mass of the water by subtracting (gross − tare = net). Using the equation below and the data given in Table 1.6, obtain the volume of water delivered and therefore the volume of your pipet.

$$\text{density} = \frac{\text{mass}}{\text{volume}} \qquad d = \frac{m}{V}$$

Normally, density is given in units of grams per milliliter (g/mL) for liquids, grams per cubic centimeter (g/cm^3) for solids, and grams per liter (g/L) for gases. Repeat this procedure in triplicate—that is, deliver and determine the mass of exactly 10 mL of water three separate times.

GIVE IT SOME THOUGHT

a. Should the mass of your three trials be exactly the same?

b. If they are not identical, does this mean that the balance is not operating properly? Explain.

EXAMPLE 1.3

Through the procedure given above, a mass of 10.0025 g was obtained for the water delivered by one 10 mL pipet at 22 °C. What is the volume delivered by the pipet?

SOLUTION: From the density equation given above, you know the following:

$$V = \frac{m}{d}$$

For the mass, substitute the value of 10.0025 g. For the density, consult Table 1.6. At 22 °C, the density is 0.997770 g/mL. The calculation is as follows:

$$V = \frac{10.0025 \text{ g}}{0.997770 \text{ g/mL}} = 10.0249 \text{ mL}$$

That number must be rounded off to 10.02 because the pipet's volume can be determined only to within a precision of ±0.02 mL.

TABLE 1.6 Density of Pure Water at Various Temperatures

T (°C)	*d* (g/mL)	T (°C)	*d* (g/mL)
15	0.999099	22	0.997770
16	0.998943	23	0.997538
17	0.998774	24	0.997296
18	0.998595	25	0.997044
19	0.998405	26	0.996783
20	0.998203	27	0.996512
21	0.997992	28	0.996232

The *precision* of a measurement is a statement about the internal agreement among repeated results; it is a measure of the reproducibility of a given set of results (Section 1.6). The arithmetic mean (average) of the results is usually taken as the "best" value. The simplest measure of precision is the *average deviation from the mean.* The average deviation is calculated by determining the mean of the measurements, calculating the deviation of each measurement from the mean and averaging the deviations (treating each as a positive quantity). Study Example 1.4; then using your own experimental results, calculate the mean volume delivered by your 10 mL pipet. Also calculate for your three trials the individual deviations from the mean and state your pipet's volume with its average deviation.

EXAMPLE 1.4

The following volumes were obtained for the calibration of a 10 mL pipet: 10.15 mL, 10.12 mL, and 10.00 mL. Calculate the mean value and the average deviation from the mean.

SOLUTION:

$$\text{mean} = \frac{10.15 + 10.12 + 10.00}{3} = 10.09$$

Deviations from the mean: $|\text{value} - \text{mean}|$

$$|10.15 - 10.09| = 0.06$$

$$|10.12 - 10.09| = 0.03$$

$$|10.00 - 10.09| = 0.09$$

Average deviation from the mean

$$= \frac{0.06 + 0.03 + 0.09}{3} = 0.06$$

Therefore, the reported value is 10.09 ± 0.06 mL.

GIVE IT SOME THOUGHT

a. Why does this reported value include +/− 0.06?

b. Is this needed for every experimentally reported value? Explain.

E. Measuring the Density of Antifreeze

Determine the mass of a dry 50 mL flask to the nearest 0.1 mg and record its mass. Using your pipet, measure a 10 mL sample of antifreeze solution into the 50 mL flask, determine the mass of the flask and its contents, and record this mass. Repeat these measurements two more times to get an indication of the precision of your measurements. Use the measured mass and volume to calculate the density of the antifreeze for each measurement. Using the three values for the density, calculate the mean density and the average deviation from the mean for your determinations. Dispose of the antifreeze as instructed.

Name ______________________________ Desk ______________________________

Date ______________ Laboratory Instructor ______________________________

Basic Laboratory Techniques | 1 Pre-lab Questions

You should be able to answer the following questions before beginning this experiment.

1. What are the basic units of electric current, time, amount of substance, and luminous intensity in the SI system?

2. What prefix do the following abbreviations represent: (a) f, (b) n, (c) G, (d) c, and (e) p?

3. What is the number of significant figures in each of the following measured quantities: (a) 351 g, (b) 0.0100 mL, (c) 1.010 mL, and (d) 3.72×10^{-3} cm?

4. What is the length in millimeters of a crystal of copper sulfate that is 0.180 in. long?

5. Perform the following conversions: (a) 72.3 mg to g, (b) 6.0×10^{-10} m to mm, and (c) 325 mm to μm.

6. DNA is approximately 2.5 nm in length. If an average man is 5 ft 10 in. tall, how many DNA molecules could be stacked to extend from the ground to the top of the head of an average man?

7. A liquid has a volume of 3.70 L. What is its volume in mL? In cm^3?

8. Why should you never determine the mass of a hot object?

9. Why is it necessary to calibrate a thermometer and volumetric glassware?

10. What is precision?

11. What is the density of an object with a mass of 1.583 g and a volume of 0.2009 mL?

12. Determining the mass of an object three times gave the following results: 8.2 g, 8.1 g, and 8.0 g. Find the mean mass and the average deviation from the mean.

13. Normal body temperature is 37.0 °C. What is the corresponding Fahrenheit temperature?

14. What is the mass in kilograms of 700 mL of a substance that has a density of 1.23 g/mL?

15. An object has a mass of exactly 5 g on an analytical balance that has an accuracy of 0.1 mg. To how many significant figures should this mass be recorded?

Name ______________________ Desk ______________

Date ______________ Laboratory Instructor ______________________

REPORT SHEET | EXPERIMENT

Basic Laboratory Techniques | 1

A. The Meterstick

Length of this lab book	________ in.	________ cm	________ mm	________ m
Width of this lab book	________ in.	________ cm	________ mm	________ m

Using an equation (including units), show that the above measurements are equivalent.

Area of this lab book (show calculations) ________ cm^2

B. The Graduated Cylinder

Volume of water in graduated cylinder ________ mL

Volume of water contained in largest test tube ________ mL

C. The Thermometer and Its Calibration

Observed temperature of water-and-ice mixture ________ °C

Temperature of boiling water ________ °C

Observed atmospheric pressure ________ kPa

True (corrected) temperature of boiling water ________ °C

Thermometer correction ________ °C

D. Using the Balance to Calibrate Your 10 mL Pipet

Temperature of water used in pipet ________ °C

Corrected temperature ________ °C

	Trial 1	*Trial 2*	*Trial 3*	
Mass of Erlenmeyer plus ~10 mL H_2O (gross mass)	________	________	________	g
Mass of Erlenmeyer (tare mass)	________	________	________	g
Mass of ~10 mL of H_2O (net mass)	________	________	________	g

Volume delivered by 10 mL pipet (show calculations) ________ ________ ________ mL

Mean volume delivered by 10 mL pipet (show calculations) ____________ mL

	Trial 1	*Trial 2*	*Trial 3*
Individual deviations from the mean	________	________	________

Average deviation from the mean (show calculations) ________ mL

Volume delivered by your 10 mL pipet ____________ mL ± __________ mL

E. Measuring the Density of Antifreeze

Temperature of antifreeze ________ °C

	Trial 1	*Trial 2*	*Trial 3*	
Mass of flask + antifreeze	________	________	________	g
Mass of empty flask	________	________	________	g
Mass of antifreeze	________	________	________	g
Density of antifreeze (show calculations)	________	________	________	g

Mean (average) density

Average deviation from the mean (show calculations)

QUESTIONS

1. Identify each of the following as measurements of length, area, volume, mass, density, time, or temperature: (a) ns, (b) 10.0 kg/m^3, (c) 1.2 pm, (d) 750 km^2, (e) 83 K, and (f) 4.0 mm^3.

2. Carry out the following operations and express the answer with the appropriate number of significant figures and units: (a) (5.231 mm)(6.1 mm), (b) 72.3 g/1.5 mL, (c) 12.21 g + 0.0132 g, and (d) 31.03 g + 12 mg.

3. Drug medications are often prescribed on the basis of body mass. The adult dosage of Elixophyllin, a drug used to treat asthma, is 6 mg/kg of body mass. Calculate the dose in milligrams for a 73 kg person.

4. A man who is 1.78 m tall weighs 73 kg. What is his height in centimeters and his mass in grams?

5. Determine the boiling point of water at 89.6 kPa.

6. A pipet delivers 9.98 g of water at 19 °C. What volume does the pipet deliver?

7. A pipet delivers 10.4 mL, 10.2 mL, 10.8 mL, and 10.6 mL in consecutive trials. Find the mean volume and the average deviation from the mean.

8. A 141 mg sample was placed on a watch glass that has a mass of 9.203 g. What is the mass of the watch glass and sample in grams?

9. (a) Using the defined freezing and boiling points of water, make a plot of degrees Fahrenheit versus degrees Celsius on the graph paper provided.

   (b) Determine the Celsius equivalent of 40 °F using your graph. The relationship between these two temperature scales is linear (that is, it is of the form $y = mx + b$). Consult Appendix C regarding linear relationships and determine the equation that relates degrees Fahrenheit to degrees Celsius.

   (c) Compute the Celsius equivalent of 40 °F using this relationship.

Thermometer Calibration Curve

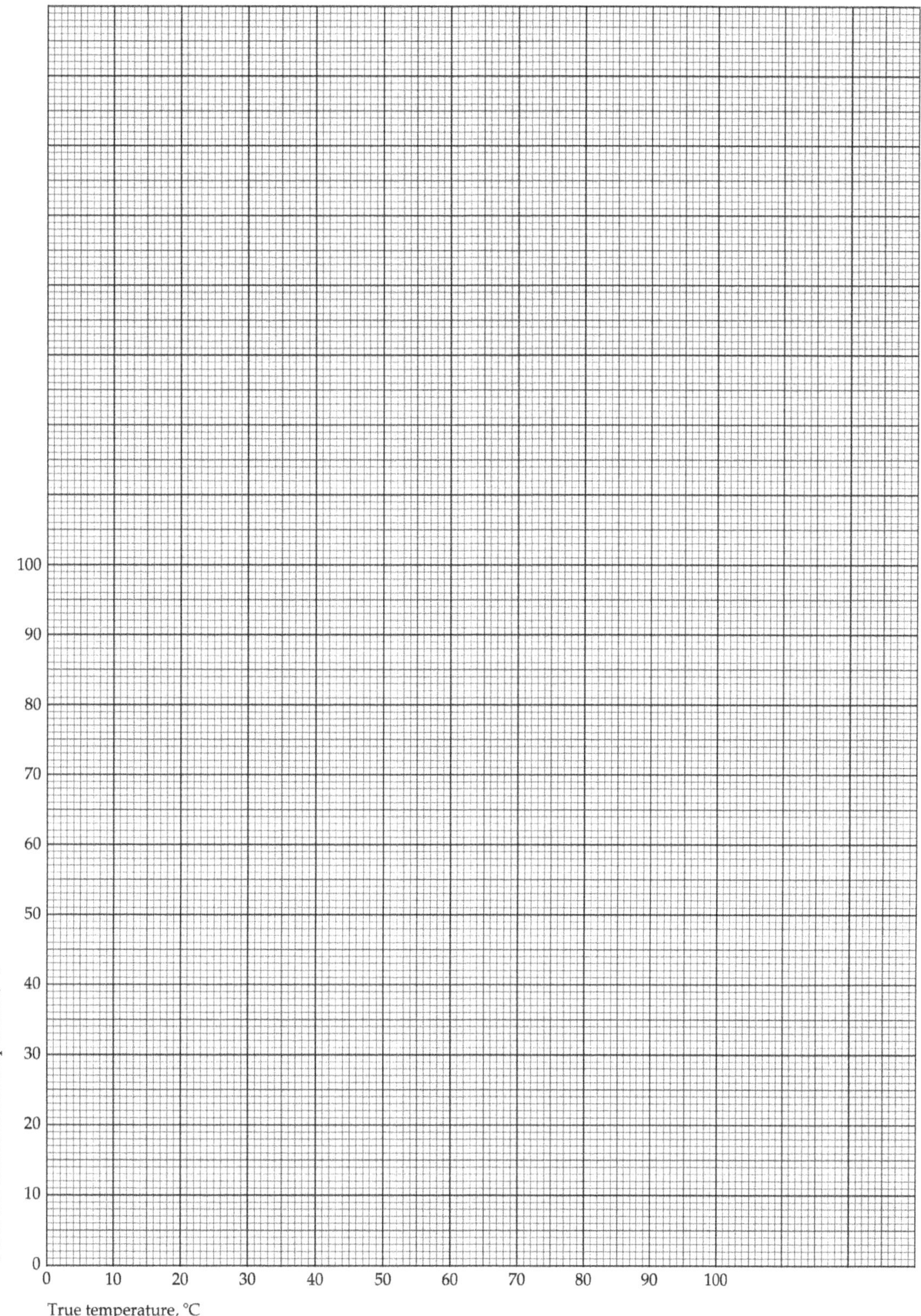

Fahrenheit—Celsius Graph

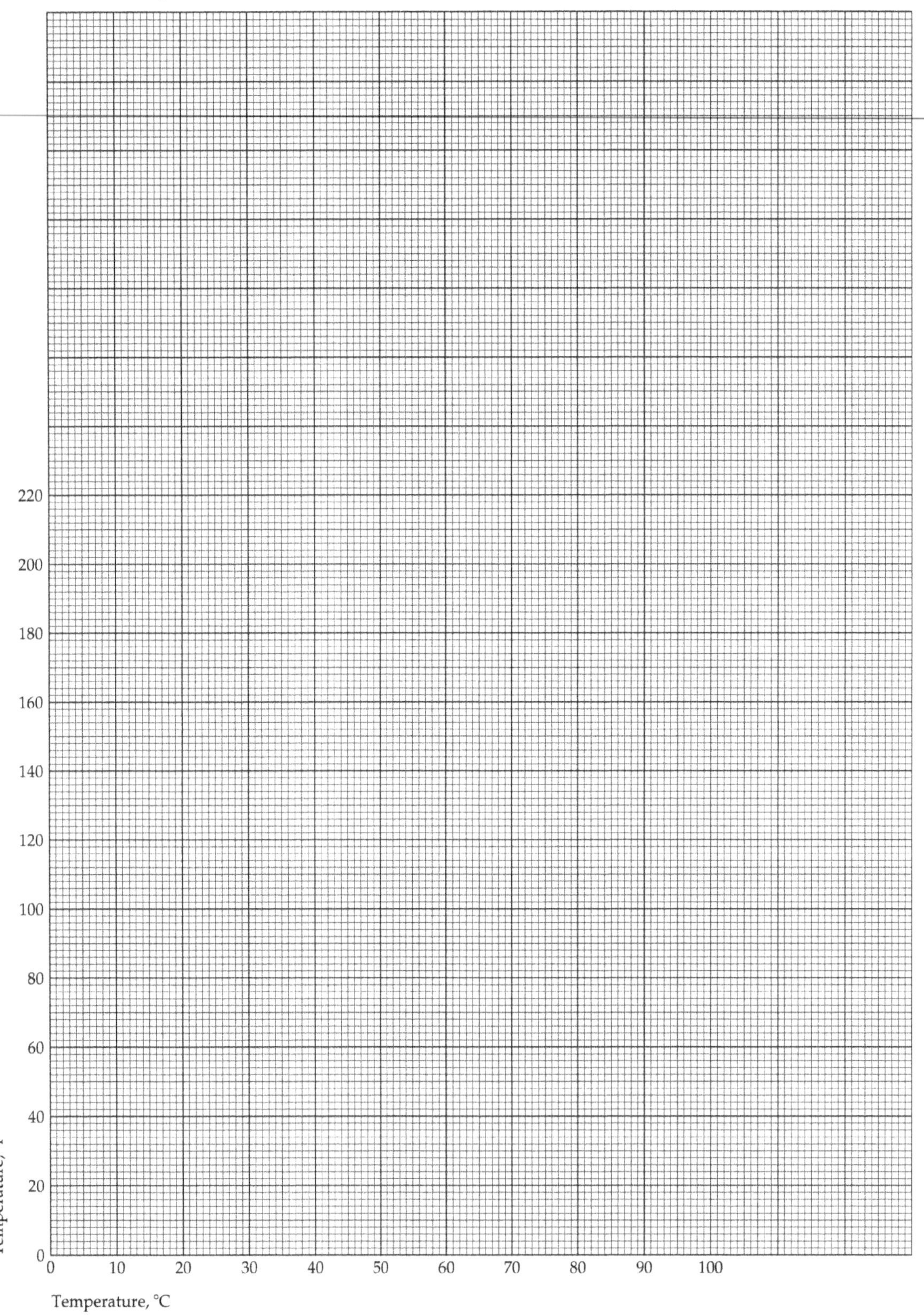

Experiment

Identification of Substances by Physical Properties

OBJECTIVE

To become acquainted with procedures used in evaluating physical properties and the use of these properties to identify substances.

APPARATUS AND CHEMICALS

Apparatus

balance
50 mL beakers (2)
250 mL beaker
50 mL Erlenmeyer flask
10 mL graduated cylinder
large test tubes (2)
small test tubes (6)
test-tube rack
10 mL pipet
ring stand and ring
wire gauze
no. 3 two-hole stopper with one of the holes slit to the side or a buret clamp
Bunsen burner and hose
stirring rod
dropper
boiling chips
thermometer
spatula
small watch glass
capillary tubes (5)
tubing with right-angle bend
utility clamp
two-hole stopper
small rubber bands (or small sections of 6.4 mm rubber tubing)
rubber bulb for pipet

Chemicals

ethyl alcohol
toluene
soap solution
boiling chips
cyclohexane
naphthalene
two unknowns (one liquid; one solid)

DISCUSSION

Properties are those characteristics of a substance that enable you to identify it and to distinguish it from other substances. Direct identification of some substances can readily be made simply by examining them. For example, you see color, size, shape, and texture and can smell odors and discern a variety of tastes. Thus, copper can be distinguished from other metals on the basis of its color.

Physical properties are those properties that can be observed without altering the composition of the substance (Section 1.3). Whereas it is difficult to assign definitive values to properties such as taste, color, and odor, other physical properties, such as melting point, boiling point, solubility, density, viscosity, and refractive index, can be expressed quantitatively. For example, the melting point of copper is 1085 °C and its density is 8.96 g/cm^3. As you may realize, a specific combination of properties is unique to a given substance, thus making it possible to identify most substances just by carefully determining several properties. This is so important that large books have been compiled listing characteristic properties of many known substances. Many scientists, most notably several German scientists during the latter part of the nineteenth century and early part of the twentieth century, spent their entire lives gathering data of this sort. Two complete references of this type that are available today are The Chemical Rubber Company's *Handbook of Chemistry and Physics* and *Lange's Handbook of Chemistry.*

In this experiment, you will use the following properties to identify a substance whose identity is unknown to you: solubility, density, melting point, and boiling point. The *solubility* of a substance in a solvent at a specified temperature is the maximum mass of that substance that dissolves in a given volume (usually 100 mL or 1000 mL) of a solvent. It is tabulated in handbooks in terms of grams per 100 mL of solvent; the solvent is usually water.

Density is an important physical property and is defined as the mass per unit volume:

$$d = \frac{m}{V}$$

Melting or freezing points correspond to the temperature at which the liquid and solid states of a substance are in equilibrium. These terms refer to the *same* temperature but differ slightly in their meaning. The *freezing point* is the equilibrium temperature when approached from the liquid phase—that is, when solid begins to appear in the liquid. The *melting point* is the equilibrium temperature when approached from the solid phase—that is, when liquid begins to appear in the solid.

A liquid is said to boil when bubbles of vapor form in it, rise rapidly to the surface, and burst. Any liquid in contact with the atmosphere will boil when its vapor pressure is equal to atmospheric pressure—that is, the liquid and gaseous states of a substance are in equilibrium. Boiling points of liquids depend upon atmospheric pressure. A liquid will boil at a higher temperature at a higher pressure and at a lower temperature at a lower pressure. The temperature at which a liquid boils at 101.3 kPa is called the *normal* boiling point. To account for these pressure effects on boiling points, people have studied and tabulated data for boiling point versus pressure for many compounds. From these data, nomographs have been constructed. A *nomograph* is a set of scales for connected variables (see Figure 2.5 p. 43 for an example); these scales are placed so that a straight line connecting the known values on some scales will provide the unknown value at the straight line's intersection with other scales. A nomograph allows you to find the correction necessary to convert the normal boiling point of a substance to its boiling point at any pressure of interest.

PROCEDURE

A. Solubility

Qualitatively determine the solubility of naphthalene (mothballs) in three solvents: water, cyclohexane, and ethyl alcohol. **(CAUTION: *Cyclohexane and ethyl alcohol are highly flammable and must be kept away from open flames.*)** Determine the solubility by adding two or three small crystals of naphthalene to 2 to 3 mL (it is not necessary to measure the solute mass or the solvent volume) of each of these three solvents in separate clean, *dry* test tubes. Try to keep the amount of naphthalene and solvent the same in each case. Place a cork in each test tube and shake briefly. Cloudiness indicates insolubility. Record your conclusions on the report sheet using the abbreviations s (soluble), sp (sparingly soluble), and i (insoluble). Into each of three more clean, *dry* test tubes, place 2 or 3 mL of these same solvents and add 4 or 5 drops of toluene in place of naphthalene. Record your observations. The formation of two layers indicates immiscibility (lack of solubility). Now repeat these experiments using each of the three solvents (water, cyclohexane, and ethyl alcohol) with your solid and liquid unknowns and record your observations.

GIVE IT SOME THOUGHT

a. It is often said that "like dissolve like." Which solvents are more like toluene and naphthalene?
b. Which solvent is more like your unknown?

Save your solid and liquid unknowns for Parts B, C, and D, but dispose of the other chemicals in the marked waste container as instructed. Do not dispose of them in the sink.

GIVE IT SOME THOUGHT

a. Can you identify your unknown from solubility measurements alone?
b. If not, can you eliminate some of the unknowns from Table 2.1?

B. Density

Determine the densities of your two unknowns in the following manner.

The Density of a Solid Determine the mass of about 1.5 g of your solid unknown to the nearest 0.001 g and record the mass. Using a pipet or a wash bottle, half fill a *clean, dry* 10 mL graduated cylinder with a solvent in which your unknown is *insoluble.* Be *careful* not to get the liquid on the inside walls because you do not want your solid to adhere to the cylinder walls when you add it in a subsequent step. Read and record this volume to the nearest 0.1 mL. Add the measured mass of the solid to the liquid in the cylinder, being careful not to lose any of the material in the transfer process and ensuring that all of the solid is beneath the surface of the liquid. Carefully tapping the sides of the cylinder with your fingers will help settle the material to the bottom. Do not be concerned about a few crystals that do not settle, but if a large quantity of the solid resists settling, add one or two drops of a soap solution and continue tapping the cylinder with your fingers. Now read the new volume to the nearest 0.1 mL. The difference in these two volumes is the volume of your solid (Figure 2.1). Calculate the density of your solid unknown.

You may recall that by measuring the density of metals in this way, Archimedes proved to the king that the charlatan alchemists had in fact not transmuted lead into gold. Archimedes did this after observing that he weighed less in the bathtub than he did normally by an amount equal to the weight of the fluid displaced. According to legend, upon making his discovery, Archimedes emerged from his bath and ran naked through the streets shouting "Eureka!" (I have found it.).

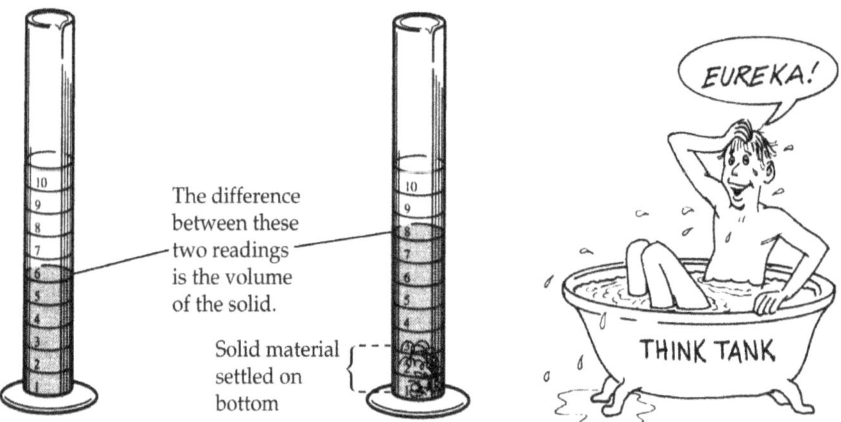

▲**FIGURE 2.1** Determination of the volume of a solid by difference using a graduated cylinder.

The Density of a Liquid Determine the mass of a clean, *dry* 50 mL Erlenmeyer flask to the nearest 0.0001 g. Obtain at least 15 mL of the unknown liquid in a clean, *dry* test tube. Using a 10 mL pipet, pipet exactly 10 mL of the unknown liquid into the 50 mL Erlenmeyer flask and quickly determine the mass of the flask containing the 10 mL of unknown to the nearest 0.0001 g. Using the calibration value for your pipet, if you calibrated it, and the mass of this volume of unknown, calculate its density. Record your results and show how (with units) you performed your calculations. *Save* the liquid for your boiling-point determination.

GIVE IT SOME THOUGHT

With the density and solubility results, narrow down the options of your unknown in Table 2.1. With this information, upon what temperatures for the boiling point and melting point should you focus to identify your unknown?

C. Melting Point of Solid Unknown

Obtain a capillary tube and a small rubber band. Seal one end of the capillary tube by carefully heating the end in the edge of the flame of a Bunsen burner until the end *completely* closes. Rotating the tube during heating will help you avoid burning yourself (Figure 2.2).

On a clean watch glass, pulverize a small portion of your solid unknown sample with the end of a test tube. Partially fill the capillary with your unknown by gently tapping the pulverized sample with the open end of the capillary to force some of the sample inside. Drop the capillary into a glass tube about 38 to 50 cm in length with the sealed end down to pack the sample into the bottom of the capillary tube. Repeat this procedure until the sample column is roughly 5 mm in height. Now set up a melting-point apparatus as illustrated in Figure 2.3.

Place the rubber band about 5 cm above the bulb on the thermometer and out of the liquid. Carefully insert the capillary tube under the rubber band with the closed end at the bottom. Place the thermometer with attached capillary into the beaker of water so that the sample is covered by water, the

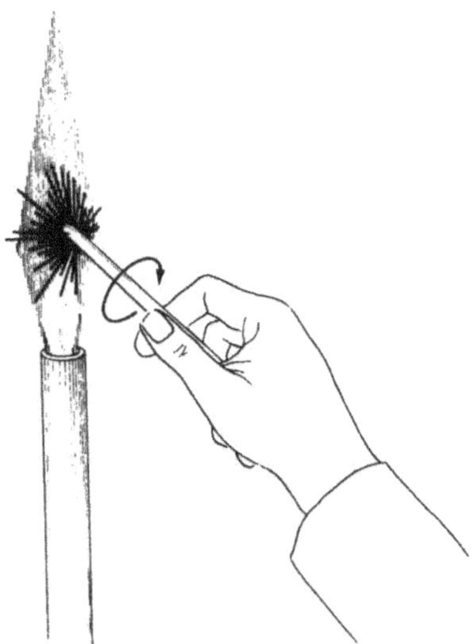

▲FIGURE 2.2 Sealing one end of a capillary tube.

thermometer does not touch the bottom of the beaker, and the open end of the capillary tube is above the surface of the water. Heat the water slowly while gently agitating the water with a stirring rod. Use boiling chips in the water bath for even boiling. Observe the sample in the capillary tube while you are doing this. At the moment the solid melts, record the temperature. Also record the melting-point range, which is the temperature range between the temperature at which the sample begins to melt and the temperature at which all of the sample has melted. Using your thermometer-calibration curve (from Experiment 1), correct these temperatures to the true temperatures and record the melting point and melting-point range. These temperatures may differ by only 1 °C or less. If the range is greater than this the rate of heating was too high and the determination should be repeated using a new sample.

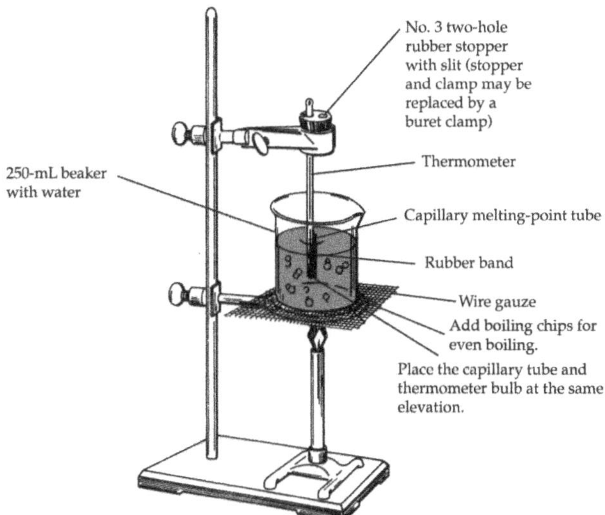

▲FIGURE 2.3 Apparatus for melting-point determination.

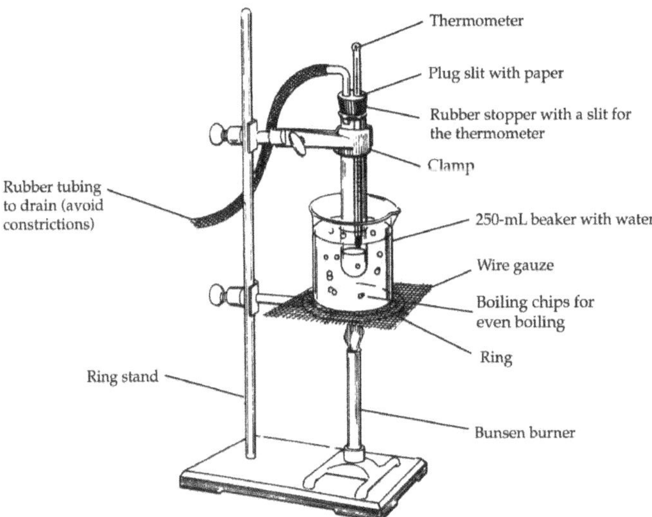

▲**FIGURE 2.4** Apparatus for boiling-point determination.

D. Boiling Point of Liquid Unknown

To determine the boiling point of your liquid unknown, put about 3 mL of the material you used to determine the density into a clean, dry test tube. Fit the test tube with a two-hole rubber stopper that has one slit; insert your thermo meter into the hole with the slit and one of your right-angle-bend glass tubes into the other hole, as shown in Figure 2.4. Add one or two small boiling chips to the test tube to ensure even boiling of your sample. Position the thermometer so that it is about 1 cm above the surface of the unknown liquid. Clamp the test tube in the ring stand and connect to the right-angle-bend tubing a length of rubber tubing that reaches to the sink. Assemble your apparatus as shown in Figure 2.4. **(CAUTION: *Make sure there are no constrictions in the rubber tubing. Your sample is flammable. Keep it away from open flames.*)**

Heat the water gradually and watch for changes in temperature. The temperature will become constant at the boiling point of the liquid. Record the observed boiling point. Correct the observed boiling point to the true boiling point at room atmospheric pressure using your thermometer-calibration curve. The normal boiling point (b.p. at 1 atm = 101.3 kPa) can now be calculated (see Example 2.1) using the nomograph provided in Figure 2.5. Your boiling-point correction should not be more than +5 °C.

EXAMPLE 2.1

What will be the boiling point of ethyl alcohol at 86.7 kPa when its normal boiling point at 101.3 kPa is known to be 78.3 °C?

SOLUTION: The answer is easily found by consulting the nomograph in Figure 2.5. A straight line drawn from 78.3 °C on the left scale of normal boiling points through 86.7 kPa on the pressure scale intersects the temperature correction scale at 4 °C. Therefore,

$$\text{normal b.p.} - \text{correction} = \text{observed b.p.}$$

$$78.3\ °\text{C} - 4.0\ °\text{C} = 74.3\ °\text{C}$$

Similar calculations can be done for the compounds in Table 2.1 at any pressure listed on the nomograph in Figure 2.5. In this experiment, you will observe a boiling point at a pressure other than at 101.3 kPa and you want to know its normal boiling point. To estimate its normal boiling point, assume, for example, that your observed boiling point is 57.0 °C and the observed pressure is 86.7 kPa. Use your observed boiling point of 57.0 °C as if it were the normal boiling point and find the correction for a pressure of 86.7 kPa. Using the nomograph, you can see that the correction is 3.8 °C. You would then *add* this correction to your observed boiling point to obtain an approximate normal boiling point:

$$57.0\ °C + 3.8\ °C = 60.8\ °C, \text{ or } 61\ °C$$

By consulting Table 2.1, you can find the compound that best fits your data; in this example, the data are for chloroform.

E. Unknown Identification

Your unknowns are substances contained in Table 2.1. Compare the properties you have determined for your unknowns with those in the table. Identify your unknowns and record your results.

Dispose of your unknowns in the appropriate marked waste containers. Clean the glassware containing the water insoluble solids and liquids by using acetone rinse and put the rinse into the waste container and then allow the glassware to air dry.

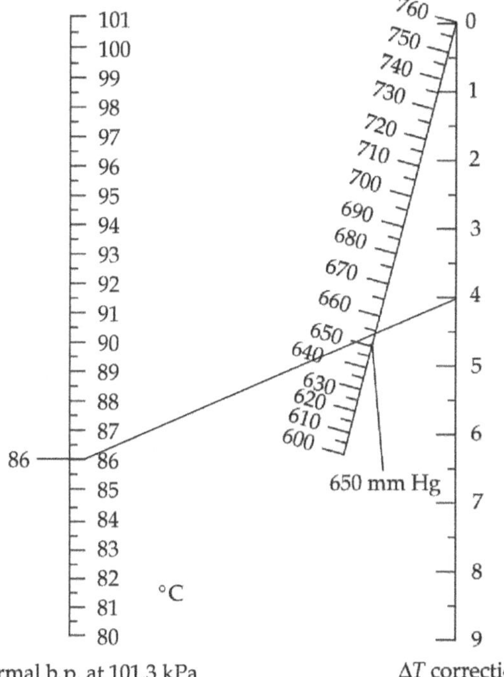

▲FIGURE 2.5 Use of nomograph for boiling-point correction for ethyl alcohol to 86.7 kPa.

TABLE 2.1 Physical Properties of Selected Pure Substances

Substance	Density (g/mL)	Melting point (°C)	Boiling point (°C)	Solubility[a] in Water	Solubility[a] in Cyclohexane	Solubility[a] in Ethyl alcohol
Acetanilide	1.22	114	304	sp	sp	s
Acetone	0.79	−95	56	s	s	s
Benzophenone	1.15	48	306	i	s	s
Bromoform	2.89	8	150	i	s	s
2,3-Butanedione	0.98	−2.4	88	s	s	s
t-Butyl alcohol	0.79	25	83	s	s	s
Cadmium nitrate · $4H_2O$	2.46	59	132	s	i	s
Chloroform[b]	1.49	−63.5	61	i	s	s
Cyclohexane	0.78	6.5	81.4	i	s	s
p-Dibromobenzene	1.83	86.9	219	i	s	s
p-Dichlorobenzene	1.46	53	174	i	s	s
m-Dinitrobenzene	1.58	90	291	i	s	s
Diphenyl	0.99	70	255	i	s	s
Diphenylamine	1.16	53	302	i	s	s
Diphenylmethane	1.00	27	265	i	s	s
Ether, ethyl propyl	1.37	−79	64	s	s	s
Hexane	0.66	−94	69	i	s	s
Isopropyl alcohol	0.79	−98	83	s	s	s
Lauric acid	0.88	43	225	i	s	s
Magnesium nitrate · $6H_2O$[c]	1.63	89	330[c]	s	i	s
Methyl alcohol	0.79	−98	65	s	i	s
Methylene chloride[b]	1.34	−97	40.1	i	s	s
Naphthalene	1.15	80	218	i	s	sp
α-Naphthol	1.10	94	288	i	i	s
Phenyl benzoate	1.23	71	314	i	s	s
Propionaldehyde	0.81	−81	48.8	s	i	s
Sodium acetate · $3H_2O$	1.45	58	123	s	i	sp
Stearic acid	0.85	70	291	i	s	sp
Thymol	0.97	52	232	sp	s	s
Toluene	0.87	−95	111	i	s	s
p-Toluidine	0.97	45	200	sp	s	s
Zinc chloride	2.91	283	732	s	i	s

[a]s = soluble; sp = sparingly soluble; i = insoluble.
[b]Toxic. Most organic compounds used in the lab are toxic.
[c]Boils with decomposition.

Name ______________________________ Desk ______________

Date ______________ Laboratory Instructor ______________________

Identification of Substances by Physical Properties | 2 | Pre-lab Questions

Before beginning this experiment in the laboratory, you should be able to answer the following questions.

1. List five physical properties.

2. A 2.40 g sample of an unknown has a volume of 4.23 cm^3. What is the density of the unknown?

3. Are the substances thymol and diphenyl solids or liquids at room temperature? (See Table 2.1)

4. Could you determine the density of chloroform using water? Why or why not? (See Table 2.1)

5. What would be the boiling point of methylene chloride at 89.3 kPa?

6. Why do you calibrate thermometers and pipets?

7. Is phenyl benzoate soluble in water? In ethyl alcohol? (See Table 2.1.)

8. When water and bromoform are mixed, two layers form. Is the bottom layer water or bromoform? (See Table 2.1.)

9. What solvent would you use to determine the density of diphenylamine? (See Table 2.1)

10. The density of a solid with a melting point of 42° to 44 °C was determined to be 0.87 ± 0.02 g/mL. What is the solid?

11. The density of a liquid whose boiling point is 55° to 57 °C was determined to be 0.77 ± 0.05 g/mL. What is the liquid?

12. Which has the greater volume—10 g of hexane or 10 g of toluene? What is the volume of each?

Name ______________________________ Desk ______________

Date ______________ Laboratory Instructor ______________________

Liquid unknown no. ______________

Solid unknown no. ______________

REPORT SHEET | EXPERIMENT 2

Identification of Substances by Physical Properties

A. Solubility

	Water	*Cyclohexane*	*Ethyl alcohol*
Naphthalene	________	________	________
Toluene	________	________	________
Liquid unknown	________	________	________
Solid unknown	________	________	________

B. Density

Solid

Final volume of liquid in cylinder ________ mL

Initial volume of liquid in cylinder ________ mL

Volume of solid ________ mL

Mass of solid ________ g

Density of solid ________ g/mL
(show calculations)

Liquid

Volume of liquid ________ mL

Volume of liquid corrected for the pipet correction ________ mL

Mass of 50 mL Erlenmeyer plus 10 mL of unknown ________ g

Mass of 50 mL Erlenmeyer ________ g

Mass of liquid ________ g

Density ________ g/mL
(show calculations)

C. Melting Point of Solid Unknown

Observed melting point ______________ °C

Corrected (apply thermometer correction to obtain) ______________ °C

Observed melting-point range ______________ °C

Corrected (apply thermometer correction to obtain) ______________ °C

D. Boiling Point of Liquid Unknown

Barometric pressure ______________ hPa

Observed ______________ °C

Corrected (apply thermometer correction to obtain) ______________ °C

Estimated true (normal) b.p. (apply pressure correction to obtain) ______________ °C

E. Unknown Identification

Solid unknown ________________________

Liquid unknown ________________________

QUESTIONS

1. Is diphenylmethane a solid or a liquid at room temperature?
2. What solvent would you use to determine the density of diphenyl?
3. Convert your densities to kg/L and compare those values with the ones in g/mL.
4. If air bubbles were trapped in your solid beneath the liquid level in your density determination, what error would result in the volume measurement? What would be the effect of this error on the calculated density?
5. A liquid unknown was found to be insoluble in water and soluble in cyclohexane and alcohol; the unknown was found to have a boiling point of 58 °C at 89.3 kPa. What is the substance? What could you do to confirm your answer?

6. A liquid that has a density of 0.80 ± 0.01 g/mL is soluble in cyclohexane. What liquid might this be?

7. What is the boiling point of cyclohexane at 82.7 kPa?

Consult a handbook or the Internet for the following questions and specify the source used.

---

8. Rhodium is an expensive metal known for its catalytic properties. What are its density and melting point?

9. What are the colors of $CuSO_4$ and $CuSO_4 \cdot 5H_2O$?

10. What are the formula, molar mass, and color of potassium dichromate (VI)?

NOTES AND CALCULATIONS

Experiment 3

Separation of the Components of a Mixture

OBJECTIVE

To become familiar with the methods of separating substances from one another using decantation, extraction, and sublimation techniques.

APPARATUS AND CHEMICALS

Apparatus

balance
Bunsen burner and hose
tongs
evaporating dishes (2)
watch glass
50 or 100 mL graduated cylinder
clay triangles (2) or wire gauze (2)
ring stands (2)
iron rings (2)
glass stirring rods

Chemicals

unknown mixture of sodium chloride, ammonium chloride, and silicon dioxide

DISCUSSION

Most of the matter people encounter in everyday life consists of mixtures of different substances. Mixtures are combinations of two or more substances in which each substance retains its own chemical identity and therefore its own properties. Whereas pure substances have fixed compositions, the composition of mixtures can vary (Section 1.2). For example, a glass of sweetened tea may contain a little or a lot of sweetener. The substances making up a mixture are called *components*. Mixtures such as cement, wood, rocks, and soil do not have the same composition, properties, and appearance throughout the mixture. Such mixtures are called *heterogeneous*. Mixtures that are uniform in composition, properties, and appearance throughout are called *homogeneous*. Such mixtures include sugar water and air. Homogeneous mixtures are also called solutions. Mixtures are characterized by two fundamental properties:

- Each of the substances in the mixture retains its chemical identity.
- Mixtures are separable into these components by physical means.

If one of the substances in a mixture is preponderant—that is, if its amount far exceeds the amounts of the other substances in the mixture—you usually call this mixture an impure substance and speak of the other substances in the mixture as impurities.

The preparation of compounds usually involves their separation or isolation from reactants or other impurities. Thus, the separation of mixtures into their components and the purification of impure substances are common problems. You are probably aware of everyday problems of this sort. For example, drinking water usually begins as a mixture of silt, sand, dissolved salts, and water. Because water is by far the largest component in this mixture, it is usually called impure water. How is it purified? The separation of the components of mixtures is based upon the fact that each component has different physical properties. The components of mixtures are always pure substances, either compounds or elements, and each pure substance possesses a unique set of properties. The properties of every sample of a pure substance are identical at a specific temperature and pressure. This means that once you have determined that a sample of sodium chloride (table salt), NaCl, is water-soluble and a sample of silicon dioxide (sand), SiO_2, is not, you realize that all samples of sodium chloride are water-soluble and all samples of silicon dioxide are not.

Likewise, every crystal of a pure substance melts at a specific temperature and a given pressure, and every pure substance boils at a specific temperature and a given pressure.

Although numerous physical properties can be used to identify a particular substance, you will be concerned in this experiment merely with the separation of the components and not with their identification. The methods you will use for the separation depend upon differences in physical properties, and they include the following:

1. *Decantation.* This is the process of separating a liquid from a solid (sediment) by gently pouring the liquid from the solid so as not to disturb the solid (Figure 3.1).
2. *Filtration.* This is the process of separating a solid from a liquid by means of a porous substance—a filter—which allows the liquid but not the solid to pass through (see Figure 3.1). Common filter materials are paper, layers of charcoal, and sand. Silt and sand can be removed from drinking water by this process.

GIVE IT SOME THOUGHT

a. When would it be best to use decantation over filtration?
b. When would you want to use filtration rather than decantation?

3. *Extraction.* This is the separation of a substance from a mixture by preferentially dissolving that substance in a suitable solvent. By this process, a soluble compound is usually separated from an insoluble compound.

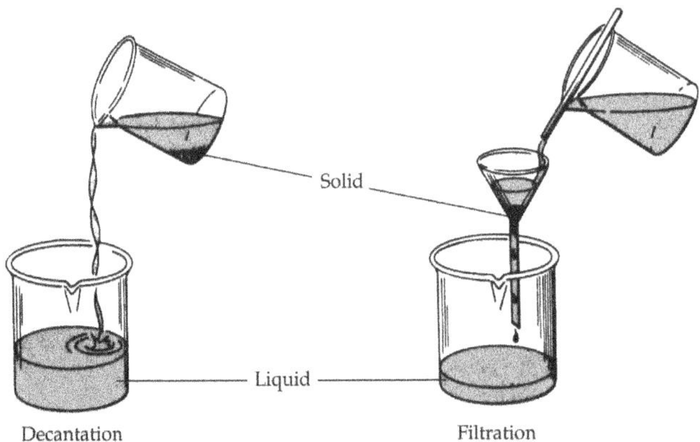

▲**FIGURE 3.1** Illustration of the processes of decantation and filtration.

4. *Sublimation.* This is the process in which a solid passes directly to the gaseous state and back to the solid state without the appearance of the liquid state. Not all substances possess the ability to be sublimed. Iodine, naphthalene, dry ice (CO_2), and ammonium chloride (NH_4Cl) are common substances that easily sublime under normal laboratory conditions.

PROCEDURE

The mixture you will separate contains three components: NaCl, NH_4Cl, and SiO_2. Their separation will be accomplished by heating the mixture to sublime the NH_4Cl, extracting the NaCl with water, and drying the remaining SiO_2, as illustrated in the scheme shown in Figure 3.2. Carefully determine the mass of a clean, dry evaporating dish to the nearest 0.01 g. Then obtain from your instructor a 2 to 3-g sample of the unknown mixture in the evaporating dish. Write the unknown number on your report sheet. If you obtain your unknown from a bottle, shake the bottle to make the sample mixture as uniform as possible. Determine the mass of the evaporating dish containing the sample and calculate the sample mass to the nearest 0.001 g.

Place the evaporating dish containing the mixture on a clay triangle (or wire gauze), ring, and ring-stand assembly *in the hood* as shown in Figure 3.3. Heat the evaporating dish with a burner until white fumes no longer form (a total of about 15 min). Heat carefully to avoid spattering, especially when liquid is present. Occasionally shake the evaporating dish gently, using crucible tongs during the sublimation process.

Allow the evaporating dish to cool until it reaches room temperature; then determine the mass of the evaporating dish with the contained solid. NEVER PUT A HOT OR WARM OBJECT ON A BALANCE! The loss in mass represents the amount of NH_4Cl in your mixture. Calculate the mass loss.

Add 25 mL of water to the solid in this evaporating dish and stir gently for 5 min. Then, determine the mass of another clean, dry evaporating dish and

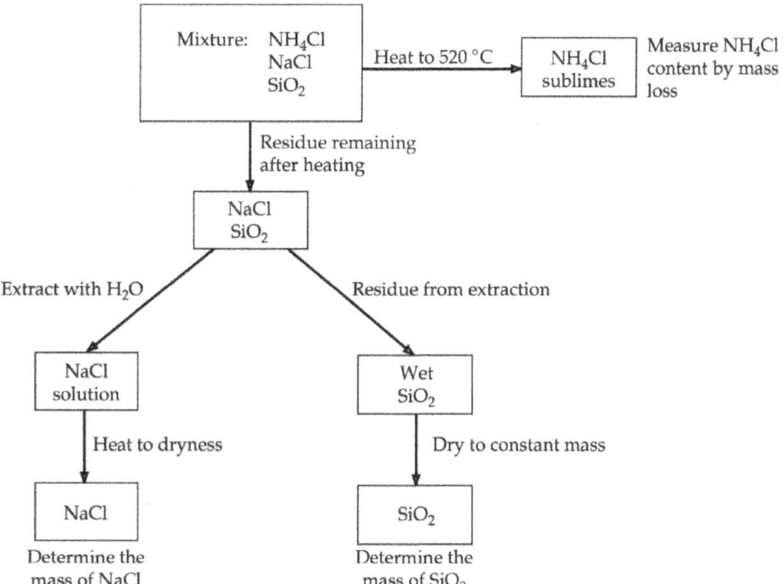

▲**FIGURE 3.2** Flow diagram for the separation of the components of a mixture.

GIVE IT SOME THOUGHT

a. What property allows you to separate NH_4Cl from NaCl and SiO_2 ?
b. Is this property a chemical or physical property?
c. What property allows you to separate NaCl from SiO_2 ?
d. Is this a chemical or physical property?

watch glass. Decant the liquid carefully into the second evaporating dish, *for which you know the mass* being careful not to transfer any of the solid into the second evaporating dish. Add 10 mL more of water to the solid in the first evaporating dish, stir, and decant this liquid into the second evaporating dish as before. Repeat with still another 10 mL of water. This process extracts the soluble NaCl from the sand. You now have two evaporating dishes—one containing wet sand, and the other containing a solution of sodium chloride.

Carefully place the evaporating dish containing the sodium chloride solution on the clay triangle on the ring stand. Begin heating the solution gently to evaporate the water. Take care to avoid boiling or spattering, especially when liquid is present. Near the end of the process, cover the evaporating dish with the watch glass whose mass was determined with this evaporating dish and reduce the heat to prevent spattering. While the water is evaporating, if you have another Bunsen burner available, you may proceed to dry the SiO_2 in the other evaporating dish as explained in the next paragraph. When you have dried the sodium chloride completely, no more water will condense on the watch glass and it too will be dry. Let the evaporating dish and watch glass cool to room temperature on a wire gauze and determine the mass. The difference between this mass and the mass of the empty evaporating dish and watch glass is the mass of the NaCl. Calculate this mass.

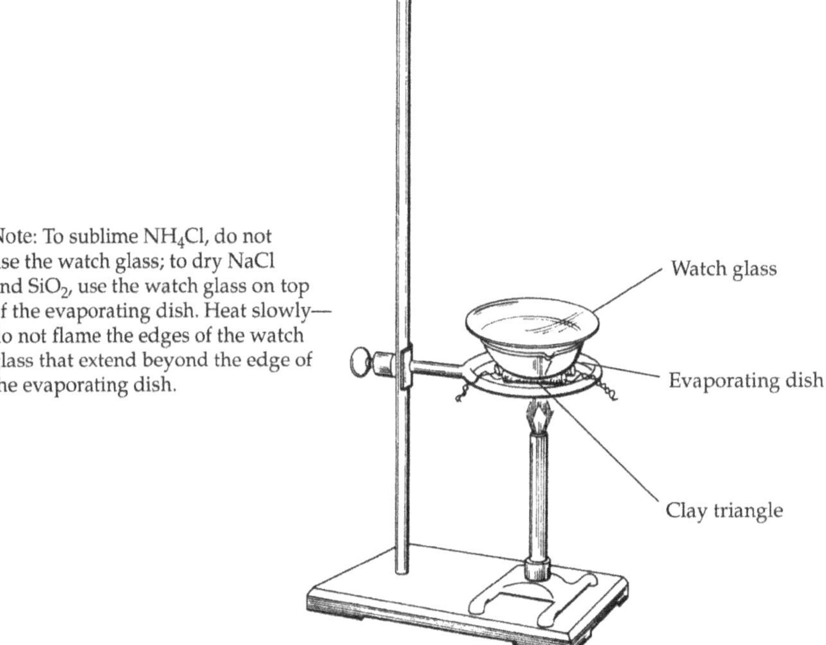

Note: To sublime NH_4Cl, do not use the watch glass; to dry NaCl and SiO_2, use the watch glass on top of the evaporating dish. Heat slowly—do not flame the edges of the watch glass that extend beyond the edge of the evaporating dish.
Watch glass
Evaporating dish
Clay triangle

▲**FIGURE 3.3** Apparatus for heating of mixture.

GIVE IT SOME THOUGHT

How will the sum of the percentage of each substance in the mixture be influenced if you do not dry your sample properly?

Place the evaporating dish containing the wet sand on the clay triangle on the ring stand and cover the evaporating dish with a clean, dry watch glass. Heat slowly at first until the lumps break up and the sand appears dry. Then heat the evaporating dish to dull redness, maintaining this heat for 10 min. Take care not to overheat the dish; otherwise, it will crack. When the sand is dry, remove the heat and let the dish cool to room temperature. Determine the mass of the dish after it has cooled to room temperature. The difference between this mass and the mass of the empty dish is the mass of the sand. Calculate this mass. Dispose of the sand in the marked container.

Calculate the percentage of each substance in the mixture using an approach similar to that shown in Example 3.1.

The accuracy of this experiment is such that the combined total of your three components should be approximately 99%. If it is less than 99%, your technique allowed for unnecessary losses of one or more of the products. If it is more than 100%, you have not sufficiently dried the sand and salt.

EXAMPLE 3.1

What is the percentage of SiO_2 in a 2.56 g sample mixture if 1.25 g of SiO_2 has been recovered?

SOLUTION: The percentage of each substance in such a mixture can be calculated as follows:

$$\%\text{component} = \frac{\text{mass component in grams}}{\text{mass sample in grams}} \times 100\%$$

Therefore, the percentage of SiO_2 in this particular sample mixture is as follows:

$$\%\,\text{SiO}_2 = \frac{1.25\text{ g} \times 100\%}{2.56\text{ g}} = 48.8\%$$

NOTES AND CALCULATIONS

Name ______________________ Desk ______________________

Date ______________ Laboratory Instructor ______________________

Separation of the Components of a Mixture | 3 | Pre-lab Questions

Before beginning this experiment in the laboratory, you should be able to answer the following questions.

1. Classify each of the following as a pure substance or a mixture; if it is a mixture, state whether it is heterogeneous or homogeneous: (a) concrete, (b) tomato juice, (c) marble, (d) seawater, and (e) iron.

   a) Mixture-hetero b) mixture-homo c) pure substance

   d) Mixture-homo e) pure substance

2. Suggest a way to determine whether a colorless liquid is pure water or a salt solution without tasting it.

   Heat it to evaporation, and observe the remaining materials

3. What distinguishes a mixture from a pure substance?

   A pure substance has a fixed composition, and a mixture's composition varies.

4. Define the process of deposition.

   Deposition is when a solid is heated directly into a gas.

5. How do decantation and filtration differ? Which should be faster?

   Filtration uses a filter, where decantation is gently pouring a liquid from the solid. Filtration should be faster.

6. Why should you never place a hot object on the pan of a balance?

   The reaction could still be occuring and the pan could be damaged.

7. How does this experiment illustrate the principle of conservation of matter?

Quantity can never be added or removed: all lost mass is accounted for in lost hydrogen.

8. A mixture was found to contain 6.02 g of Fe, 2.45 g of lime, and 0.42 g of gypsum. What percentage of iron is in the mixture? 6.02 + 2.45 + .42 = 8.89

$\left(\frac{6.02}{8.89}\right) \times 100 = 67.7\%$ Fe → 68%

9. How could you separate a mixture of NaCl and acetone? NaCl + C_3H_6O →

You could separate it by adding water.

10. How could you separate zinc chloride from SiO_2?

Zinc chloride can be separated from SiO_2 by dissolving the ZnCl with water - silicone dioxide would be insoluble.

11. A student found that her mixture was 13% NH_4Cl, 18% NaCl, and 75% SiO_2. Assuming that her calculations are correct, what did she most likely do incorrectly in her experiment?

Her measurements were most likely inaccurate, or she didn't leave the mixture on the evaporating dish for long enough

12. Why is the NaCl extracted with water three times as opposed to only once?

Each extraction only yields about 50% of the NaCl from the water.

Name ______________________ Desk ______________

Date ______________ Laboratory Instructor ______________________

Unknown no. ______________

REPORT SHEET | EXPERIMENT

Separation of the Components of a Mixture

3

A. Determination of Mass Percent of Ammonium Chloride

Mass of evaporating dish and original sample ______ g

Mass of evaporating dish ______ g

Mass of original sample ______ g

Mass of evaporating dish after subliming NH_4Cl ______ g

Mass of NH_4Cl ______ g

Percent of NH_4Cl (show calculations) ______ %

B. Determination of Mass Percent of Sodium Chloride

Mass of evaporating dish, watch glass, and NaCl ______ g

Mass of evaporating dish and watch glass ______ g

Mass of NaCl ______ g

Percent of NaCl (show calculations) ______ %

C. Determination of Mass Percent of Silicon Dioxide

Mass of evaporating dish and SiO_2 ______ g

Mass of evaporating dish ______ g

Mass of SiO_2 ______ g

Percent of SiO_2 (show calculations) ______ %

D. Determination of Percent Recovery

Mass of original sample ____________ g

Mass of determined $(NH_4Cl + NaCl + SiO_2)$ ____________ g

Differences in these weights ____________ g

Percent recovery of matter $= \dfrac{\text{g matter recovered}}{\text{g original sample}} \times 100\% =$ ____________ %

Account for your errors.

QUESTIONS

1. Could the separation in this experiment have been done in a different order? For example, if the mixture was first extracted with water and then both the extract and the insoluble residue were heated to dryness, could you determine the amounts of NaCl, NH_4Cl, and SiO_2 originally present? Why or why not?

Consult a handbook to answer these questions.

2. How could you separate barium sulfate, $BaSO_4$, from NaCl?

3. How could you separate magnesium chloride, $MgCl_2$, from silver chloride, AgCl?

4. How could you separate tellurium dioxide, TeO_2, from SiO_2?

5. How could you separate lauric acid from α-naphthol? (See Table 2.1.)

Chemical Reactions

OBJECTIVE

To observe typical chemical reactions, identify some of the products and summarize the chemical changes in terms of balanced chemical equations.

APPARATUS AND CHEMICALS

Apparatus

Bunsen burner and hose
16 cm test tube
thistle tube or long-stem funnel
droppers (2)
crucible and cover
glass tubing
ring stand, iron ring, wire triangle
tongs

Chemicals

0.1 *M* sodium oxalate, $Na_2C_2O_4$
10 *M* NaOH
1 *M* K_2CrO_4
mossy zinc
6 *M* $NH_3(aq)$*
conc. HNO_3
0.1 *M* $NaHSO_3$ (freshly prepared)
10 cm of copper wire (diameter: 1.0 to 1.6 mm)
0.1 *M* $BaCl_2$
powdered sulfur
0.1 *M* $KMnO_4$
0.1 *M* $Pb(NO_3)_2$
6 *M* HCl
6 *M* H_2SO_4
3 *M* $(NH_4)_2CO_3$
$KMnO_4$
Na_2CO_3
Na_2SO_3
ZnS

DISCUSSION

Chemical equations represent what transpires in a chemical reaction (Section 3.1). For example, the equation

$$2KClO_3(s) \xrightarrow{\Delta} 2KCl(s) + 3O_2(g)$$

means that potassium chlorate, solid $KClO_3$, decomposes upon heating (Δ is the symbol used for heat) to yield solid potassium chloride, KCl, and oxygen gas, O_2. *Before* an equation can be written for a reaction, someone must establish what the products are. How does that person decide what these products are? Products are identified by their chemical and physical properties as well as by analyses. That oxygen rather than chlorine gas is produced in the above reaction can be established by the fact that oxygen is a colorless, odorless gas. Chlorine, on the other hand, is a pale yellow-green gas with an irritating odor.

In this experiment, you will observe that in some cases, gases are produced, precipitates are formed, or color changes occur during the reactions. These are all indications that a chemical reaction has occurred. To identify some of the products of the reactions, consult Table 4.1, which lists some of the properties of the substances that can be formed in these reactions.

*Reagent bottle may be labeled 6 *M* NH_4OH.

TABLE 4.1 Properties of Reaction Products

Water-soluble solids	Water-insoluble solids	Manganese oxyanions	Gases
KCl: white (colorless solution)	CuS: very dark blue or black	MnO_4^-: purple solution	H_2: colorless; odorless
NH_4Cl: white (colorless solution)	Cu_2S: black	MnO_4^{2-}: dark green solution	NO_2: brown; pungent odor (TOXIC)
$KMnO_4$: purple	$BaCrO_4$: yellow	MnO_4^{3-}: dark blue solution	NO: colorless; slight pleasant odor
$MnCl_2$: pink (very pale)	$BaCO_3$: white		CO_2: colorless; odorless
$Cu(NO_3)_2$: colorless	$PbCl_2$: white		Cl_2: pale yellow-green; pungent odor (TOXIC)
	MnO_2: black or brown		SO_2: colorless; choking odor (as from matches) (TOXIC)
			H_2S: colorless; rotten-egg odor (TOXIC)

PROCEDURE

A. A Reaction between the Elements Copper and Sulfur

PERFORM THIS EXPERIMENT IN THE HOOD WITH A PARTNER. Obtain about a 5 cm length of copper wire and note its properties. Observe that its surface is shiny, that it can be easily bent, and that it has a characteristic color. Make a small coil of the wire by wrapping it around your pencil and place the wire coil in a crucible. Add sufficient powdered sulfur to completely cover the copper in the crucible. Cover and place the crucible on a clay triangle on an iron ring for heating. (See Figure 3.3) THIS APPARATUS MUST BE SET UP AND ALL THE PROCEDURE PERFORMED IN THE HOOD because some sulfur will burn to form noxious sulfur dioxide. Heat the crucible with a Bunsen burner initially with low heat on all sides; then use the hottest flame to heat the bottom of the crucible to red heat. Continue heating until no more smoking occurs, indicating that all the sulfur is burned off. Using the crucible tongs, remove the crucible from the clay triangle without removing the cover and place it on a heat-resistant pad or wire gauze, *not on the desktop*, to cool. After the crucible has cooled, remove the cover and inspect the substance. Note its properties. Record your answers on the Report Sheet.

GIVE IT SOME THOUGHT

a. In this reaction, what is the limiting reactant?
b. Which reactant is in excess?

1. Does the substance resemble copper?
2. Is it possible to bend the substance without breaking it? Handle it with tongs or forceps—not your fingers.
3. What color is it?
4. Has a reaction occurred?

Copper(II) sulfide, CuS, is insoluble in aqueous ammonia, $NH_3(aq)$ (that is, does not react with NH_3), whereas copper(I) sulfide, Cu_2S, dissolves (that is, does react) to give a blue solution with NH_3. Place a small portion of your product in a test tube and add 2 mL of 6 M NH_3 in the hood. Heat gently with a Bunsen burner.

5. Does your product react with NH_3?
6. Suggest a possible formula for the product.
7. Write a reaction showing the formation of your proposed product:

$$Cu(s) + S_8(s) \longrightarrow ?$$

Waste Disposal Instruction The copper compounds and the acids are toxic and should be handled with care. Avoid spilling any solution and immediately clean up spills that occur (using paper towels). If you spill any solution on your hands, wash them immediately. After completing each series of reactions and before moving on to the next series, dispose of the contents of your test tubes in the designated receptacles. Do not wash the contents down the sink. Acids should be neutralized with sodium bicarbonate.

B. Oxidation–Reduction Reactions

Many metals react with acids to liberate hydrogen and form the metal salt of the acid (Section 4.4). The noble metals do not react with acids to produce hydrogen. Some of the unreactive metals do react with nitric acid, HNO_3; however, in these cases, gases that are oxides of nitrogen rather than hydrogen are formed.

Add a small piece of zinc to a test tube containing 2 mL of 6 *M* HCl and note what happens.

8. Record your observations.
9. Suggest possible products for the observed reaction: $Zn(s) + HCl(aq) \longrightarrow ?$

Place a 2.5 cm piece of clean shiny copper wire in a clean test tube. Add 2 mL of 6 *M* HCl and note whether a reaction occurs.

10. Record your observations.
11. Is Cu an active or an inactive metal?

WHILE HOLDING A CLEAN TEST TUBE IN THE HOOD, place a 2.5 cm piece of copper wire in the test tube and add 1 mL of concentrated nitric acid, HNO_3.

12. Record your observations.
13. Is the gas colored? If so, what color is it?
14. Suggest a formula for the gas.
15. After the reaction has proceeded for 5 min, carefully add 5 mL of water. Based on the color of the solution, what substance is present?

GIVE IT SOME THOUGHT

a. What is the oxidation state of manganese in $KMnO_4$?

b. Based on this oxidation state, why is $KMnO_4$ an excellent oxidizing agent?

Potassium permanganate, $KMnO_4$, is an excellent oxidizing agent in acidic media. The permanganate ion is purple and is reduced to the manganous ion, Mn^{2+}, which has a very faint pink color. Place 1 mL of 0.1 *M* sodium oxalate, $Na_2C_2O_4$, in a clean test tube. Add 10 drops of 6 *M* sulfuric acid. Mix thoroughly. To the resulting solution add 1 or 2 drops of 0.1 *M* $KMnO_4$ and stir. If there is no obvious indication that a reaction has occurred, warm the test tube gently in a hot water bath (see Appendix J).

16. Record your observations. Was the $KMnO_4$ reduced to Mn^{2+} ?

Place 3 mL of 0.1 *M* sodium hydrogen sulfite, $NaHSO_3$, solution in a test tube. Add 1 mL of 10 *M* sodium hydroxide, NaOH, solution and stir. To the mixture in the test tube add 1 drop of 0.1 *M* $KMnO_4$ solution.

17. Record your observations. Was the $KMnO_4$ reduced? Identify the manganese compound formed.

Add additional 0.1 *M* $KMnO_4$ solution, one drop at a time, and observe the effect of each drop until you have added 10 drops.

18. Record your observations.
19. Suggest why the effect of additional potassium permanganate changes as more is added.

WHILE HOLDING A TEST TUBE IN THE FUME HOOD, add one or two crystals of potassium permanganate, $KMnO_4$, to 1 mL of 6 *M* HCl.

20. Record your observations.
21. Note the color of the gas evolved.
22. Based on the color of the gas, what is the gas?

Dispose of the solution in an appropriate container. All of the above Mn containing solution as well as the copper ones should be disposed of in the correct manner.

C. Metathesis Reactions

Additional observations are needed before equations can be written for the reactions above, but some of the products can be identified (Section 4.2). The remaining reactions are simple, and you will be able, from available information, not only to identify products, but also to write equations. A number of reactions may be represented by equations of the following type:

$$AB + CD \longrightarrow AD + CB$$

These are called exchange, double replacement, or *metathesis*–reactions. This type of reaction involves the exchange of ions between interacting substances. The following is a specific example:

$$NaCl(aq) + AgNO_3(aq) \longrightarrow AgCl(s) + NaNO_3(aq)$$

Place a small sample of sodium carbonate, Na_2CO_3, in a test tube and add several drops of 6 *M* HCl.

23. Record your observations.
24. Note the odor and color of the gas that forms (see safety instructions).
25. What is the evolved gas?
26. Write an equation for the reaction $HCl(aq) + Na_2CO_3(s) \longrightarrow ?$ (NOTE: In this reaction, the products must have H, Cl, Na, and O atoms in new combinations but no other elements can be present.)

Note that H_2CO_3 and H_2SO_3 readily decompose as follows:

$$H_2CO_3(aq) \longrightarrow H_2O(l) + CO_2(g)$$

$$H_2SO_3(aq) \longrightarrow H_2O(l) + SO_2(g)$$

IN THE HOOD, repeat the same test with sodium sulfite, Na_2SO_3.

27. Record your observations.
28. What is the gas?
29. Write an equation for the following reaction (note the similarity to the equation above): $HCl(aq) + Na_2SO_3(s) \longrightarrow ?$

IN THE HOOD, repeat this test with zinc sulfide, ZnS.

30. Record your observations.
31. What is the gas?
32. Write an equation for the reaction $HCl(aq) + ZnS(s) \longrightarrow ?$

To 1 mL of 0.1 *M* lead nitrate, $Pb(NO_3)_2$, solution in a clean test tube add a few drops of 6 *M* HCl.

33. Record your observations.
34. What is the precipitate?
35. Write an equation for the reaction $Pb(NO_3)_2(aq) + HCl(aq) \longrightarrow ?$

To 1 mL of 0.1 *M* barium chloride, $BaCl_2$, solution add 2 drops of 1 *M* potassium chromate, K_2CrO_4, solution.

36. Record your observations.
37. What is the precipitate?
38. Write an equation for the reaction $BaCl_2(aq) + K_2CrO_4(aq) \longrightarrow ?$

To 1 mL of 0.1 *M* barium chloride, $BaCl_2$, solution add several drops of 3 *M* ammonium carbonate, $(NH_4)_2CO_3$, solution in a test tube.

39. What is the precipitate?
40. Write an equation for the reaction $(NH_4)_2CO_3(aq) + BaCl_2(aq) \longrightarrow ?$

After the precipitate has settled somewhat, carefully decant (that is, pour off) the excess liquid. Add 1 mL of water to the test tube, shake it, allow the precipitate to settle, and again carefully pour off the liquid. To the remaining solid, add several drops of 6 *M* HCl.

41. Record your observations.
42. Note the odor.
43. What is the evolved gas? (Recall the reaction in step 26 of this experiment.)

Dispose of solutions containing Pb, Cr, and Ba in appropriate containers.

NOTES AND CALCULATIONS

Name ______________________________ Desk ______________

Date ______________ Laboratory Instructor ______________________

Chemical Reactions | 4 Pre-lab Questions

Before beginning this experiment in the laboratory, you should be able to answer the following questions.

1. Before you can write a chemical equation, what must you know?

2. What observations might you make that suggest that a chemical reaction has occurred?

3. How could you distinguish between NaCl and $PbCl_2$?

4. Define metathesis reactions. Give an example.

5. What is a precipitate?

6. Balance these equations:

$$KBrO_3(s) \xrightarrow{\Delta} KBr(s) + O_2(g)$$

$$MnBr_2(aq) + AgNO_3(aq) \longrightarrow Mn(NO_3)_2(aq) + AgBr(s)$$

7. How could you distinguish between the gases O_2 and H_2?

8. Using water, how could you distinguish between the white solids $Pb(OH)_2$ and $NaHCO_3$?

9. Write equations for the decomposition of $H_2CO_3(aq)$ and $H_2SO_3(aq)$.

Name ______________________________ Desk ______________

Date ______________ Laboratory Instructor ______________________

Partner's Name ______________________

REPORT SHEET | EXPERIMENT

Chemical Reactions | 4

A. A Reaction between the Elements Copper and Sulfur

1. ______________________________

2. ______________________________

3. ______________________________

4. ______________________________

5. ______________________________

6. ______________________________

7. $Cu(s) + S_8(s) \xrightarrow{\Delta}$ ______________________________

B. Oxidation–Reduction Reactions

8. ______________________________

9. $Zn(s) + HCl(aq) \longrightarrow$ ______________________________

10. ______________________________

11. ______________________________

12. ______________________________

13. ______________________________

14. ______________________________

15. ______________________________

16. ______________________________

17. ______________________________

18. ______________________________

19. ______________________________

20. ______________________________

21. ______________________________

22. ______________________________

C. Metathesis Reactions

23. ______________________________

24. ______________________________

25. ______________________________

26. $HCl(aq) + Na_2CO_3(s) \longrightarrow$ ______________________________

27. ______________________________

28. ______________________________

29. $HCl(aq) + Na_2SO_3(s) \longrightarrow$ ______________________________

30. ______________________________

31. ______________________________

32. $HCl(aq) + ZnS(s) \longrightarrow$ ______________________________

33. ______________________________

34. ______________________________

35. $Pb(NO_3)_2(aq) + HCl(aq) \longrightarrow$ ______________________________

36. ______________________________

37. ______________________________

38. $BaCl_2(aq) + K_2CrO_4(aq) \longrightarrow$ ______________________________

39. ______________________________

40. $(NH_4)_2CO_3(aq) + BaCl_2(aq) \longrightarrow$ ______________________________

41. ______________________________

42. ______________________________

43. ______________________________

QUESTIONS

1. Complete and balance the following chemical reactions:

   $2HCl(aq) + Pb(NO_3)_2(aq) \longrightarrow$

   $2HI(aq) + K_2SO_3(s) \longrightarrow$

   $Pb(NO_3)_2(aq) + 2\ KCl(aq) \longrightarrow$

   $Ba(NO_3)_2(aq) + Na_2SO_4(aq) \longrightarrow$

   $K_2CO_3(aq) + Ba(NO_3)_2(aq) \longrightarrow$

   $HCl(aq) + AgNO_3(aq) \longrightarrow$

2. How could you separate gold from a mixture of zinc and gold?
3. Using a hydrochloric acid solution, how could you determine whether a white powder was zinc sulfide or silver nitrate? Write balanced equations.

   $ZnS(s) + 2HCl(aq) \longrightarrow$

   $AgNO_3(s) + HCl(aq) \longrightarrow$

NOTES AND CALCULATIONS

Experiment

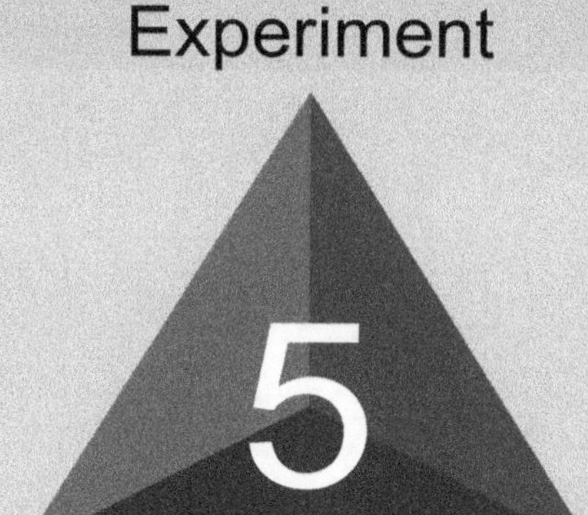

Chemical Formulas

OBJECTIVE

To become familiar with chemical formulas and how they are obtained.

APPARATUS AND CHEMICALS

Apparatus

balance
Bunsen burner and hose
50 mL graduated cylinder
wire gauze
crucible and cover
boiling chips
250 mL beaker
evaporating dish
ring stand and two iron rings
stirring rod
clay triangle

Chemicals

granular zinc
powdered sulfur
copper wire
6 *M* HCl

DISCUSSION

Chemists use an abbreviated notation to indicate the exact chemical composition of compounds (chemical formulas). You then use these chemical formulas to indicate how new compounds are formed by chemical combinations of other compounds (chemical reactions). However, before you can learn how chemical formulas are written, you must first become acquainted with the symbols used to denote the elements from which these compounds are formed.

Symbols and Formulas

One or two letters are used (with the first letter capitalized) to denote a chemical element (Sections 2.5 and 2.6). These symbols are derived, as a rule, from the first two letters or first part of the element's (Latin) name.

Many elements are found in nature in molecular form; that is, two or more of the same type of atom are tightly bound together. The resultant "package" of atoms, or *molecule* as it is termed, behaves in many ways as a single distinct object or unit. For example, the oxygen normally found in air consists of molecules that contain two oxygen atoms. This molecular form of oxygen is represented by the chemical formula O_2. The subscript in the formula shows that two oxygen atoms are present in each oxygen molecule.

Compounds that are composed of molecules are called *molecular compounds*, and they may contain more than one type of atom. For example, a molecule of water consists of two hydrogen atoms and one oxygen atom and is represented by the chemical formula H_2O. The absence of a subscript on the O implies that there is one oxygen atom per water molecule. Another compound composed of these same elements but in different proportions is hydrogen peroxide, H_2O_2. The physical and chemical properties of these two compounds are very different. You shouldn't be surprised, for they are two different substances.

Chemical formulas that indicate the *actual* numbers and types of atoms in a molecule are called *molecular formulas*, whereas chemical formulas that indicate only the *relative* numbers of atoms of a type in a molecule are called *empirical formulas*. The subscripts in an empirical formula are always the smallest whole-number ratios. For example, the molecular formula for hydrogen peroxide is H_2O_2, whereas its empirical formula is HO. The molecular formula for glucose is $C_6H_{12}O_6$; its empirical formula is CH_2O. For many substances, the molecular formula and empirical formula are identical, as is the case for water, H_2O, and sulfuric acid, H_2SO_4.

Atomic Weights

It is important to know something about masses of atoms and molecules (Section 2.4). Using a mass spectrometer, you can measure the masses of individual atoms with a high degree of accuracy. For example, the hydrogen-1 atom has a mass of 1.6735×10^{-24} g and the oxygen-16 atom has a mass of 2.656×10^{-23} g. Because it is cumbersome to express such small masses in grams, a unit called the *atomic mass unit*, or u, is used. An u equals 1.66054×10^{-24} g. Most elements occur as mixtures of isotopes. The average atomic mass of each element expressed in u is also known as its *atomic weight*. The atomic weights of the elements listed in the table of elements and in the periodic table inside the front and back covers of this book, are in u.

Formula and Molecular Weights

The *formula weight* of a substance is merely the sum of the atomic weights of all atoms in its chemical formula (Section 3.3). For example, nitric acid, HNO_3, has a formula weight of 63.0 u.

$$\begin{aligned} \text{FW} &= (\text{AW of H}) + (\text{AW of N}) + 3(\text{AW of O}) \\ &= 1.0\text{ u} + 14.0\text{ u} + 3(16.0\text{ u}) \\ &= 63.0\text{ u} \end{aligned}$$

If the chemical formula of a substance is its molecular formula, the formula weight is also called the *molecular weight*. For example, the molecular formula for formaldehyde is CH_2O. Therefore, the molecular weight of formaldehyde is as follows:

$$\begin{aligned} \text{MW} &= 12.0\text{ u} + 2(1.0\text{ u}) + 16.0\text{ u} \\ &= 30.0\text{ u} \end{aligned}$$

For ionic substances such as NaCl that exist as three-dimensional arrays of ions, it is not appropriate to speak of molecules. Similarly, the terms *molecular weight* and *molecular formula* are inappropriate for these ionic substances. It is correct to speak of their formula weight, however. Thus, the formula weight of NaCl is as follows:

$$\begin{aligned} \text{FW} &= 23.0\text{ u} + 35.5\text{ u} \\ &= 58.5\text{ u} \end{aligned}$$

Percentage Composition from Formulas

New compounds are made in laboratories every day, and the formulas of these compounds must be determined. The compounds are often analyzed for their *percentage composition* (that is the percentage by mass of each element present in the compound). The percentage composition is useful information in establishing the formula for the substance. If the formula of a compound is known, calculating its percentage composition is a straightforward matter. In general, the percentage of an element in a compound is given by the following formula:

$$\frac{\text{(number of atoms of element)(AW)}}{\text{FW of compound}} \times 100\%$$

where AW = atomic weight and FW = formula weight

If you want to know the percentage composition of formaldehyde, CH_2O, whose formula weight is 30.0 u, you proceed as follows:

$$\%\text{C} = \frac{12.0\ \text{u}}{30.0\ \text{u}} \times 100\% = 40.0\%$$

$$\%\text{H} = \frac{2(1.0\ \text{u})}{30.0\ \text{u}} \times 100\% = 6.7\%$$

$$\%\text{O} = \frac{16.0\ \text{u}}{30.0\ \text{u}} \times 100\% = 53.3\%$$

The Mole

Even the smallest samples used in the laboratory contain an enormous number of atoms. A drop of water contains about 2×10^{21} water molecules! The unit the chemist uses for dealing with such a large number of atoms, ions, or molecules is the *mole*, abbreviated mol. Just as the unit *dozen* refers to 12 objects, the mole refers to a collection of 6.02×10^{23} objects. This number is called Avogadro's number. Thus, a mole of water molecules contains 6.02×10^{23} H_2O molecules and a mol of sodium contains 6.02×10^{23} Na atoms. The mass (in grams) of 1 mol of a substance is called its *molar mass*. The molar mass (in grams) of any substance is numerically equal to its formula weight. Thus:

One CH_2O molecule has a mass of 30.0 u; 1 mol CH_2O has a mass of 30.0 g and contains 6.02×10^{23} CH_2O molecules.

One Na atom has a mass of 23.0 u; 1 mol Na has a mass of 23.0 g and contains 6.02×10^{23} Na atoms.

It is a simple matter to calculate the number of moles of any substance whose mass and formula are known. For example, suppose you have 946 mL (1 quart in the English system of units) of rubbing alcohol (generally isopropyl alcohol) and know its density to be 0.785 g/mL and want to know how many moles of isopropyl alcohol this is.

Now you can calculate the mass:

$$946 \text{ mL} \times 0.785 \text{ g/mL} = 743 \text{ g}$$

Next, you need the chemical formula for isopropyl alcohol. This is C_3H_7OH. Therefore, the molecular weight is as follows:

Weight carbon	$3 \times 12.0 = 36.0$ u
Weight hydrogen	$8 \times 1.0 = 8.0$ u
Weight oxygen	$1 \times 16.0 = 16.0$ u
Molecular weight	$C_3H_7OH = 60.0$ u

Hence, on the gram mole scale using the molar mass as the conversion factor

$$\text{moles of } C_3H_7OH = (743 \text{ g } C_3H_7OH)\left(\frac{1 \text{ mol } C_3H_7OH}{60.0 \text{ g } C_3H_7OH}\right) = 12.4 \text{ mol.}$$

Thus, 946 mL of rubbing alcohol contains 12.4 mol of isopropyl alcohol. It should now be apparent to you how much information is contained in a chemical formula.

Empirical Formulas from Analyses

The empirical formula for a substance tells you the relative number of atoms of each element in the substance. Thus, the formula H_2O indicates that water contains 2 hydrogen atoms for each oxygen atom. This ratio applies on the molar level as well; thus, 1 mol of H_2O contains 2 mol of H atoms and 1 mol of O atoms. Conversely, the ratio of the number of moles of each element in a compound gives the subscripts in a compound's empirical formula. Thus, the mole concept provides a way of calculating the empirical formula of a chemical substance. This is shown in the following example.

Example 5.1

While you are working in a hospital laboratory, a patient complaining of severe stomach cramps and labored respiration dies within minutes of being admitted. Relatives of the patient tell you that he may have ingested rat poison. Therefore, you have his stomach pumped to verify this and to determine the cause of death. One of the more logical things to do would be to attempt to isolate the agent that caused death and perform chemical analyses on it. Assume that this was done, and the analyses showed that the isolated chemical compound contained, by weight, 60.0% potassium, 18.5% carbon, and 21.5% nitrogen. What is the chemical formula for this compound?

SOLUTION: One simple and direct way of making the necessary calculations is as follows: Assume that you had 100 g of the compound. This 100 g would contain the following:

$$(100 \text{ g})(0.600) = 60.0 \text{ g potassium}$$
$$(100 \text{ g})(0.185) = 18.5 \text{ g carbon}$$
$$(100 \text{ g})(0.215) = 21.5 \text{ g nitrogen}$$

Divide each of these masses by the appropriate molar mass to obtain the number of moles of each element in the 100 g.

$$60.0 \text{ g K}\left(\frac{1 \text{ mol K}}{39.0 \text{ g K}}\right) = 1.54 \text{ mol K}$$

$$18.5\text{ g C}\left(\frac{1\text{ mol C}}{12.0\text{ g C}}\right) = 1.54\text{ mol C}$$

$$21.5\text{ g N}\left(\frac{1\text{ mol N}}{14.0\text{ g N}}\right) = 1.54\text{ mol N}$$

Then divide each number by 1.54 (the smallest number of moles) to determine the simplest whole-number ratio of moles of each element. (In general, after determining the number of moles of each element, you determine the simplest whole-number ratio by dividing each number of moles by the smallest number of moles. In this example, all of the numbers are the same.)

$$K = \frac{1.54}{1.54} = 1.00$$

$$C = \frac{1.54}{1.54} = 1.00$$

$$N = \frac{1.54}{1.54} = 1.00$$

The ratio obtained in this case is 1.00, and you conclude that the formula is KCN. This is the simplest, or empirical, formula because it uses as subscripts the smallest set of integers to express the correct ratios of atoms present. Because KCN is a common rat poison, you may justifiably conclude that the relatives' suggestion of rat poison ingestion as the probable cause of death is correct.

Molecular Formulas from Empirical Formulas

The formula obtained from percentage composition is *always* the empirical formula. You can obtain the molecular formula from the empirical formula if you know the molecular weight of the compound. *The subscripts in the molecular formula of a substance are always a whole-number multiple of the corresponding subscripts in its empirical formula*. The multiple is found by comparing the formula weight of the empirical formula with the molecular weight. For example, suppose you determined the empirical formula of a compound to be CH_2O. Its formula weight is as follows:

$$\text{FW} = 12.0\text{ u} + 2(1.0\text{ u}) + 16.0\text{ u} = 30.0\text{ u}$$

Suppose the experimentally determined molecular weight is 180 u. Then the molecule has six times the mass (180 u/30.0 u = 6.00); therefore, it must have six times as many atoms as the empirical formula. The subscripts in the empirical formula must be multiplied by 6 to obtain the molecular formula: $C_6H_{12}O_6$.

In this experiment, you will determine the empirical formulas of two chemical compounds. One is copper sulfide, which you will prepare according to the following chemical reaction:

$$x\text{Cu}(s) + y\text{S}(s) \longrightarrow \text{Cu}_x\text{S}_y(s)$$

The other is zinc chloride, which you will prepare according to this chemical reaction:

$$x\text{Zn}(s) + y\text{HCl}(aq) \longrightarrow \text{Zn}_x\text{Cl}_y(s) + \frac{y}{2}\text{H}_2(g)$$

The objective is to determine the combining ratios of the elements (that is, to determine x and y) and to balance the chemical equations given above.

PROCEDURE

A. Zinc Chloride

Clean and dry your evaporating dish and place it on the wire gauze resting on the iron ring. Heat the dish with your Bunsen burner, gently at first, then more strongly, until all of the condensed moisture has been driven off. This should require heating for about 5 min. Allow the dish to cool to room temperature on a wire gauze (do not place the hot dish on the countertop) and determine its mass. Record the mass of the empty evaporating dish to the nearest 0.01 g.

Obtain a sample of granular zinc from your laboratory instructor and add about 0.5 g of it to the dried and massed evaporating dish. Determine the mass of the evaporating dish containing the zinc and record the total mass to the nearest 0.01 g. Calculate the mass of the zinc.

Slowly, and with constant swirling, add 15 mL of 6 *M* HCl to the evaporating dish containing the zinc. A vigorous reaction will ensue, and hydrogen gas will be produced. **(CAUTION:** ***No flames are permitted in the laboratory while this reaction is taking place because hydrogen gas is explosive.*****)** If any undissolved zinc remains after the reaction ceases, add an additional 5 mL of acid. Continue to add 5 mL portions of acid as needed until all of the zinc has dissolved. **(CAUTION:** ***Zinc chloride is caustic and must be handled carefully to avoid contact with your skin. Should you come in contact with zinc chloride, immediately wash the area with copious amounts of water.*****)**

Set up a steam bath as illustrated in Figure 5.1 using a 250 mL beaker and place the evaporating dish on the steam bath. Heat the evaporating dish carefully on the steam bath until most of the liquid has disappeared. Then remove the steam bath and heat the dish on the wire gauze. During this last stage of heating, the flame must be carefully controlled; otherwise, spattering and some loss of product will occur. **(CAUTION:** ***Do not heat to the point that the compound melts. If you do, some of it will be lost due to sublimation.*****)** Leave the compound looking somewhat pasty while hot.

Allow the dish to cool to room temperature and determine its mass. Record the mass. After this first mass determination, gently heat the dish again. Cool it and determine its mass again. If masses do not agree within 0.02 g, repeat the

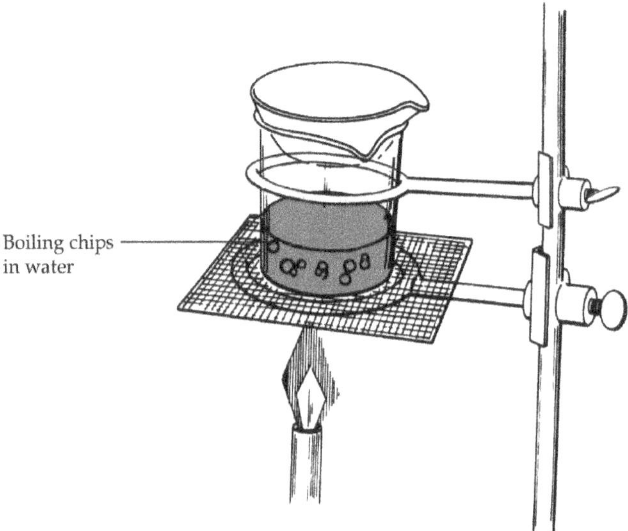

▲**FIGURE 5.1** Steam bath.

heating and mass determination until two successive masses agree within the specified precision. This is known as *drying to constant mass* and is the only way to make sure all of the moisture is driven off. Zinc chloride is very deliquescent (rapidly absorbs moisture from the air); so the mass should be determined as soon as possible after cooling.

Calculate the mass of zinc chloride. The difference in mass between the zinc and zinc chloride is the mass of chlorine. Calculate the mass of chlorine in zinc chloride. From this information, you can readily calculate the empirical formula for zinc chloride and balance the chemical equation for its formation. Perform these operations on the report sheet.

B. Copper Sulfide

Support a clean, dry porcelain crucible and cover on a clay triangle and dry by heating to a dull red in a Bunsen flame, as illustrated in Figure 5.2. Allow the crucible and cover to cool to room temperature and determine their mass. Record the mass to the nearest 0.01 g.

Place 1.5 to 2.0 g of tightly wound copper wire or copper turnings in the crucible and determine the mass of the copper, crucible, and lid to the nearest 0.01 g; record your results. Calculate the mass of copper.

In the hood, add sufficient sulfur to cover the copper, place the crucible with the cover in place on the triangle and heat the crucible gently until sulfur ceases to burn (blue flame) at the end of the cover. Do not remove the cover while the crucible is hot. Finally, heat the crucible to dull redness for about 5 min.

Allow the crucible to cool to room temperature. This will take about 10 min. Then determine the mass with the cover in place. Record the mass. Again cover the contents of the crucible with sulfur and repeat the heating procedure. Allow the crucible to cool and determine the mass again. Record the mass. If the last two mass values do not agree to within 0.02 g, the chemical reaction between the copper and sulfur is incomplete. If you find this to be the case, add more sulfur and repeat the heating and determining the mass until a constant mass is obtained.

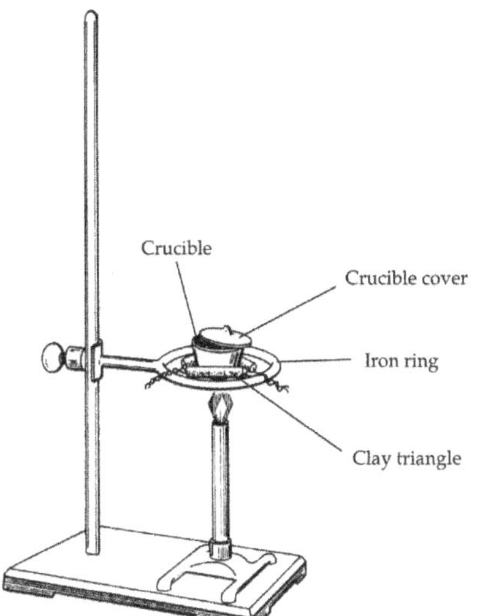

▲**FIGURE 5.2** Setup for copper sulfide determination.

Calculate the mass of copper sulfide obtained. The difference in mass between the copper sulfide and copper is the mass of sulfur in copper sulfide. Calculate this mass. From this information, you can obtain the empirical formula for copper sulfide and balance the chemical equation for its production. Perform these operations on your report sheet.

Waste Disposal Instructions All chemicals must be disposed of in the appropriately labeled containers.

Name ____________________ Desk ____________________

Date ____________________ Laboratory Instructor ____________________

Quiz Friday 10/16
→ separation of components & mixture

Chemical Formulas | 5 | Pre-lab Questions

Before beginning this experiment in the laboratory, you should be able to answer the following questions.

1. Give the chemical symbols for the following elements: (a) mercury, (b) tellurium, (c) platinum, (d) bromine, and (e) neon.

   a) Mg b) Te c) Pt d) Br e) Ne

2. What are the formula weights of (a) MgS and (b) $CaCO_3$?

   a) Mg -
   S -
   b) $CaCO_3$

3. Define the term *compound*.

   A compound is 2 or more molecules bonded together - can be the same or different element

4. The molecular formula of glucose is $C_6H_{12}O_6$. What is its empirical formula? 1:2:1 ???

5. Balance the following equation: $Al + O_2 \rightarrow Al_2O_3$. $2Al + 3O_2 \rightarrow Al_2O_3$

6. What is the percentage composition of each element in $PbCO_3$?

   Pb = (207.2) = 207.2
   C = (12) 12
   O_3 = (16×3) + 48
   267.2

   (207.2/267.2)×100 = 77.5% Pb = 78%
   (12/267.2)×100 = 4.49% C = 5%
   (48/267.2)×100 = 17.96% O = 18%

7. In 7 g of a compound made of aluminum and oxygen, 3.29 g is oxygen. What is the empirical formula for the substance?

8. What is the law of definite proportions?

9. How do empirical and molecular formulas differ?

10. What is the mass in grams of 1.73 mol CaI_2?

11. Calculate the number of moles of
(a) sulfur dioxide, SO_2 in 32 g of sulfur dioxide

(b) sulfuric acid, H_2SO_4 in 4.90 g of sulfuric acid

12. A 5.325 g sample of methyl benzoate, a compound in perfumes, was found to contain 3.758 g of carbon, 0.316 g of hydrogen, and 1.251 g of oxygen. What is the empirical formula? If its molar mass is about 130 g/mol, what is its molecular formula?

13. An analysis of an oxide of nitrogen with a molecular weight of 92.02 u has a percent composition of 69.57% oxygen and 30.43% nitrogen. What are the empirical and molecular formulas for this nitrogen oxide? Complete and balance the equation for its formation from the elements nitrogen and oxygen.

14. How many sodium atoms are in 0.1310 g of sodium?

Name Grace Rademacher Desk ______

Date 10/9/20 Laboratory Instructor Prof. Briguglio

REPORT SHEET | EXPERIMENT

Chemical Formulas | 5

A. Zinc Chloride

1. Mass of evaporating dish and zinc 45.476 g
2. Mass of evaporating dish 44.985 g
3. Mass of zinc 0.491 g
4. Mass of evaporating dish and zinc chloride:
   - first determination 46.137 g
   - second determination 46.129 g
   - third determination ______ g
5. Mass of zinc chloride 1.152 g
6. Mass of chlorine in zinc chloride .661 g

44.985 + .491 = 45.476

Diff. is < .02 = DRY

1.152 − .491 =

7. Empirical formula for zinc chloride (show calculations) $ZnCl_2$

   $Zn_{(s)} + 2HCl_{(aq)} \rightarrow ZnCl_{2(aq)} + H_{2(g)}$

   Zn^{+2} Cl^{-1} Cl^{-1}

8. Balanced chemical equation for the formation of zinc chloride from zinc and HCl

   $Zn_{(s)} + 2HCl_{(aq)} \rightarrow ZnCl_{2(aq)} + H_{2(g)}$

B. Copper Sulfide

1. Mass of crucible, cover, and copper ______ g
2. Mass of crucible and cover ______ g
3. Mass of copper ______ g
4. Mass of crucible, cover, and copper sulfide:
   - first determination ______ g
   - second determination ______ g
   - third determination ______ g
5. Mass of copper sulfide ______ g

6. Mass of sulfur in copper sulfide ___________ g

7. Empirical formula for copper sulfide
(show calculations)

____

8. Balanced chemical equation for the formation of copper sulfide from copper and sulfur

QUESTIONS

1. Can you determine the molecular formula of a substance from its percent composition?

2. Given that zinc chloride has a formula weight of 136.28 u, what is its formula?

3. Can you determine the atomic weights of zinc or copper by the methods used in this experiment? If so, how? What additional information is necessary to do this?

4. How many grams of zinc chloride could be formed from the reaction of 3.57 g of zinc with excess HCl?

5. Magnesium reacts directly with nitrogen gas to form magnesium nitride. In an experiment conducted, it was found that 0.36 g of magnesium produced 0.50 g of magnesium nitride. (a) How many moles of nitrogen combined with 0.36 g of magnesium? (b) What is the ratio of nitrogen atoms to magnesium atoms in magnesium nitride? (c) What is the empirical formula of magnesium nitride?

6. When copper(I) sulfide is partially roasted in air (reaction with O_2), copper(I) sulfite is formed first. Subsequently, upon heating, the copper sulfite thermally decomposes to copper(I) oxide and sulfur dioxide. Write balanced chemical equations for these two reactions.

NOTES AND CALCULATIONS

Experiment 6

Chemical Reactions of Copper and Percent Yield

OBJECTIVE

To gain familiarity with some basic laboratory procedures, some chemistry of a typical transition element, and the concept of percent yield.

APPARATUS AND CHEMICALS

Apparatus

balance	Bunsen burner and hose
250 mL beakers (2)	100 mL graduated cylinder
evaporating dish	weighing paper
stirring rod	boiling chips
towel	ring stand and two iron rings
wire gauze	

Chemicals

0.5 g piece of copper wire (diameter: 1.0 or 1.3 mm)	conc. HNO_3
6 *M* H_2SO_4	3.0 *M* NaOH
	granular zinc
methanol	acetone
aluminum foil cut in 2.5 cm squares	conc. HCl

DISCUSSION

Most chemical syntheses involve the separation and purification of the desired product from unwanted contaminants. Common methods of separation are filtration, sedimentation, decantation, extraction, and sublimation. This experiment is designed to be a quantitative evaluation of your individual laboratory skills in carrying out some of these operations. At the same time, you will become acquainted with two fundamental types of chemical reactions called metathesis reactions and redox reactions. Metathesis reactions are reactions in which two substances react through an exchange of their component ions: $AX + BY \rightarrow AY + BX$ (Section 4.2). Redox reactions are reactions in which certain atoms undergo changes in their oxidation states (Section 4.4). By means of these reactions, you will carry out several chemical transformations involving copper and its compounds. Finally, you will recover the copper sample with maximum efficiency. The chemical reactions involved are the following:

$$Cu(s) + 4HNO_3(aq) \longrightarrow Cu(NO_3)_2(aq) + 2NO_2(g) + 2H_2O(l) \qquad \text{Redox} \qquad [1]$$

$$Cu(NO_3)_2(aq) + 2NaOH(aq) \longrightarrow Cu(OH)_2(s) + 2NaNO_3(aq) \qquad \text{Metathesis} \qquad [2]$$

$$Cu(OH)_2(s) \xrightarrow{\Delta} CuO(s) + H_2O(g) \qquad \text{Dehydration} \qquad [3]$$

$CuO(s) + H_2SO_4(aq) \longrightarrow CuSO_4(aq) + H_2O(l)$	Metathesis	[4]
$CuSO_4(aq) + Zn(s) \longrightarrow ZnSO_4(aq) + Cu(s)$	Redox	[5]
$3CuSO_4(aq) + 2Al(s) \longrightarrow Al_2(SO_4)_3(aq) + 3Cu(s)$	Redox	[6]

GIVE IT SOME THOUGHT

After performing this series of reactions, how would you expect the mass of Cu(*s*) from step [1] to compare with the mass of Cu(*s*) from step [6]?

Each of these reactions tend to proceed to completion. Metathesis reactions proceed to completion whenever one of the products is removed from the solution, such as in the formation of a gas or an insoluble substance.

This is the case for reactions [1], [2], and [3], where in reaction [1] a gas and in reaction [2] an insoluble precipitate are formed. In reaction [3], some of the water is boiled off, but copper(II) hydrate decomposes above 60–80 °C. (Reactions [5] and [6] proceed to completion because copper is more difficult to oxidize than either zinc or aluminum.)

Reactants that are completely consumed in a reaction are termed limiting reagents because they determine, or limit, the amount of product that can be formed. In this experiment, you will allow copper to react with an excess of nitric acid. Copper is the limiting reagent and limits, or determines, the amount of copper nitrate that can be formed (Section 3.7).

The amount of product that is calculated to form when all of the limiting reagent reacts is called the theoretical yield. The amount of product actually obtained in a reaction is termed the actual yield. Percent yield of a reaction relates to the actual yield and the theoretical (calculated) yield as follows:

$$\%\ \text{yield} = \frac{\text{mass of actual yield}}{\text{mass of theoretical yield}} \times 100\%$$

The object of this experiment is to recover all of the copper with which you began. This is the test of your laboratory skills.

The percent yield of the copper can be expressed as the ratio of the recovered mass to initial mass multiplied by 100, as follows:

$$\%\ \text{yield} = \frac{\text{recovered mass of Cu}}{\text{initial mass of Cu}} \times 100\%$$

PROCEDURE

Determine the mass of approximately 0.500 g of 16- or 18-gauge copper wire to the nearest 0.0001 g and record its mass (1). Place it in a 250-mL beaker. *In the hood*, add 4 or 5 mL of concentrated HNO_3 to the reaction beaker. **(CAUTION:** ***Be careful not to get any of the nitric acid on yourself. If you do, wash it off immediately with copious amounts of water. The gas produced in this reaction is toxic, and the reaction must be performed in the hood.*****)** After the reaction is complete, add 100 mL of distilled H_2O to the reaction beaker. Describe the reaction as to color change, evolution of a gas, and change in temperature (exothermic or endothermic) on the report sheet (6).

GIVE IT SOME THOUGHT

a. What product is formed in this step?
b. Where does this fit in with the series of reactions on pages 87–88?
c. Why can you allow the gas produced to evolve without collecting it?

Add 30 mL of 3.0 *M* NaOH to the solution in your beaker with stirring and describe the reaction on the report sheet (7). Add two or three boiling chips and carefully heat the solution—while stirring with a stirring rod—just to the boiling point. Describe the reaction on your report sheet (8).

GIVE IT SOME THOUGHT

a. What product is formed in this step?
b. Where does this fit in with the series of reactions on pages 87–88?

Allow the black CuO to settle; then decant the supernatant liquid. Add about 200 mL of very hot distilled water, stir, and allow the CuO to settle. Decant once more. What are you removing by the washing and decantation (9)?

Add 15 mL of 6.0 *M* H_2SO_4. **(CAUTION: *This is a corrosive acid.*)** What copper compound is present in the beaker now (10)?

Your instructor will tell you whether you should use zinc or aluminum for the reduction of Cu(II) in the following step.

A. Reduction with Zinc

In the hood, add 2.0 g of zinc metal (granules of about 0.6 mm diameter) all at once and stir until the supernatant liquid is colorless. Describe the reaction on your report sheet (11). What is present in solution (12)? When gas evolution has become *very* slow, heat the solution gently (but do not boil) and allow it to cool. What gas is formed in this reaction (13)? How do you know (14)?

B. Reduction with Aluminum

In the hood, add several 2.5 cm squares of aluminum foil and a few drops of concentrated HCl. Continue to add pieces of aluminum until the supernatant liquid is colorless. Describe the reaction on your report sheet (11). What is present in solution (12)? What gas is formed in this reaction (13)? How do you know (14)?

For Either Reduction Method When gas evolution has ceased, decant the solution, transfer the precipitate to a porcelain evaporating dish of known mass, and record its mass on the report sheet (3). Wash the precipitated copper with about 5 mL of distilled water, allow it to settle, decant the solution, and repeat the process. What are you removing by washing (15)? Wash the precipitate with about 5 mL of methanol. **(CAUTION: *Keep the methanol away from flames—it is flammable! Methanol is also extremely toxic. Avoid breathing the vapors as much as possible.*)** Allow the precipitate to settle and decant the methanol. Finally, wash the precipitate with about 5 mL of acetone. **(CAUTION: *Keep the***

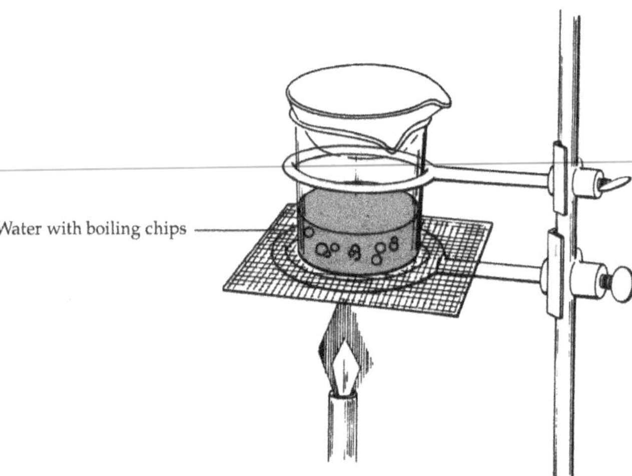

▲**FIGURE 6.1** Steam bath.

***acetone away from flames—it is extremely flammable!*)** Allow the precipitate to settle and decant the acetone from the precipitate. **(CAUTION: *The methanol and acetone washes should be disposed of in a waste container for organic liquids.*)** Prepare a steam bath as illustrated in Figure 6.1 and dry the product on your steam bath for at least 5 min. Wipe the bottom of the evaporating dish with a towel, remove the boiling chips, determine the mass of the evaporating dish plus copper, and record its mass (2). Calculate the final mass of copper (4). Compare the mass with your initial mass and calculate the percent yield (5). What color is your copper sample (16)? Is it uniform in appearance (17)? Suggest possible sources of error in this experiment (18).

Dispose of all the chemicals in the designated receptacles.

Name Grace Rademacher Desk

Date Laboratory Instructor

Chemical Reactions of Copper and Percent Yield | 6 Pre-lab Questions

Before beginning this experiment in the laboratory, you should be able to answer the following questions.

1. Give an example, ~~other than the ones listed in this experiment,~~ of redox and metathesis reactions.

Redox: $Cu_{(s)} + 4HNO_{3(aq)} \rightarrow Cu(NO_3)_{2(aq)} + 2NO_{2(g)} + 2H_2O_{(l)}$

Metathesis: $Cu(NO_3)_2 + 2NaOH_{(aq)} \rightarrow Cu(OH)_{2(s)} + 2NaNO_{3(aq)}$

2. When will reactions proceed to completion?

Reactions will proceed to completion when all of the solute is dissolved

3. Define *percent yield* in general terms.

Percent yield is the amount of a substance which is recovered after a reaction that releases gas as a reactant

4. Name six methods of separating materials.

- sedimentation
- decantation
- filtration
- evaporation
- crystallization
- distillation

5. Give criteria in terms of temperature changes for exothermic and endothermic reactions.

AB + CD → AD + CB (↑ energy, ↓ energy) exothermic

AB + CD → AD + CB (↓ energy, ↑ energy) endothermic

energy diff = enthalpy change

ΔH

6. If 3.35 g of $Cu(NO_3)_2$ are obtained from the reaction of 2.25 g of Cu with excess HNO_3, what is the percent yield of the reaction? $Cu + 2HNO_3 \rightarrow Cu(NO_3)_2 + 2H$

$2.25\,g\,Cu \times \frac{1\,mol\,Cu}{65.55\,g\,Cu} = .0354\,mol\,Cu$ mol Cu = mol $Cu(NO_3)_2$

~~$.0354\,mol\,Cu(NO_3)_2 \times \frac{187.56\,g\,Cu(NO_3)_2}{1\,mol\,Cu(NO_3)_2}$~~ $= 6.64$ mol $Cu(NO_3)_2$

% yield: $\left(\frac{3.35\,g\,Cu(NO_3)_2}{6.64\,g\,Cu(NO_3)_2}\right) \times 100 = 50.5\%$ yield

7. What two liquids are flammable in this experiment?

Hydrogen and ________ are flammable

8. Define the term *limiting agent*.

Limiting agent is the reactant that will be consumed completely in the rxn

9. What is the maximum percent yield in any reaction?

100%

10. What are meant by the terms *decantation* and *filtration*?

Decantation is the process of pouring out the liquid in a solution, leaving the solids behind. Filtration is the separation of solids and liquids with a filter

11. When $Cu(OH)_2(s)$ is heated, copper(II) oxide and water are formed. Write a balanced equation for the reaction.

$Cu(OH)_{2(s)} \rightarrow CuO_{(s)} + H_2O_{(l)}$

12. When sulfuric acid and copper(II) oxide are allowed to react, copper(II) sulfate and water are formed. Write a balanced equation for this reaction.

$\overset{+1}{H_2}\overset{-2}{SO_4} + \overset{+2}{Cu}\overset{-2}{O} \rightarrow H_2O + CuSO_4$

Name Grace Rademacher Desk

Date 10/9/20 Laboratory Instructor Prof. Briguglio

Complete all

REPORT SHEET | EXPERIMENT

Chemical Reactions of Copper and Percent Yield | 6

1. Initial mass of copper 0.553 g
2. Mass of copper and evaporating dish 46.417 g
3. Mass of evaporating dish 45.864 g
4. Mass of recovered copper 0.504 g
5. Percent yield (show calculations) 91%

$\frac{\text{Recovered copper (g)}}{\text{O.g (g)}} \times 100$

$\left(\frac{.504 g}{.553 g}\right) \times 100 = 91.139\%$

6. Describe the reaction $Cu(s) + HNO_3(aq) \longrightarrow$. $Cu_{(s)} + 4HNO_{3(aq)} \rightarrow Cu(NO_3)_{2(aq)} + 2NO_{2(g)} + 2H_2O_{(l)}$

This is an oxidation-reduction equation. Copper metal dissolved, producing NO_2

7. Describe the reaction $Cu(NO_3)_2(aq) + NaOH(aq) \longrightarrow$. $Cu(NO_3)_2 + 2NaOH \rightarrow Cu(OH)_2 + 2NaNO_3$ (aq, aq, s)

Double displacement, metathesis

8. Describe the reaction $Cu(OH)_2(s) \xrightarrow{\Delta}$. $CuO_2 + HOH$

Dehydration rxn $Cu(OH_2) \xrightarrow{heat} CuO + HOH$: balanced

9. What are you removing by this washing? Hydroxide and $NaNO_3$ which didn't react
10. What copper compound is present in the beaker? $CuSO_4$ (copper II sulfate)
11. Describe the reaction $CuSO_4(aq) + Zn(s)$ or ~~$CuSO_4(aq) + Al(s)$~~. Oxidation-reduction
12. What is present in solution? SO_4 ions and Zn ions are present
13. What is the gas? Hydrogen
14. How do you know? Acid + metal = release of H. $Zn_{(s)} + H_2SO_4 \rightarrow ZnSO_{4(aq)} + H_{2(g)}$
15. What are you removing by washing? Zn^{+2}, SO_4^{-2} or Al^{+3} and SO_4^{-2}
16. What color is your copper sample? It was brown w/ a hint of red
17. Is it uniform in appearance? Other than boiling stones, no
18. Suggest possible sources of error in this experiment. Not all of the copper fell out of solution, not able to get all of the copper from the beaker

$CuSO_4$
63.546 + 32.059 + (15.999)4

QUESTIONS

$CuSO_4 + Zn \rightarrow ZnSO_4 + Cu$ — Cu as limiting reagent

1. When zinc (or aluminum) was allowed to react with the copper sulfate, what was the limiting reagent?

.553 g Cu × (1 mol Cu / 63.546 g Cu) × (1 mol $CuSO_4$ / 1 mol Cu) × (159.601 g $CuSO_4$ / 1 mol $CuSO_4$) = 1.38 g $CuSO_4$ .5689 Zn

2. If your percent yield of copper was greater than 100%, what are three plausible errors you may have made?

My percent yield of Copper was 91%.

3. Potassium carbonate reacts readily with hydrochloric acid to form potassium chloride, water and carbon dioxide.

   (a) Write a balanced chemical equation for this reaction.

   $K_2CO_3 + 2HCl \rightarrow 2KCl + H_2CO_3$

   K: +1, CO_3: −2, Cl: −1, H: +1

   (b) Potassium carbonate is highly soluble in water. Calculate the mass of potassium carbonate required to make 250 cm³ of 1.50 mol/dm³ solution.

   Concentration = mols solute / Vol. of Sol. (L) = 1.5 mol/dm³ / .25 L = 6

   6g potassium carbonate is required

4. Consider the precipitation reaction between lead(II) nitrate solution and potassium iodide solution:

$$Pb(NO_3)_2(aq) + 2\ KI(aq) \rightarrow PbI_2(s) + 2\ KNO_3(aq)$$

1:2:1:2

Suppose 20.0 g of lead(II) nitrate and 20.0 g of potassium iodide were dissolved separately to form 250 cm³ of lead(II) nitrate solution and 250 cm³ of potassium iodide solution. (a) What is the limiting reactant? (b) Determine the mass of lead(II) iodide (PbI_2) that can be obtained from this reaction?

20g $Pb(NO_3)_2$ → 250cm³ $Pb(NO_3)_2$ (aq)
20g KI → 250 KI (aq)

Pb - 207.2
N - 14.006
O - 15.999
K - 39.098
I - 126.9

a) K is the limiting reactant

b) 20g $Pb(NO_3)_2$ × (1 mole / 331.2 g $Pb(NO_3)_2$) × (166 g KI / 2 mol KI) × (1 mol PbI_2 / 166 g KI) × (461 g PbI_2 / 1 mol PbI_2)

13.92 g of lead (II) iodide can be obtained

207.2 + 2(14.006) + 6(15.999) = 331.206

63.546
+15.999
79.545

5. Molarity, abbreviated M, is defined as the concentration of a solution expressed as moles of solute per liter of solution. How many milliliters of 4.00 *M* H_2SO_4 are required to react with 1.60 g of CuO according to Equation [4]?

Concentration = mols sol / vol of sol(L) = .02 / .004 = 5M.02

1.6g CuO × 1 mol CuO / 79.545g CuO = .02

.02 mol H_2SO_4 / 4.00 mol/L = .00503 L H_2SO_4

.00503 × 1000 = 5.03 mL H_2SO_4 required

6. If 2.00 g of Zn is allowed to react with 2.00 g of $CuSO_4$, according to Equation [5], how many grams of Zn will remain after the reaction is complete?

2.00g Zn × 1mol Zn / 65.37g Zn = 0.0306mol Zn

2.00g $CuSO_4$ × 1mol $CuSO_4$ / 159.5g $CuSO_4$ = 0.0125 mol $CuSO_4$

.0306 − .0125 = .0181

.0181 mol Zn × 65.37gZn / 1molZn = 1.18gZn will remain after the reaction

NOTES AND CALCULATIONS

Chemicals in Everyday Life: What Are They, and How Do We Know?

Experiment

OBJECTIVE

To observe some reactions of common substances found around the home and to learn how to identify them.

APPARATUS AND CHEMICALS

Apparatus

150 mL beaker	7 cm test tubes (6)
medicine droppers (3)	red and blue litmus paper

Chemicals

household ammonia	chemical fertilizer
household bleach (chlorine)	table salt
baking soda	vinegar
chalk	Epsom salts
mineral oil	NaI
1 *M* NH_4Cl	8 *M* NaOH
$(NH_4)_2CO_3$	18 *M* H_2SO_4
$Ba(OH)_2$ (sat. soln.)	0.1 *M* $AgNO_3$
3 *M* HNO_3	solid unknown containing CO_3^{2-}, Cl^-, SO_4^{2-}, or I^-
0.2 *M* $BaCl_2$	

DISCUSSION

One important aspect of chemistry is the identification of substances. The identification of minerals—for example, fool's gold as opposed to genuine gold—was and still is of great importance to prospectors. The rapid identification of a toxic substance ingested by an infant may expedite the child's recovery, or it may be the determining factor in saving a life. Substances are identified by the use of instruments, by reactions characteristic of the substance, or by both. Reactions that are characteristic of a substance are frequently referred to as a *test.* For example, you may test for oxygen with a glowing splint; if the splint bursts into flame, oxygen is probably present. You test for chloride ions by adding silver nitrate to an acidified solution. The formation of a white precipitate suggests the presence of chloride ions. Because other substances may yield a white precipitate under these conditions, you "confirm" the presence of chloride ions by observing that this precipitate dissolves in ammonium hydroxide while the other white precipitates do not. The area of chemistry concerned with identification of substances is termed *qualitative analysis.*

In this experiment, you will perform tests on or with substances you are apt to encounter in everyday life, such as table salt, bleach, smelling salts, and baking soda. You probably don't think of these as "chemicals," but they are, even though you don't refer to them at home by their chemical names (which are sodium chloride, sodium hypochlorite, ammonium carbonate, and sodium bicarbonate, respectively); instead, you use their trade names. After observing some reactions of these household chemicals, you will partially identify an unknown. Your task will be to determine whether the substance contains the carbonate (CO_3^{2-}), chloride (Cl^-), sulfate (SO_4^{2-}), or iodide (I^-) ion.

(CAUTION: ***Even though household chemicals may appear innocuous, NEVER mix them unless you are certain you know what you are doing. Innocuous chemicals, when combined, may produce severe explosions or other hazardous reactions.*****)**

PROCEDURE

A. Household Ammonia

Obtain 1 mL of household ammonia in a 150 mL beaker. Hold a dry piece of red litmus paper over the beaker, being careful not to touch the paper with the sides of the beaker or the solution. Record your observations on the report sheet (1). Repeat the operation using a piece of red litmus paper that has been moistened with distilled water. Record your observations on the report sheet (2).

GIVE IT SOME THOUGHT

Do you note any difference in the time required for the litmus to change colors or the intensity of the color change?

Ammonium salts are converted to ammonia, NH_3, by the action of strong bases. Hence, you can test for the ammonium ion, NH_4^+, by adding sodium hydroxide, NaOH, and noting the familiar odor of NH_3 or by using red litmus paper. The *net* reaction is as follows:

$$NH_4^+(aq) + OH^-(aq) \rightleftharpoons NH_3(g) + H_2O(l)$$

NH_3 gas is released from this solution and it reacts with the moist litmus to change its color.

GIVE IT SOME THOUGHT

What happens to red litmus when it comes in contact with ammonia?

Place about 1 mL of 1 *M* NH_4Cl, ammonium chloride, in a test tube and hold a moist piece of red litmus in the mouth of the tube. Record your observations (3). Now add about 1 mL of 8 *M* NaOH, mix and repeat the test. (Do not allow the litmus to touch the sides of the tube because it may come in contact with NaOH, which will turn the litmus blue.) If the litmus does not change color, gently warm the test tube, but do not boil the solution. Record your observations (4).

You may suspect that ordinary garden fertilizer contains ammonium compounds. Confirm your suspicions by placing solid fertilizer, an amount about the size of a pea, in a test tube; add 1 mL of 8 *M* NaOH and test as above using moist litmus paper. Does the fertilizer contain ammonium salts (5)?

What is the active ingredient in smelling salts? Hold a moist piece of red litmus paper over the mouth of an open jar of ammonium carbonate,

$(NH_4)_2CO_3$. Carefully fan your hand over the jar and see if you can detect a familiar odor. Test with moist red litmus paper to confirm. Record your observations on the report sheet (6). Most ammonium salts are stable; for example, the ammonium chloride solution that you tested above should not have had any effect on litmus paper *before* you added the sodium hydroxide. However, $(NH_4)_2CO_3$ is quite unstable and decomposes to ammonia and carbon dioxide:

$$(NH_4)_2CO_3(s) \xrightarrow{\Delta} 2NH_3(g) + CO_2(g) + H_2O(g)$$

Smelling salts contain ammonium carbonate that has been moistened with ammonium hydroxide.

GIVE IT SOME THOUGHT

Will the reactants and products have the same effect on litmus, or do you expect different colors from reactants to products?

Use the designated containers to dispose of the chemicals used in this part as well as those used in Parts B, C, D, and E.

B. Baking Soda, $NaHCO_3$

Substances that contain the carbonate ion, CO_3^{2-}, react with acids to liberate carbon dioxide, CO_2, which is a colorless and odorless gas. Carbon dioxide, when released from baking soda by acids (for example, those present in lemon juice and sour milk), helps a cake rise:

$$NaHCO_3(s) + H^+(aq) \longrightarrow CO_2(g) + H_2O(l) + Na^+(aq)$$

Place in a small, dry test tube an amount of solid baking soda about the size of a small pea. **(CAUTION: *Concentrated H_2SO_4 causes severe burns. Do not get it on your skin. If you come in contact with it, immediately wash the area with copious amounts of water.*)** Then add 1 or 2 drops of 18 *M* H_2SO_4 and notice what happens. Record your observations on the report sheet (7). Repeat this procedure, but use vinegar instead of the sulfuric acid. Record your observations on the report sheet (8).

A confirmatory test for CO_2 is to allow it to react with $Ba(OH)_2$, barium hydroxide, solution. A white precipitate of $BaCO_3$, barium carbonate, is produced:

$$CO_2(g) + Ba(OH)_2(aq) \longrightarrow BaCO_3(s) + H_2O(l)$$

Many substances, such as eggshells, oyster shells, and limestone, contain the carbonate ion. To determine whether common blackboard chalk contains the carbonate ion, place a small piece of chalk in a dry test tube and add a few drops of 2 *M* HCl. Test the escaping gas for CO_2 by carefully holding a drop of $Ba(OH)_2$, suspended from the tip of a medicine dropper or a wire loop, a short distance into the mouth of the test tube. Clouding of the drop is due to the formation of $BaCO_3$ and proves the presence of carbonate. **(*NOTE: Breathing on the drop will cause it to cloud because your breath contains* CO_2.)** Record your observations on the report sheet (9). Dispose of all waste in designated containers.

C. Table Salt, NaCl

Chloride salts react with concentrated sulfuric acid to liberate hydrogen chloride, which is a pungent and colorless gas that turns moist blue litmus paper red:

$$Cl^- + H_2SO_4(l) \rightarrow HCl(g) + HSO_4^-$$

This reaction will occur whether the substance is $BaCl_2$, KCl, or $ZnCl_2$; the only requirement is that the salt be a chloride. For KCl, the complete equation is as follows:

$$KCl(s) + H_2SO_4(l) \rightarrow HCl(g) + KHSO_4(s)$$

GIVE IT SOME THOUGHT

Why can these chlorides be used interchangeably?

Another reaction characteristic of the chloride ion is its reaction with silver nitrate to form silver chloride, AgCl, a white, insoluble substance:

$$Cl^-(aq) + AgNO_3(aq) \longrightarrow AgCl(s) + NO_3^-(aq)$$

GIVE IT SOME THOUGHT

According to the solubility rules, what other cations could precipitate out the $Cl^-(aq)$?

Place in a small, dry test tube an amount of sodium chloride about the size of a small pea and add 1 or 2 drops of 18 *M* H_2SO_4. (**CAUTION:** ***Concentrated*** H_2SO_4 ***causes severe burns. Do not get it on your skin. If you come in contact with it, immediately wash the area with copious amounts of water.***) Carefully note the color and odor of the escaping gas by fanning the gas with your hand toward your nose. DO NOT PLACE YOUR NOSE DIRECTLY OVER THE MOUTH OF THE TEST TUBE. Record your observations on the report sheet (10). Complete the equation $H_2SO_4 + 2NaCl(s) \longrightarrow ?$ on the report sheet (11).

Place a small amount (about the size of a pea) of NaCl in a small test tube and add 15 drops of distilled water and 1 drop of 3 *M* HNO_3. Then add 3 or 4 drops of 0.1 *M* $AgNO_3$ and mix the contents. Record your observations on the report sheet (12). Why should you use distilled water for this test (13)? Confirm your answer by testing tap water for chloride ions: Add 1 drop of 3 *M* HNO_3 to about 2 mL of tap water and then add 3 drops of 0.1 *M* $AgNO_3$. Does this test indicate the presence of chloride ions in tap water? Record your answer on the report sheet (14).

Sodium ions impart a yellow color to a flame. When potatoes boil over on a gas stove or campfire, a burst of yellow flames appears because of the presence of sodium ions. Simply handling a utensil contaminates it sufficiently with sodium ions from the skin so that when the utensil is placed in a hot flame, a yellow color will appear. Obtain a few crystals of table salt on the tip of a clean spatula and place the tip in the flame of your burner for a moment. Record your observations on the report sheet (15).

D. Epsom Salts, $MgSO_4 \cdot 7\,H_2O$

Epsom salts are used as a purgative, and solutions of this salt are used to soak tired, aching feet. The following tests are characteristic of the sulfate ion, SO_4^{2-}. Place a small quantity of Epsom salts in a small, dry test tube. Add 1 or 2 drops of 18 M H_2SO_4. **(CAUTION: *Concentrated H_2SO_4 causes severe burns. Do not get it on your skin. If you come in contact with it, immediately wash the area with copious amounts of water.*)** Record your observations on the report sheet (16). Note the difference in the behavior of this substance toward sulfuric acid compared with the behavior of baking soda toward sulfuric acid. Dispose of all waste in designated containers.

Place some Epsom salts (an amount the size of a small pea) in a small test tube and dissolve it in 1 mL of distilled water. Add 1 drop of 3 M HNO_3, then 1 or 2 drops of 0.2 M $BaCl_2$. Record your observations on the report sheet (17). Barium sulfate is a white insoluble substance that forms when barium chloride is added to a solution of any soluble sulfate salt, such as Epsom salts, as follows:

$$SO_4^{2-}(aq) + BaCl_2(aq) \longrightarrow BaSO_4(s) + 2\,Cl^-(aq)$$

E. Bleach and Iodide

Commercial bleach is usually a 5% solution of sodium hypochlorite, NaOCl. This solution behaves as though only chlorine, Cl_2, were dissolved in it. Because this solution is fairly concentrated, you must avoid direct contact with your skin and eyes. The element chlorine, Cl_2, behaves differently from the chloride ion. Chlorine is a pale yellow-green gas with an irritating odor, is slightly soluble in water, and is toxic. It is capable of liberating the element iodine, I_2, from iodide salts:

$$Cl_2(aq) + 2\,I^-(aq) \longrightarrow I_2(aq) + 2\,Cl^-(aq)$$

Iodine gives a reddish-brown color to water; it is more soluble in mineral oil than in water, and it imparts a violet color to mineral oil. Thus, chlorine can be used to identify iodide salts, or iodide salts can be used to detect chlorine.

In a small test tube, dissolve a small amount (about the size of a pea) of sodium iodide, NaI, in 1 mL of distilled water; add 5 drops of bleach. Note the color; then add several drops of mineral oil, shake, and allow the layers to separate, which takes about 20 sec. Note that the mineral oil is the top layer. Record your observations on the report sheet (18). Dispose of all waste in designated containers.

Another reaction characteristic of iodides is that they form a pale yellow precipitate when treated with silver nitrate solution:

$$I^-(aq) + AgNO_3(aq) \longrightarrow AgI(s) + NO_3^-(aq)$$

Dissolve a small amount of sodium iodide in 1 mL of distilled water and add a drop of 3 M HNO_3; then add 3 or 4 drops of 0.1 M $AgNO_3$ solution. Record your observations on the report sheet (19).

Solid iodide salts react with concentrated sulfuric acid by instantly turning dark brown, with the slight evolution of a gas that fumes in moist air and with the appearance of violet fumes of iodine. **(CAUTION: *Concentrated H_2SO_4 causes severe burns. Do not get it on your skin. If you come in contact with it,***

***immediately wash the area with copious amounts of water.*)** Place a small amount (about the size of a pea) of sodium iodide in a small, dry test tube and *in the hood*, add 1 or 2 drops of 18 *M* H_2SO_4. Record your observations on the report sheet (20). Dispose of all waste in designated containers.

TABLE 7.1 Reaction of Solid Salts with H_2SO_4

Ion	Reaction
CO_3^{2-}	Colorless, odorless gas, CO_2, evolved
Cl^-	Colorless, pungent gas, HCl, evolved, which turns blue litmus red
SO_4^{2-}	No observable reaction*
I^-	Violet vapors of I_2 formed

F. Unknown

Your solid unknown will contain only one of the following ions: carbonate, chloride, sulfate, or iodide. Table 7.1 summarizes the behavior of these ions toward H_2SO_4.

GIVE IT SOME THOUGHT

a. What test should you perform first?

b. What clues will the first test give you about identifying your unknown?

Place a small amount of your unknown (save some for further tests) in a small, dry test tube and add a drop of 18 *M* H_2SO_4. Record your observations and the formula for the unknown ion on the report sheet (21). What additional test might you perform to help identify the ion? Consult with your instructor *before* doing the test. Record your answer on the report sheet.

---

*Concentrated H_2SO_4 will liberate heat when it reacts with $MgSO_4 \cdot 7H_2O$ because of the H_2O present, but not because of the presence of SO_4^{2-}.

Name ______________________________ Desk ______________

Date ______________ Laboratory Instructor ______________________

Chemicals in Everyday Life: What Are They, and How Do We Know? | 7 | Pre-lab Questions

Before beginning this experiment in the laboratory, you should be able to answer the following questions.

1. Why is it unwise to haphazardly mix household chemicals or other chemicals?

2. How could you detect the presence of the NH_4^+ ion?

3. How could you detect the presence of the CO_3^{2-} ion?

4. How could you detect the presence of the Cl^- ion?

5. How could you detect the presence of the SO_4^{2-} ion?

6. How could you detect the presence of the I^- ion?

7. How could you detect the presence of the Ag^+ ion?

8. Complete and balance the following equations:

$$LiCl(s) + H_2SO_4(l) \longrightarrow$$

$$NH_4^+(aq) + OH^-(aq) \rightleftharpoons$$

$$AgNO_3(aq) + I^-(aq) \longrightarrow$$

$$NaHCO_3(s) + H^+(aq) \longrightarrow$$

9. Why should distilled water be used when conducting chemical tests?

10. Assume that you had a mixture of solid Na_2CO_3 and NaCl. Could you use only aqueous $AgNO_3$ to determine whether NaCl was present? Explain.

11. Assume that you had a mixture of solid Na_2CO_3 and NaCl. How could you show the presence of both carbonate and chloride in this mixture?

12. How could you show the presence of both iodide and sulfate in a mixture? Consult Appendix C for help.

Name ______________________________ Desk ______________

Date ______________ Laboratory Instructor ______________________

Unknown no. or letter ______________

REPORT SHEET | EXPERIMENT 7

Chemicals in Everyday Life: What Are They, and How Do We Know?

A. Household Ammonia

1. Effect of household ammonia on dry litmus ______________
2. Effect of household ammonia on moist litmus ______________
3. Effect of NH_4Cl on litmus ______________
4. Effect of $NH_4Cl + NaOH$ on litmus ______________
5. Fertilizers contain ammonium salts: Yes ______ ______
6. Smelling salts ______________

B. Baking Soda, $NaHCO_3$

7. Baking soda + H_2SO_4 ______________
8. Baking soda + vinegar ______________
9. Chalk contains carbonate ion: Yes ______ No ______

C. Table Salt, NaCl

10. Effect of H_2SO_4 on table salt ______________
11. $H_2SO_4(l) + 2NaCl(s) \longrightarrow$ ______________
12. Effect of $AgNO_3(aq)$ on table salt solution ______________
13. Why use distilled water? ______________
14. Chloride ions in tap water: Yes ___ No ______
15. Salt in flame ______________

D. Epsom Salts, $MgSO_4 \cdot 7H_2O$

16. Effect of H_2SO_4 on Epsom salts ______________
17. $BaCl_2$ + Epsom salts ______________

E. Bleach

18. Bleach + NaI(aq) ______________________________

19. $AgNO_3(aq)$ + NaI(aq) ______________________________

20. Effect of H_2SO_4 on NaI(s) ______________________________

____

F. Unknown

21. Unknown ion ______________________________

22. Confirmatory test ______________________________

____

QUESTIONS

1. How could you distinguish sodium bromide from sodium iodide?

2. How could you distinguish solid barium chloride from solid barium sulfate?

3. Do you think that washing soda, Na_2CO_3, could be used for the same purpose as baking soda, $NaHCO_3$? Will Na_2CO_3 react with HCl? Write the chemical equation. Write the chemical equation for the reaction of $NaHCO_3$ with HCl.

4. Sodium benzoate is a food preservative. What are its formula and its solubility in water? (Consult a handbook or the Internet.) Source: *Handbook of Chemistry and Physics.*

5. Citric acid is often found in soft drinks. What is its melting point? (Consult a handbook or the Internet.) Source: *Handbook of Chemistry and Physics.*

6. *p*-Phenylenediamine (also named 1,4-diaminobenzene) dyes hair black. Is this substance a liquid or solid at room temperature? (Consult a handbook or the Internet.) Source: *Handbook of Chemistry and Physics.*

7. Household vinegar is a 5% solution of acetic acid. Consult your textbook or Appendix E and give the formula for acetic acid.

8. Consult the label on a can of Drano (not liquid) and write the chemical formula for the contents. (There are also small pieces of aluminum inside the can.)

NOTES AND CALCULATIONS

Experiment

Gravimetric Analysis of a Chloride Salt

OBJECTIVE

To illustrate typical techniques used in gravimetric analysis by quantitatively determining the amount of chloride in an unknown.

APPARATUS AND CHEMICALS

Apparatus

balance
250 mL beakers (6)
Bunsen burner and hose
funnels (3)
funnel support
plastic wash bottle
graduated cylinders (2), 10 and 100 mL
ring stand, iron ring, and wire gauze
stirring rods (3)
rubber policeman (3)
shark skin filter paper (3)
watch glasses (3)
weighing paper

Chemicals

unknown chloride sample
0.5 *M* $AgNO_3$
6 *M* HNO_3
acetone
distilled water

DISCUSSION

Quantitative analysis is that aspect of analytical chemistry concerned with determining *how much* of one or more constituents is present in a particular sample of material. Information such as percentage composition is essential to establishing formulas for compounds (Sections 3.3 and 3.5). Two common quantitative methods used in analytical chemistry are gravimetric and volumetric analysis. *Gravimetric analysis* derives its name from the fact that the constituent being determined can be isolated in some weighable form. *Volumetric analysis*, on the other hand, derives its name from the fact that the method used to determine the amount of a constituent involves measuring the volume of a reagent. Usually, gravimetric analyses involve the following steps:

1. Drying and then accurately determining the mass of representative samples of the material to be analyzed
2. Dissolving the samples
3. Precipitating the constituent in the form of a substance of known composition by adding a suitable reagent
4. Isolating the precipitate by filtration
5. Washing the precipitate to free it of contaminants
6. Drying the precipitate to a constant mass (to obtain an analytically weighable form of known composition)
7. Calculating the percentage of the desired constituent from the masses of the sample and precipitate

Although the techniques of gravimetric analysis are applicable to a variety of substances, they will be illustrated here with an analysis that also incorporates a number of other techniques. Chloride ion may be quantitatively precipitated from solution by the addition of silver ion according to the following ionic equation:

$$Ag^{+}(aq) + Cl^{-}(aq) \longrightarrow AgCl(s) \qquad [1]$$

GIVE IT SOME THOUGHT

Is the formation of $AgCl(s)$ consistent with the solubility rules?

Silver chloride is quite insoluble (only about 0.0001 g of AgCl dissolves in 100 mL of H_2O at 20 °C); hence, the addition of silver nitrate solution to an aqueous solution containing chloride ion precipitates AgCl quantitatively. The precipitate can be collected on a filter paper, dried, and then the mass determined. From the mass of the AgCl obtained, the amount of chloride in the original sample can then be calculated.

This experiment also illustrates the concept of stoichiometry. *Stoichiometry* is the determination of the proportions in which chemical elements combine and the mass relations in any chemical reaction (Section 3.1). In this experiment, stoichiometry specifically means the mole ratio of the substances entering into and resulting from the combination of Ag^{+} and Cl^{-}. In the reaction of Ag^{+} and Cl^{-} in Equation [1], 1 mol of chloride ions reacts with 1 mol of silver ions to produce 1 mol of silver chloride. Thus,

$$\text{moles Cl}^{-} = \text{moles AgCl} = \frac{\text{grams AgCl}}{\text{molar mass of AgCl}}$$

$$\begin{aligned}\text{grams Cl in sample} &= (\text{moles Cl}^{-})(\text{atomic weight Cl}) \\ &= \frac{(\text{atomic weight Cl})(\text{grams AgCl})}{\text{molar mass of AgCl}} \\ &= \frac{(35.45 \text{ g Cl})(\text{grams AgCl})}{143.32 \text{ g AgCl}} \\ &= (0.2473 \text{ g Cl/g AgCl})(\text{grams AgCl})\end{aligned}$$

The number 0.2473 is called a gravimetric factor. It converts grams of AgCl into grams of Cl. Gravimetric factors are used repeatedly in analytical chemistry and are tabulated in handbooks. The percentage of Cl in the sample can be calculated according to the following formula:

$$\%\text{ Cl in sample} = \frac{(\text{grams Cl in sample})(100)\%}{\text{gram sample}}$$

EXAMPLE 8.1

In a gravimetric chloride analysis, it was found that 0.2516 g AgCl was obtained from an unknown that had a mass of 0.1567 g. What is the percent of chloride in this sample?

SOLUTION: The mass of Cl in the sample is as follows:

$$\begin{aligned}\text{g Cl} &= (0.2473 \text{ g Cl/g AgCl}) \times (0.2516 \text{ g AgCl}) \\ &= 0.06222 \text{ g Cl}\end{aligned}$$

This calculation uses the gravimetric factor given above.

$$\% \text{ Cl} = \frac{(0.06222 \text{ g Cl})(100)\%}{(0.1567 \text{ g sample})}$$
$$= 39.71\%$$

or using dimensional analysis, $\text{g Cl} = 0.2516 \text{ g AgCl}\left(\frac{1 \text{ mol AgCl}}{143.32 \text{ g AgCl}}\right)\left(\frac{1 \text{ mol Cl}}{1 \text{ mol AgCl}}\right)\left(\frac{35.45 \text{ g Cl}}{1 \text{ mol Cl}}\right) = 0.06222 \text{ g Cl}$

$$\text{and } \% \text{ Cl} = \frac{(0.06222 \text{ g Cl})(100)\%}{(0.1567 \text{ g sample})} = 39.71\%$$

PROCEDURE

Obtain an unknown and record its number on your report sheet. On a piece of weighing paper, determine the mass to the nearest 0.0001 g of about 0.2 to 0.4 g of your unknown sample. Transfer the sample quantitatively to a clean 250 mL beaker (do not determine the mass of the beaker) and use a pencil to label the beaker #1. Record the sample mass. Add 150 mL of distilled water and 1 mL of 6 *M* HNO_3 to the beaker. Repeat with samples 2 and 3, labeling the beakers #2 and #3, respectively. Using a different glass rod for each solution, stir until all of the sample has dissolved. Leave the stirring rods in the beakers. Do not place them on the desktop.

While stirring one of the solutions, add about 20 mL of 0.5 *M* $AgNO_3$ solution. Place a watch glass over the beaker. Warm the solution gently with your Bunsen burner and keep it warm for 5 to 10 min. Do not boil the solution.

GIVE IT SOME THOUGHT

If this amount of $AgNO_3$ is added, what is the limiting reagent?

Obtain a filter paper (three will be needed) and determine the mass of it accurately after it has been folded and torn, not before. Fold the paper as

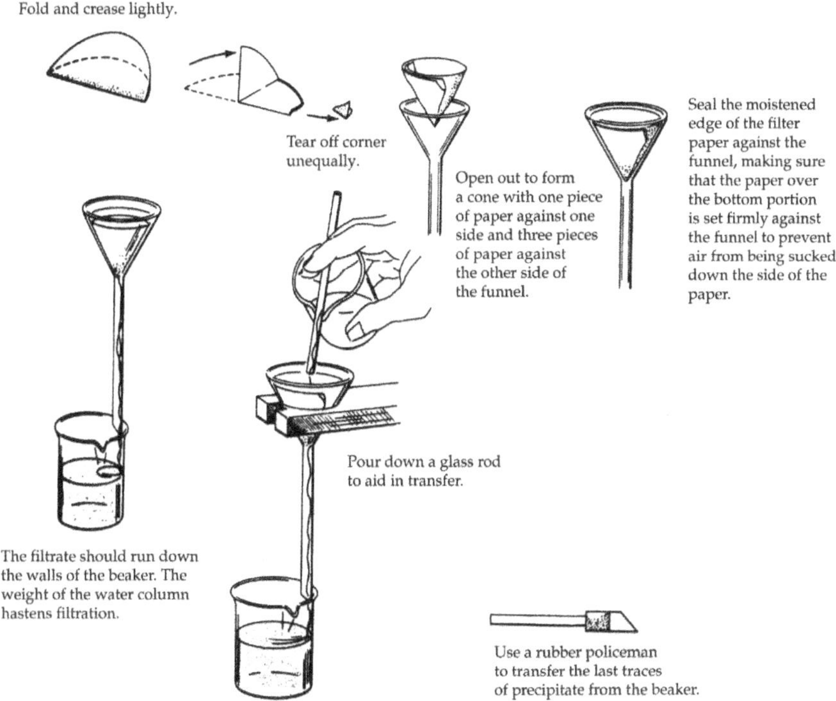

Fold and crease lightly.
Tear off corner unequally.
Open out to form a cone with one piece of paper against one side and three pieces of paper against the other side of the funnel.
Seal the moistened edge of the filter paper against the funnel, making sure that the paper over the bottom portion is set firmly against the funnel to prevent air from being sucked down the side of the paper.
Pour down a glass rod to aid in transfer.
The filtrate should run down the walls of the beaker. The weight of the water column hastens filtration.
Use a rubber policeman to transfer the last traces of precipitate from the beaker.

▲**FIGURE 8.1** Filter paper use.

illustrated in Figure 8.1 and fit it into a glass funnel. Make sure you open the filter paper in the funnel so that one side has three pieces of paper and one side has one piece of paper against the funnel—not two pieces on each side. *Why?* Your instructor will demonstrate this for you. Wet the paper with distilled water to hold it in place in the funnel. Completely and quantitatively transfer the precipitate and all of the warm solution from the beaker onto the filter, using a rubber policeman (your laboratory instructor will show you how to use it) and a wash bottle to remove the last traces of precipitate. The level of solution in the filter funnel should be *below* the top edge of the filter paper. Wash the precipitate on the filter paper with two or three 5 mL portions of water from the wash bottle. Finally, pour three 5 mL portions of acetone through the filter. **(CAUTION:** ***Acetone is highly flammable! Keep it away from open flames.*****)** Remove the filter paper, place it on a numbered watch glass, and store it in your locker until the next lab period.

Repeat the above processes with your other two samples, making sure you have numbered your watch glasses so that you can identify the samples. The precipitated AgCl must be kept out of bright light because it is photosensitive and slowly decomposes in the presence of light as follows:

$$2\text{AgCl}(s) \xrightarrow{hv} 2\text{Ag}(s) + \text{Cl}_2(g)$$

In this equation, hv is a symbol for electromagnetic radiation; here it represents radiation in the visible and ultraviolet regions of the spectrum. This is the reaction used by Corning to make photosensitive sunglasses. *In the next lab period*, when the AgCl is thoroughly dry, determine the mass of the filter papers and AgCl and calculate the mass of AgCl. Using these data, calculate the percentage of chloride in your original sample. Dispose of the filter papers and AgCl as instructed by your instructor.

Standard Deviation and Error Analysis

As a means of estimating the precision of your results, calculate the standard deviation. Before you see an illustration of this, however, study the definitions of the following terms:

Accuracy: a measure of how closely individual measurements agree with the correct (true) value.

Precision: the closeness of agreement among several measurements of the same quantity; the reproducibility of a measurement.

Error: the difference between the true result and the determined result.

Determinate errors: errors in method or performance that can be discovered and eliminated.

Indeterminate errors: random errors, which are incapable of discovery but can be treated by statistics.

Mean: arithmetic mean or average (μ), where

$$\mu = \frac{\text{sum of results}}{\text{number of results}}$$

For example, if an experiment's results are 1, 3, and 5,

$$\mu = \frac{1+3+5}{3} = 3$$

Median: the midpoint of the results for an odd number of results and the average of the two middle results for an even number of results (m). For example, if an experiment's results are 1, 3, and 5, $m = 3$. If results are 1.0, 3.0, 4.0, and 5.0,

$$m = \frac{3.0 + 4.0}{2} = 3.5$$

The scatter about the mean or median—that is, the deviations from the mean or median—can be used to measure precision. Thus, the smaller the deviations, the more reproducible or precise the measurements. The relative average deviation is the average deviation divided by the mean.

EXAMPLE 8.2

Assuming that an experiment's results are 1.0, 2.0, 3.0, and 4.0, calculate the mean, the deviations from the mean, the average deviation from the mean, and the relative average deviation from the mean.

SOLUTION: The mean is calculated as follows:

$$\mu = \frac{1.0 + 2.0 + 3.0 + 4.0}{4} = \frac{10.0}{4} = 2.5$$

The deviations from the mean are as follows:

$$|2.5 - 1.0| = 1.5$$
$$|2.5 - 2.0| = 0.5$$
$$|2.5 - 3.0| = 0.5$$
$$|2.5 - 4.0| = 1.5$$

The symbol | | means absolute value; so all deviations are positive. Therefore, the average deviation from the mean is as follows:

$$\frac{1.5 + 0.5 + 0.5 + 1.5}{4} = 1.0$$

The relative average deviation from the mean is calculated by dividing the average deviation from the mean by the mean. Thus,

$$\text{relative deviation} = \frac{1.0}{2.5} = 0.40$$

This can be expressed as 40%, 400 parts/thousand (ppt), or 40,000 parts/million (ppm). Note that if the mean were larger (for example, 100 instead of 2.5) and the average deviation were still 1.0, the relative deviation would be 1.0/100 or 1.0% (10 ppt). In this case, the precision is better because the relative deviation is smaller.

EXAMPLE 8.3

Assuming that an experiment's results are 1.0, 1.5, 2.0, and 2.5, calculate the mean, the deviations from the mean, the average deviation from the mean, and the relative average deviation from the mean.

SOLUTION: The mean is calculated as follows:

$$\mu = \frac{1.0 + 1.5 + 2.0 + 2.5}{4}$$
$$= \frac{7.0}{4} = 1.75, \text{ or } 1.8 \text{ to two significant figures}$$

The deviations from the mean are as follows:

$$|1.8 - 1.0| = 0.8$$
$$|1.8 - 1.5| = 0.3$$

$$|1.8 - 2.0| = 0.2$$
$$|1.8 - 2.5| = 0.7$$

Therefore, the average deviation from the mean is as follows:

$$\frac{0.8 + 0.3 + 0.2 + 0.7}{4} = 0.5$$

The relative average deviation from the mean is as follows:

$$\frac{0.5}{1.8} = 0.3$$
$$= 30\%, \text{ or } 300 \text{ ppt, or } 30{,}000 \text{ ppm}$$

Obviously, the data in Example 8.3 are internally more consistent than the data in Example 8.2 and hence are more precise because the deviations are smaller. Thus, the average deviation and relative average deviation can be used to measure precision.

Standard deviation (s), which is related to statistics and is a better measure of precision, is calculated using the following formula:

$$s = \sqrt{\frac{\text{sum of the squares of the deviations from the mean}}{\text{number of observations} - 1}}$$
$$= \sqrt{\frac{\Sigma_i |\chi_i - \mu|^2}{N - 1}}$$

where s = standard deviation from the mean, χ_i = members of the set, μ = mean, and N = number of members in the set of data. The symbol Σ_i means to sum over the members. The relative standard deviation is the standard deviation divided by the mean.

EXAMPLE 8.4

An experiment's results are 1, 3, and 5. Calculate the mean, the deviations from the mean, the standard deviation, and the relative standard deviation for the data.

SOLUTION: The mean is as follows:

$$\mu = \frac{1 + 3 + 5}{3} = 3$$

The deviations from the mean are as follows:

$$|\chi_i - \mu| = \text{deviation}$$
$$|1 - 3| = 2$$
$$|3 - 3| = 0$$
$$|5 - 3| = 2$$

$$s = \sqrt{\frac{2^2 + 0^2 + 2^2}{3 - 1}}$$
$$= \sqrt{\frac{4 + 0 + 4}{2}}$$
$$= \sqrt{\frac{8}{2}}$$
$$= \sqrt{4} = 2$$

The results of this experiment could be reported as 3 ± 2. The relative standard deviation is as follows:

$$\frac{2}{3} = 0.7, \text{ or } 70\%$$

EXAMPLE 8.5

The results of an experiment are 2.100, 2.110, and 2.105. Calculate the mean, the deviations from the mean, the standard deviation, and the relative standard deviation.

SOLUTION: The mean is as follows:

$$\mu = \frac{2.100 + 2.110 + 2.105}{3} = 2.105$$

The deviations from the mean are as follows:

$$|2.105 - 2.100| = 0.005$$
$$|2.105 - 2.110| = 0.005$$
$$|2.105 - 2.105| = 0.000$$

Therefore, the standard deviation is as follows:

$$\begin{aligned} s &= \sqrt{\frac{(0.005)^2 + (0.005)^2 + (0.000)^2}{2}} \\ &= \sqrt{\frac{5 \times 10^{-5}}{2}} \\ &= 0.005 \end{aligned}$$

The results could be reported as 2.105 ± 0.005. The relative standard deviation is

$$\frac{0.005}{2.105} = 0.002, \text{ or } 0.2\%$$

Obviously, the data in Example 8.5 are more precise although not necessarily more accurate than the data in Example 8.4 because both the deviations and standard deviation are smaller in Example 8.5.

Calculate the standard deviation of your data and record the results on your report sheet.

The standard deviation may be used to determine whether a result should be retained or discarded. As a rule of thumb, you should discard any result that is more than two standard deviations from the mean. For example, if you had a result of 49.65% and had determined that your percentage of chloride was 49.25 ± 0.09%, you should discard the result (49.65%). This is because $s = 0.09$ and $|49.25 - 49.65| = 0.40$, which is greater than 2 × 0.09. You should discard the result because it is more than two standard deviations from the mean.

NOTES AND CALCULATIONS

Name ______________________ Desk ______________

Date ______________ Laboratory Instructor ______________________

Gravimetric Analysis of a Chloride Salt | 8 | Pre-lab Questions

Before beginning this experiment in the laboratory, you should be able to answer the following questions.

1. What is the fundamental difference between gravimetric and volumetric analysis?

2. What does *stoichiometry* mean?

3. Why should silver chloride be protected from light? Will your result be high or low if you don't protect your silver chloride from light?

4. Can you eliminate indeterminate errors from your experiment?

5. Does standard deviation give a measure of accuracy or precision?

---

6. Why don't you open your folded filter paper so that two pieces touch each side of the funnel?

7. If your silver chloride undergoes extensive photodecomposition before you determine its mass will your results be high or low?

8. Assuming that an experiment's result are 5.5, 5.6, and 6.3, find the mean, the average deviation from the mean, the standard deviation from the mean, and the relative deviation from the mean.

9. What is meant by the term *gravimetric factor*?

Name ______________________________ Desk ______________

Date ______________ Laboratory Instructor ______________________

Unknown no. or letter ______________

REPORT SHEET | EXPERIMENT

Gravimetric Analysis of a Chloride Salt | 8

	Trial 1	*Trial 2*	*Trial 3*
Mass of sample	________ g	________ g	________ g
Mass of filter paper + AgCl	________	________	________
Mass of filter paper	________	________	________
Mass of AgCl	________	________	________
Mass of Cl in original sample (show calculations)	________	________	________
Percent chloride in original sample (show calculations)	________	________	________

Average percent chloride (show calculations) ________

Standard deviation (show calculations) ________

Relative standard deviation expressed as a percent (show calculations) ________

Do any of your results differ from the mean by more than two standard deviations? ________

Reported percent chloride ________ ± ________ %

QUESTIONS

1. The following percentages of chloride were found: 32.52%, 32.14%, 32.61%, and 32.75%.

   (a) Find the mean, the standard deviation, and the relative standard deviation.

   (b) Can any result be discarded?

2. Barium can be analyzed by precipitating it as $BaSO_4$ and determining the mass of the precipitate. When a 0.269 g sample of a barium compound was treated with excess H_2SO_4, 0.0891 g of $BaSO_4$ formed. What percentage of barium is in the compound?

3. What percentage of sodium is in pure table salt?

4. How many milligrams of sodium are contained in 4.50 g of NaCl?

5. An impure sample of table salt that weighed 0.8421 g when dissolved in water and treated with excess $AgNO_3$ formed 2.044 g of AgCl. What percentage of NaCl is in the impure sample?

6. List at least three sources of error in this experiment.

Solubility and the Effect of Temperature on Solubility

Experiment

OBJECTIVE

To determine the solubility of a compound as a function of temperature.

APPARATUS AND CHEMICALS

Apparatus

25 × 200 mm test tube (outer)
20 × 150 mm test tube (inner)
Two hole stopper, one side slit
Thermometer, alcohol
Wire stirring loop
800 mL beaker (×1), 250 mL beaker (×1)
Ring stand
Clamp, 3 prong
Magnetic stirring plate
Top loading 0.001g balance or 0.0001g analytical balance
10 mL and 5 mL pipet and pipet bulb
Evaporating dish
Bunsen burner and hose
Weigh boats
Watch glass

Chemicals

KCl
oxalic acid dihydrate, $H_2C_2O_4{\cdot}2H_2O$
Ice

WORK IN GROUPS OF TWO OR THREE, DEPENDING ON EQUIPMENT AVAILABILITY, BUT ANALYZE THE DATA INDIVIDUALLY

DISCUSSION

The solubility of a substance (the solute) in a solvent depends on the nature of both the solvent and the substance and the interactions between them: solute–solute, solvent–solvent, and solute–solvent. For ionic substances in water, there is an upper limit to the amount of solute that may be dissolved. The solubility may be quantified as the amount of a solute that may be dissolved in a specific amount of solvent. Typical units of solubility are mass of solute in grams/100 g solvent. The solubility may also be expressed as a mole fraction (section 13.4): moles solute/(moles solute + moles solvent).

The value of the solubility depends on the temperature. For many substances, the solubility of a solid solute increases as the temperature increases. A representative set of solubilities of ionic compounds in water as a function of temperature is shown in Figure 9.1. Note the different temperature dependence of the solubility for $Ce_2(SO_4)_3$, which exhibits a decrease in solubility with an increase in temperature. There is no linear relationship or mathematical equation that governs the temperature dependence of solubilities of substances. These solubility curves are determined experimentally.

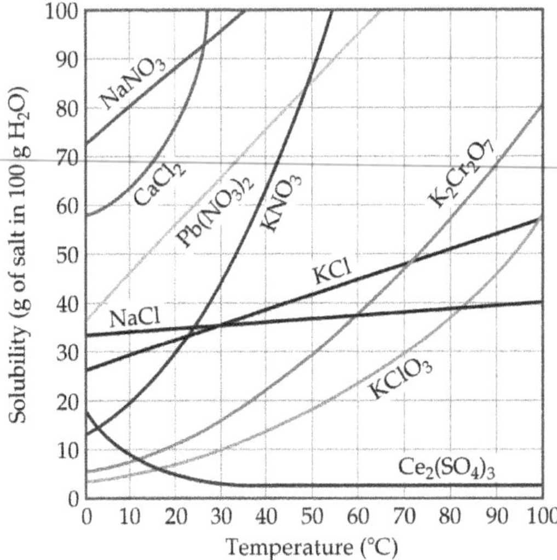

▲FIGURE 9.1 Solubility of selected ionic compounds in water as a function of temperature.

The energetics of solution formation require examination of several competing processes. In an ionic compound $MX(s) \rightarrow M^+(aq) + X^-(aq)$, the ionic bonds between M and X must be broken, which require energy. The ions are then surrounded by water molecules in the process of solvation (hydration, in the case of water as the solvent), which evolves energy as a result of the formation of the ion–dipole interactions. The water molecules also need to be separated. The overall process (ΔH_{soln}) may be exothermic or endothermic, as shown in Figure 9.2, depending on the value of the enthalpy of each step.

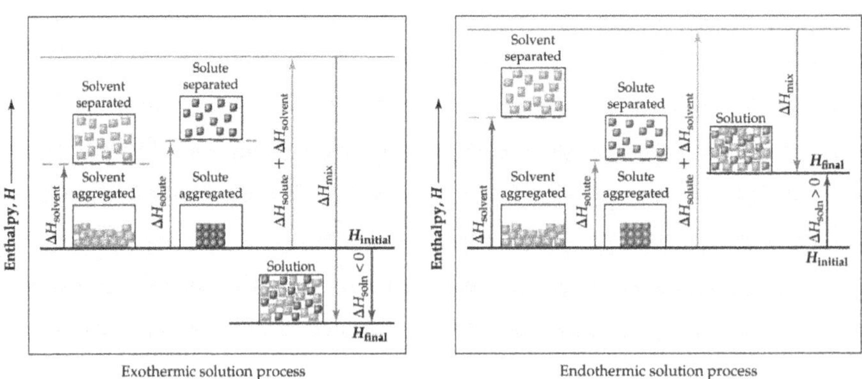

▲FIGURE 9.2 Enthalpy changes in the solution process.

The process of a solid and a solvent forming a solution is termed dissolution, whereas the reverse process is termed crystallization.

$$\text{solid} + \text{water} + \text{heat} \rightleftharpoons \text{solution (endothermic)}$$

$$\text{solid} + \text{water} \rightleftharpoons \text{solution} + \text{heat (exothermic)}$$

The solubility gives an upper limit on the amount of solute that may be dissolved in a solution at a particular temperature. A solution that contains the maximum amount of dissolved solid is termed **saturated**, whereas a solution with less than the maximum amount of dissolved solid is denoted **unsaturated** (Section 13.2). Solutions that contain more than the amount of solute to form a saturated solution are **supersaturated.**

EXAMPLE 9.1

A saturated solution contains 4.60 g barium nitrate in 15.00 mL water at 25.0°C. What is the solubility in g solute/100 g H_2O? Determine the mole fraction of $Ba(NO_3)_2$ at 25.0 °C. The density of water depends on the temperature and is given in Table 1.6.

SOLUTION: The density of water is obtained from Table 1.6.

$$15.00 \text{ mL H}_2\text{O} \times \frac{0.997044 \text{ g H}_2\text{O}}{1 \text{ mL}} = 14.96 \text{ g H}_2\text{O}$$

$$100 \text{ g H}_2\text{O} \times \frac{4.60 \text{ g Ba(NO}_3)_2}{14.96 \text{ gH}_2\text{O}} = 30.8 \text{ g Ba(NO}_3)_2 \text{ in } 100 \text{ g H}_2\text{O}$$

Calculate moles of solute and solvent.

$$4.60 \text{ g Ba(NO}_3)_2 \times \frac{1 \text{ mol Ba(NO}_3)_2}{261.34 \text{ g Ba(NO}_3)_2} = 0.0176 \text{ mol Ba(NO}_3)_2$$

$$14.96 \text{ g H}_2\text{O} \times \frac{1 \text{ mol H}_2\text{O}}{18.015 \text{ g H}_2\text{O}} = 0.8302 \text{ mol H}_2\text{O}$$

$$\text{mole fraction Ba(NO}_3)_2 = \frac{0.0176 \text{ mol}}{0.0176 \text{ mol} + 0.8302 \text{ mol}} = 0.0208$$

PROCEDURE

A. Determination of Solubility above Ambient Temperature

Prepare a water bath with a sufficient level to cover the bottom one half of a test tube when it is placed into the water bath. Begin heating the water bath but avoid boiling. Construct the experimental apparatus similar to the one shown in Figure 9.3. Ensure the test tubes, two-hole stopper, stirrer, and thermometer are clean and dry. Assemble the experimental apparatus by placing the stirrer through one hole of the stopper and the thermometer through the side slit of the stopper. Tare the balance and then determine the mass of the apparatus. Obtain approximately 6 g of the solid oxalic acid dihydrate in a weight boat, and then transfer into the inner test tube. Insert the two-hole stopper with stirrer and thermometer into the outer test tube, and then determine the mass of the apparatus with the oxalic acid. Obtain about 100 mL distilled water and measure the temperature. Remove the stopper and add 10.00 mL of distilled water to the test tube using a pipet.

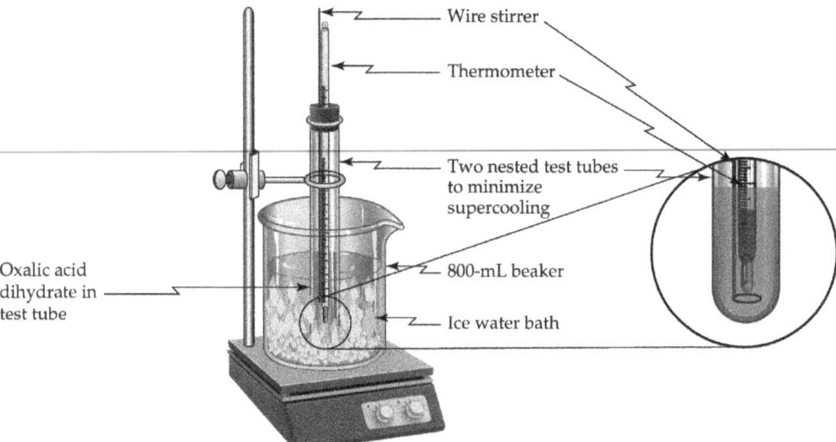

▲**FIGURE 9.3** Experimental apparatus to measure the variation of solubility with temperature.

Clamp the test tube in the heated water bath, ensuring the bottom of the test tube is not touching the beaker. Continue heating, with periodic stirring, until all of the oxalic acid is dissolved. Remove the apparatus from the water bath and clamp higher on the ring stand.

Periodically stir the solution while cooling to maintain a uniform temperature and concentration. Watch for the formation of crystals and record the temperature at which the crystals first form. Deliver by pipet an additional 5.00 mL distilled water, then reheat to dissolve and cool to determine the temperature at which crystals first form. Repeat the addition of 5.00 mL distilled water to the test tube, dissolve with adequate stirring and heating, and measure the temperature of crystal formation for the one additional volume. After the third data point collection, transfer the solution to the waste container. Clean and dry the experimental apparatus. Determine the concentration in mass solute/100g H_2O and the mol fraction.

B. Determination of Solubility below Ambient Temperature

Measure the mass of the dry apparatus. Add approximately 1.8 g of the oxalic acid dihydrate, whose mass has been determined and recorded to the inner test tube. Deliver by pipet 10.00 mL distilled water, then stir and heat as needed to completely dissolve the solid. Remove the test tube from the hot water bath and watch for formation of crystals on cooling. Record the temperature at which crystals first form. It may be necessary to cool the test tube in an ice bath. Deliver by pipet an additional 5.00 mL distilled water, then reheat to dissolve and cool to determine the temperature at which crystals first form. Repeat the addition of 5.00 mL distilled water to the test tube, dissolve with adequate stirring and heating, and measure the temperature of crystal formation for the one additional volume. After the third data point collection, transfer the solution to the waste container. Thoroughly clean the glassware, stir wire, and thermometer by scrubbing with a brush while washing with soap and hot water in the sink. Dry the experimental apparatus. Determine the concentration in mass solute/100g H_2O and the mol fraction.

C. Determination of the Solubility of an Unknown in Water at Ambient Temperature

Record the ambient temperature of the room. Obtain and determine the mass of an empty evaporation dish. Assemble an iron ring and ring stand. Obtain about 15 mL of the unknown saturated solution. Deliver by pipet 10.00 mL of the assigned saturated solution to the evaporation dish. Place the burner beneath the iron ring and *gently* heat the solution to dryness. A thick paste will form after most of the water has evaporated. Gentle and slow heating is required to avoid sample loss. Cool the evaporating dish and then loosely cover with a watch glass in the last stages of heating to prevent powder from being lost from the dish. After cooling to room temperature, obtain the mass of the evaporation dish and solid ionic compound. Determine the solubility in *g* solute/100 g H_2O. Using the solubility in *g* solute/100 g H_2O, temperature, and Figure 9.1 determine the most likely identity of the solid.

GIVE IT SOME THOUGHT

If a aqueous saturated solution were left uncovered overnight, would the concentration of the dissolved solute change?

CALCULATIONS

Determine the mass of solute per 100 g water.

Use your pipet calibration from experiment #1 to determine the volume of the saturated solution.

The solubility may be calculated using the density to find the mass of solvent. If the temperature were 21 °C, the mass of water is:

$$10.00\text{ mL H}_2\text{O}\times\frac{0.997992\text{ g H}_2\text{O}}{1\text{ mL}}=9.980\text{ g H}_2\text{O}$$

If 4.241 g of solute is dissolved 10.00 mL in water, the mass of solute in 100 g H_2O is determined by:

$$100\text{ g H}_2\text{O}\times\frac{4.241\text{ g solute}}{9.980\text{ g H}_2\text{O}}=42.49\text{ g solute in g H}_2\text{O}$$

Determine the mass of solute in 100 g water for each volume for parts A and B. Prepare a graph of the mass of solute in 100 g water (as the ordinate, vertical axis) versus temperature in °C (as the abscissa, horizontal axis). Prepare an additional graph with mole fraction versus temperature in °C. Ensure correct significant figures and units are used in all calculations.

Determine the mass of solute in 100 g water for each volume for part C. Suggest the identity of the salt using Figure 9.1.

NOTES AND CALCULATIONS

Name ______________________ Desk ______________________

Date ______________________ Laboratory Instructor ______________________

Solubility and the Effect of Temperature on Solubility | 9 | Pre-lab Questions

Before beginning this experiment in the laboratory, you should be able to answer the following questions.

1. What is meant by the terms unsaturated, saturated, and supersaturated?

2. Why is the temperature at which crystals first form recorded instead of the temperature at which the crystals stop forming?

3. A saturated solution contains 7.0 g NaCl in 20.0 mL water at 25°C. Determine the solubility in g solute/100 g H_2O.

NOTES AND CALCULATIONS

Name ______________________ Desk ______________

Date ______________ Laboratory Instructor ______________________

REPORT SHEET | EXPERIMENT

Solubility and the Effect of Temperature on Solubility

9

A. Determination of Solubility Above Ambient Temperature

Temperature ______________

Mass of apparatus ______________

Mass of apparatus with oxalic acid dehydrate ______________

Mass of oxalic acid dihydrate used ______________

Total Volume (mL)	Crystal formation Temperature (°C)	Concentration mass solute/100 g H_2O	Mole fraction

B. Determination of Solubility below Ambient Temperature

Temperature ______________

Mass of oxalic acid dihydrate used ______________

Total Volume (mL)	Crystal formation Temperature (°C)	Concentration mass solute/100 g H_2O	Mole fraction

Show your calculations for the solution concentrations and mole fractions.

C. Determination of the Solubility of a Saturated Solution at Ambient Temperature

Temperature (°C)	______________
Volume of saturated solution	______________
Mass of evaporating dish	______________
Mass of evaporating dish with solid	______________
Mass of solid	______________
Mass of water	______________
g solid/100 g water	______________
Proposed identity of solid	______________

QUESTIONS

1. Does the solubility of cerium(III) sulphate solute increase or decrease with temperature?

2. Is there a simple relationship that governs the temperature dependence of the solubility of compounds in aqueous solutions?

3. Write an equation for the solubility equilibrium for oxalic acid dihydrate and include heat on the reactant or product side.

4. A saturated solution contains 6.51 g Na_2SO_4 in 33.30 mL of water at 16.0 °C. Determine the solubility in g solute/100 g H_2O. Determine the mole fraction of solute at 16.0 °C.

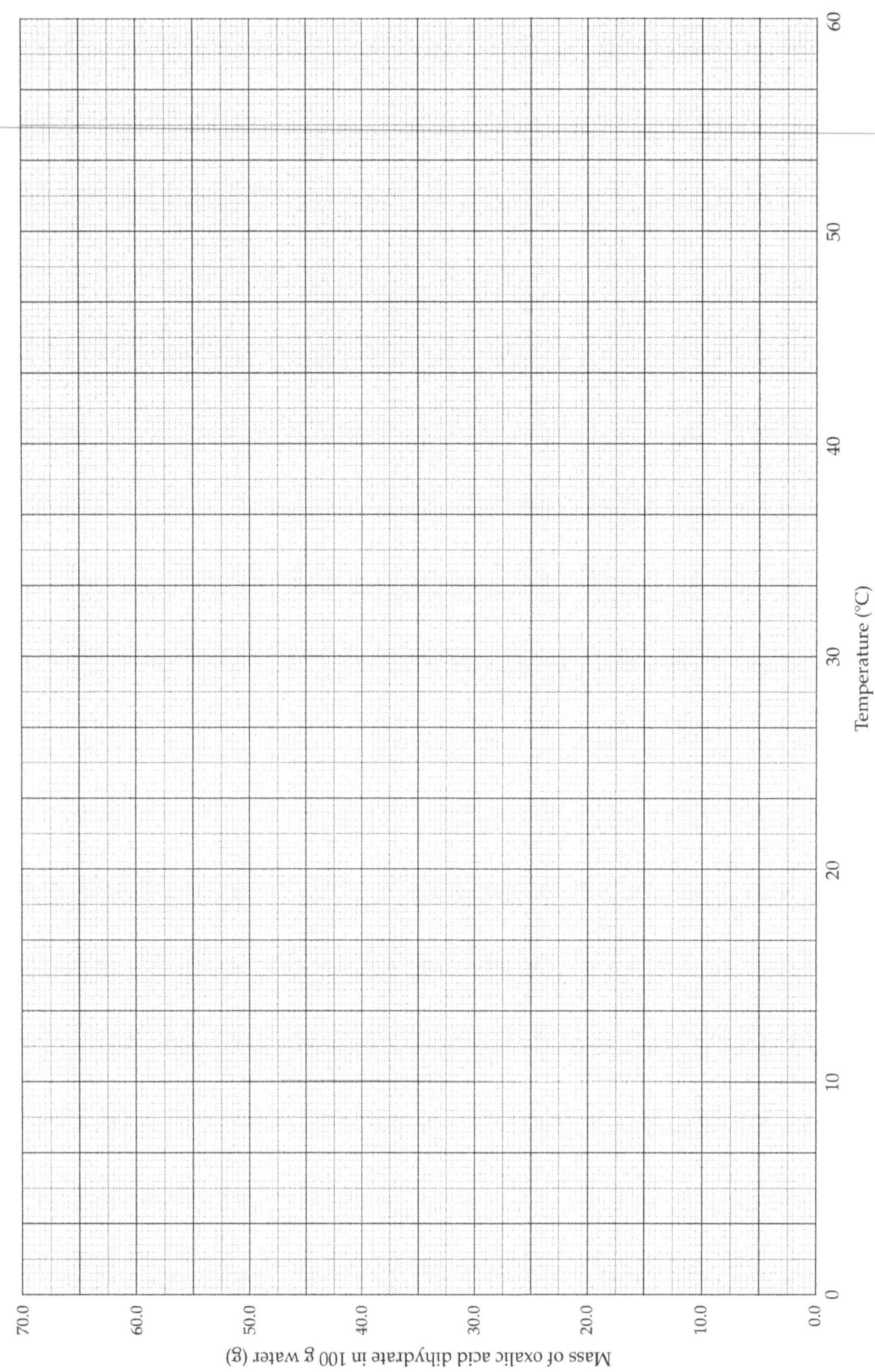

0
10
20
30
40
50
60
Temperature (°C)
0.0
10.0
20.0
30.0
40.0
50.0
60.0
70.0
Mass of oxalic acid dihydrate in 100 g water (g)

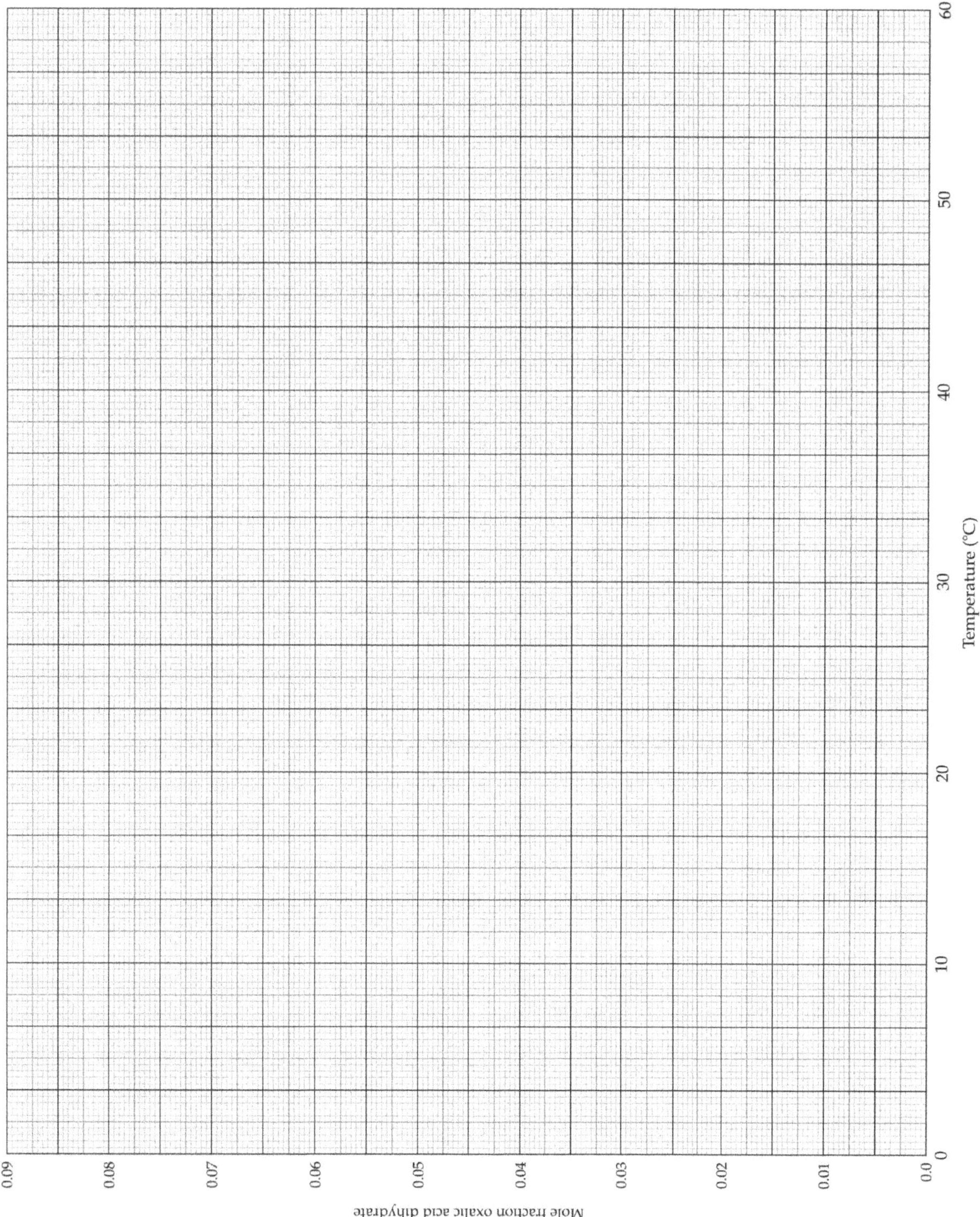
Temperature (°C)
0
10
20
30
40
50
60
Mole fraction oxalic acid dihydrate
0.0
0.01
0.02
0.03
0.04
0.05
0.06
0.07
0.08
0.09

NOTES AND CALCULATIONS

Paper Chromatography: Separation of Cations and Dyes

Experiment

OBJECTIVE

To become acquainted with chromatographic techniques as a method of separation (purification) and identification of substances.

APPARATUS AND CHEMICALS

Apparatus

Petri or evaporating dishes (2)
paper chromatography strips
600 mL beaker
capillary pipets (6)
15 cm glass stirring rod
metric ruler
10 cm watch glasses (2)
12.5-cm Whatman No.1 filter paper
scissors
50 or 100 mL graduated cylinder
paper clips (3)

Chemicals

0.5 *M* $Cu(NO_3)_2$
0.5 *M* $Ni(NO_3)_2$
15 *M* NH_3
1% dimethylglyoxime in ethanol
black felt-tip pen
0.5 *M* $Fe(NO_3)_3$
solvent (90% acetone, 10% 6 *M* HCl, freshly prepared)
isopropyl alcohol
unknown solutions

DISCUSSION

Chromatography (from the Greek *chrōma*, for color, and *graphein*, to write) is a technique often used by chemists to separate components of a mixture (Section 1.3). In 1906, the Russian botanist Mikhail Tsvett separated color pigments present in leaves by allowing a solution of these pigments to flow down a column packed with an insoluble material such as starch, alumina (Al_2O_3), or silica (SiO_2). Because different color bands appeared along the column, he called the procedure chromatography. Color is not a requisite property to achieve separation of compounds using this procedure. Colorless compounds can be made visible by being allowed to react with other reagents to form colored products, or they can be detected by physical means. Consequently, because of its simplicity and efficiency, this technique has wide applicability for separating and identifying compounds such as drugs and natural products. Chromatography is also widely used in analyses to detect the use of steroids by competitive athletes to aid forensic scientists in criminal investigations. There are many specific chromatographic techniques currently available and many of these techniques are quite complex and require the use of sophisticated instrumentation.

The basis of chromatography is the *partitioning* (that is, separation arising from differences in solubility) of compounds between a stationary phase and a moving phase. Stationary phases such as alumina (Al_2O_3), silica (SiO_2), and

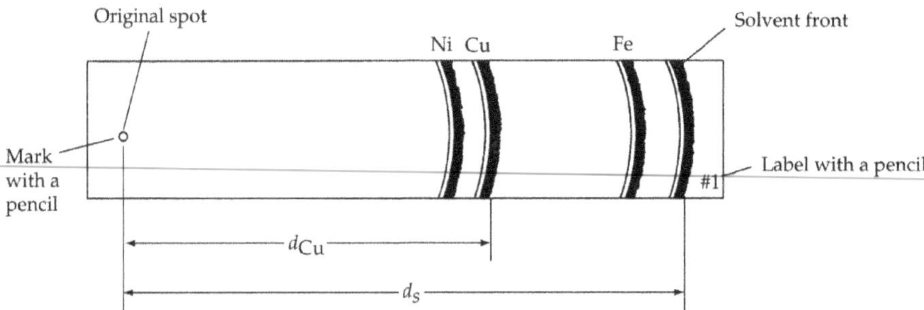

▲**FIGURE 10.1** Typical chromatogram strip.

paper (cellulose) have enormous surface areas. The molecules or ions of the substances to be separated are continuously adsorbed and then released (desorbed) into the solvent flowing over the surface of the stationary phase. This brings about a separation of the components—that is, they travel with different speeds in the moving solvent because there are generally different attractions between these components and the stationary phase. This may be described as a surface phenomenon.

In paper chromatography, a small spot of the mixture to be separated is placed on one end of a strip of paper and solvent is allowed to move up the paper through the spot by capillary action. The solvent and the various components of the mixture travel at different speeds along the paper. In this experiment, Fe^{3+}, Cu^{2+}, and Ni^{2+} will be separated with a solvent that consists of a mixture of acetone, water, and hydrochloric acid. A diagram of a portion of a typical chromatogram strip is shown in Figure 10.1.

GIVE IT SOME THOUGHT

a. Do you expect Fe^{3+}, Cu^{2+}, and Ni^{2+} to travel at the same speed in the moving solvent?

b. How will your experimental results verify this?

You can deduce the identity of components in a mixture by comparing a chromatogram of the unknown with chromatograms of standard mixtures of components suspected to be present in the unknown. An additional aid in identifying a compound is its R_f value, which is defined as the ratio of the distance traveled by a compound to the distance traveled by the solvent. For example, for Cu^{2+} (Figure 10.1),

$$R_f(\text{Cu}) = \frac{d_{\text{Cu}}}{d_s} \quad [1]$$

The R_f value of a compound is a characteristic of the compound, the support, the solvent used, and temperature, and it serves to identify the constituents of a mixture. An estimate of the relative amount of each of the constituents in a mixture can be made from the relative intensities and sizes of the various bands in a chromatogram.

The three ions studied in this experiment will be discerned in the following manner: Fe^{3+} in water imparts a red-brown rust color and thus will produce a rust-colored band on the paper. Although Cu^{2+} is blue, the color is faint and not easily detected, especially when copper is present in small amounts. In aqueous solution, however, Cu^{2+} reacts with NH_3 (from ammonium hydroxide) to form a complex ion, $[Cu(NH_3)_4]^{2+}$, which is deep blue and therefore easily observed. Finally, Ni^{2+} reacts with an organic reagent, dimethylglyoxime, to produce a strawberry-red color.

The following reactions will occur during the development process on the paper:

$$\underset{\text{Faint yellow}}{Fe^{3+}} + 3OH^- \longrightarrow \underset{\text{Rust colored}}{Fe(OH)_3}$$

$$\underset{\text{Pale blue}}{Cu^{2+}} + 4NH_3 \longrightarrow \underset{\text{Deep blue}}{[Cu(NH_3)_4]^{2+}}$$

$$\underset{\text{Pale green}}{Ni^{2+}} + 2NH_3 + 2\,(H_3C)(HON{=})C{-}C({=}NOH)(CH_3) \longrightarrow \underset{\text{Strawberry red}}{Ni(C_4H_7N_2O_2)_2} + 2NH_4^+$$

There are two simple ways of performing this experiment: by circular horizontal chromatography or by ascending chromatography using paper strips. One half of the class should perform this experiment using one technique, and the other half of the class should use the other technique. The two techniques can then be compared in terms of time required, separation, and ease of identification.

PROCEDURE

GIVE IT SOME THOUGHT

Should both techniques yield similar results?

Make six capillary pipets (5 cm long) by drawing out 6 mm glass tubing. Your instructor will demonstrate how this is done, or you can obtain the capillary tubes from your instructor.

Obtain an unknown; you will analyze this at the same time you are conducting the experiment on the known solutions. Your unknown may contain one, two, or three of the ions being studied. Record the colors of the *known and the unknown solutions* on your report sheet (1).

Do not pour the chemicals down the drain. Dispose of them in the designated waste container.

A. Horizontal Circular Technique

Cut a wick about 1 cm wide on a piece of 12.5 cm Whatman No.1 filter paper, as shown in Figure 10.2. Avoid touching the filter paper if your hands are oily. Trim about 1 cm from the end of the wick so that the filter paper will fit into the Petri dish. Mark the center of the paper with a pencil dot (do not use a

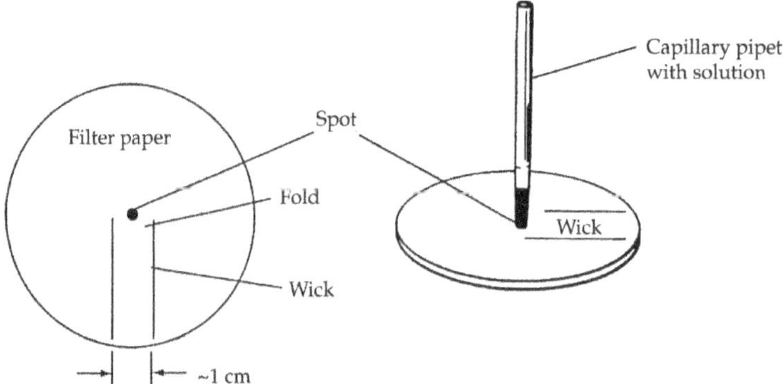

▲**FIGURE 10.2**

ballpoint pen or marker). Place the cut filter paper on top of another piece of filter paper or a paper towel, which will serve as an absorbing pad. Using the following technique, "spot" the filter paper at the center directly on the pencil dot sequentially with 1 drop of each of the known solutions of Cu^{2+}, Ni^{2+}, and Fe^{3+}, letting the spot dry completely between applications. Pour a small quantity (only 1 to 2 mL) of Cu^{2+} solution in a clean sample vial and fill your capillary pipet by dipping the end in the solution. Allow the solution to rise by capillary action. Withdraw the pipet and touch the inside of the vial with the tip of the pipet to remove the hanging drop. Spot the filter paper on the pencil dot by touching it with the capillary held perpendicular to the paper. Allow the solution to flow out of the capillary until a spot with a diameter of about 5 to 7 mm is obtained. Dry the filter paper completely by waving it in the air. In the same way, apply a single drop of the Ni^{2+} and then the Fe^{3+} solution to the same spot, making sure the paper is dry before each application. In the same manner, spot a second piece of filter paper that you have prepared for chromatography with 1 drop of your unknown solution.

Place two Petri (or evaporating) dishes on your desktop away from direct sunlight or heat. Fill the dishes with the solvent mixture to a depth of 5 to 7 mm. **(CAUTION:** ***Acetone is highly flammable! Keep it away from open flames.*****)** Carefully place the (dry) spotted filter papers on the rims of the petri dishes with the wicks bent down into the solvent (Figure 10.3). Cautiously place a 4 in. watch glass on top of each filter paper, being careful not to push the papers into the solvent. The purpose of the cover is to prevent uneven evaporation of the solvent from the paper by providing an enclosed atmosphere that is saturated with solvent vapor. These conditions permit the solvent to travel across the paper in a more uniform manner, which is necessary for effective separation of the ions. Do not disturb the systems while the chromatograms are

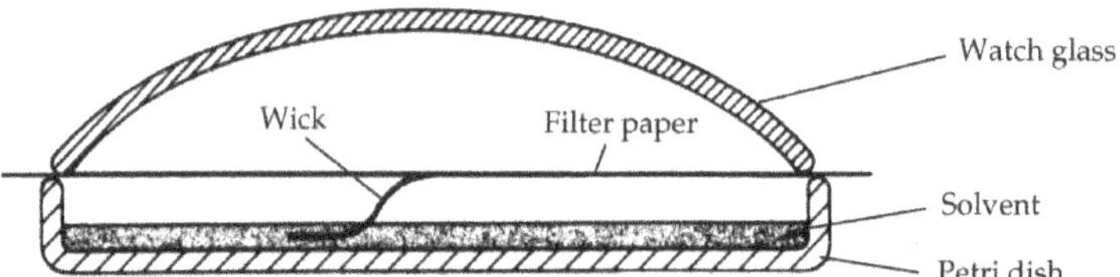

▲**FIGURE 10.3** Side view of apparatus for paper chromatography.

developing. When the solvent front has nearly reached the edge of the paper in each dish (15 to 25 min), carefully remove the watch glass and filter paper. Immediately mark the position of the solvent front with a pencil. Because the solvent evaporates quickly, this marking must be made as soon as possible. Allow each paper to dry by fanning the air with it.

Which of the known ions do you detect without resorting to the use of any other reagents (or development) (2)? Is there any difference between the ring front of this ion and the solvent front (3)? Mark the ring front of this ion.

Developing Reactions *In the hood*, pour about 5 mL of 15 *M* ammonia into a clean, shallow dish and rest the filter paper containing the knowns on top of the dish. Do not let the paper dip into the solution. What color develops (4)? What ion does this indicate (5)? Mark the ring front.

To detect the third ion, dip a new piece of filter paper into a 1% solution of dimethylglyoxime; then, using the new piece as a brush, paint the test filter paper. As an alternative procedure, you may spray the paper with a dimethylglyoxime-containing aerosol. What color develops (6)? What ion does this indicate (7)? Mark the ring front.

Measure the distance from the original spot to each of the ring fronts in millimeters (8). Calculate the R_f values for the three known ions (9) using Equation [1].

Repeat the above sequence of developing reactions with the paper containing your unknown. What ions are present in your unknown (10)? Record the distances in millimeters to the ring fronts (11) and calculate the R_f values (12).

Relative amounts of the components of a mixture can be determined by using the methods employed in this experiment. Prepare a mixture of these ions in the following manner: Mix together 5 mL of the Fe^{3+} and 5 mL of the Ni^{2+} solutions; then add 2 or 3 drops of the Cu^{2+} solution as a trace contaminant. Apply 1 drop of this mixture to a prepared piece of filter paper. Run and develop the chromatogram. Record your observations including the relative intensities of the colors (13). Compare your results with those of someone who performed the experiment using the other technique (14).

Dyes You may think that food coloring consists of one substance or that the black ink in felt-tip pens is a single substance. Spot a piece of filter paper heavily with a black felt-tip pen. Make the spot quite dark so that enough ink is transferred to the paper. Obtain a chromatogram, using as the solvent a solution prepared by diluting 10 mL of isopropyl alcohol with 5 mL of water.

GIVE IT SOME THOUGHT

a. If the ink from a felt-tip pen contains more than one substance, what should its chromatogram look like?

b. If the ink has only one substance, what should its chromatogram look like?

Attach all of the dry, developed chromatograms to your report sheet.

Dispose of all solutions in appropriate waste containers.

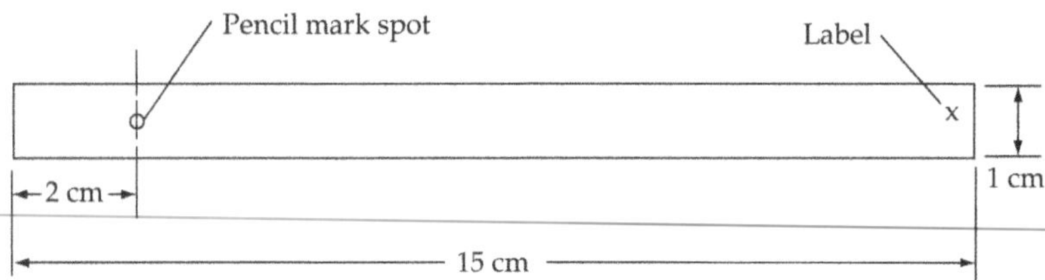

▲FIGURE 10.4

B. Ascending Strip Technique

Obtain a strip of chromatography paper about 50 cm long. Cut this into 15 cm strips and mark each strip with a pencil dot as indicated in Figure 10.4. Label as diagrammed in Figure 10.4. You will be running three chromatograms simultaneously using this technique: the knowns, the unknown, and the trace Cu^{2+}. Place the developing solvent to a depth of 10 to 12 mm in the bottom of a 600 mL beaker (Figure 10.5). Spot the strips as described previously for the circles: one strip with the three knowns, one strip with the unknown solution, and one strip with the solution containing a trace of Cu^{2+}. Attach the labeled ends of the strips to a 6 in. glass rod by folding the ends over the rod and clipping the paper together using paper clips (Figure 10.5). Place the rod with the three strips attached across the top of the beaker. Make sure the spots on the strips are completely dry before you place the strips in the beaker. Check that the strips touch neither one another nor the walls of the beaker and that the bottom of each strip is resting in the solution and that the spots are above the liquid level in the beaker. Cover the beaker carefully with a watch glass. Do not disturb the beaker while the chromatograms are developing. When the solvent front has nearly reached the union of the folded part of the paper, carefully remove the watch glass and the glass rod with the strips. Immediately mark the center of the solvent fronts with a pencil. Allow the strips to dry by fanning the air with them and proceed as directed below with the developing solutions.

Which of the known ions do you detect without resorting to the use of any other reagents (or development) (2)? Is there any difference between the spot of this ion and the solvent front (3)? Mark the spot of this ion.

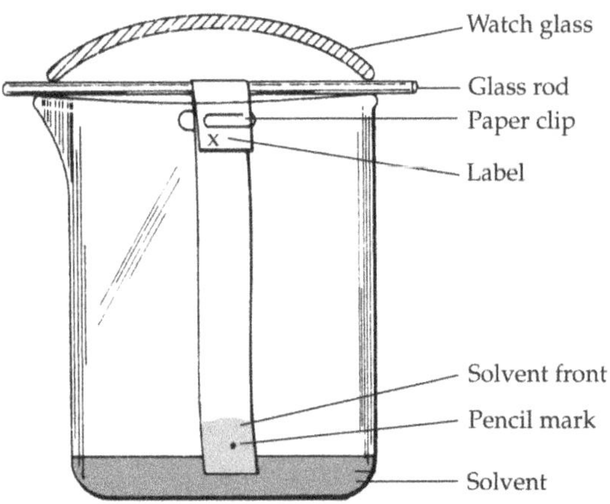

▲FIGURE 10.5

Developing Reactions *In the hood*, pour about 5 mL of 15 *M* ammonia into a clean, shallow dish and rest the filter paper containing the knowns on top of the dish. Do not let the paper dip into the solution. What color develops (4)? What ion does this indicate (5)? Mark the spot.

To detect the third ion, dip a new piece of filter paper into a 1% solution of dimethylglyoxime; then using the new piece as a brush, paint the test filter paper. As an alternative, you may spray the paper with a dimethylglyoxime-containing aerosol. What color develops (6)? What ion does this indicate (7)? Mark the ring front.

Measure the distance from the original spots to the center of each of the ring fronts in millimeters (8). Calculate the R_f values for the three known ions (9) using Equation [1].

Repeat the preceding sequence of developing reactions with paper containing your unknown. What ions are present in your unknown (10)? Record the distances in millimeters from the original spots to the center of the ring fronts (11) and calculate the R_f values (12).

Relative amounts of the components of a mixture can be determined by using the methods employed in this experiment. Prepare a mixture of these ions in the following manner: Mix together 5 mL of the Fe^{3+} solution and 5 mL of the Ni^{2+} solution; then add 2 or 3 drops of the Cu^{2+} solution as a trace contaminant. Apply 1 drop of this mixture to a prepared piece of filter paper. Run and develop the chromatogram. Record your observations (13). Compare your results with those of someone who performed the experiment using the other technique (14).

Dyes You may think that food coloring consists of one substance or that the black ink in felt-tip pens is a single substance. Spot a piece of filter paper heavily with a black felt-tip pen. Make the spot quite dark so that enough ink is transferred to the paper. Obtain a chromatogram, using as the solvent a solution prepared by diluting 10 mL of isopropyl alcohol with 5 mL water.

Attach all of the dry, developed chromatograms to your report sheet.

NOTES AND CALCULATIONS

Name ______________________________ Desk ______________________________

Date ______________________ Laboratory Instructor ______________________________

Paper Chromatography: Separation of Cations and Dyes | 10 Pre-lab Questions

Before beginning this experiment in the laboratory, you should be able to answer the following questions.

1. What does the technique of chromatography allow you to do?

2. What is the meaning and utility of an R_f value? What would you expect to influence an R_f value?

3. What are the developing reactions that allow the identification of Ni^{2+} and Cu^{2+}?

4. What forces cause the eluting solution to move along the chromatographic support material?

5. Why is it important to mark the solvent front immediately?

6. Why is it important to use only a pencil to mark the starting line of the chromatogram?

7. Would you expect changing the solvent to change the R_f value? Why or why not?

8. Suggest simple reasons why the Ni^{2+} and Fe^{3+} ions have different R_f values.

9. Is it necessary that compounds be colored to be separated by chromatography?

10. Why should the Petri dish (or beaker) be covered during development of the chromatogram?

Name ______________________________ Desk ______________________

Date ____________________ Laboratory Instructor ______________________

Unknown no. or letter ______________

REPORT SHEET | EXPERIMENT

Paper Chromatography: Separation of Cations and Dyes

10

1. Color of solutions: Cu^{2+} __________; Fe^{3+} __________; Ni^{2+} __________
2. Ion requiring no development __________________
3. Yes __________ No __________
4. Ammonia develops a __________ color.
5. The ion that causes the color of (4) is __________________.
6. Dimethylglyoxime develops a __________ color with (7) __________ ion.

8. Ring front distances for knowns in mm: solvent __________mm;

   Fe^{3+} __________mm; Cu^{2+} __________mm; Ni^{2+} __________mm
9. R_f values (show calculations at end of report): Fe^{3+} __________; Cu^{2+} __________;

   Ni^{2+} __________
10. Ions present in unknown __________________
11. Ring front distances for unknown __________ : __________ : __________ : __________

    ____
12. R_f values for unknown ____

13. Observation on relative amounts ____

____

14. Comment on the relative merits of the two techniques.

Calculations for R_f's (9) and R_f's (12):

ATTACH CHROMATOGRAMS

Molecular Geometries of Covalent Molecules: Lewis Structures and the VSEPR Model

OBJECTIVE

To become familiar with Lewis structures, the principles of the VSEPR model, and the three-dimensional structures of covalent molecules.

APPARATUS

Prentice Hall Molecular Model Set for General and Organic Chemistry or Styrofoam balls* and pipe cleaners or other appropriate model kit.

DISCUSSION

Types of Bonding Interactions

Whenever atoms or ions are strongly attached to one another, there is a *chemical bond* between them. There are three types of chemical bonds: ionic, covalent, and metallic. The term *ionic bond* refers to electrostatic forces that exist between ions of opposite charge (Section 8.2). Ions may be formed by the transfer of one or more electrons from an atom with a low ionization energy to an atom with a high electron affinity. Thus, ionic substances generally result from the interaction of metals on the far left side of the periodic table with nonmetals on the far right side of the periodic table (excluding the noble gases, group 8A). A *covalent bond* results from the sharing of electrons between two atoms (Section 8.3). The more familiar examples of covalent bonding are found in nonmetallic elements interacting with one another. This experiment illustrates the geometric (three-dimensional) shapes of molecules and ions resulting from covalent bonding among various numbers of elements and two of the consequences of geometric structure—isomers and polarity. *Metallic bonds* are found in metals such as gold, iron, and magnesium. In the metals, each atom is bonded to several neighboring atoms. The bonding electrons are relatively free to move throughout the three-dimensional structure of the metal. Metallic bonds give rise to such typical metallic properties as high electrical and thermal conductivity and luster.

Lewis Symbols

The electrons involved in chemical bonding are the *valence electrons*, those residing in the incomplete outer shell of an atom. The American chemist G. N. Lewis suggested a simple way of showing these valence electrons, which are now called Lewis electron-dot symbols, or simply Lewis symbols (Section 8.1). The **Lewis symbol** for an element consists of the chemical abbreviation for the element and a dot for each valence electron. For example, oxygen has the electron configuration [He]$2s^2 2p^4$; therefore, its Lewis symbol shows six valence electrons. The dots are placed on the four sides of the atomic abbreviation.

*Available from Snow Foam Products, Inc., 9917 W. Gidley St., El Monte, CA 91731.

Each side can accommodate up to two electrons. All four sides are equivalent; the placement of two electrons on any given side versus one electron is arbitrary. The number of valence electrons of any representative element is the same as the group number of the element in the periodic table.

The Octet Rule

Atoms often gain, lose, or share electrons to achieve the same number of electrons as the noble gas closest to them in the periodic table (Section 8.1). Because all noble gases (except He) have eight valence electrons, many atoms undergoing reactions also end up with eight valence electrons. This observation has led to the **octet rule:** *Atoms tend to gain, lose, or share electrons until they are surrounded by eight valence electrons.* An octet of electrons consists of full *s* and *p* subshells on an atom. In terms of Lewis symbols, an octet can be thought of as four pairs of valence electrons arranged around the atom, as in the configuration for Ne, which is $:\overset{..}{\underset{..}{\text{Ne}}}:$. There are many exceptions to the octet rule, but it provides a useful framework for many important concepts of bonding.

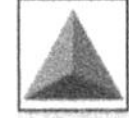

GIVE IT SOME THOUGHT

One common exception to the octet rule involves transition metal (TM) complexes. In TM complexes, the stable arrangement of electrons consists of full s, p, and d orbitals. TM atoms tend to gain, lose, or share electrons until they are surrounded by what number of electrons?

Covalent Bonding

Ionic substances have several characteristic properties (Section 8.2). They are usually brittle, have high melting points, and are crystalline solids with well-formed faces that can often be cleaved along smooth, flat surfaces. These characteristics result from electrostatic forces that maintain the ions in a rigid, well-defined, three-dimensional arrangement.

The vast majority of chemical substances do not have the characteristics of ionic materials. Most of the substances with which you come in daily contact, such as water, gasoline, oxygen, and sugar, tend to be liquids, gases, or low-melting solids. Many of them, such as charcoal lighter fluid, vaporize readily. Many are pliable in their solid form (for example, plastic bags and paraffin wax).

For the very large class of substances that do not behave like ionic substances, a different model is needed for the bonding between atoms. Lewis reasoned that atoms might acquire a noble gas electron configuration by sharing electrons with other atoms to form *covalent bonds* (Section 8.3). The hydrogen molecule, H_2, provides the simplest example of a covalent bond. When two hydrogen atoms are close to each other, electrostatic interactions occur between them. The two positively charged nuclei and the two negatively charged electrons repel each other, whereas the nuclei and electrons attract each other. The attraction between the nuclei and the electrons cause electron density to concentrate between the nuclei. As a result, the overall electrostatic interactions are attractive in nature.

In essence, the shared pair of electrons in any covalent bond acts as a kind of "glue" to bind the two hydrogen atoms together in the H_2 molecule.

Lewis Structures

The formation of covalent bonds can be represented using Lewis symbols as shown below for H_2 (Section 8.5).

$$\mathrm{H\cdot + \cdot H \rightarrow H:H}$$

The formation of a bond between two F atoms to give an F_2 molecule can be represented in a similar way.

$$:\ddot{\underset{..}{\mathrm{F}}}\cdot + \cdot\ddot{\underset{..}{\mathrm{F}}}: \rightarrow :\ddot{\underset{..}{\mathrm{F}}}:\ddot{\underset{..}{\mathrm{F}}}:$$

By sharing the bonding electron pair, each fluorine atom acquires eight electrons (an octet) in its valence shell. It thus achieves the noble-gas electron configuration of neon. The structures shown here for H_2 and F_2 are called *Lewis structures* (or Lewis electron-dot structures). When writing Lewis structures, you usually show each electron pair shared between atoms as a line (to emphasize that it is a bond) and the unshared electron pairs as dots. Writing them this way, the Lewis structures for H_2 and F_2 are:

$$\mathrm{H{-}H} \qquad :\ddot{\underset{..}{\mathrm{F}}}{-}\ddot{\underset{..}{\mathrm{F}}}:$$

The number of valence electrons for the nonmetal is the same as the group number. Therefore, you might predict that 7A elements, such as F, would form one covalent bond to achieve an octet; 6A elements, such as O, would form two covalent bonds; 5A elements, such as N, would form three covalent bonds; and 4A elements, such as C, would form four covalent bonds. For example, consider the simple hydrogen compounds of the nonmetals of the second row of the periodic table.

$$\mathrm{H{-}}\ddot{\underset{..}{\mathrm{F}}}: \qquad \mathrm{H{-}}\underset{\displaystyle |\atop \mathrm{H}}{\ddot{\mathrm{O}}}: \qquad \mathrm{H{-}}\underset{\displaystyle |\atop \mathrm{H}}{\ddot{\mathrm{N}}}\mathrm{{-}H} \qquad \mathrm{H{-}}\overset{\mathrm{H}\atop |}{\underset{|\atop \mathrm{H}}{\mathrm{C}}}\mathrm{{-}H}$$

GIVE IT SOME THOUGHT

a. Do Lewis structures tell you anything about how each molecule is oriented in three dimensions?
b. If not, would you need to expand on this model?

Thus, the Lewis model succeeds in accounting for the compounds of nonmetals, in which covalent bonding predominates.

Multiple Bonds

The sharing of one pair of electrons constitutes a single covalent bond, generally referred to simply as a *single bond*. In many molecules, atoms attain complete octets by sharing more than one pair of electrons between them.

When two electron pairs are shared, two lines (representing a *double bond*) are drawn. A *triple bond* corresponds to the sharing of three pairs of electrons. Such multiple bonding is found in CO_2 and N_2.

$$\ddot{O}=C=\ddot{O} \qquad :N\equiv N:$$

Drawing Lewis Structures

Lewis structures are useful in understanding the bonding in many compounds and are frequently used when discussing the properties of molecules. To draw Lewis structures, you follow a regular procedure.

1. *Sum the valence electrons from all atoms.* Use the periodic table as necessary to help determine the number of valence electrons on each atom. For an anion, add an electron to the total for each negative charge. For a cation, subtract an electron for each positive charge.
2. *Write the symbols for the atoms to show which atoms are attached and connect them with a single bond* (a dash representing two electrons). Chemical formulas are often written in the order in which the atoms are connected in the molecule or ion, as in HCN. When a central atom has a group of other atoms bonded to it, the central atom is usually written first, as in CO_3^{2-} and BF_3. In other cases, you may need more information before you can draw the Lewis structure.
3. *Complete the octets of the atoms bonded to the central atom.* (Remember, however, that hydrogen can have only two electrons.)
4. *Place any leftover electrons on the central atom*, even if doing so results in more than an octet.
5. *If there are not enough electrons to give the central atom an octet, try multiple bonds.* Use one or more of the unshared pairs of electrons on the atoms bonded to the central atom to form double or triple bonds.

Formal Charge

When drawing a Lewis structure, you are describing how the electrons are distributed in a molecule (or ion) (Section 8.5). In some instances, you can draw several different Lewis structures that obey the octet rule. How do you decide which one is most reasonable? One approach is to do some "bookkeeping" of the valence electrons to determine the *formal charge* of each atom in each Lewis structure. The formal charge of an atom is the charge that an atom in a molecule or ion would have if all atoms had the same electronegativity. To calculate the formal charge on any atom in a Lewis structure, you assign the electrons to the atoms as follows:

1. All of the unshared (nonbonding) electrons are assigned to the atom on which they are found.
2. Half of the bonding electrons are assigned to each atom in the bond.
3. The formal charge of each atom is calculated by subtracting the number of electrons assigned to the atom from the number of valence electrons in the neutral atom.
4. All formal charges should be reasonable (zero formal charges are preferred; otherwise, more electronegative atoms should have negative formal charges).

EXAMPLE 11.1

Draw three Lewis structures for SCN^-. Calculate formal charges and designate the preferred structure.

SOLUTION: The total number of valence electrons is 6(S) + 4(C) + 5(N) + 1 (minus 1 charge) = 16 electrons. Following are three possible Lewis structures:

$[:\ddot{\underset{\cdot\cdot}{S}}-C\equiv N:]^-$	$[:\ddot{S}=C=\ddot{N}:]^-$	$[:S\equiv C-\ddot{\underset{\cdot\cdot}{N}}:]^-$
−1 0 0	0 0 −1	+1 0 −2
I	II	III

The formal charges for structure I are calculated as follows: For the S atom there are six nonbonding electrons and one bonding electron for a total of seven. The number of valence electrons on a neutral S atom is six. Thus, the formal charge is 6 − 7= −1. For C, there are four bonding electrons and four valence electrons on a neutral C atom. Thus, the formal charge is 4 − 4 = 0. For N, there are two nonbonding electrons 6 × ½ = 3 bonding electrons, and five valence electrons on a neutral N atom. Thus, the formal charge is 5 − 5 = 0. All of the atoms have octets in each structure. Because nitrogen is more electronegative than sulfur is, structure II is preferred on the basis of formal charges. Structure III is not favored because of the large charge separation.

Bond Polarity and Dipole Moments

A covalent bond between two different kinds of atoms is usually a polar bond (Section 8.4). This is because the two different atoms have different electronegativities, which means that the electrons in the bond are not shared equally by the two bound atoms. This creates a bond dipole moment, μ, which is a charge separation, Q, over a distance, r:

$$\mu = Qr$$

The bonding electrons have an increased attraction to the more electronegative atom, thus creating an excess of electron density (δ^-) near it and a deficiency of electron density (δ^+) near the less electronegative atom. You symbolize the bond dipole (a vector) with an arrow and a cross, ↦, with the point of the arrow representing the negative end of the dipole and the cross representing the positive end of the dipole. Thus, in the polar covalent molecule H—Cl, because chlorine is more electronegative than hydrogen, the bond dipole is as illustrated below.

$$\overset{\delta^+}{H} \mapsto \overset{\delta^-}{Cl}$$

The bond dipole is a vector (has magnitude and direction), and the dipole moments of polyatomic molecules are the vector sums of the individual bond dipoles as illustrated in Figure 11.1.

Because H_2O has a molecular dipole and CCl_4 does not, H_2O is a polar molecule and CCl_4 is a nonpolar molecule.

GIVE IT SOME THOUGHT

When a bond dipole vector is drawn between two atoms, to which atom will the bond vector be drawn?

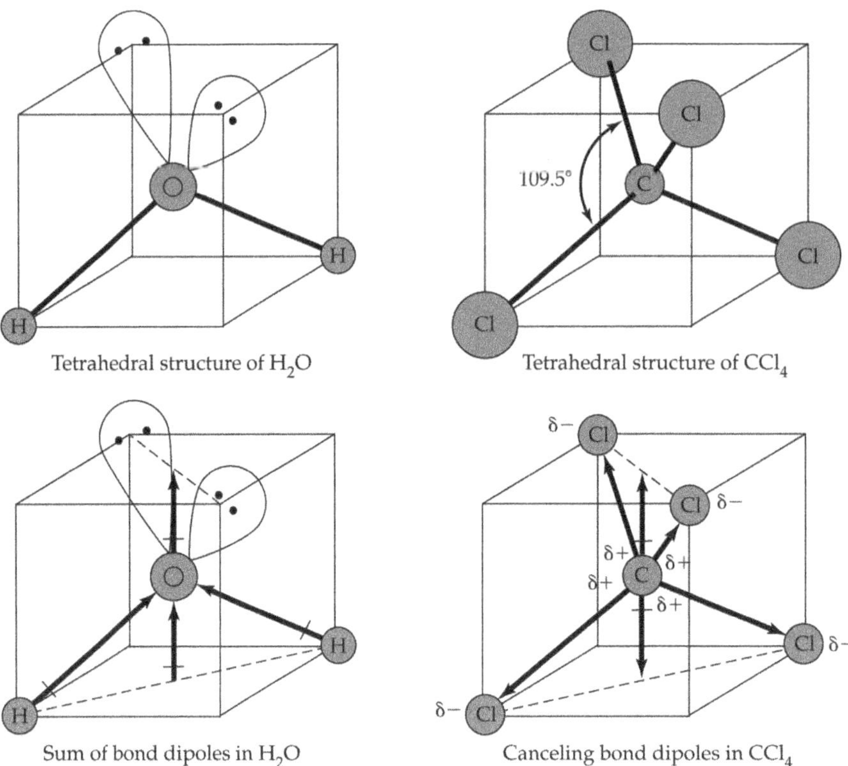

▲FIGURE 11.1 The bond dipoles in H_2O are added to give a net molecular dipole that bisects the HOH angle and points upward toward the O atom in this drawing. In contrast, because of the symmetry of CCl_4, the bond dipoles cancel and the CCl_4 molecule has no molecular dipole moment.

The VSEPR Model

In covalent molecules, atoms are bonded together by sharing pairs of valence-shell electrons (Section 9.2). Electron pairs repel one another and try to stay out of each other's way. *The best arrangement of a given number of electron pairs is the one that minimizes the repulsions among them.* This simple idea is the basis of valence shell electron pair repulsion, or the **VSEPR** model. Thus, as illustrated in Table 11.1, two electron pairs are arranged *linearly*, three pairs are arranged in a *trigonal planar* fashion, four are arranged *tetrahedrally*, five are arranged in a *trigonal bipyramidal* geometry, and six are arranged *octahedrally*. The shape of a molecule or ion can be related to these five basic arrangements of electron pairs.

Predicting Molecular Geometries When drawing Lewis structures, you encounter two types of valence shell electron pairs: **bonding pairs**, which are shared by atoms in bonds, and nonbonding pairs (or *lone pairs*), as in the Lewis structure for NH_3.

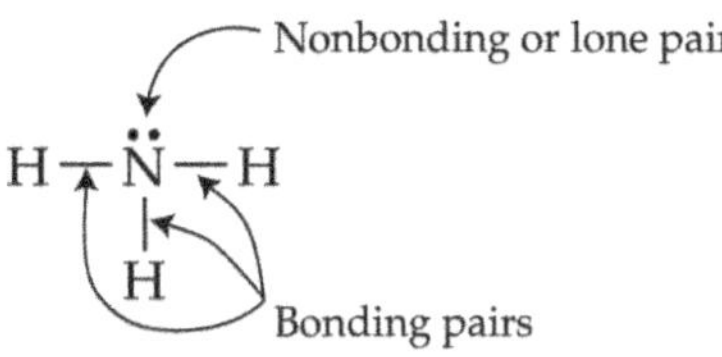

TABLE 11.1 Electron-Domain Geometries as a Function of the Number of Electron Pairs

Number of electron pairs	Arrangement of electron pairs	Electron-domain geometry	Predicted bond angles
2	180°	Linear	180°
3	120°	Trigonal planar	120°
4	109.5°	Tetrahedral	109.5°
5	90° 120°	Trigonal bipyramidal	120° 90°
6	90° 90°	Octahedral	90° 180°

GIVE IT SOME THOUGHT

How do the molecular geometries from the VSEPR model differ from those of Lewis structures?

Because there are four electron pairs around the N atom, the electron-pair repulsions will be minimized when the electron pairs point toward the vertices of a tetrahedron (see Table 11.1). The arrangement of electron pairs about the central atom of an AB_n molecule, where A is the central atom and B are the peripheral atoms, is called its **electron-domain geometry**. However, when using experiments to determine the structure of a molecule, you locate atoms, not electron pairs. The **molecular geometry** of a molecule (or ion) is the arrangement of the *atoms* in space. You can predict the molecular geometry of a molecule from its electron-domain geometry. In NH_3, the three bonding pairs point toward three vertices of a tetrahedron. Therefore, the hydrogen atoms are located at the three vertices of a tetrahedron and the lone pair is located at a fourth vertex. The arrangement of the atoms in NH_3 is thus a *trigonal pyramid*; the sequence of steps to arrive at this prediction are illustrated in Figure 11.2. It shows that the trigonal-pyramidal molecular geometry of NH_3 is a consequence of its tetrahedral electron-domain geometry. When describing the shapes of molecules, you give the molecular geometry rather than the electron-domain geometry.

Following are the steps used to predict molecular geometries with the VSEPR model:

1. Sketch the Lewis structure of the molecule or ion.
2. Count the total number of electron pairs around the central atom and arrange them in a way that minimizes electron-pair repulsions (see Table 11.1).
3. Describe the molecular geometry in terms of the angular arrangement of the bonding pairs, which corresponds to the arrangement of bound atoms.

Application of the VSEPR model to molecules that contain multiple bonds reveals that a double or triple bond has essentially the same effect on bond angles as does a single bond. This observation leads to a fourth rule:

4. A double or triple bond is treated as one bonding pair in predicting geometry.

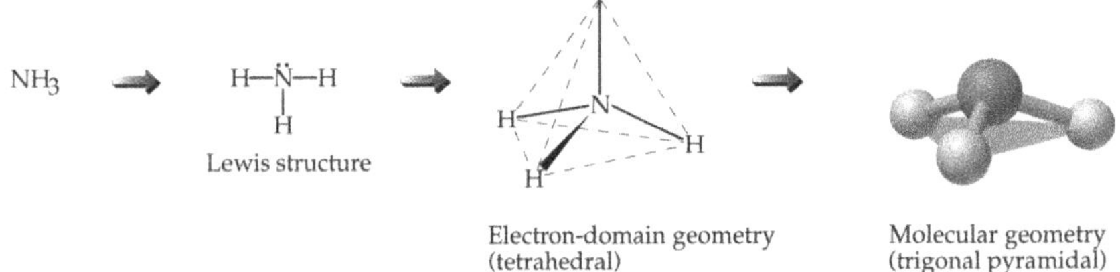

▲**FIGURE 11.2** Chemical formula, Lewis structure, electron-domain geometry, and molecular geometry of NH_3.

EXAMPLE 11.2

Using the VSEPR model, predict the molecular geometries of (a) SCl_2 and (b) NO_2^-.

SOLUTION:

a. The Lewis structure for SCl_2 is :C̤̈l—S̤̈—C̤̈l:. The central S atom is surrounded by two nonbonding electron pairs and two single bonds. Thus, the electron-domain geometry is tetrahedral, and the molecular geometry is bent.

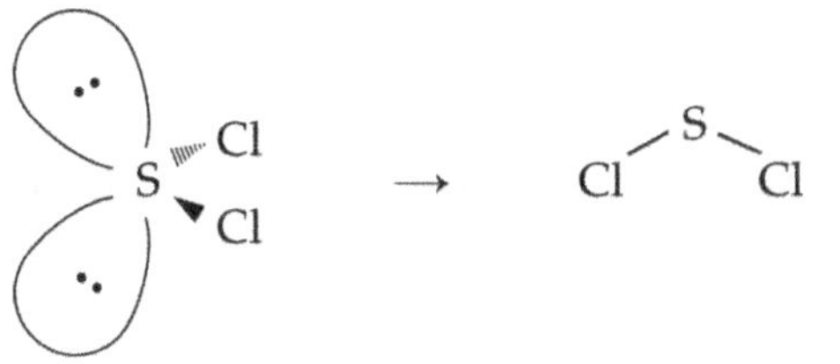

b. For NO_2^-, you can draw two equivalent resonance structures.

Because of resonance, the bonds between N and the two O atoms are of equal length. Both resonance structures show one nonbonding electron pair, one single bond, and one double bond about the central N atom. In predicting geometry, a double bond is treated as one electron pair (rule 4). Thus, the arrangement of valence shell electrons is trigonal planar. Two of these positions are occupied by O atoms; so the ion has a bent shape.

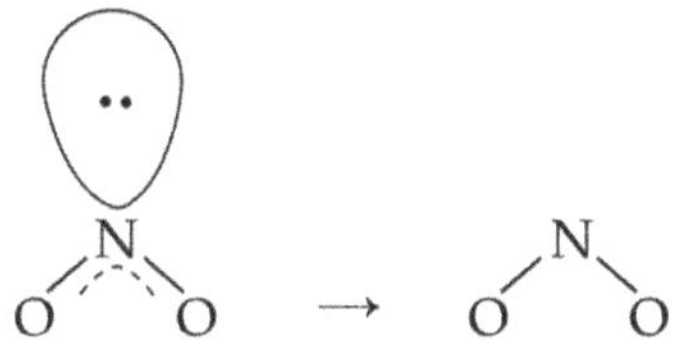

Valence Bond (VB) Theory: Covalent Bonding and Orbital Overlap

Although the VSEPR model provides a simple means for predicting molecular shapes, it does not explain why bonds exist between atoms. In 1931, Linus Pauling developed a bonding model called **valence bond theory**, based on the marriage of Lewis's notion of electron-pair bonds to the idea of atomic orbitals (Section 9.4). By extending this approach to include the ways in which atomic orbitals can mix with one another, a picture appears that corresponds nicely to the VSEPR model.

In the Lewis theory, covalent bonding occurs when atoms share electrons. Such sharing concentrates electron density between the nuclei. In the valence bond theory, the buildup of electron density between two nuclei is visualized as occurring when a valence atomic orbital of one atom merges with that of another atom. The orbitals are then said to share a region of space, or to overlap. The overlap of orbitals allows two electrons of opposite spin to share the common space between the nuclei, forming a covalent bond as shown for the H_2, HCl, and Cl_2 molecules in Figure 11.3.

▲**FIGURE 11.3** The overlap of orbitals to form covalent bonds. (a) The bond in H_2 results from the overlap of two 1*s* orbitals from two H atoms. (b) The bond in HCl results from the overlap of a 1*s* orbital of H and one of the lobes of a 3*p* orbital of Cl. (c) The bond in Cl_2 results from the overlap of two 3*p* orbitals, one from each of the two Cl atoms.

GIVE IT SOME THOUGHT

When orbitals form covalent bonds, in what location are you most likely to find an electron?

Hybrid Orbitals

Although the idea of orbital overlap allows you to understand the formation of covalent bonds, it is not always easy to extend these ideas to polyatomic molecules (Section 9.5). You need to understand both the formation of electron-pair bonds and the observed geometry. Consider the molecule BeF_2:

$$:\ddot{\underset{..}{F}}—Be—\ddot{\underset{..}{F}}:$$

The VSEPR model predicts, correctly, that this molecule is linear. Beryllium has a $1s^2 2s^2$ electron configuration with no unpaired electrons. Which orbitals on Be overlap with orbitals on F to form bonds? If the electron configuration of Be were promoted to $1s^2 2s^1 2p^1$, a Be *s* and a *p* orbital could be used for bonding. You can also answer the question by "mixing" the 2*s* orbital with one of the 2*p* orbitals on Be to generate two new orbitals as shown in Figure 11.4.

Like *p* orbitals, each of the new *sp* orbitals has two lobes. The two new orbitals are identical in shape, but their large lobes point in opposite directions. Two

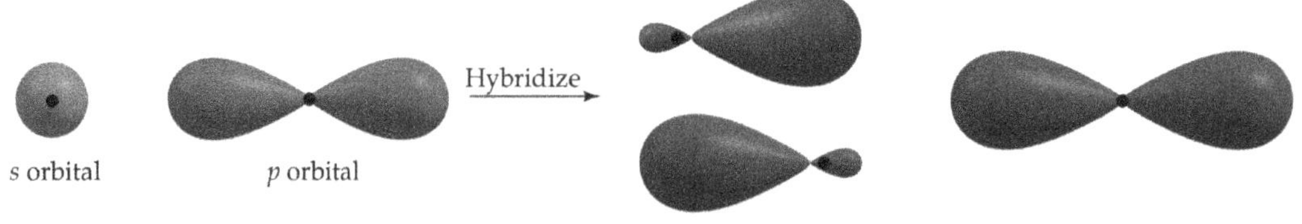

▲**FIGURE 11.4** One *s* orbital and one *p* orbital can hybridize to form two equivalent *sp* hybrid orbitals. The two hybrid orbitals have their large lobes pointing in opposite directions, 180° apart.

GIVE IT SOME THOUGHT

a. Would the shape of the hybrid orbitals be the same if one *s* orbital and two *p* orbitals mixed?

b. What if one *s* orbital and three *p* orbitals mixed?

hybrid orbitals have been created, orbitals formed by mixing two or more atomic orbitals on an atom, a procedure called hybridization. Other types of hybrid orbitals and their geometric arrangements are shown in Table 11.2.

TABLE 11.2 Geometrical Arrangements Characteristic of Hybrid Orbital Sets

Atomic orbital set	Hybrid orbital set	Geometry	Examples
s,p	Two *sp*	180° Linear	BeF_2, $HgCl_2$
s,p,p	Three sp^2	120° Trigonal planar	BF_3, SO_3
s,p,p,p	Four sp^3	109.5° Tetrahedral	CH_4, NH_3, H_2O, NH_4^+

In this experiment, based on molecular models, you will draw Lewis structures for molecules and ions, predict their geometric structures from the VSEPR model and valence bond theory, predict some of their properties, and build molecular models.

NOTES AND CALCULATIONS

Name ______________________________ Desk ______________

Date ______________ Laboratory Instructor ______________________

Molecular Geometries of Covalent Molecules: Lewis Structures and the VSEPR Model | 11 Pre-lab Questions

Before beginning this experiment in the laboratory, you should be able to answer the following questions.

1. Distinguish among ionic, covalent, and metallic bonding.

2. Which of the following molecules possess polar covalent bonds: H_2, N_2, HCl, HCN, and CO_2?

3. Which of the molecules in question 2 have molecular dipole moments?

4. What are the favored geometrical arrangements for AB_n molecules for which the A atom has 2, 3, 4, 5, and 6 pairs of electrons in its valence shell?

5. How many equivalent orbitals are involved in each of the following sets of hybrid orbitals: sp, sp^2, sp^3d, and sp^3d^2?

6. Define the term *formal charge.*

7. Calculate the formal charges of all of the atoms in O_2, SO_2, and SO_3.

NOTES AND CALCULATIONS

Name Grace Rademacher Desk

Date Laboratory Instructor Prof Briguglio

REPORT SHEET | EXPERIMENT

Molecular Geometries of Covalent Molecules: Lewis Structures and the VSEPR Model

11

1. Using an appropriate set of models, make molecular models of the compounds listed below and complete the table.

Molecular formula	No. of bond pairs (bp)	No. of lone pairs on central atom (lp)	Hybridization of central atoms	Molecular geometry	Bond angle(s)	Dipole moment (yes or no)
$BeCl_2$	2	0	sp	linear	180°	no
BF_3	3	0	sp^2	trigonal planar	120°	no
$SnCl_2$	2	1	sp^2	bent	120°	yes
CH_4	4	0	sp^3	tetrahedral	109.5°	no
NH_3	3	1	sp^3	trigonal pyramidal	109.5°	yes
H_2O	2	2	sp^3	bent	109.5°	yes
PCl_5	5	0	sp^3d	trigonal bipyramidal	120/90°	no
SF_4	4	1	sp^3d	saw horse	120/90°	yes
BrF_3	3	2	sp^3d	T-shaped	90/180°	yes
XeF_2	2	3	sp^3d	linear	180°	no
SF_6	6	0	sp^3d^2	octahedral	90°	no
IF_5	5	1	sp^3d^2	square pyramid	90°	yes
XeF_4	4	2	sp^3d^2	square planar	90°	no

2. From your models of SF_4, BrF_3, and XeF_4, deduce whether different atom arrangements, called geometrical isomers, are possible; if so, sketch them below. Indicate the preferred geometry for each case and suggest a reason for your choice. Indicate which structures have dipole moments and show their direction.

	Molecule	*Dipole moment*	*Preferred geometry*	*Reason*
(a)	SF_4			
(b)	BrF_3			
(c)	XeF_4			

3. Using the Lewis structure predict the geometrical structures of the following ions and state the hybridization of the central atom.

Ion	*Structure*	*Central atom hybridization*
N_3^-		

Ion	*Structure*	*Central atom hybridization*
CO_3^{2-}		
NO_3^{-}		
BF_4^{-}		

4. Because lone pairs are spread over a larger region of space than bonding pairs, lone pair–lone pair interactions are greater than lone pair–bonding pair interactions, which are in turn larger than bonding pair–bonding pair interactions. Using this notion, suggest how the following species would distort from regular geometries.

   (a) OF_2

   (b) SCl_2

   (c) PF_3

5. There are several families of hydrocarbons, among which are alkanes, alkenes, and alkynes. The parents of each family are CH_4 (methane), $H_2C=CH_2$ (ethene), and $H—C\equiv C—H$ (ethyne), respectively. Predict the geometries of these molecules, give the hybridization of carbon in each molecule, and suggest whether they are polar or nonpolar molecules.

Molecule	*C-hybridization*	*Polar (yes or no)*
CH_4		

Molecule	*C-hybridization*	*Polar (yes or no)*
C_2H_4		
C_2H_2		

6. Calculate the formal charges of all atoms in NH_3, NO_2^-, and NO_3^-.

NH3

N needs 5
H needs 1

FC: 0

NO2-

N: 5-4 = +1
O1: 6-7 = -1 } 1 double bond
O2: 6-7 = -1 → F.C.

NO2- FC: 1-

NO3-

N: 5
3 O: 18
+ 1: 1
24
- 6
18
- 18
0

5-3 = 2
6-7 = -1 } 1 doub
6-7 = -1
6-7 = -1

NO3- FC: 1-

Experiment

Atomic Spectra and Atomic Structure

OBJECTIVE

To gain some understanding of the relationship between emission (line) spectra and atomic structure.

APPARATUS AND CHEMICALS

Apparatus

spectroscope with illuminated scale or diffraction grating
hydrogen lamp
mercury-vapor lamp
high-voltage power supply with lamp holder
Nichrome wire loop
Bunsen burner and hose
wood splints

Chemicals

6 *M* HCl
0.1 *M* $CaCl_2$
0.1 *M* LiCl
unknown 1 solution (mixture containing two or more cations from above)
0.1 *M* NaCl
0.1 *M* $SrCl_2$
0.1 *M* KCl
0.1 *M* $BaCl_2$
unknown 2 solution (one cation from above)

INTRODUCTION

A person's present understanding of atomic structure has come from studying the properties of light or radiant energy and emission and absorption spectra. Radiant energy is characterized by two variables: its wavelength, λ, and its frequency, v. The wavelength and frequency are related to each other by the relation

$$v\lambda = c \qquad [1]$$

where c is the speed of light, 3.00×10^8 meters per second, or m/s; wavelength is usually expressed in nanometers, or nm (10^{-9} m); and frequency is given in cycles per second, or hertz (Hz) (Section 6.1).

EXAMPLE 12.1

What frequency corresponds to a wavelength of 500 nm?

SOLUTION:

$$v = \frac{c}{\lambda} = \left(\frac{3.00 \times 10^8 \text{ m/s}}{500 \text{ nm}}\right)\left(\frac{10^9 \text{ nm}}{1 \text{ m}}\right)$$

$$= 6.00 \times 10^{14} \text{ s}^{-1}, \text{ or } 6.00 \times 10^{14} \text{ Hz}$$

Radiation of different wavelengths affects matter differently. Infrared radiation may cause a "heat burn," visible and near-ultraviolet light may cause a sunburn or suntan, and X-rays may cause tissue damage or even cancer. Some wavelength units for various types of radiation are given in Table 12.1.

TABLE 12.1 Wavelength Units for Electromagnetic Radiation

Unit	Symbol	Length (m)	Type of radiation
Ångstrom	Å	10^{-10}	X-ray
Nanometer	nm	10^{-9}	ultraviolet, visible light
Micrometer	μm	10^{-6}	infrared
Millimeter	mm	10^{-3}	infrared
Centimeter	cm	10^{-2}	microwaves
Meter	m	1	TV, radio

A particular source of radiant energy may emit a single wavelength, as in the light from a laser, or many different wavelengths, as in the radiations from an incandescent lightbulb or a star. Radiation composed of a single wavelength is termed *monochromatic*. When the radiation from a source such as the sun or other stars is separated into its components, a *spectrum* is produced. This separation of radiations of differing wavelengths can be achieved by passing the radiation through a prism. Each component of the polychromatic radiation is bent to a different extent by the prism, as shown in Figure 12.1. This rainbow of colors, containing light of all wavelengths, is called a *continuous spectrum.* The most familiar example of a continuous spectrum is the rainbow, produced by the dispersion of sunlight by raindrops or mist.

It was found that by placing a narrow slit between the source of radiation and the prism, the quality of the spectrum was improved and the monochromatic components were more sharply resolved. An instrument used for studying line spectra is called a *spectroscope.* A spectroscope (Figure 12.2) contains the following: a slit for admitting a narrow collimated beam of light, a prism for dispersing the light into its components, an eyepiece for viewing the spectrum, and an illuminated scale against which the spectrum may be viewed (Section 6.3).

Most substances will emit light energy if heated to a high enough temperature. For example, a fireplace poker will glow red if left in the flame for several minutes. Similarly, neon gas will emit bright red light when excited with a sufficiently high electrical voltage. When energy is absorbed by a substance, electrons within the atoms of the substance may be excited to positions of higher potential energy, farther away from the nucleus. When these electrons return to their normal, or ground-state, position, energy is emitted. Usually, some of the energy emitted occurs in the visible region of the electromagnetic spectrum (4000 to 7000 Å, or 400 to 700 nm).

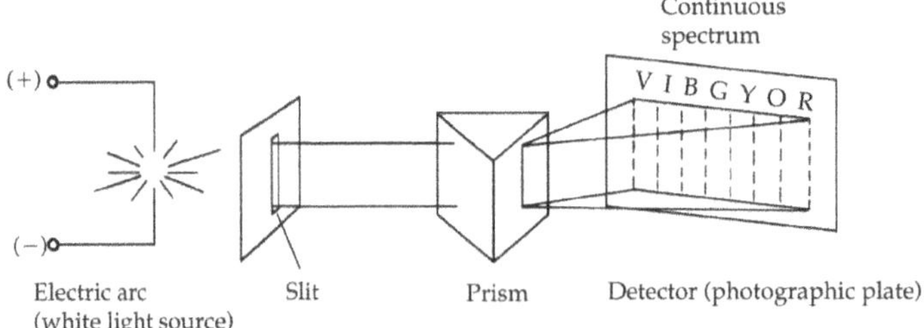

▲**FIGURE 12.1** The spectrum of white light. When light from the sun or from a high-intensity incandescent bulb is passed through a prism, the component wavelengths are spread out into a continuous rainbow spectrum, where VIBGYOR represents; violet, indigo, blue, green, yellow, orange, and red, respectively.

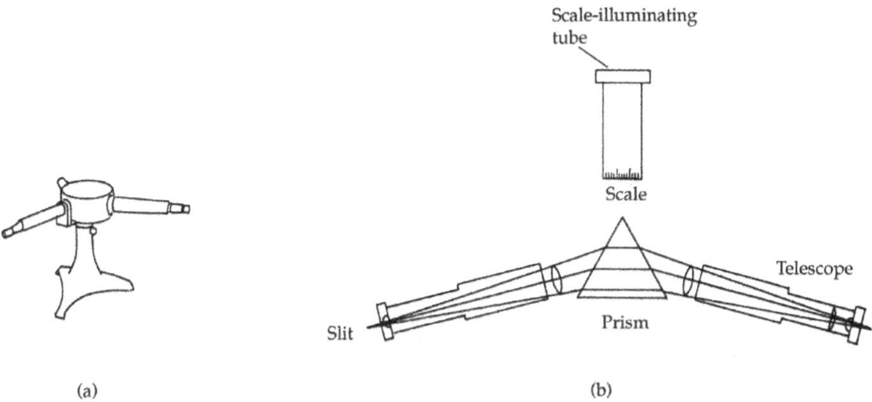

▲FIGURE 12.2 (a) A sketch of the spectroscope to be used. (b) A diagram showing its component parts. When viewed through the telescope, the spectrum will appear superimposed on the numeric scale.

Excited atoms do not emit a continuous spectrum; rather, they emit radiation at certain discrete, well-defined fixed wavelengths. For example, if you have ever spilled table salt (NaCl) into a flame, you have seen the characteristic yellow emission of excited sodium atoms. If the light emitted by the atoms of a particular element is viewed in a spectroscope, only certain bright-colored lines are seen in the spectrum. See Figure 12.3, which illustrates the emission spectrum of hydrogen.

The observation that a given excited atom emits radiation at only certain fixed wavelengths indicates that the atom can undergo energy changes only of certain fixed, definite amounts. An atom does not emit continuous radiation, but rather energy corresponding to definite regular changes in the energies of its component electrons. The experimental demonstration of bright-line atomic emission spectra implied a regular fixed electronic microstructure for the atom and led to the Bohr model for the hydrogen atom.

In this experiment, you will obtain emission spectra for a few elements and explain your results in terms of the Bohr model. You will use a spectroscope like the one illustrated in Figure 12.2 to obtain your data (Section 6.3).

CALIBRATION OF THE SPECTROSCOPE

The scale of the spectroscope has arbitrary divisions and must be calibrated with a known spectrum before the spectrum of an unknown may be obtained. Calibration is accomplished by viewing the emission spectrum of mercury

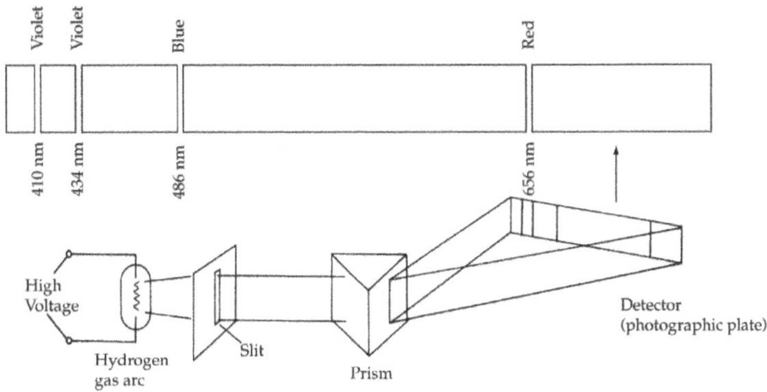

▲FIGURE 12.3 The emission spectrum of hydrogen.

because the emission wavelengths for mercury are precisely known. You will record the positions on the spectroscope scale of the mercury emission lines and prepare a calibration curve by plotting the positions against the known wavelengths of the lines, which are as follows:

Violet:	404.7 nm
Blue:	435.8 nm
Green:	546.1 nm
Yellow:	579.0 nm

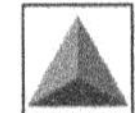

GIVE IT SOME THOUGHT

All four lines in this spectrum are a different color. What does that tell you about their energies?

Turn on the illuminated scale of the spectroscope and look through the eyepiece to make sure the scale is visible but not so brightly lit that the mercury spectral lines are obscured when the mercury lamp is turned on. With the power supply unplugged, position the power supply and mercury lamp directly in front of the slit opening of the spectroscope. **(CAUTION: *The power supply develops several thousand volts. Do not touch any portion of the power supply, wire leads, or lamps unless the power supply is unplugged from the wall outlet. In addition to visible light, the lamps may emit ultraviolet radiation. Ultraviolet radiation is damaging to your eyes. Wear your safety glasses at all times during this experiment because they will absorb some of the ultraviolet radiation. Do not look directly at any of the lamps while they are illuminated.* DO NOT LET THE POWER SUPPLY OR LAMP TOUCH THE SPECTROSCOPE.)** With your instructor's permission, turn on the power supply; then turn on the power supply switch to illuminate the mercury lamp. Look into the eyepiece and adjust the slit opening to maximize the brightness and sharpness of the emission lines on the scale. If necessary, adjust the position of the illuminated scale so that the numbered divisions can be easily read but do not obscure the mercury spectral lines. Once the slit and scale have been adjusted, do not move them during the rest of the experiment. Record on your report sheet the color and location of the mercury lines on the numbered scale for each line in the visible spectrum of mercury. On the graph paper provided, plot the observed scale reading versus the known wavelength for each line. You will use this calibration curve to determine the wavelengths of the emission lines for other atoms.

DISCUSSION

A. Emission Spectrum of Atomic Hydrogen

Atoms absorb and emit radiation with characteristic wavelengths. This was one of the observations that led the Danish physicist Niels Bohr to develop a model for the structure of the hydrogen atom. Within this model, the electron of the hydrogen atom moves about the central proton in a circular orbit. Only orbits of certain radii having certain energies are allowed. In the absence of radiant energy, an electron in an atom remains indefinitely in one of the allowed energy states or orbits. When electromagnetic energy impinges upon the atom, the atom may absorb energy, and in the process, an electron will be promoted from one energy state to another. The frequency of energy absorbed is related to the energy difference:

$$\Delta E = h\nu = \frac{hc}{\lambda} \qquad [2]$$

In the Bohr model, the radius of the orbit is related to the principal quantum number, n, where n is a positive integer:

$$\text{radius} = n^2(5.3\times10^{-11}\,\text{m}) \qquad [3]$$

The energy of the electron is also related to n:

$$E_n = -R_H\left(\frac{1}{n^2}\right) \qquad [4]$$

GIVE IT SOME THOUGHT

a. Is the preference for an electron to be in a lower energy or higher energy state?

b. Would the preference be for an electron to have a large or small principle quantum number?

Thus, as n increases, the electron moves farther from the nucleus and its energy increases. The constant R_H in Equation [4] is called the *Rydberg constant*; it has the value 2.18×10^{-18} J (Section 6.3).

EXAMPLE 12.2

What is the energy of a hydrogen electron when $n = 3$? When $n = 2$?

SOLUTION:

$$E_3 = (-2.18\times10^{-18}\text{ J})\left(\frac{1}{3^2}\right)$$
$$= -2.42\times10^{-19}\text{ J}$$
$$E_2 = (-2.18\times10^{-18}\text{ J})\left(\frac{1}{2^2}\right)$$
$$= -5.45\times10^{-19}\text{ J}$$

The orbital radii and energies are illustrated for $n = 1, 2$ and 3 in Figure 12.4.

According to Bohr's theory, if an electron were to move from an outer orbit to an inner orbit, a photon of light should be emitted having the energy:

$$\Delta E = E_{\text{inner}} - E_{\text{outer}} = -R_H\left(\frac{1}{n^2_{\text{inner}}} - \frac{1}{n^2_{\text{outer}}}\right) \qquad [5]$$

Thus, from Example 12.2, an electron moving from $n = 3$ to $n = 2$ would emit light of energy:

$$[(-5.45)-(-2.42)](10^{-19}\text{J}) = -3.03\times10^{-19}\text{J}$$

The wavelength of this photon is given by the Planck relation:

$$\lambda = \frac{hc}{\Delta \text{E}} \qquad [6]$$

See Example 12.3.

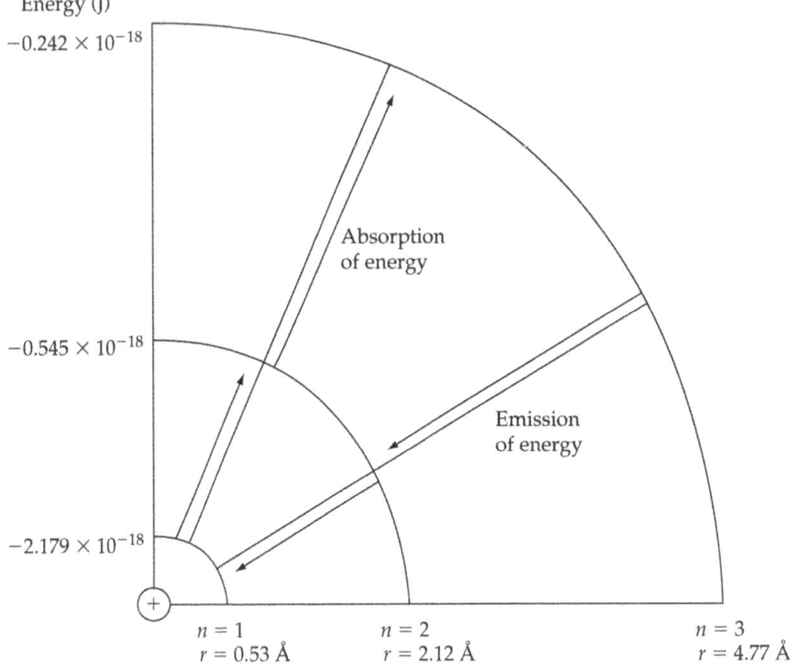

▲**FIGURE 12.4** Radii and energies of the three lowest energy orbits in the Bohr model of hydrogen. The arrows refer to transitions of the electron from one allowed energy state to another. When the transition takes the electron from a lower to a higher energy state, absorption occurs. When the transition is from a higher to a lower energy state, emission occurs.

EXAMPLE 12.3

Calculate the wavelength of light emitted for the $n = 3 \longrightarrow n = 2$ transition.

SOLUTION:

$$\lambda = \frac{(6.63 \times 10^{-34}\ \text{J-s})(3.00 \times 10^{8}\ \text{m/s})}{3.03 \times 10^{-19}\ \text{J}}$$
$$= 6.56 \times 10^{-7}\ \text{m}$$
$$= 656\ \text{nm}$$

By similar calculations, Bohr predicted wavelengths for the hydrogen emission spectrum that agreed exactly with the experimental values. He even predicted emission wavelengths in the infrared and ultraviolet regions of the spectrum that had not yet been measured but were later confirmed.

Using the spectroscope, you will measure the emission wavelengths in the visible region for hydrogen and assign those wavelengths to their corresponding transitions using calculations similar to those in Examples 12.2 and 12.3.

GIVE IT SOME THOUGHT

What is the difference between emission and absorption?

PROCEDURE

Turn on the illuminated scale of the spectroscope and look through the eyepiece to make sure the scale is visible but not so brightly lit that the hydrogen spectral lines are obscured when the hydrogen lamp is turned on. With the power supply unplugged, position the power supply and hydrogen lamp directly in front of the slit opening of the spectroscope. **(CAUTION: *The power supply develops several thousand volts. Do not touch any portion of the power supply, wire leads, or lamps unless the power supply is unplugged from the wall outlet. In addition to visible light, the lamps may emit ultraviolet radiation. Ultraviolet radiation is damaging to your eyes. Wear your safety glasses at all times during this experiment because they will absorb some of the ultraviolet radiation. Do not look directly at any of the lamps while they are illuminated.* DO NOT LET THE POWER SUPPLY OR LAMP TOUCH THE SPECTROSCOPE.)** With your instructor's permission, turn on the power supply; then turn on the power supply switch to illuminate the hydrogen lamp. You should have adjusted the slit and the illuminated scale in the calibration step. Should they require further adjustment, adjust the slit to maximize the line intensity and sharpness. On your report sheet, record the color and location of the hydrogen lines on the numbered scale for each line in the visible spectrum of hydrogen. You should easily observe the red, blue-green, and violet lines. A second faint violet line may also be visible if the room and scale illumination are not too bright.

GIVE IT SOME THOUGHT

Arrange the colored lines in order of increasing energy.

Use the calibration curve you constructed to determine the wavelengths of the lines in the hydrogen emission spectrum. Using your textbook or a handbook, find the true wavelengths of these lines and calculate the percent error in your determination of the wavelength of each line:

$$\%\ \text{error} = \frac{(\text{true value} - \text{experimental value})}{(\text{true value})} \times 100\% \qquad [7]$$

Use Equations [5] and [6] to calculate the wavelengths in nanometers for the $n = 3 \longrightarrow n = 2$, $n = 4 \longrightarrow n = 2$, $n = 5 \longrightarrow n = 2$, and $n = 6 \longrightarrow n = 2$ transitions. How do these calculated values compare with your experimental values? Assign the transitions.

GIVE IT SOME THOUGHT

Arrange these transitions in order of increasing energy. How does this ordering allow you to assign the transitions to the colored line in the spectrum?

DISCUSSION

B. Emission Spectra of Group 1A and Group 2A Elements

Because the energies of the electrons in the atoms of different elements are different, the emission spectrum of each element is unique. The emission spectrum may be used to detect the presence of an element in both a qualitative and quantitative way. A number of common metallic elements emit light strongly in the visible region, allowing them to be detected with a spectroscope. For these elements, the emissions are so intense that the elements may often be recognized by the gross color they impart to a flame. For example, lithium ions impart a red color to a flame; sodium ions, a yellow color; potassium ions, a violet color; calcium ions, a brick-red color; strontium ions, a bright red color; and barium ions, a green color. If you examine the emission spectra of these ions with a spectroscope, you will find that as with mercury and hydrogen, the emission spectra are composed of a series of lines. The series is unique for each metal. Consequently, a flame into which both lithium and strontium, for example, had been placed would be red, and you could not tell with the naked eye that both ions were present. However, with the aid of a spectroscope, you could detect the presence of both ions.

GIVE IT SOME THOUGHT

a. Assuming that different metals result in different flame tests, what does that tell you about the energy required to excite or emit an electron?

b. How does this relate to Einstein's explanation of the photoelectric effect (Section 6.2)?

PROCEDURE

You will obtain the emission spectra of the ions of each of the elements listed in the preceding discussion. You will use a Bunsen burner as an excitation source and observe the gross color imparted to the flame by the ions. With this information, you will determine the contents of two unknown solutions—one containing only one of these metal ions and the other containing a mixture of two or more of these metal ions.

There are three ways to introduce the metal ions into the flame. First, you can soak wood splints in each salt solution. Then you can ignite the splints using a Bunsen burner. If you soak the wood splints sufficiently, this procedure will produce a substantial burst of color. Second, you can use a wire loop to pick up a drop of the metal-ion containing solution and place the drop in the flame for vaporization. Although this method is simple and inexpensive, it produces only a brief burst of color before the sample evaporates completely. Third, you can introduce a fine mist of sample into the flame by using a spray bottle. This method produces a longer-lived emission that is therefore easier to see. Your instructor will tell you which method to use.

WORK IN PAIRS FOR THIS PART OF THE EXPERIMENT.

If you use the wire-loop method, obtain several 15 cm lengths of Nichrome wire and about 10 mL of 6 *M* HCl. Bend the last 6 mm of each wire in a small circular loop to use in picking up the sample solutions. Dip the loop in the 6 *M* HCl solution to remove any oxides that are present, rinse the loop in distilled water, and heat the loop in the hottest part of the flame until no color is imparted to the flame by the wires.

If you are using the sprayer, make sure it produces a *fine* mist. If the sprayer nozzle is adjustable, try adjusting it to create a very fine mist. If the nozzle cannot be adjusted, ask your instructor how to clean it so that you get the mist you want.

Set up a Bunsen burner directly in front of the slit of the spectroscope but at a sufficient distance to avoid damage to the spectroscope. Have your instructor check the placement before you ignite the burner. Ignite the burner and adjust the fuel/air mix of the flame so that it is as hot as possible. Adjust the illuminated scale of the spectroscope so that you can determine approximate positions of the emission lines. It is not necessary to make exact measurements of the emission wavelengths.

With either method of introduction (loop, mist, or wood splint), introduce the metal-ion solution into the flame and note the gross color of the flame. Record your observations on your report sheet. Then while looking through the eyepiece of the spectroscope, have your partner introduce the metal-ion solution into the flame; note the color, intensity, and approximate scale position of the brightest lines in the emission spectrum of the metal ions. Repeat with each of the other metal-ion containing solutions. If you use the wire-loop method, use a new loop for each solution or clean the wire loop between samples with HCl and distilled water and place the wire loop in the flame until it imparts no color to the flame.

After obtaining the emission spectrum of each of the known metal-ion solutions, obtain the emission spectrum of a single metal ion containing solution unknown, and by matching the colors, intensities, and positions of the lines in the emission spectrum to those of a known metal ion, identify your unknown. Record your results on your report sheet.

Obtain an unknown mixture and its emission spectrum as described. Then compare the color, intensity, and positions of the brightest lines in this spectrum with those of the knowns. In this way, *determine which metal ions are present in the unknown mixture*. Record your results on your report sheet.

NOTES AND CALCULATIONS

Name ______________________________ Desk ______________

Date ______________ Laboratory Instructor ______________________

Atomic Spectra and Atomic Structure | 12 Pre-lab Questions

Before beginning this experiment in the laboratory, you should be able to answer the following questions.

1. Name the colors of visible light beginning with that of highest energy (shortest wavelength).

2. Distinguish between absorption and emission of energy.

3. A system proposed by the U.S. Navy for underwater submarine communication, called ELF (for extremely low frequency), operates with a frequency of 76 Hz. What is the wavelength of this radiation in meters? In miles? (1 mile = 1.61 km)

4. What is the energy in joules of the frequency given in question 3?

5. Yellow and blue light have wavelengths of 579 nm and 436 nm, respectively. Which light has the higher frequency—yellow or blue? Which light has the higher energy—yellow or blue?

6. Ba emits radiation at 553.6 nm. Could a spectroscope be used to detect this emission?

7. If calcium emits radiation at 622 nm, what color will boron impart to a flame?

8. From the wavelengths and colors given for the mercury emission spectrum in this experiment, construct a graphical representation of the mercury emission spectrum as it would appear on the scale of a spectroscope.

Name Grace Rademacher Desk ______________

Date ______________ Laboratory Instructor ______________

Single-ion unknown no. ______________ Unknown mixture no. ______________

REPORT SHEET | EXPERIMENT

Atomic Spectra and Atomic Structure | 12

Calibration of Spectroscope

Lines observed in emission spectrum of mercury

Color	*Position on scale*	*Known wavelength*
Violet	4.5	404.7 nm
Blue	4.3	435.8 nm
Green	5.7	546.0 nm
Yellow	6.0	579.0 nm

A. Emission Spectrum of Atomic Hydrogen

Lines observed in emission spectrum of hydrogen

Color	*Position on scale*	*-2 Observed *Wavelength from calibration curve*	*Assignment*	*1 Expected
violet	4.5	410 nm	n=6	410.1
blue	5.6	435 nm	n=5	434.0
green	5.2	485 nm	n=4	486.1
red	6.7	660 nm	n=3	656.2

To make the assignments of the observed transitions, use Equations [5] and [6] to calculate the wavelengths of the following:

$n = 6 \longrightarrow n = 2$ transition

$n = 5 \longrightarrow n = 2$ transition

↓ 4.34 x 10^-45

→ 4.11 x 10^-45

% error

Violet: (410.1 - 410) / 410.1 x 100 = 0.024 % error

blue: (434 - 435) / 434 x 100 = 0.23 % error

green: (486.1 - 485) / 486.1 x 100 = 0.23 % error

Red: (656.2 - 660) / 656.2 x 100 = 0.579 % → 0.58 % error

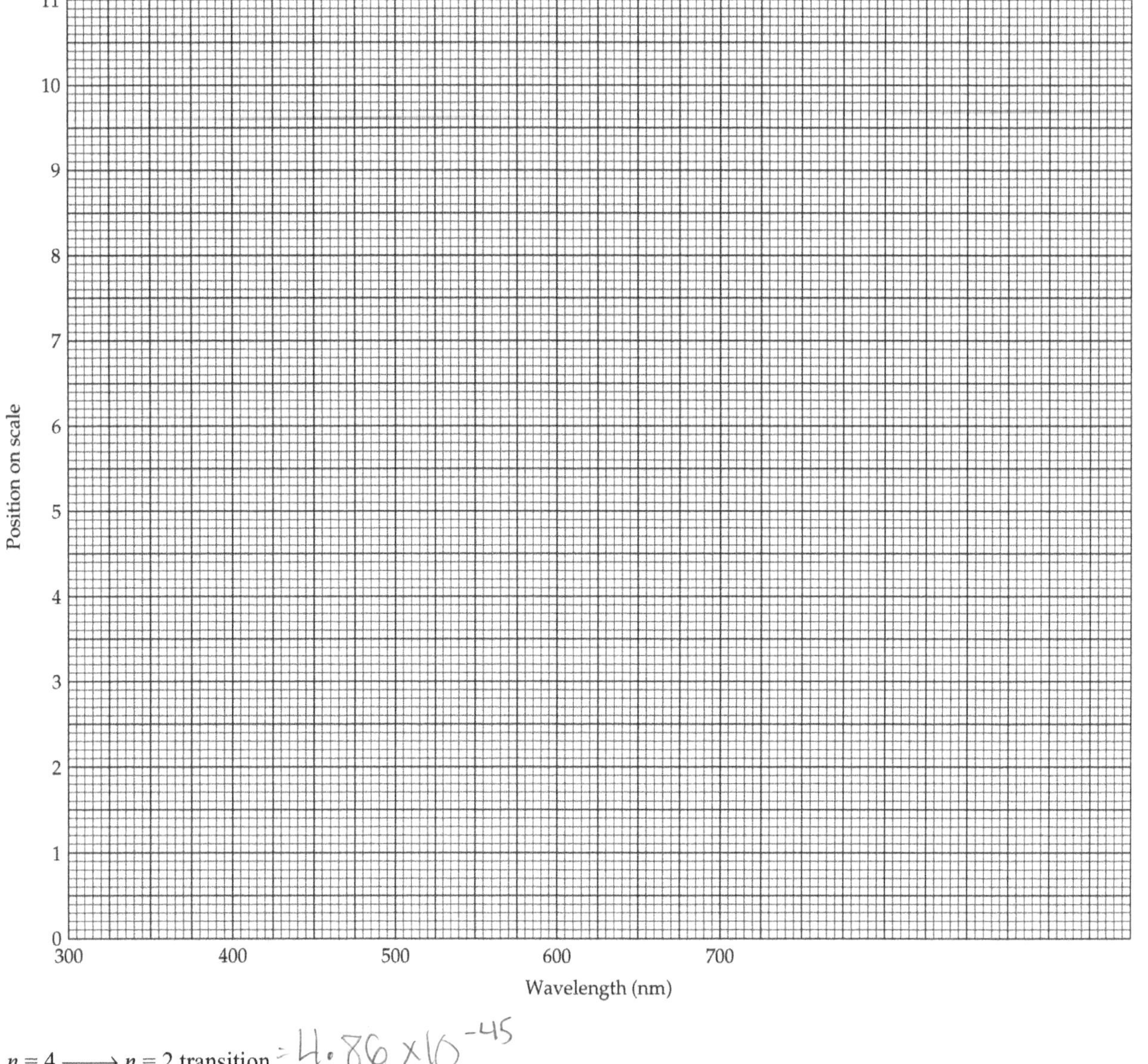

$n = 4 \longrightarrow n = 2$ transition = 4.86 x 10^{-45}

$n = 3 \longrightarrow n = 2$ transition = 6.56 x 10^{-45}

Consult your text or a handbook to find the accepted values for the emission spectrum of hydrogen and calculate the percent errors of your experimental results.

Accepted wavelength	*Observed wavelength*	*Percent error*
410.1	410	0.024%
434.0	435	0.23%
486.1	485	0.23%
656.2	660	0.58%

Percent error calculations – On page 1

line 1 n=6 → n=2

$\frac{(6.63\times10^{-34})(3.00\times10^{8})}{4.84\times10^{-19}} = 4.11\times10^{-45}$

line 2 n=5 → n=2

$\frac{(6.63\times10^{-34})(3.00\times10^{8})}{4.58\times10^{-19}} = 4.34\times10^{-45}$

line 3 n=4 → n=2

$\frac{(6.63\times10^{-34})(3.00\times10^{8})}{4.09\times10^{-19}} = 4.86\times10^{-45}$

line 4 n=3 → n=2

$\frac{(6.63\times10^{-34}\,J/s)(3.00\times10^{8}\,m/s)}{3.03\times10^{-19}\,J} = 6.56\times10^{-45}$

B. Emission Spectra of Group 1A and Group 2A Elements

Which method did you use to introduce the metal ion into the Bunsen burner flame? online method

1.

Known metal ions	*Gross flame color*	*Position and colors of the lines observed on the spectroscope scale*
Li	Red	red wavelength · med
Na	Yellow	none
K	Purple	green wavelength - medium
Ca	Red	green wavelengths - large
Sr	Red	green wavelength > Ca
Ba	Green	None

2.

Unknowns	*Gross flame color*	*Position and colors of the lines observed on the spectroscope scale*
Single-ion unknown	Green	530 · 570
~~Mixture~~		

What is the identity of the single-ion unknown? Ba

~~What ions are present in the mixture?~~

QUESTIONS

1. What is the purpose of the slit in the spectroscope?

So that we can observe the virtual image and the meter and so that the illumination will show.

2. Why is the spectroscope scale illuminated?

The spectroscope scale is illuminated so that one can know exactly where the lines are falling.

3. Why was the emission spectrum of mercury used to calibrate the spectroscope?

Mercury was used because its emission are known and identifyable

4. Could the emission spectrum of some other element be used to calibrate the spectroscope?

Yes, any spectrum can be used for calibration as long as they have a clearly identifyable wavelength

5. In addition to the spectral lines you observed in the emission spectrum of hydrogen, several other lines are also present in other regions of the spectrum. Calculate the wavelengths of the $n = 4 \longrightarrow n = 1$ and $n = 4 \longrightarrow n = 3$ transitions and indicate in which regions of the spectrum these transitions would occur. (Section 6.1)

$n_1 = 4 \quad n_2 = 1$

$\frac{1}{\lambda} = 1.097 \times 10^7$

$= 9.72 \times 10^{-8}$ m

UV Spectrum

$n_1 = 4 \quad n_2 = 3$

$\frac{1}{\lambda} = 1.097 \times 10^7$

$= 1.88 \times 10^{-6}$ m

= Infrared spectrum

6. Of the metal ions tested, sodium produces the brightest and most persistent color in the flame. Do you think potassium could be detected visually in the presence of sodium by burning this mixture in a flame? Could you detect both with a spectroscope?

7. The minimum energy required to break the oxygen–oxygen bond in O_2 is 495 kJ/mol. What is the longest wavelength of radiation that possesses the necessary energy to break the O—O bond? What type of electromagnetic radiation is this?

Behavior of Gases: Molar Mass of a Vapor

OBJECTIVE

To observe how changes in temperature and pressure affect the volume of a fixed amount of a gas; to determine the molar mass of a gas from knowing its mass, temperature, pressure, and volume.

APPARATUS AND CHEMICALS

Apparatus

gas law demonstration apparatus
balance
125 mL Erlenmeyer flask
250 mL graduated cylinder (one per class)
rubber band
wire gauze
boiling chips
barometer
Bunsen burner and hose
600 mL beaker
5 cm square of aluminum foil
pins
ring stand and two iron rings
utility clamp
thermometer

Chemicals

volatile unknown liquid

DISCUSSION

The Effect of Pressure on the Volume of a Gas

The effect of pressure on the volume of a gas can be determined by using a gas buret, as shown in Figure 13.1. The volume of the buret is graduated in terms of cubic centimeters (cm^3) or milliliters (mL). When the stopcock is opened, air can enter the buret, and the level of mercury will be equal in both tubes. If the stopcock is then closed, a fixed volume of air will be trapped in the buret at prevailing atmospheric pressure. Raising the leveling bulb increases the pressure on the gas; the new pressure on the gas corresponds to the prevailing atmospheric pressure plus the height the mercury in the leveling bulb is above the level of mercury in the buret. Experiments show that when the pressure is doubled, the volume is halved and when the pressure is halved, the volume is doubled. From such experiments, you may conclude that at constant temperature, the volume of a given amount of gas is *inversely* proportional to the pressure. This is *Boyle's law*, which Boyle enunciated in 1662 (Section 10.3). This may be expressed mathematically as

$$\frac{V_1}{V_2} = \frac{P_2}{P_1}$$

or

$$V = \frac{k}{P} \qquad [1]$$

where V_1 is the volume at pressure P_1, V_2 is the volume at pressure P_2, and k is a proportionality constant.

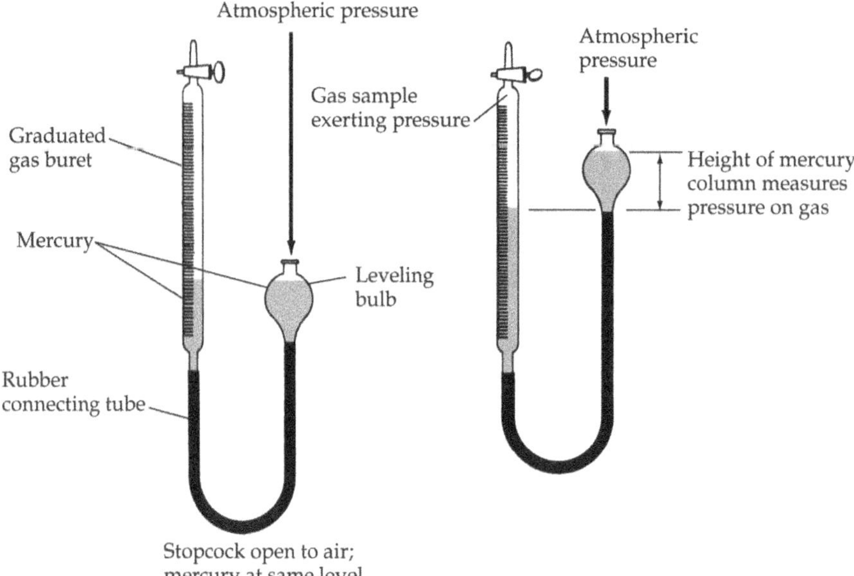

▲**FIGURE 13.1** A gas buret.

By means of Boyle's law, you can calculate the volume of a gas at any pressure, provided you know the volume at a given pressure. Example 13.1 is illustrative.

EXAMPLE 13.1

If 100 mL of a gas is enclosed in the buret at 101.3 kPa, what volume would the gas occupy at 202.6 kPa at the same temperature?

SOLUTION: You could let $V_1 = 100\text{ mL}$, $P_1 = 101.3\text{ kPa}$, and $P_2 = 202.6\text{ kPa}$; substitute these values into Equation [1]; and solve for V_2. A better approach is to reason through the problem. Because the pressure increases from 101.3 kPa to 202.6 kPa, the new volume must be *less* than 100 mL. The volume decreases in the same ratio the pressure increases. Therefore, the new volume is as follows:

$$V_2 = 100\text{ mL} \times \frac{101.3\text{ kPa}}{202.6\text{ kPa}} = 50.0\text{ mL}$$

If the volume had been multiplied by the fraction 202.6 kPa/101.3 kPa, a fraction larger than 1, the answer would have been larger than 100 mL. And as you know, this would be incorrect.

The Effect of Temperature on the Volume of a Gas

The relation between temperature and volume of a given sample of gas can be studied in a gas buret while the pressure is held constant. You are familiar with the fact that gases expand when heated. Experiments have shown that at constant pressure, the volume of a given mass of a gas is directly proportional to the Kelvin (or absolute) temperature. This is known as *Charles's law* (Section 10.3).

To convert a Celsius temperature to the Kelvin (or absolute) scale, add 273°. Thus, 20 °C = (20° + 273°) = 293 K.

Charles's law may be expressed mathematically as follows:

$$\frac{V_1}{V_2} = \frac{T_1}{T_2}$$

or

$$V = k'T \qquad [2]$$

where V_1 is the volume at temperature T_1, V_2 is the volume at temperature T_2, and k' is a proportionality constant. An application of this law is illustrated in Example 13.2.

EXAMPLE 13.2

Suppose a sample of oxygen has a volume of 100 mL at a temperature of 20 °C. What will be its volume at 100 °C at the same pressure?

SOLUTION: Because the temperature increases, the volume must increase. The volume at 100 °C equals the original volume multiplied by a fraction made up of the Kelvin temperatures, which is larger than 1.

$$V \text{ at } 100\ °\text{C} = 100\ \text{mL} \times \frac{373\ \text{K}}{293\ \text{K}} = 127\ \text{mL}$$

Because gases can occupy any volume, you gain the most information when you compare them under the same conditions. Standard conditions of temperature and pressure, designated as STP, are often used for this purpose. Standard temperature is 0 °C, and standard pressure is 101.3 kPa, or 1 atmosphere (1 atm).

Problems involving changes in both temperature and pressure are worked by combining the effects due to each of these changes. Example 13.3 is illustrative.

EXAMPLE 13.3

Suppose a sample of oxygen at 100 °C occupies a volume of 100 mL at 93.3 kPa. What will be its volume at STP?

SOLUTION:

$$\begin{aligned} V_1 &= 100\ \text{mL} \\ V_2 &= ? \\ T_1 &= 100° + 273° = 373\ \text{K} \\ T_2 &= 0° + 273° = 273\ \text{K} \\ P_1 &= 93.3\ \text{kPa} \\ P_2 &= 101.3\ \text{kPa} \end{aligned}$$

The volume at STP will be equal to the original volume multiplied by a fraction made up of the two temperatures and another fraction made up of the two pressures. The temperature fraction will be less than 1 because the temperature decreases; this results in a decrease in volume. The pressure fraction will be less than 1 because the pressure increases; an increase in pressure also causes the volume to decrease.

$$\begin{aligned} \text{Volume at STP} &= 100\ \text{mL} \times \frac{273\ \text{K}}{373\ \text{K}} \times \frac{93.3\ \text{kPa}}{101.3\ \text{kPa}} \\ &= 67.4\ \text{mL} \end{aligned}$$

Ideal Gas Law

It is possible to relate the four variables of a gas—pressure, volume, temperature, and number of moles—in one equation, which is referred to as the *ideal gas law* (Section 10.4). Before doing this, you must recognize *Avogadro's law*, which states that *equal volumes of all gases, at the same conditions of temperature and pressure, contain the same number of molecules*. Thus, 6.022×10^{23} molecules (Avogadro's number) of any gaseous substance should occupy the same volume under the same conditions. One mole of an ideal gas at STP occupies 22.4 L, a value known as the *molar volume*.

According to Charles's law, the volume of a fixed number of moles, *n*, of a gas at a constant pressure is directly proportional to the Kelvin temperature.

$$V \alpha T \quad \text{at constant } P \text{ and } n$$

According to Boyle's law, the volume of a fixed number of moles of a gas at constant temperature is inversely proportional to the pressure.

$$V \alpha \frac{1}{P} \quad \text{at constant } T \text{ and } n$$

And according to Avogadro's law,

$$V \alpha n \quad \text{at constant } P \text{ and } T$$

If a quantity is proportional to two or more variables, it is proportional to the product of those variables.

$$V \alpha T \times \frac{1}{P} \times n \qquad [3]$$

The proportionality symbol, α, in Equation [3] can be replaced by an equal sign by introducing a proportionality constant, *R*.

$$V = R \times T \times \frac{1}{P} \times n$$

or

$$PV = nRT \qquad [4]$$

Equation [4] is the ideal gas law. *R*, the ideal gas constant, can be calculated by considering 1 mol of an ideal gas at STP.

$$R = \frac{P \times V}{n \times T} = \frac{1.013 \times 10^5 \text{ pa} \times 22.4 \text{ L}}{1 \text{ mol} \times 273 \text{ K}} = 8{,}310 \text{ L} - \text{Pa/mol} - \text{K}$$

or, in other units,

$$R = \frac{8.314 \text{ nm}}{\text{mol} - \text{K}} = 8.314 \text{ J/mol} - \text{K}$$

The number of moles of a substance (*n*) equals its mass in grams, *m*, divided by the number of grams per mole (that is, its molar mass, $n = m/M$ (Section 10.5). Making this substitution into Equation [4] gives

$$PV = \left(\frac{m}{M}\right) RT \qquad [5]$$

The units of *P*, *V*, *T*, and *m* must, of course, be expressed in units consistent with the value of *R*. Example 13.4 illustrates how this equation can be used to calculate the molar mass of a gas.

EXAMPLE 13.4

A gaseous sample weighing 0.896 g was found to occupy a volume of 524 mL at 97.3 kPa and 28 °C. What is the molar mass of the gas?

SOLUTION: Solving Equation [5] for molar mass and substituting the appropriate values and using the appropriate value of *R* that is consistent with the given units, you find

$$M = \frac{mRT}{PV} \qquad [6]$$

$$= \frac{0.896\ \text{g} \times 8310\ \text{L} - \text{Pa/mol} - \text{K} \times 301\ \text{K}}{97300\ \text{Pa} \times 0.524\ \text{L}}$$

$$= 44.0\ \text{g/mol}$$

The first portion of this experiment will be a demonstration of Charles's and Boyle's laws. The laboratory instructor will use a gas buret to vary the pressure, temperature, and volume of a sample of air.

In the second portion of this experiment, using Equation [6], you will determine the molar mass of a volatile liquid. A small quantity of liquid sample is placed in a flask of known mass and vaporized to expel all of the air from the flask, leaving it filled with the vapor at a known temperature (temperature of boiling water) and atmospheric pressure. The flask and vapor is then cooled so that the vapor condenses. The mass of the flask plus condensed vapor plus air is then determined. The mass of air, being nearly identical before and after, cancels out and allows you to determine the mass of the vapor. The above data, in conjunction with the volume of the flask, permit calculation of the molar mass.

PROCEDURE

A. Demonstration Experiment

Use the report sheet to record the data collected by the laboratory instructor. Calculate the volumes by means of the appropriate gas laws and compare them with the observed volumes. Express the differences as percent error.

B. Molar Mass of a Vapor

Record the number of your unknown liquid (1). Place a small square of aluminum foil over the mouth of a clean, dry 125 mL Erlenmeyer flask. Fold the foil loosely around the neck and secure with a rubber band (Figure 13.2). Use a pin to make a very small hole in the center of the foil. Determine the mass of the flask, foil cap, and rubber band (2).

Place about 350 mL of water and some boiling chips in a 600 mL beaker and begin heating it to bring the water to a boil. While the water is heating, remove the foil from the Erlenmeyer flask and place about 2 mL of your unknown liquid in the flask; then replace the foil and rubber band. Clamp the flask at the top of the neck. After the water has been brought to a boil, record its temperature (3) and barometric pressure (4). If you calibrated your thermometer in Experiment 1, apply its correction. Insert the Erlenmeyer flask as far as possible into the boiling water, holding it by hand using the clamp. It is not necessary to clamp the flask to the ring stand. After about 4 min, remove the flask from the boiling water and examine it to see whether all of the unknown liquid has vaporized (including any that condensed in the neck). If it has not, reinsert the flask into the boiling water for a few minutes. After all of the liquid has vaporized, use the clamp to remove the flask and set it aside to cool.

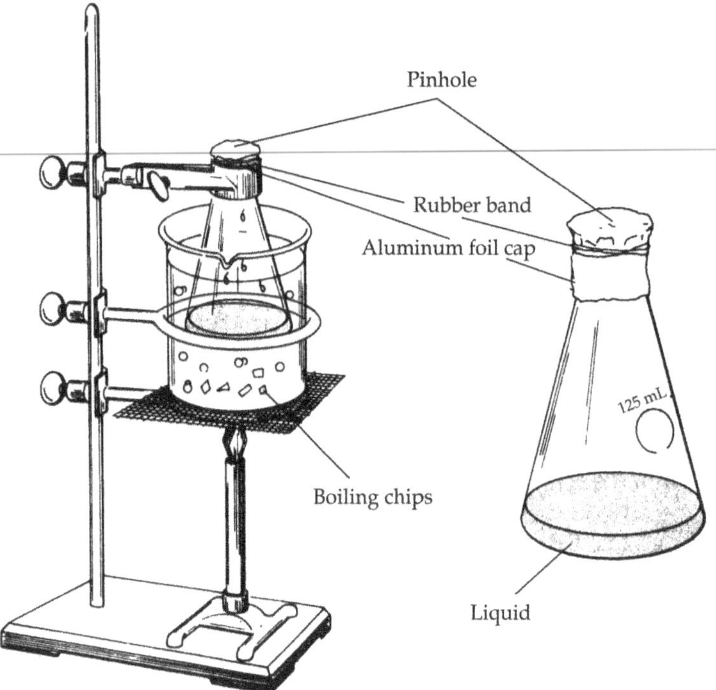

▲**FIGURE 13.2** Experimental apparatus.

Note: The clamp holding the flask may be held by hand instead of being secured to the ring stand.

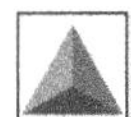

GIVE IT SOME THOUGHT

What role will temperature and barometric pressure play in this experiment?

After the flask has cooled to room temperature, wipe it dry and remove any water that adheres to the aluminum. Determine the mass of the flask, cap, rubber band, and condensed unknown liquid (5). Calculate the mass of the condensed liquid (6).

Remove the cap and fill the flask completely with water. Measure the volume by pouring the water into a large graduated cylinder. Dispose of the condensed liquid in an appropriate waste container.

Calculate the molar mass of the unknown using Equation [6].

Name ______________________________ Desk ______________

Date ______________ Laboratory Instructor ______________________

Behavior of Gases: Molar Mass of a Vapor | 13 Pre-lab Questions

Before beginning this experiment in the laboratory, you should be able to answer the following questions.

1. How does the volume of an ideal gas at constant pressure change as the temperature increases?

2. How does the volume of an ideal gas at constant temperature change as the pressure increases?

3. How does the pressure of an ideal gas at constant temperature and volume change as the number of molecules increase?

4. An inflated bicycle tyre contains 670 cm^3 of air at an internal pressure of 6 bar and a temperature of 20 °C. Use the ideal gas equation to calculate the amount of air in the tyre in moles. (1 bar = 100 kPa)

5. Show by mathematical equations how to determine the molar mass of a volatile liquid by measurement of the pressure, volume, temperature, and mass of the liquid.

6. At 338 K, pure PCl_5 gas present in a flask has a pressure of 25.5 kPa. At 480 K, the gas dissociates completely into PCl_3 (gas) and Cl_2 (gas). What is the pressure in the flask at 480 K?

7. A sample of nitrogen occupies a volume of 200 mL at 60 °C and 66.7 kPa of pressure. What will its volume be at STP?

8. Consider Figure 13.1. If the height of the mercury column in the leveling bulb is 10 mm greater than that in the gas buret and atmospheric pressure is 670 mm, what is the pressure on the gas trapped in the buret (1 mm Hg = 0.1333 kPa)?

9. Consider Figure 13.1. If the level of the mercury in the leveling bulb is lowered, what happens to the volume of the gas in the gas buret?

10. Show that Boyle's law, Charles's law, and Avogadro's law can be derived from the ideal gas law.

11. Methane burns in oxygen to produce CO_2 and H_2O.

$$CH_4(g) + 2O_2(g) \longrightarrow 2H_2O(l) + CO_2(g)$$

If 0.50 L of gaseous CH_4 is burned at STP, what volume of O_2 is required for complete combustion? What volume of CO_2 is produced?

12. Calculate the density of N_2 at STP using (a) the ideal gas law and (b) the molar volume and molar mass of N_2. How do the densities compare?

Name ______________________________ Desk ______________

Date ______________ Laboratory Instructor ______________________

Unknown no. ______________

REPORT SHEET | EXPERIMENT

Behavior of Gases: Molar Mass of a Vapor | 13

A. Demonstration Experiment

1. Boyle's law—effect of pressure at constant temperature

	Trial 1	*Trial 2*
First pressure, barometer reading, kPa	______	______
Difference in mercury levels (+ or −), kPa	______	______
Second pressure, kPa	______	______
First volume, mL	______	______
Second volume, measured, mL	______	______
Second volume, calculated, mL	______	______
Percent error	______	______
(show calculations)		

Trial 1 *Trial 2*

2. Charles's law—effect of temperature at constant pressure

First temperature	______ °C =	______ K
Second temperature	______ °C =	______ K
First volume, mL		______
Second volume, measured, mL		______
Second volume, calculated, mL		______
Percent error		______

(show calculations)

B. Molar Mass of a Vapor

1. Unknown liquid number ____________
2. Mass of flask + cap + rubber band ____________ g
3. Temperature of boiling water ____________ °C
   Correction ____________ °C
   Corrected temperature ____________ °C
4. Barometric pressure ____________ kPa
5. Mass of flask + rubber band + cap + condensed vapor ____________ g
6. Mass of condensed liquid ____________ g
7. Volume of flask ____________ mL
8. Molar mass of vapor ____________ g/mol
   (show calculations)

QUESTIONS

1. If an insufficient amount of liquid unknown had been used, how would this have affected the value of the experimental molar mass?

2. What are the major sources of error in your determination of the molar mass?

3. If the flask were not wiped completely dry, how would this derived mass of the gas affect the molar mass?

4. Isobutyl alcohol has a boiling point of 108 °C. How would you modify the procedure used in this experiment to determine its molar mass?

GAS LAW PROBLEMS

1. There are two gas cylinders. Gas cylinder A, containing nitrogen gas, occupies a volume of 48.2 dm^3 at 25 °C and 846 kPa pressure. Gas cylinder B, of unknown volume, contains helium gas at 25 °C and 962 kPa pressure. When the two cylinders are connected and the gases are mixed, the pressure in the cylinders is found to be 883 kPa. What is the volume of gas cylinder B?

2. A sample of gas of mass 2.82 g occupies a volume of 639 mL at 27 °C and 101.3 kPa pressure. What is the molar mass of the gas?

3. A gas is placed in a storage tank at a pressure of 3.04 MPa at 20.3 °C. As a safety device, a small metal plug in the tank is made of a metal alloy that melts at 130 °C. If the tank is heated, what is the maximum pressure that will be attained in the tank before the plug melts and releases gas?

4. Forty liters (40 L) of a gas were collected over water when the barometer read 82.9 kPa and the temperature was 20 °C. What volume would the dry gas occupy at standard conditions? [HINT: Consider Dalton's law of partial pressures (Section 10.6).]

5. Five moles (5.0 mol) of hydrogen gas at 0 °C are forced into a steel cylinder with a volume of 2.0 L. What is the pressure of the gas, in kPa, in the cylinder?

6. What is the density of He at STP? Why do helium-filled balloons rise in air?

7. What volume in milliliters will 6.5 g of CO_2 occupy at STP?

8. If 48.0 g of O_2 and 4.4 g of CO_2 are placed in a 10.0 L container at 21 °C, what is the pressure of the mixture of gases?

9. A mixture of cyclopropane gas and oxygen is used as an anesthetic. Cyclopropane contains 85.7% C and 14.3% H by mass. At 50.0 °C and 99.7 kPa pressure, 1.56 g cyclopropane has a volume of 1.00 L. What is the molecular formula of cyclopropane?

Experiment

Determination of *R*: The Gas Law Constant

OBJECTIVE

To understand how well real gases obey the ideal gas law and to determine the ideal gas constant, *R*.

APPARATUS AND CHEMICALS

Apparatus

balance	barometer
Bunsen burner and hose	glass tubing with 60-degree bends (2)
test tube	and straight pieces (2)
250 mL beaker	125 mL Erlenmeyer flask
250 mL wide-mouth bottle	rubber tubing
rubber stoppers (2)	thermometer
pinch clamp	Styrofoam cups
clamp	ring stand

Chemicals

$KClO_3$	MnO_2

DISCUSSION

Most gases obey the ideal gas equation, $PV = nRT$, under ordinary conditions (that is, near room temperature and atmospheric pressure). Small deviations from this law are observed, however, because real gas molecules are finite in size and exhibit mutual attractive forces. The van der Waals equation,

$$\left(P + \frac{n^2 a}{V^2}\right)(V - nb) = nRT$$

where *a* and *b* are constants characteristic of a given gas, takes into account these two causes for deviation and is applicable over a much wider range of temperatures and pressures than is the ideal gas equation. The term *nb* in the expression $(V - nb)$ is a correction for the finite volume of the molecules; the correction to the pressure by the term n^2a/V^2 takes into account the intermolecular attractions (Section 10.9).

GIVE IT SOME THOUGHT

a. How is this equation different from the ideal gas equation?

b. What does this equation account for that the ideal gas equation does not?

In this experiment, you will determine the numerical value of the gas law constant *R*, in its common units of L-kPa/mol-K. This will be done using both the ideal gas law and the van der Waals equation together with measured values of pressure (*P*), temperature (*T*), volume (*V*), and number of moles (*n*) of an enclosed sample of oxygen. Then you will perform an error analysis on the experimentally determined constant.

The oxygen will be prepared by the decomposition of potassium chlorate, with manganese dioxide used as a catalyst.

$$2KClO_3(s) \xrightarrow[\Delta]{MnO_2(s)} 2KCl(s) + 3O_2(g)$$

If the mass of the $KClO_3$ is accurately determined before and after the oxygen has been driven off, the mass of the oxygen can be obtained by difference. The oxygen can be collected by displacing water from a bottle, and the volume of gas can be determined from the volume of water displaced. The pressure of the gas may be obtained through use of Dalton's law of partial pressures, the vapor pressure of water, and atmospheric pressure. Dalton's law states that the pressure of a mixture of gases in a container is equal to the sum of the pressures that each gas would exert if it were present alone (Section 10.6).

$$P_{total} = \sum_i P_i$$

Because this experiment is conducted at atmospheric pressure, $P_{total} = P_{atmospheric}$. Hence,

$$P_{atmospheric} = P_{O_2} + P_{H_2O\ vapor}$$

PROCEDURE

Add a small amount of MnO_2 (about 0.02 g) and approximately 0.3 g of $KClO_3$ to a test tube and accurately determine the mass to the nearest 0.0001 g. Your instructor will demonstrate how to insert the glass tubing into the rubber stoppers. Be extremely careful to follow his or her instructions. Assemble the apparatus illustrated in Figure 14.1, but do not attach the test tube. Make sure tube *B* does not extend below the water level in the bottle. Fill glass tube *A* and the rubber tubing with water by loosening the pinch clamp, attaching a rubber bulb to tube *B*, and applying pressure through it. Close the clamp when the tube is filled.

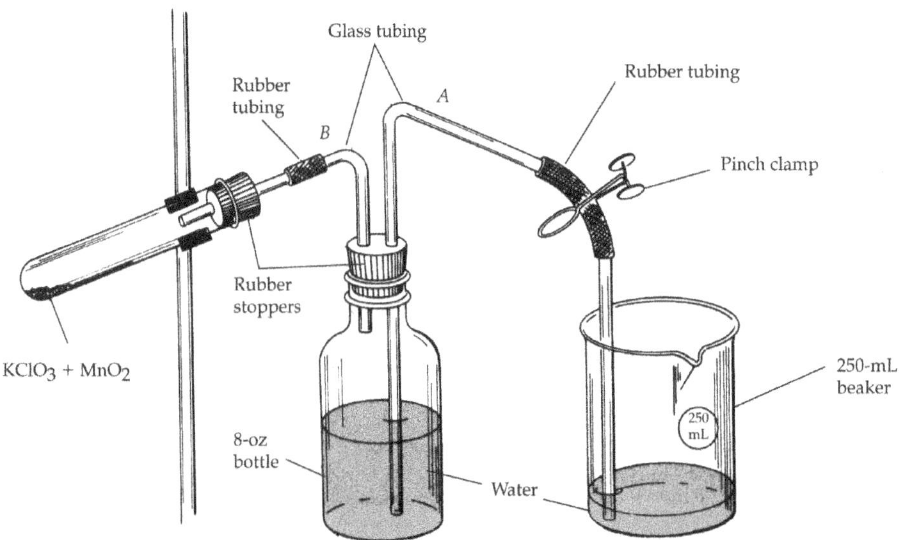

▲FIGURE 14.1 Apparatus for determining *R* (8 oz bottle = 250 mL bottle).

GIVE IT SOME THOUGHT

Why can't tube *B* extend below the water level in the bottle?

Mix the solids in the test tube by rotating the tube, making sure none of the mixture is lost from the tube, and attach tube *B* as shown in Figure 14.1. **(CAUTION: *When you attach the test tube, make sure that none of the* $KClO_3$ *and* MnO_2 *comes in contact with the rubber stopper; otherwise, a severe explosion may result. Make certain that the clamp holding the test tube is secure so that the test tube does not move.*)**

Fill the beaker about half full of water, insert glass tube *A*, open the pinch clamp, and lift the beaker until the levels of water in the bottle and beaker are identical. Then close the clamp, discard the water in the beaker, and dry the beaker. The purpose of equalizing the levels is to produce atmospheric pressure inside the bottle and test tube.

Set the beaker with tube *A* in it on the desk and open the pinch clamp. A little water will flow into the beaker, but if the system is airtight and has no leaks, the flow will soon stop and tube *A* will remain filled with water. If this is not the case, check the apparatus for leaks and start over. Keep in the beaker the water that has flowed into it; at the end of the experiment, the water levels will be adjusted and this water will flow back into the bottle.

Heat the lower part of the test tube gently (**make certain the pinch clamp is open**) so that a slow but steady stream of gas is produced, as evidenced by the flow rate of water into the beaker. When the rate of gas evolution slows considerably, increase the rate of heating, and heat until no more oxygen is evolved. Allow the apparatus to cool to room temperature, making sure the end of the glass tube in the beaker is always below the surface of the water. Equalize the water levels in the beaker and the bottle as before and close the clamp. Determine the mass of a 125 mL Erlenmeyer flask* to the nearest 0.01 g and empty the water from the beaker into the flask*. Determine the mass of the flask* with the water in it. Measure the temperature of the water and using the density of water in Table 14.1, calculate the volume of the water displaced. This is equal to the volume of oxygen produced. Remove the test tube from the apparatus and accurately determine the mass of the tube and its contents. The difference in mass between this and the original mass of the tube plus MnO_2 and $KClO_3$ is the mass of the oxygen produced.

GIVE IT SOME THOUGHT

How is the volume of water displaced equal to the volume of oxygen produced?

Record the barometric pressure. The vapor pressure of water at various temperatures is also given in Table 14.1.

---

*Or Styrofoam cup. The volume of water may also be measured directly but less accurately with a graduated cylinder.

TABLE 14.1 Density and Vapor Pressure of Pure Water at Various Temperatures

Temperature (°C)	Density (*d*) (g/mL)	Temperature (°C)	H_2O vapor pressure (kPa)
15	0.999099	15	1.70
16	0.998943	16	1.81
17	0.998774	17	1.93
18	0.998595	18	2.06
19	0.998405	19	2.19
20	0.998203	20	2.33
21	0.997992	21	2.47
22	0.997770	22	2.63
23	0.997538	23	2.81
24	0.997296	24	2.98
25	0.997044	25	2.17
26	0.996783		
27	0.996512		
28	0.996232		

Waste Disposal Instructions $KClO_3$ is a powerful oxidizing agent and must not be disposed of in a waste basket! Do not attempt to clean out the residue that remains in the test tube. Return the test tube to your instructor or follow his or her instructions for disposal of its contents.

Calculate the gas law constant, *R*, from your data, using the ideal gas equation. Calculate *R* using the van der Waals equation $(P + n^2a/V^2)(V - nb) = nRT$ (for O_2, $a = 1.360\ L^2\ atm/mol^2$ and $b = 31.83\ cm^3/mol$). Make sure you keep your units straight.

Error Analysis

Determine the maximum and minimum values of *R* consistent with the experimental reliability of your data from the ideal gas law.

$$R = \frac{PV}{nT} = \frac{(32.00\ \text{g/mol})\ PV}{mT}$$

GIVE IT SOME THOUGHT

Before performing any calculations, what do you expect *R* to be?

Assume that the reliabilities for the various measured quantities in this experiment are as follows:

$$P = \pm 0.01\ \text{kPa} \qquad T = \pm 1\ °\text{C}$$

$$V = \pm 0.0001\ \text{L} \qquad m = \pm 0.0001\ \text{g}$$

To determine the maximum value of *R*, use the maximum values that the pressure and volume may have and the minimum values that the mass and temperature may have. Similarly, calculate the minimum value of *R* from the minimum values of *P* and *V* and the maximum values for *m* and *T*. Determine the average value of *R* and assign an uncertainty range to this average value.

EXAMPLE 14.1

Assume that the measured quantities were as follows: $P = 94.06$ kPa $T = 20$ °C, $V = 242.9$ mL, and $m = 0.3002$ g. What would be the maximum and minimum values of R, the average value of R, and the uncertainty range assigned to this average value?

SOLUTION: First, put the measured quantities into proper units as follows:

$$P = 94.06 \text{ kPa}$$
$$V = 242.9 \text{ mL} = 0.2429 \text{ L}$$
$$m = 0.3002 \text{ g}$$
$$T = (20 + 273) \text{ K} = 293 \text{ K}$$

Therefore,

$$\text{Maximum } R = \frac{[94.07 \text{ kPa}](0.2430 \text{ L})(32.00 \text{ g/mol})}{(0.3001 \text{ g})(292 \text{ K})} = 8.35 \text{ L-Pa/mol-K}$$

$$\text{Minimum } R = \frac{[94.05 \text{ kPa}](0.2428 \text{ L})(32.00 \text{ g/mol})}{(0.3003 \text{ g})(294 \text{ K})} = 8.28 \text{ L-Pa/mol-K}$$

Therefore, the average value for R is as follows:

$$\text{Average } R = \frac{8.35 + 8.28}{2} \text{ L-Pa/mol-K} = 8.31 \text{ L-Pa/mol-K}$$

Note that the minimum and maximum values of R differ from the average by 0.04. Consequently, the uncertainty in R can be written as ±0.04 L-Pa/mol-K and the data would be reported as

$$R = (8.31 \pm 0.04) \text{ L-kPa/mol-K}$$

NOTES AND CALCULATIONS

Name ______________________ Desk ______________

Date ____________ Laboratory Instructor ______________________

Determination of *R*: The Gas Law Constant | 14 Pre-lab Questions

Before beginning this experiment in the laboratory, you should be able to answer the following questions.

1. State whether the behavior of methylamine (CH_3NH_2) would be less ideal than that of argon.

2. Calculate the value of *R* in L-Pa/mol-K by assuming that an ideal gas occupies 22.4 L/mol at STP.

3. Why do you equalize the water levels in the bottle and the beaker?

4. Why does the vapor pressure of water contribute to the total pressure in the bottle?

5. What is the value of an error analysis?

6. Suggest reasons why real gases might deviate from the ideal gas law on the molecular level.

7. At present, automobile batteries are sealed. When lead storage batteries discharge, they produce hydrogen. Suppose the void volume in a battery is 100 mL at 101.3 kPa of pressure and 25 °C. What would be the pressure increase if 0.05 g H_2 were produced by the discharge of the battery? Does this present a problem? Explain. Why were sealed lead storage batteries not used in the past?

8. Why is the corrective term to the volume subtracted and not added to the volume in the van der Waals equation?

9. A sample of a pure gas at 20 °C and 89.3 kPa occupied a volume of 562 cm^3. How many moles of gas does this represent? (HINT: Use the value of *R* that you found in question 2.)

10. A certain compound containing only carbon and hydrogen was found to have a vapor density of 2.550 g/L at 100 °C and 101.3 kPa. If the empirical formula of this compound is CH, what is the molecular formula of the compound?

11. Which gas would you expect to behave more like an ideal gas—Ne or HBr? Why?

12. Why must the $KClO_3$–MnO_2 mixture be kept away from a rubber stopper?

Name Grace Rademacher Desk ______

Date ______ Laboratory Instructor Professor Briguglio

$2KClO_{3(s)} \xrightarrow{MnO_2(s)} 2KCl(s) + 3O_2(g)$

REPORT SHEET | EXPERIMENT

Determination of *R*: The Gas Law Constant

14

$KClO_3$ = 0.303 g
MnO_2 = 0.026 g
empty tube = 19 g

1. Mass of test tube + $KClO_3$ + MnO_2 19.329 g
2. Mass of test tube + contents after reaction 19.206 g
3. Mass of oxygen produced 0.158 g
4. Mass of 125 mL flask* + water 177.034 g
5. Mass of 125 mL flask* 87.847 g
6. Mass of water 89.187 g
7. Temperature of water 21.5° C
8. Density of water 0.713 g/mL
9. Volume of water 125.087 = volume of O_2 gas 125.087 g/mL
10. Barometric pressure ______
11. Vapor pressure of water 3.27 KPa
12. Pressure of O_2 gas (show calculations) 0.57987 atm

    $P = \frac{nRT}{V}$ $\frac{3(0.082057)(294.65K)}{125.087 g/mL}$ P = 0.57987

13. Gas law constant, *R*, from ideal gas law (show calculations) R = 0.082507

    $PV = nRT$ $R = \frac{PV}{nT}$ $\frac{(0.57987)(125.087)}{3(294.65 K)} = 0.082054$

14. *R* from the van der Waals equation (show calculations) R = 4.06208

    $P = \frac{RT}{V-b} - \frac{a}{V^2}$ $P = \frac{RT}{V}$ $R1 = \frac{T}{PV}$ $\frac{294.65}{0.57987 \cdot 125.087} = 4.062$

15. Accepted value of *R* 0.082507 (source of *R* value) Ideal Gas Law

*Or Styrofoam cup.

16. Uncertainty in *R* (show calculations) ____________________

QUESTIONS

1. Does your value of *R* agree with the accepted value within your uncertainty limits?

2. Discuss possible sources of error in the experiment. Indicate the ones that you believe are most important.

3. Which gas would you expect to deviate more from ideality—H_2 or HBr? Explain your answer.

4. How does the solubility of oxygen in water affect the value of *R* you determined? Explain your answer.

5a. Use the van der Waals equation to calculate the pressure exerted by 1.000 mol of Cl_2 in 22.41 L at 0.0 °C. The van der Waals constants for Cl_2 are $a = 658\ L^2\text{-kPa/mol}^2$ and $b = 0.0562$ L/mol.

P+ (6.49 L² atm/mol × (1.00 mol)² / (22.41 L)²) = 6.49 L² atm/mol² (22.41 L − 1.000 × 0.0562 mol/4 mol)

P = 1.00268 atm − 0.012922 atm = 0.990 atm

= 1.000 mol × 0.082057 L atm/mol K × 273.15 K

(P+ 6.49 L² atm/mol² × 1.000 mol² / 502.2081 L²)(22.3538 L) = 22.4138696 L atm

(P+ 0.0129229298 atm) = 22.4138696 L atm / 22.3538 L = 1.00268722 atm

5b. Which factor is the major cause for deviation from ideal behavior—the volume of the Cl_2 molecules or the attractive forces between them?

6. 20.83 g of a gas occupies 4.167 L at 79.97 kPa at 30.0 °C. What is its molecular weight?

7. 5.600 g of solid CO_2 is put in an empty sealed 4.00 L container at a temperature of 300 K. When all the solid CO_2 turns into gas, what will be the pressure in the container?

8. A gas consisting of only carbon and hydrogen has an empirical formula of CH_2. The gas has a density of 1.65 g/L at 27.0 °C and 97.9 kPa. Determine the molecular weight and the molecular formula of the gas.

9. The gauge pressure in an automobile tire reads 220 kPa in the winter at 0 °C. The gauge reads the difference between the tire pressure and the atmospheric pressure (101 kPa). In other words, the tire pressure is the gauge reading plus 101 kPa. If the same tire were used in the summer at 43 °C and no air had leaked from the tire, what would be the tire gauge reading in the summer? (Note: For automobile tires, usually the unit bar is used, 1 bar = 100 kPa.)

NOTES AND CALCULATIONS

Experiment

Activity Series

OBJECTIVE

To become familiar with the relative activities of metals in chemical reactions.

APPARATUS AND CHEMICALS

Apparatus

small test tubes* (13)

test tube rack

Chemicals

0.2 *M* $Ca(NO_3)_2$

0.2 *M* $Mg(NO_3)_2$

0.2 *M* $Zn(NO_3)_2$

0.2 *M* $Fe(NO_3)_3$

0.2 *M* $FeSO_4$

0.2 *M* $SnCl_4$

0.2 *M* $CuSO_4$

6 *M* HCl

7 small pieces each of calcium, magnesium, zinc, iron wool, tin, and freshly cleaned copper

DISCUSSION

Chemical elements are usually classified by their properties into three groups: metals, nonmetals, and metalloids. Most of the known elements are metals. Their physical properties include high thermal and electrical conductivity, high luster, malleability (ability to be pounded flat without shattering), and ductility (ability to be drawn out into a fine wire). All common metals are solids at room temperature except mercury, which is a liquid. The periodic table illustrated in Figure 15.1 shows the three classifications of the elements.

All elements to the left of the shaded area except hydrogen are metals. Those to the right are nonmetals. Those in the shaded area have intermediate properties and are called semimetals or metalloids. Families or groups of elements consist of elements in vertical columns in the periodic table. Elements within a group or family (called congeners) have similar chemical properties because they have similar valence electronic structures; that is, the number of valence electrons (electrons in the outermost shell) is the same for all members of a family or group. For historical reasons, most of the groups have names, some often referred to by their names. These are the following:

1. Group 1A, called *alkali metals* because they react with oxygen to form bases
2. Group 2A, called *alkaline earth metals* because their presence makes soils alkaline
3. Group 3A, no common name
4. Group 4A, no common name

*A spot plate may be used in place of test tubes.

Periods — horizontal rows

Elements arranged in order of increasing atomic number

Steplike line divides metals from nonmetals

Groups — vertical columns containing elements with similar properties

	1A 1	2A 2	3B 3	4B 4	5B 5	6B 6	7B 7	8B 8	8B 9	8B 10	1B 11	2B 12	3A 13	4A 14	5A 15	6A 16	7A 17	8A 18
1	1 H																	2 He
2	3 Li	4 Be											5 B	6 C	7 N	8 O	9 F	10 Ne
3	11 Na	12 Mg											13 Al	14 Si	15 P	16 S	17 Cl	18 Ar
4	19 K	20 Ca	21 Sc	22 Ti	23 V	24 Cr	25 Mn	26 Fe	27 Co	28 Ni	29 Cu	30 Zn	31 Ga	32 Ge	33 As	34 Se	35 Br	36 Kr
5	37 Rb	38 Sr	39 Y	40 Zr	41 Nb	42 Mo	43 Tc	44 Ru	45 Rh	46 Pd	47 Ag	48 Cd	49 In	50 Sn	51 Sb	52 Te	53 I	54 Xe
6	55 Cs	56 Ba	71 Lu	72 Hf	73 Ta	74 W	75 Re	76 Os	77 Ir	78 Pt	79 Au	80 Hg	81 Tl	82 Pb	83 Bi	84 Po	85 At	86 Rn
7	87 Fr	88 Ra	103 Lr	104 Rf	105 Db	106 Sg	107 Bh	108 Hs	109 Mt	110 Ds	111 Rg	112 Cn	113 Nh	114 Fl	115 Mc	116 Lv	117 Ts	118 Og

57 La	58 Ce	59 Pr	60 Nd	61 Pm	62 Sm	63 Eu	64 Gd	65 Tb	66 Dy	67 Ho	68 Er	69 Tm	70 Yb
89 Ac	90 Th	91 Pa	92 U	93 Np	94 Pu	95 Am	96 Cm	97 Bk	98 Cf	99 Es	100 Fm	101 Md	102 No

Metals
Metalloids
Nonmetals

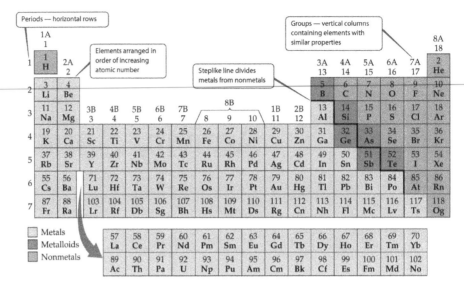

▲**FIGURE 15.1** Periodic table of the elements.

5. Group 5A, called *pnictides*, from the Greek word meaning choking suffocation
6. Group 6A, called *chalcogens*, from Greek roots meaning ore former
7. Group 7A, called *halogens*, from Greek roots meaning salt former
8. Group 8A, called *rare*, *noble*, or *inert gases* because they are rare and were thought to be unreactive

Those most frequently referred to by group name are the alkali metals, the alkaline earth metals, the halogens, and the rare gases.

The three broad categories of the elements also have somewhat similar chemical properties. For example, metals, as compared with the other elements, have relatively low ionization energies and enter into chemical combination with nonmetals by *losing* electrons to become cations. This can be symbolized by the following equation:

$$\mathrm{M} \longrightarrow \mathrm{M}^{n+} + ne^-$$

Nonmetals, as compared with metals, have relatively high electron affinities and enter into chemical combination with metals by *gaining* electrons to become anions. This can be symbolized by the following equation:

$$\mathrm{X} + ne^- \longrightarrow \mathrm{X}^{n-}$$

Specific examples of these types of reactions can be divided into several useful categories, which the following examples illustrate.

Electron-Transfer Reactions

1 REACTIONS WITH OXYGEN

$$2\mathrm{Mg}(s) + \mathrm{O}_2(g) \longrightarrow 2\mathrm{MgO}(s)$$

In this reaction, magnesium is oxidized by oxygen, which is reduced by magnesium (Section 4.4). This can be better illustrated by breaking down the reaction into the following fictitious although helpful steps:

$$2(Mg \longrightarrow Mg^{2+} + 2e^-) \qquad \text{oxidation}$$

$$O_2 + 4e^- \longrightarrow 2O^{2-} \qquad \text{reduction}$$

$$2Mg + O_2 \longrightarrow 2MgO \qquad \text{oxidation-reduction (redox) reaction}$$

In *oxidation*, the oxidized element loses electrons and becomes more positive. In *reduction*, the reduced element gains electrons and becomes more negative. Oxidation is always associated with a concomitant reduction and in a balanced reaction the number of electrons lost must equal the number of electrons gained.

2 REACTIONS WITH WATER

$$2Na(s) + 2H_2O(l) \longrightarrow 2NaOH(aq) + H_2(g)$$

$$Ca(s) + 2H_2O(l) \longrightarrow Ca(OH)_2(aq) + H_2(g)$$

The *ionic* equations for these reactions better illustrate the electron-transfer process.

$$2Na(s) + 2H_2O(l) \longrightarrow 2Na^+(aq) + 2OH^-(aq) + H_2(g)$$

$$Ca(s) + 2H_2O(l) \longrightarrow Ca^{2+}(aq) + 2OH^-(aq) + H_2(g)$$

3 REACTIONS WITH ACIDS

$$Zn(s) + 2HCl(aq) \longrightarrow ZnCl_2(aq) + H_2(g)$$

or

$$Zn(s) + 2H^+(aq) + 2Cl^-(aq) \longrightarrow Zn^{2+}(aq) + 2Cl^-(aq) + H_2(g)$$

Because the chloride ion is merely a spectator—that is, it does not participate in the reaction—it may be omitted, yielding the *net ionic equation* (Section 4.2).

$$Zn(s) + 2H^+(aq) \longrightarrow Zn^{2+}(aq) + H_2(g)$$

or simply

$$Zn + 2H^+ \longrightarrow Zn^{2+} + H_2$$

4 ELECTRON TRANSFER AMONG METALS

$$Zn(s) + Cu(NO_3)_2(aq) \longrightarrow Zn(NO_3)_2(aq) + Cu(s)$$

or

$$Zn(s) + Cu^{2+}(aq) \longrightarrow Zn^{2+}(aq) + Cu(s)$$

or simply

$$Zn + Cu^{2+} \longrightarrow Zn^{2+} + Cu$$

Note once again in the ionic equation, the spectator ion (NO_3^-) has been omitted because it takes no active part in the reaction and serves only to provide electrical neutrality (Section 4.4). Therefore, any other soluble salt of copper(II), such as chloride, sulfate, or acetate, could perform the same function.

To achieve the lowest energy level for the system, the more active metal of a pair will lose electrons to the more passive metal or will react more vigorously with water, acids, or oxygen. In some cases, no reaction will occur. Without prior knowledge, you have no way to predict these events.

PROCEDURE

A. Reactions of Metals with Acid

To each of six test tubes containing 0.5 mL of dilute 6 *M* HCl, add a small piece of the metals Ca, Cu, Fe, Mg, Sn, and Zn, one metal to each tube. Observe the test tubes and note any changes that occur (such as whether a gas is evolved, whether the gas evolution is vigorous and what color changes occurred). Enter your observations on the report sheet and write both complete and ionic equations for each reaction noted. After completing each series of reactions, dispose of the contents of your test tubes in the designated waste containers.

B. Reactions of Metals with Solutions of Metal Ions

(WORK IN PAIRS FOR THIS STEP.) Add a small piece of calcium metal to each of seven test tubes containing about 0.5 mL of 0.2 *M* $Ca(NO_3)_2$, 0.2 *M* $CuSO_4$, 0.2 *M* $FeSO_4$, 0.2 *M* $Fe(NO_3)_3$, 0.2 *M* $Mg(NO_3)_2$, 0.2 *M* $SnCl_4$, and 0.2 *M* $Zn(NO_3)_2$ solutions. Note any reaction that occurs by observing whether a color change occurs on the surface of the metal or in the solution or whether a gas is evolved. Pay close attention to whether the Ca dissolves (because there may be other side reactions). Record your observations on the report sheet. Write both complete and ionic equations for any reaction that occurs. After completing each series of reactions, dispose of the contents of your test tubes in the designated waste containers.

Repeat the preceding process by adding a small piece of copper to another 0.5 mL of each of the metal-cation solutions. Do the same for iron, magnesium, tin, and zinc and record your observations. Pay close attention to whether the metal dissolves. Write both complete molecular and ionic equations for all reactions.

C. Relative-Activity Series

From the information in the table you constructed in Part B, you can rank these six metals according to their relative chemical reactivities. One of the metals will replace all others in solution. For example, if calcium metal is oxidized by solutions containing cations of each of the other metals, it is most reactive. One of the other metals that will replace all but calcium is the next most reactive. Finally, one of the metals will not replace any of the other metal cations from solution. Therefore, it is least reactive. List the metals on your report sheet in terms of decreasing reactivity starting with the most reactive (1) and ending with the least reactive (6).

GIVE IT SOME THOUGHT

Are your results from this experiment consistent with the activity series from your textbook?

Name ______________________________ Desk ______________________________

Date ______________________ Laboratory Instructor ______________________________

Activity Series | 15 Pre-lab Questions

Before beginning this experiment in the laboratory, you should be able to answer the following questions.

1. What distinguishes a metal from a nonmetal?

2. What does ionization energy measure?

3. What does electron affinity measure?

4. Why must oxidation be accompanied by a reduction?

5. How do you determine the relative reactivities of metals?

6. Complete and balance the following:

$$Pb^{2+}(aq) + Mg(s) \longrightarrow$$

$$Al(s) + Fe_2O_3(s) \longrightarrow$$

$$Mg(s) + HCl(aq) \longrightarrow$$

7. Balance the following reactions and identify the species that have been oxidized and the species that have been reduced.

Reaction	**Species oxidized**	**Species reduced**
$Cl_2(g) + I^-(aq) \longrightarrow I_2(s) + Cl^-(aq)$		
$WO_2(s) + H_2(g) \longrightarrow W(s) + H_2O(l)$		
$Ca(s) + H_2O(l) \longrightarrow H_2(g) + Ca(OH)_2(s)$		
$Al(s) + O_2(g) \longrightarrow Al_2O_3(s)$		

8. Assuming that the following redox reactions are found to occur spontaneously, identify the more active metal in each reaction.

Reaction	**More active metal**
$2Li(s) + Cu^{2+}(aq) \longrightarrow 2Li^+(aq) + Cu(s)$	
$Cr(s) + 3V^{3+}(aq) \longrightarrow 3V^{2+}(aq) + Cr^{3+}(aq)$	
$Cd(s) + 2Ti^{3+}(aq) \longrightarrow 2Ti^{2+}(aq) + Cd^{2+}(aq)$	

Name ______________________________ Desk ______________

Date ______________ Laboratory Instructor ______________________

REPORT SHEET | EXPERIMENT

Activity Series | 15

A. Reactions of Metals with Acid

Metal	Reaction with HCl	Observation	Equations*
Ca			
Cu			
Mg			
Fe			
Sn			
Zn			
Example: Co			

* For Fe and Sn assume the lower oxidation state of plus two.

B. Reactions of Metals with Solutions of Metal Ions

Metal ions / Metal	Ca^{2+}	Cu^{2+}	Fe^{3+}	Fe^{2+}	Mg^{2+}	Sn^{4+}	Zn^{2+}	Al^{3+}
Ca								Yes, calcium dissolves
Cu								N.R.
Fe								N.R.
Mg								Yes, magnesium dissolves
Sn								N.R.
Zn								N.R.
Example: Al		Yes, aluminum turns brown	Yes, turns aluminum dark	Yes, aluminum turns dark; solution is colorless	N.R.	Yes, aluminum turns dark; solution is colorless	Yes, aluminum turns dark	

Example:

$$2Al(s) + 3Zn(NO_3)_2(aq) \longrightarrow 2Al(NO_3)_3(aq) + 3Zn(s)$$
$$2Al(s) + 3Zn^{2+}(aq) \longrightarrow 2Al^{3+}(aq) + 3Zn(s)$$

Complete equation	*Net ionic equation*

C. Relative-Activity Series

Most reactive — Least reactive

1. ______ 2. ______ 3. ______ 4. ______ 5. ______ 6. ______

QUESTIONS

1. Which of these six metals should be most reactive toward oxygen?
2. Which of the oxides would you expect to be thermally unstable and decompose according to the equation?

   $2MO \xrightarrow{\Delta} 2M + O_2$

3. Sodium is slightly less reactive than calcium. Predict the outcome of the following reactions:

   $Na + H_2O \longrightarrow$

   $Na + O_2 \longrightarrow$

   $Na + HCl \longrightarrow$

   $Na + Ca^{2+} \longrightarrow$

4. From the data in Table B, rank the activity of aluminum.

5. For each of the following reactions, indicate which substance is oxidized and which is reduced. Which substance is the oxidizing agent, and which is the reducing agent?

	Substance oxidized	*Substance reduced*	*Oxidizing agent*	*Reducing agent*
$2Al(s) + 3Cl_2(g) \longrightarrow 2AlCl_3(s)$	______	______	______	______
$8H^+(aq) + MnO_4^-(aq) + 5Fe^{2+}(aq) \longrightarrow 5Fe^{3+}(aq) + Mn^{2+}(aq) + 4H_2O(l)$	______	______	______	______
$FeS(s) + 3NO_3^-(aq) + 4H^+(aq) \longrightarrow 3NO(g) + SO_4^{2-}(aq) + Fe^{3+}(aq) + 2H_2O(l)$	______	______	______	______
$Zn(s) + 2HCl(aq) \longrightarrow ZnCl_2(aq) + 2H_2(g)$	______	______	______	______

Experiment 16

Electrolysis, the Faraday Constant, and Avogadro's Number

OBJECTIVE

To determine the values for the Faraday constant and Avogadro's number by electrolysis.

APPARATUS AND CHEMICALS

Apparatus

DC source of electricity	ammeter
insulated copper wires (2)	timer or watch
50 mL buret	clamp
250 mL beaker	thermometer
barometer	glass stirring rod
ring stand	millimeter ruler or meterstick

Chemicals

3 *M* H_2SO_4

DISCUSSION

The passage of an electric current through a solution is accompanied by chemical reactions at the electrodes. Oxidation (loss of electrons) occurs at the anode; reduction (gain of electrons) occurs at the cathode. The amount of reaction that occurs at the electrodes is directly proportional to the number of electrons transferred (Sections 20.5 and 20.9). The Faraday constant is defined as the total charge carried by Avogadro's number of electrons; in other words, 1 faraday represents the charge on 1 mol of electrons.

In this experiment, you will determine the value of the faraday by measuring the amount of charge required to reduce 1 mol of H^+ ions according to the reaction

$$2H^+(aq) + 2e^- \longrightarrow H_2(g) \quad \text{or} \quad 1H^+(aq) + 1e^- \longrightarrow \tfrac{1}{2}H_2(g)$$

Electric charge is conveniently measured in coulombs (Section 20.9). A coulomb, *C*, is the amount of electrical charge that passes a point in a circuit when a current of 1 *A* (ampere), flows for 1 *s* (second)

$$1\,\text{C} = 1\,\text{A} \times 1\,\text{s}$$

Therefore, the number of coulombs passing through a solution in a cell can be obtained by multiplying the current in amperes by the time in seconds for which it flows. The charge on the electron can also be measured in coulombs and is equal to 1.60×10^{-19} C.

EXAMPLE 16.1

A current of 3.00 A was passed through a solution of sulfuric acid for exactly 20.0 min. How many electrons and how many coulombs were passed through the solution?

SOLUTION:

$$\text{coulombs} = 3.00\ \text{A} \times 20.0\ \text{min} \times \frac{60\ \text{s}}{\text{min}} \times \frac{1\ \text{C}}{\text{A-s}}$$
$$= 3.60 \times 10^3\ \text{C}$$

$$\text{electrons} = \frac{3.60 \times 10^3\ \text{C}}{1.60 \times 10^{-19}\ \text{C/electron}}$$
$$= 2.25 \times 10^{22}\ \text{electrons}$$

From Equation [1], note that one hydrogen ion is reduced for every electron passed through the solution and that one molecule of H_2 is produced for every two electrons. Thus, in Example 16.1, the 2.25×10^{22} electrons would produce 1.125×10^{22} molecules of H_2.

If you were to measure the volume, pressure, and temperature of the hydrogen gas associated with the electrolysis in Example 16.1, you could calculate the number of moles of H_2 produced using the relation $n = PV/RT$. Example 16.2 illustrates how this information can be used to calculate Avogadro's number.

EXAMPLE 16.2

A current of 3.00 A was passed through a solution of sulfuric acid for 20 min. The hydrogen produced was collected over water at 20 °C and 87.66 kPa and was found to occupy a volume of 534 mL. How many moles of H_2 were produced, and what is the value of Avogadro's number, *N*?

SOLUTION: At 20 °C the vapor pressure of H_2O is 2.33 kPa. Therefore,

$$P_{\text{total}} = P_{H_2O} + P_{H_2}$$
$$P_{H_2} = 87.66\ \text{kPa} - 2.33\ \text{kPa}$$
$$= 85.33\ \text{kPa}$$

Solving the ideal gas equation for *n* gives the following:

$$n = \frac{PV}{RT} = \frac{[(8.533 \times 10^4\ \text{Pa})](0.534\ \text{L})}{(8.314 \times 10^3\ \text{L-Pa/mol-K})(293\ \text{K})}$$
$$= 0.0187\ \text{mol}\ H_2$$

Because two moles of electrons are required for each mole of H_2 produced,

$$\text{mol electrons} = (0.0187\ \text{mol}\ H_2) \times (2\ \text{mol electrons/mol}\ H_2)$$
$$= 0.0374\ \text{mol electrons}$$

Avogadro's number is the number of electrons in 1 mol of electrons. From the previous example,

$$2.25 \times 10^{22}\ \text{electrons} = 0.0374\ \text{mol electrons}$$

Therefore,

$$N = \frac{2.25 \times 10^{22}\ \text{electrons}}{0.0374\ \text{mol}}$$
$$= 60.2 \times 10^{22}\ \text{electrons/mol}$$
$$= 6.02 \times 10^{23}\ \text{electrons/mol}$$

Faraday's constant (abbreviated F), is defined as the number of coulombs that is equivalent to 1 mol of electrons. Thus, in the previous examples, the 3600 C corresponds to the charge associated with 0.0374 mol electrons, or

$$\frac{3.60\times10^3\ \text{C}}{0.0374\ \text{mol}} = 96{,}300\ \text{C/mol electrons}$$

This corresponds very closely to the accepted value for the faraday, which is

$$1\,F = 96{,}500\ \text{C/mol electrons}$$

Also note from Example 16.2 that (1.125×10^{22}) molecules of H_2 were produced and represent 0.0187 mol H_2.

$$0.0187\ \text{mol H}_2 = (1.125\times10^{22})\ \text{molecules H}_2$$

or

$$1\ \text{mol H}_2 = \frac{1.125\times10^{22}\ \text{molecules H}_2}{0.0187\ \text{mol H}_2} = 6.02\times10^{23}\ \text{molecules H}_2/\text{mol H}_2$$
$$= N$$

PROCEDURE

Assemble the apparatus illustrated in Figure 16.1. Obtain a DC source with an attached ammeter. In a 250 mL beaker, add 100 mL of distilled water and then slowly add 50 mL of dilute (3 *M*) sulfuric acid. Stir with a glass rod to mix well. Fill a 50 mL buret with this solution and invert it in the solution, holding it in place with a ring stand and clamp. Attach the copper wire cathode to the negative terminal of the DC source and place the other end uninsulated into the inverted mouth of the buret. Make sure the entire bare part of the wire is inside the buret; otherwise, some of the H_2 generated will not be collected in the buret. The anode electrode should be hung over the edge of the beaker and immersed in the acid solution, with the other end attached to the positive electrode of the DC source. The top of the solution in the buret should be within the graduated region of the buret so that the volume may be accurately measured, as illustrated in Figure 16.1.

Record the time, turn on the DC source, and record the ammeter reading. During electrolysis, be careful not to move the electrodes because this may change the current. It is important to maintain a steady current throughout the duration of the electrolysis. If current does fluctuate, it may be necessary to use an average value. Continue the electrolysis until at least 20 mL of hydrogen has been collected. Note the time the electrolysis is stopped. Measure the volume of H_2 collected.

Measure the height of the water column in mm in the buret above the solution in your beaker. Also measure the temperature of the acid solution and obtain the barometric pressure and the vapor pressure of water at the solution temperature. Note the following:

$$P_{H_2} = P_{\text{barometric}} - P_{H_2O\ \text{column}} - P_{H_2O\ \text{vapor}}$$
$$P_{H_2O\ \text{column}} = (\text{density of H}_2\text{O})\times\text{gravity}\times(\text{height of H}_2\text{O})$$
$$= (\text{height of H}_2\text{O in mm})\times9.81\ \text{Pa}$$

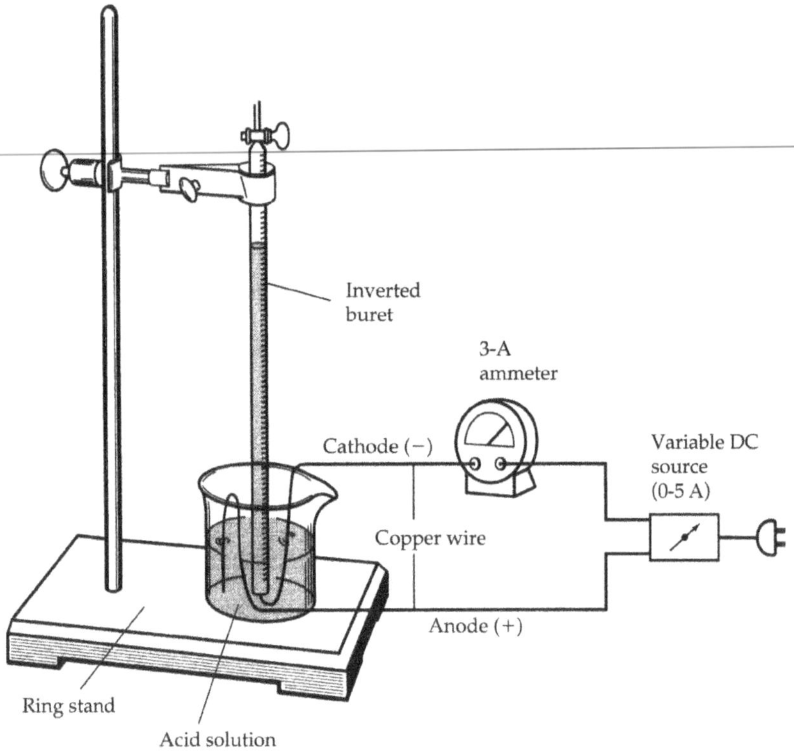

▲FIGURE 16.1 Apparatus for electrolysis experiment.

GIVE IT SOME THOUGHT

What do you expect for your result of the Faraday constant?

If time permits, repeat the experiment.

Dispose of the sulfuric acid in the designated waste containers.

Name ______________________________ Desk ______________

Date ______________ Laboratory Instructor ______________________

Electrolysis, the Faraday Constant, and Avogadro's Number | 16 Pre-lab Questions

Before beginning this experiment in the laboratory, you should be able to answer the following questions.

1. In this experiment, what type of chemical reaction is occurring to produce O_2 at the anode?

2. Define the term *Faraday's constant.*

3. What process is occurring at the cathode?

4. Why do you include the height of the water column in the buret in your calculation of the pressure?

5. Why is H_2SO_4 present in the electrolysis solution?

6. If a current of 6 A is drawn from a solar panel for three weeks, how many faradays are involved?

7. When an aqueous NaCl solution is electrolyzed, how many faradays need to be transferred at the anode to release 0.150 mol of Cl_2 gas?

$$2Cl^-(aq) \longrightarrow Cl_2(g) + 2e^-$$

8. How long must a current of 2.4 A pass through a sulfuric acid solution to liberate 0.200 L of H_2 gas at STP?

9. How much silver was in the solution if all of the silver was removed as Ag metal by electrolysis for 0.40 hr with a current of 1.00 mA $(1\ mA = 10^{-3}\ A)$?

10. Electrolysis of molten NaCl is done in a Downs cell operating at 7.0 volts (V) and 4.0×10^4 A. How much Na(s) and $Cl_2(g)$ can be produced in four hours in such a cell?

Name Grace Rademacher Desk

Date 04/27/21 Laboratory Instructor Prof. Towle

REPORT SHEET | EXPERIMENT

Electrolysis, the Faraday Constant, and Avogadro's Number | 16

1. Final volume in buret 29.5 mL
2. Initial volume in buret 50.00 mL
3. Volume of hydrogen 20.5 mL
4. Temperature of solution 25.5°C
5. Height of water column 340 mm
6. Barometric pressure 760 kPa
7. Vapor pressure of H_2O (see Appendix L) 25.2 kPa
8. Pressure of H_2 709.8 kPa (show calculations) $P_{H_2} = P_{baro} - P_{H_2O\ col} - P_{H_2O\ vapor}$

$H_2 = 760\ pKa - 25\ kPa - 25.2\ kPa = 709.8\ kPa$ / 101.325

$P_{H_2O col} = \frac{340 mm}{13.6 \frac{mm H_2O}{mm Hg}} = 25\ kPa = 7.0052\ atm$

9. Moles of H_2 produced (show calculations) 0.00587

$n = \frac{PV}{RT}$ $\frac{(7.0052)(.0205)}{(0.0821)(298.15)} = n$

~~10. Time reaction started~~

~~11. Time reaction terminated~~

12. Elapsed time 4 min 47 s
13. Current 0.5 A
14. Number of coulombs passed 143.5 C $C = (0.5)(287s)$ $1C = 1A \times 1s$
15. Value for faraday 48,976 Accepted value $96500 C/e^-$ (show calculations)

$\frac{(7.0052)(0.0205 L)}{(0.0821)(298.65)} = 0.00585$

$F = \frac{(2 \times 287)}{(2 \times 0.00586)} = 48976.12$

16. Value for Avogadro's number 2.6×10^{25} Accepted value 6.022×10^{23} (show calculations)

$N = \left(\frac{2}{0.00587} \times \frac{12223.17}{1.6 \times 10^{-19}\ C/e} \right) = 2.6 \times 10^{25}$ $= 7.64 \times 10^{22}\ e^-$

$\left(\frac{\#\ of\ e^-}{mol\ of\ e^-} \right) \left(\frac{Faraday}{1.6 \times 10^{-19}} \right)$

QUESTIONS

1. Discuss the major sources of error in this experiment.

The biggest source of error in this experiment would be human error in calculations.

2. Calculate the percentage error in F and N.

F

$\%\ error = \left(\frac{48976 - 96500}{96500} \right) 100$

49% error :(

N

$\%\ error: \left(\frac{2.6 \times 10^{25} - 6.022 \times 10^{23}}{6.022 \times 10^{23}} \right)$

$= (4.218 \times 10^{47}) 100 = 4.218 \times 10^{49}$

4.218×10^{49} % error ?:(

3. If the amperage in your electrolysis cell were increased by a factor of 2, what effect would this have on the time required to produce the same amount of hydrogen?

If amperage increased by a factor of 2, the time required to produce the same amount of hydrogen would decrease by a ½ factor

4. Electrolysis of an NaCl solution with a current of 2.00 A for a period of 200 s produced 59.6 mL of Cl_2 at 86.7 kPa pressure and 27 °C. Calculate the value of the Faraday's Constant from these data.

Temp = 300.15 K

Pressure = $\frac{86.7}{101} = 0.856$ atm

$n = \frac{PV}{RT}$

$\frac{(0.856)(0.0596)}{(0.0821)(300.15\ K)} = 0.0021$ mol

$F = \frac{(2 \times 200)}{(2 \times 0.0021)} = 95238.1\ C$

5. Why are different products obtained when molten NaCl and aqueous NaCl are electrolyzed? Predict the products in each case.

Experiment

Electrochemical Cells and Thermodynamics

OBJECTIVE

To become familiar with some fundamentals of electrochemistry, including the Nernst equation, by constructing electrochemical (voltaic) cells and measuring their potentials at various temperatures. The quantities ΔG, ΔH, and ΔS are calculated from the temperature variation of the measured emf.

APPARATUS AND CHEMICALS

Apparatus

DC voltmeter or potentiometer (to measure mV)
emery cloth
600 mL beaker
thermometer
glass stirring rods (3)
Bunsen burner and hose
clamps (2)
alligator clips and lead wires (2 sets)
50 mL test tubes (3)
glass U-tubes (to fit large test tubes) (3)
cotton
ring stand, two iron rings, and wire gauze
test tube clamps (2)

Chemicals

1 *M* $Pb(NO_3)_2$
1 *M* $SnCl_2$
lead, tin, and copper strips or wire
1 *M* $Cu(NO_3)_2$
0.1 *M* KNO_3
ice
agar

DISCUSSION

BACKGROUND

Electrochemistry is that area of chemistry that deals with the relations between chemical changes and electrical energy. It is primarily concerned with oxidation-reduction phenomena. Chemical reactions can be used to produce electrical energy in cells that are referred to as *voltaic*, or galvanic, cells (Section 20.3). Electrical energy, on the other hand, can be used to bring about chemical changes in what are termed *electrolytic* cells (Section 20.9). In this experiment, you will investigate some of the properties of voltaic cells.

In principle, any spontaneous redox reaction can be used to produce electrical energy. This task can be accomplished by means of a voltaic cell, a device in which electron transfer takes place through an external circuit or pathway rather than directly between reactants. One such spontaneous reaction occurs when a strip of zinc is immersed in a solution containing Cu^{2+}. As the reaction proceeds, the blue color of the $Cu^{2+}(aq)$ ions begins to fade and metallic copper deposits on the zinc strip. At the same time, the zinc begins to dissolve. The redox reaction that occurs is given in Equation [1].

$$Zn(s) + Cu^{2+}(aq) \longrightarrow Zn^{2+}(aq) + Cu(s) \quad [1]$$

Figure 17.1 shows a voltaic cell that utilizes the same reaction. The two solid metal strips connected by the external circuit are called electrodes (Figure 20.5). The electrode at which oxidation occurs is called the *anode*, and the electrode at which reduction occurs is called the *cathode.* The voltaic cell may be regarded as two "half-cells," one corresponding to the oxidation half-reaction and the other to the reduction half-reaction. Recall that a substance that loses electrons is said to be oxidized and a substance that gains electrons is said to be reduced. In the example below, Zn is oxidized and Cu^{2+} is reduced.

Anode (oxidation half-reaction) $Zn(s) \longrightarrow Zn^{2+}(aq) + 2e^-$

Cathode (oxidation half-reaction) $Cu^{2+}(aq) + 2e^- \longrightarrow Cu(s)$

Because Zn^{2+} ions are formed in one compartment and Cu^{2+} ions are depleted in the other compartment, a salt bridge is used to maintain electrical neutrality by allowing the migration of ions between these compartments.

The cell voltage, or electromotive force (*emf*), is indicated on the voltmeter in units of volts. The cell emf is also called the cell potential. The magnitude of the emf is a quantitative measure of the driving force or thermodynamic tendency for the reaction to occur. In general, the emf of a voltaic cell depends on the substances that make up the cell as well as on their concentration and temperature. Hence, it is a common practice to compare *standard cell potentials*, symbolized by $E°_{cell}$ (Section 20.4). These potentials correspond to cell voltages under standard conditions: gases at 101.3 kPa (1 atm) pressure; solutions at 1 *M* concentration and at 25 °C. For the Zn/Cu voltaic cell in Figure 17.1, the standard cell potential at 25 °C is 1.10 V.

$$Zn(s) + Cu^{2+}(aq, 1\,M) \longrightarrow Zn^{2+}(aq, 1\,M) + Cu(s) \qquad E°_{cell} = 1.10\text{ V}$$

Recall that the superscript ° denotes standard state conditions (Section 20.4).

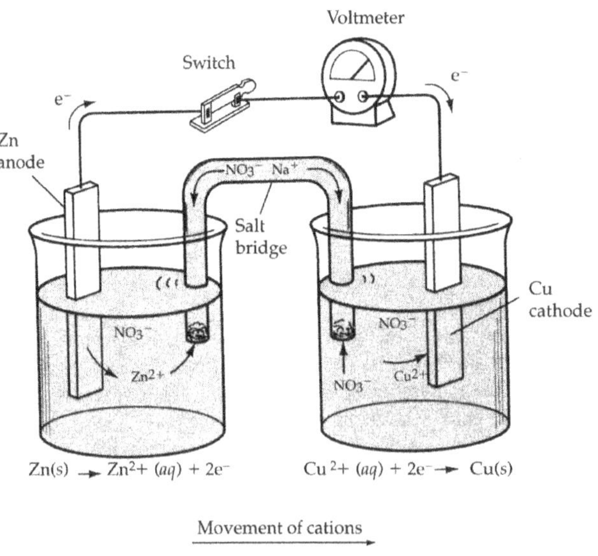

▲**FIGURE 17.1** A complete and functioning voltaic cell using a salt bridge to complete the electrical circuit.

The cell potential is the difference between two electrode potentials, one associated with the cathode and the other associated with the anode. By convention, the potential associated with each electrode is chosen to be the potential for reduction to occur at that electrode. Thus, standard electrode potentials are tabulated for reduction reactions, and they are denoted by the symbol, E°_{red}. The cell potential is given by the standard reduction potential of the cathode reaction, E°_{red} (cathode), *minus* the standard reduction potential of the anode reaction, E°_{red} (anode) as follows:

$$E^{\circ}_{cell} = E^{\circ}_{red}\,(\text{cathode}) - E^{\circ}_{red}\,(\text{anode}) \qquad [2]$$

Because it is not possible to directly measure the potential of an isolated half-cell reaction, the standard hydrogen reduction half-reaction, in which $H^{+}(aq)$ is reduced to $H_2(g)$ under standard conditions, has been selected as a reference (Section 20.4). It has been assigned a standard reduction potential of exactly 0 V.

$$2H^{+}(aq,\ 1\ M) + 2e^{-} \longrightarrow H_2(g, 101.3\ \text{kPa}) \qquad E^{\circ}_{red} = 0\ \text{V}$$

An electrode designed to produce this half-reaction is called the standard hydrogen electrode (SHE). Figure 17.2 shows a voltaic cell using a SHE and a standard Zn^{2+}/Zn electrode. The spontaneous reaction occurring in this cell is the oxidation of Zn and the reduction of H^{+}.

$$Zn(s) + 2H^{+}(aq) \longrightarrow Zn^{2+}(aq) + H_2(g)$$

The standard cell potential for this cell is 0.76 V. By using the defined standard reduction potential of H^{+} ($E^{\circ}_{red} = 0$ V) and Equation [2], you can determine the standard reduction potential for the Zn^{2+}/Zn half-reaction as follows:

$$E^{\circ}_{cell} = E^{\circ}_{red}(\text{cathode}) - E^{\circ}_{red}(\text{anode})$$

$$0.76\ \text{V} = 0\ \text{V} - E^{\circ}_{red}(\text{anode})$$

$$E^{\circ}_{red}(\text{anode}) = -0.76\ \text{V}$$

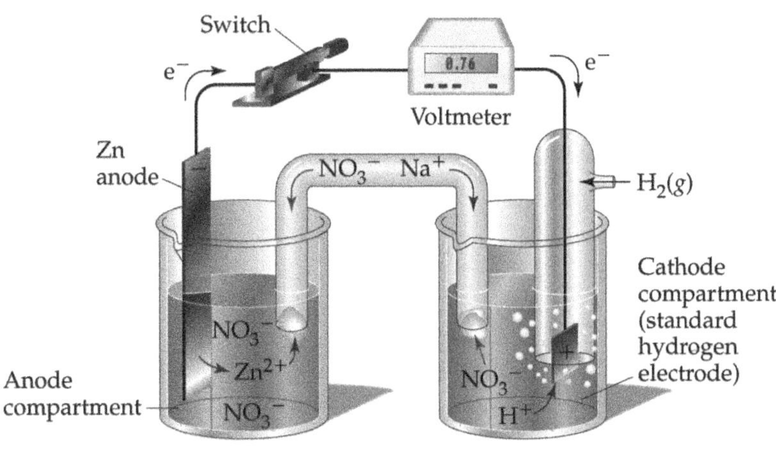

▲**FIGURE 17.2** Voltaic cell using a standard hydrogen electrode.

Thus, a standard reduction potential of −0.76 V can be assigned to the reduction of Zn^{2+} to Zn as follows:

$$Zn^{2+}(aq,\ 1\ M) + 2e^- \longrightarrow Zn(s) \qquad E^\circ_{red} = -0.76\ V$$

Notice that the reaction is written as a reduction even though it is "running in reverse" as an oxidation in the cell in Figure 17.2. Whenever you assign a potential to a half-cell reaction, you write the reaction as a reduction reaction.

Standard reduction potentials for other half-reactions can be established in a manner similar to that used for the $Zn^{2+} \mid Zn$ half-reaction. The table in Appendix H lists some standard reduction potentials. Example 17.1 illustrates how this method can be used to determine the standard reduction potential for the $Cu^{2+} \mid Cu$ half-reaction.

Example 17.1

The cell in Figure 17.1 may be represented by the following cell notation:

$$Zn \mid Zn^{2+}(aq) \parallel Cu^{2+}(aq) \mid Cu$$

The single bar represents the phase separation of the electrode from the solution. The double bar represents the salt bridge. The cell notation is generally written as | Anode || Cathode |. Given that E°_{cell} for this cell is 1.10 V, the Zn electrode is the anode, and the standard reduction potential of Zn^{2+} is −0.76 V, calculate the E°_{red} for the reduction of Cu^{2+} to Cu.

$$Cu^{2+}(aq,\ 1\ M) + 2e^- \longrightarrow Cu(s)$$

SOLUTION: Use Equation [2] and the information provided. See Figure 17.3.

$$E^\circ_{cell} = E^\circ_{red}(\text{cathode}) - E^\circ_{red}(\text{anode})$$

$$1.10\ V = E^\circ_{red}(\text{cathode}) - (-0.76\ V)$$

$$E^\circ_{red}(\text{cathode}) = 1.10\ V - 0.76\ V = 0.34\ V$$

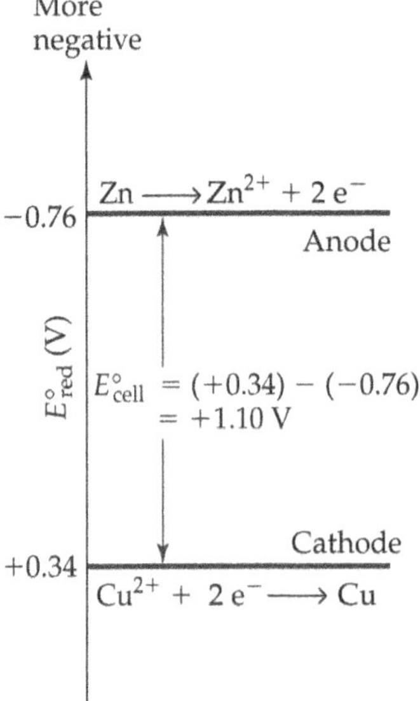

▲**FIGURE 17.3** Half-cell potentials and standard cell potential for the Zn-Cu voltaic cell.

The free-energy change, ΔG, associated with a chemical reaction is a measure of the driving force or spontaneity of the process. If the free-energy change of a process is negative, the reaction will occur spontaneously in the direction indicated by the chemical equation (Section 20.5).

The cell potential of a redox process is related to the free-energy change as follows:

$$\Delta G = -nFE \qquad [3]$$

In this equation, F is Faraday's constant, the electrical charge on 1 mol of electrons is

$$1F = 96{,}485 \frac{\text{C}}{\text{mol e}^-} = 96{,}485 \frac{\text{J}}{\text{V-mol e}^-}$$

and n represents the number of moles of electrons transferred in the reaction. When both reactants and products are in their standard states, Equation [3] takes the following form:

$$\Delta G^\circ = -nFE^\circ \qquad [4]$$

Example 17.2

Calculate the standard free-energy change associated with the redox reaction $2Ce^{4+} + Tl^{+} \longrightarrow 2Ce^{3+} + Tl^{3+}$ ($E^\circ = 0.450$ V). Would this reaction occur spontaneously under standard conditions?

SOLUTION:

$$\begin{aligned} \Delta G^\circ &= -nFE^\circ \\ &= -(2 \text{ mol e}^-)\left(\frac{96{,}485 \text{ J}}{\text{V-mol e}^-}\right)(0.450 \text{ V}) \\ &= -86.9 \times 10^3 \text{ J} \\ &= -86.9 \text{ kJ} \end{aligned}$$

Because $\Delta G^\circ < 0$, this reaction would occur spontaneously.

The standard free-energy change of a chemical reaction is also related to the equilibrium constant for the reaction as follows:

$$\Delta G^\circ = -RT \ln K \qquad [5]$$

where R is the gas law constant (8.314 J/K-mol) and T is the temperature in Kelvin. Consequently, E° is also related to the equilibrium constant. From Equations [4] and [5], it follows that

$$\begin{aligned} -nFE^\circ &= -RT \ln K \\ E^\circ &= \frac{RT}{nF} \ln K \end{aligned} \qquad [6]$$

When $T = 298$ K, $\ln K$ is converted to $\log K$, and the appropriate values of R and F are substituted, Equation [6] becomes

$$E^\circ = \frac{0.0592 \text{ V}}{n} \log K \qquad [7]$$

You can see from this relation that the larger K is, the larger the standard cell potential will be.

In practice, most voltaic cells are not likely to be operating under standard state conditions. It is possible, however, to calculate the cell emf, E, under non-standard state conditions with a knowledge of $E°$, temperature, and concentrations of reactants and products as follows:

$$E = E° - \frac{0.0592\text{ V}}{n}\log Q \qquad [8]$$

Q is called the reaction quotient; it has the form of an equilibrium constant expression, but the concentrations used to calculate Q are not equilibrium concentrations. The relationship given in Equation [8] is referred to as the Nernst equation (see Example 17.3) (Section 20.6).

Now consider the operation of the cell shown in Figure 17.1 in more detail. Earlier you saw that the reaction

$$\text{Cu}(aq)^{2+} + \text{Zn}(s) \longrightarrow \text{Zn}(aq)^{2+} + \text{Cu}(s)$$

is spontaneous. Consequently, it has a positive electrochemical potential ($E° = 1.10$ V) and a negative free energy ($\Delta G° = -nFE°$). As this reaction occurs, Cu^{2+} will be reduced and deposited as copper metal onto the copper electrode. The electrode at which reduction occurs is called the cathode. Simultaneously, zinc metal from the zinc electrode will be oxidized and go into solution as Zn^{2+}. The electrode at which oxidation occurs is called the anode. Effectively, then, electrons will flow in the external wire from the zinc electrode through the voltmeter to the copper electrode and be given up to copper ions in solution. These copper ions will be reduced to copper metal and plate out on the copper electrode. Concurrently, zinc metal will give up electrons to become Zn^{2+} ions in solution. These Zn^{2+} ions will diffuse through the salt bridge into the copper solution and replace the Cu^{2+} ions that are being removed. See Figure 17.4.

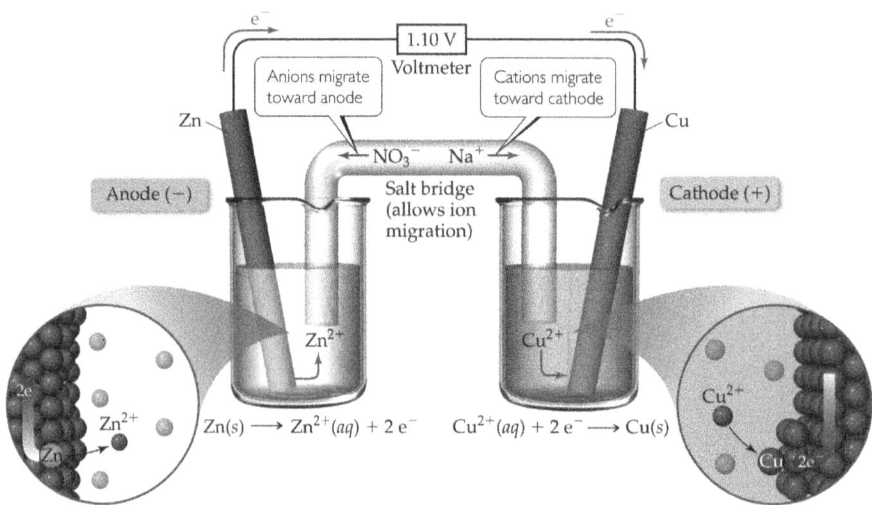

▲**FIGURE 17.4** Voltaic cell indicating movement of electrons and ions.

EXAMPLE 17.3

Calculate the cell potential for the cell

$$Zn \,|\, Zn^{2+}(0.60\ M) \,||\, Cu^{2+}(0.20\ M) \,|\, Cu$$

given the following:

$$Cu(aq)^{2+} + Zn(s) \longrightarrow Cu(s) + Zn(aq)^{2+} \qquad E° = 1.10\ V$$

(HINT: Recall that Q includes expressions for species in solution but not for pure solids.)

SOLUTION:

$$\begin{aligned} E &= E° - \frac{0.0592\ V}{n}\log\frac{[Zn^{2+}]}{[Cu^{2+}]} \\ &= 1.10\ V - \frac{0.0592\ V}{2}\log\frac{[0.60]}{[0.20]} \\ &= (1.10 - 0.014)\ V \\ &= 1.086\ V \\ &= 1.09\ V \end{aligned}$$

You can see that changes in concentrations have small effects on the cell emf because of the log term in the above equation.

A list of the properties of electrochemical cells and some definitions of related terms are given in Table 17.1.

Chemists have developed a shorthand notation for electrochemical cells, as shown in Example 17.1. The notation for the Cu-Zn cell that explicitly shows concentrations is as follows:

$$Zn \,|\, Zn^{2+}(xM) \,||\, Cu^{2+}(yM) \,|\, Cu$$

Anode (oxidation) Cathode (reduction)

In this notation, the anode (oxidation half-cell) is written on the left and the cathode (reduction half-cell) is written on the right.

Your objective in this experiment is to construct a set of three electrochemical cells and to measure their cell potentials. With a knowledge of two half-cell potentials and the cell potentials obtained from your measurements, you will calculate the other half-cell potentials and the equilibrium constants for the reactions. By measuring the cell potential as a function of temperature, you may also determine the thermodynamic constants, ΔG, ΔH, and ΔS, for the reactions. This can be done with the aid of Equation [9].

$$\Delta G = \Delta H - T\Delta S \qquad [9]$$

ΔG may be obtained directly from measurements of the cell potential using the following relationship:

$$\Delta G = -nFE$$

TABLE 17.1 Summary of Properties of Electrochemical Cells and Some Definitions

Voltaic cells: $E > 0$, $\Delta G < 0$; reaction spontaneous, K large (greater than 1)
Electrolytic cells: $E < 0$, $\Delta G > 0$; reaction nonspontaneous, K small (less than 1)
Anode: electrode at which oxidation occurs
Cathode: electrode at which reduction occurs
Oxidizing agent: species accepting electrons to become reduced
Reducing agent: species donating electrons to become oxidized

A plot of ΔG versus temperature in Kelvin will give $-\Delta S$ as the slope and ΔH as the intercept. A more accurate measure of ΔH can be obtained, however, by substituting ΔG and ΔS back into Equation [9] and calculating ΔH.

EXAMPLE 17.4

For the voltaic cell

$$\mathrm{Pb}\,|\,\mathrm{Pb}^{2+}(1\,M)\,||\,\mathrm{Cu}^{2+}(1\,M)\,|\,\mathrm{Cu}$$

the following data were obtained:

$$E = 0.464\text{ V} \qquad T = 298\text{ K}$$
$$E = 0.468\text{ V} \qquad T = 308\text{ K}$$
$$E = 0.473\text{ V} \qquad T = 318\text{ K}$$

Calculate ΔG, ΔH, and ΔS for this cell.

SOLUTION:

$$\Delta G = -nFE$$

At 298 K,

$$\Delta G = -(2\text{ mol e}^-)(96{,}485\text{ J/V-mol e}^-)(0.464\text{ V})(1\text{ kJ}/1000\text{ J})$$
$$= -89.5\text{ kJ}$$

At 308 K,

$$\Delta G = -(2\text{ mol e}^-)(96{,}485\text{ J/V-mol e}^-)(0.468\text{ V})(1\text{ kJ}/1000\text{ J})$$
$$= -90.3\text{ kJ}$$

At 318 K,

$$\Delta G = -(2\text{ mol e}^-)(96{,}485\text{ J/V-mol e}^-)(0.473\text{ V})(1\text{ kJ}/1000\text{ J})$$
$$= -91.3\text{ kJ}$$

Because $\Delta G = \Delta H - T\Delta S$, a plot of ΔG versus T in Kelvin gives $\Delta H = -64.2$ kJ/mol and $\Delta S = 85.0$ J/mol-K.

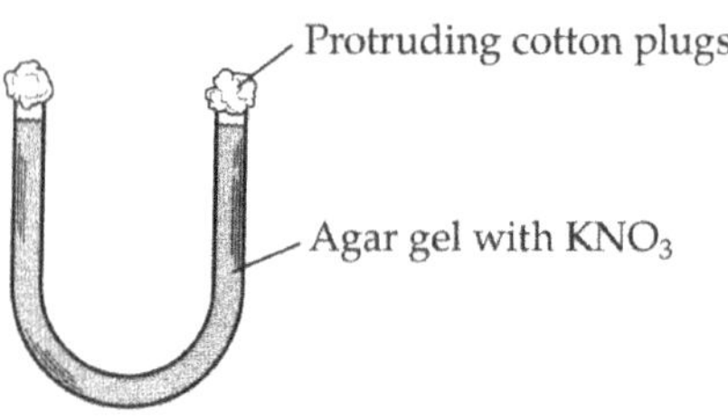

▲**FIGURE 17.5** U-tube salt bridge.

PROCEDURE

Before setting up pairs of half-cells, make a complete cell of the type shown in Figure 17.1 in the following manner illustrated in Figure 17.6: Place 30 mL of 1 *M* $Pb(NO_3)_2$ and 30 mL of 1 *M* $Cu(NO_3)_2$ in separate large test tubes. Obtain a lead strip and a copper strip and clean the surfaces of each with emery cloth or sandpaper. Boil 100 mL of 0.1 *M* KNO_3. Remove this solution from the heat and add to the boiling solution 1 g of agar, stirring constantly until all of the agar dissolves. Invert a U-tube and fill the U-tube with this solution before it cools, leaving about a 12.7 mm of air space at each end of the U-tube as shown in Figure 17.5. The cotton plugs must protrude from the ends of the U-tube. Construct two additional agar-filled U-tubes in the same manner. Place a U-tube in the test tubes as a salt bridge, as shown in Figure 17.6.

Insert the lead strip into the $Pb(NO_3)_2$ solution and the copper strip into the $Cu(NO_3)_2$ solution. Obtain a voltmeter and attach the positive lead to the copper strip and the negative lead to the lead strip using alligator clips. Read the voltage. Check that the alligator clips make good contact with the metal strips. Record this voltage and the temperature of the cells on your report sheet. If your measured potential is negative, reverse the wire connection. Now construct the following cells and measure their voltages in the same manner.

$$\mathrm{Sn}\,|\,\mathrm{Sn}^{2+}(1\,M)\,||\,\mathrm{Cu}^{2+}(1\,M)\,|\,\mathrm{Cu}$$

$$\mathrm{Pb}\,|\,\mathrm{Pb}^{2+}(1\,M)\,||\,\mathrm{Sn}^{2+}(1\,M)\,|\,\mathrm{Sn}$$

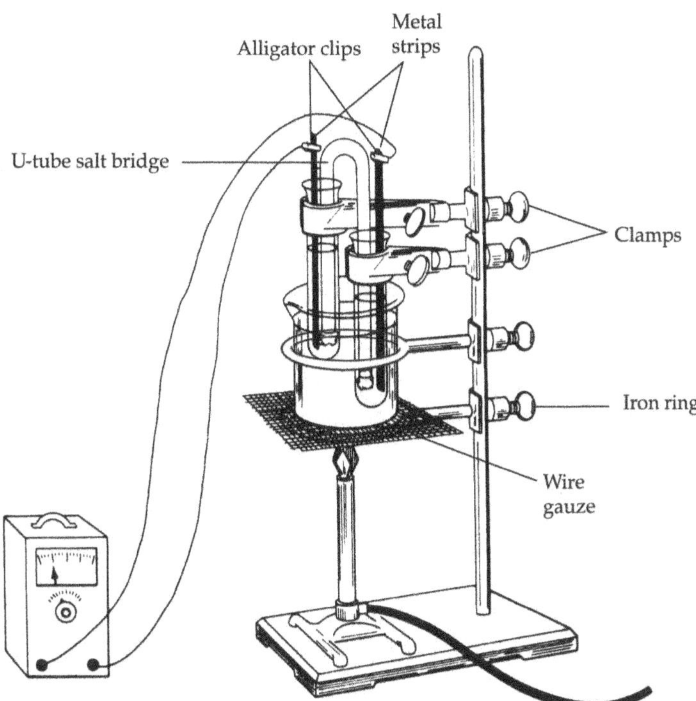

▲FIGURE 17.6 Electrochemical reaction apparatus.

GIVE IT SOME THOUGHT

a. In the cells you constructed, which half-reaction occurs at the anode?
b. Which half reaction occurs at the cathode?
c. Explain your choice.
d. In each of these cells, do you observe a spontaneous reaction?
e. How can you observe this experimentally?

GIVE IT SOME THOUGHT

Consider the cell illustrated in Figure 17.4

a. Do the concentrations of $Cu^{2+}(aq)$ and $Zn^{2+}(aq)$ increase, decrease, or remain the same as the reaction proceeds?
b. Does the mass of the Cu(s) and Zn(s) increase, decrease, or remain the same as the reaction proceeds?
c. Why do the concentrations and masses change?

Record the voltage and temperature of each cell on your report sheet. From the measured voltages, calculate the half-cell potentials for the lead and tin half-cells and the equilibrium constants for these two reactions. In these calculations, use $E° = 0.34$ V for the $Cu^{2+} \mid Cu$ couple.

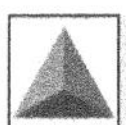

GIVE IT SOME THOUGHT

Describe a voltaic cell consistent with Figure 17.1 by sketching each voltaic cell. In your sketch, label the anode and cathode compartments and indicate the half-reaction occurring at the anode and the cathode, the overall cell reaction, the overall cell potential, the direction of electron flow, and the ions present in the anode and cathode compartments.

Now choose any one of the three cells and measure the cell potential as a function of temperature as follows: Insert the metal strips into each of two 50 mL test tubes containing the respective 1 M cation solution and place the test tubes in a 600 mL beaker containing about 200 mL of distilled water.

Place the U-tube into the two test tubes and connect the metallic strips to the voltmeter as before (see Figure 17.6). Because the voltage changes are of the order of 30 mV for the temperature range you will study, make certain the voltmeter you use is sensitive enough to detect these small changes.

Begin heating the water in the 600 mL beaker, using your Bunsen burner. Make sure the test tubes are clamped firmly in place. *Do not move* any part of the cell; if you do, the voltage will fluctuate. Heat the cell to approximately 70 °C. Measure the temperature and record it on your report sheet. Determine the cell potential at this temperature and record it on your report sheet. Remove the Bunsen burner and record the temperature and voltage at 15° intervals as the cell cools to room temperature. Finally, replace the beaker of water with a beaker containing an ice-water mixture, being careful not to move the test tubes and their contents. After the cell has been in the ice-water mixture for about 10 min, measure the temperature of the ice-water mixture in the 600 mL beaker and record it and the cell potential. You have determined the cell potential at various temperatures. Calculate ΔG for the cell at each of these temperatures and plot ΔG versus temperature on the graph paper provided. The slope of the plot is $-\Delta S$. From the values of ΔG and ΔS, calculate ΔH at 298 K. If time permits, determine the temperature dependence of E for another cell.

GIVE IT SOME THOUGHT

a. Which variable would you plot on the x- and y-axis on your graph?

b. Rearrange Equation [9]$(\Delta G = \Delta H - T\Delta S)$ into $y = mx + b$. Which of the following graphs should yours resemble?

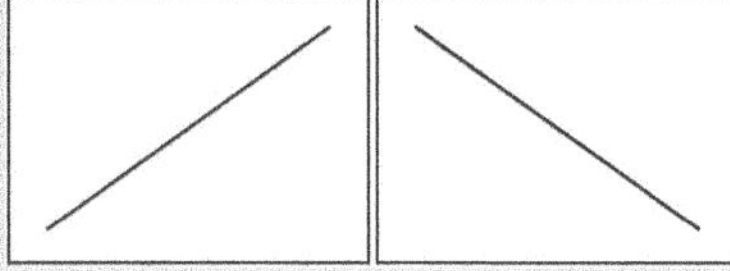

Dispose of the solutions in the test tubes and the agar containing U-tubes in the designated waste container.

NOTES AND CALCULATIONS

Name ______________________________ Desk ______________

Date ______________ Laboratory Instructor ______________________

Electrochemical Cells and Thermodynamics | 17 Pre-lab Questions

Before beginning this experiment in the laboratory, you should be able to answer the following questions.

1. Define the following terms: *faraday*, *salt bridge*, *anode*, *cathode*, *voltaic cell*, and *electrolytic cell.*

2. Write a chemical equation for the reaction that occurs in the following cell: $Cu \mid Cu^{2+}(aq) \parallel Ag^{+}(aq) \mid Ag$.

3. Given the following E's, calculate the standard-cell potential for the cell in question 2.

$$Cu^{2+}(aq) + 2e^{-} \longrightarrow Cu(s) \qquad E° = +0.34\text{ V}$$
$$Ag^{+}(aq) + e^{-} \longrightarrow Ag(s) \qquad E° = +0.80\text{ V}$$

4. Calculate the voltage of the following cell:

$$Zn \mid Zn^{2+}(0.10\ M) \parallel Cu^{2+}(0.20\ M) \mid Cu$$

5. Calculate the cell potential, the equilibrium constant, and the free-energy change for

$$Ca(s) + Mn^{2+}(aq)(1\ M) \rightleftharpoons Ca^{2+}(aq)(1\ M) + Mn(s)$$

given the following $E°$ values:

$$Ca^{2+}(aq) + 2e^{-} \longrightarrow Ca(s) \qquad E° = -2.87\text{ V}$$
$$Mn^{2+}(aq) + 2e^{-} \longrightarrow Mn(s) \qquad E° = -1.18\text{ V}$$

6. Would you normally expect $\Delta H°$ to be positive or negative for a voltaic cell? Justify your answer.

---

7. Predict whether the following reactions are spontaneous.

$$2Mn^{3+}(aq) + 2Cl^{-}(aq) \longrightarrow 2Mn^{2+}(aq) + Cl_2(g)$$

$Mn^{3+} + e \longrightarrow Mn^{2+}$ $\quad E^{\theta} = +1.49$

$Cl_2 + 2e \longrightarrow 2Cl^{-}$ $\quad E^{\theta} = +1.36$

$$Cu^{+}(aq) + Fe^{3+}(aq) \longrightarrow Cu^{2+}(aq) + Fe^{2+}(aq)$$

$Fe^{3+} + e \longrightarrow Fe^{2+}$ $\quad E^{\theta} = +0.77$

$Cu^{2+} + e \longrightarrow Cu^{+}$ $\quad E^{\theta} = +0.15$

8. Identify the oxidizing agents and reducing agents in the reactions in question 7.

Name Grace Rademacher Desk

Date 4/20/21 Laboratory Instructor Towle

REPORT SHEET | EXPERIMENT

Electrochemical Cells and Thermodynamics | 17

	Shorthand cell designation	Temperature (°C)	E cell (measured)	ΔG (calculated)	K_{eq} (calculated)
1.	$Fe\|Fe^{2+}\|\|Cu^{2+}\|Cu$	25°C	0.778	-300.31 kJ	2.8×10^{-53}
2.	$Pb\|Pb^{2+}\|\|Cu^{2+}\|Cu$	25°C	0.480	-185.28 kJ	3.47×10^{-33}
3.	$Fe\|Fe^{2+}\|\|Pb^{2+}\|Pb$	25°C	0.298	-115.03 kJ	7.06×10^{-21}

Show calculations for $\Delta G°$ and K_{eq} for an exemplary pair. For $Pb\,|\,Pb^{2+}\,||\,Sn^{2+}\,|\,Sn$:

ΔG

1) $-(4)(96500)(0.778) = -300308$ J | $\ln K = \frac{\Delta G}{-RT} = \frac{-300308}{(8.3144)(298.15)} = e^{-121}$

2) $-(4)(96500)(0.48) = -185280$ J | 2) $-185280/((8.3144)(298.15)) = e^{-74.74}$

3) $-(4)(96500)(0.298) = -115028$ J | 3) $-115028/((8.3144)(298.15)) = e^{-46.4}$

	Half-cell equation	E half-cell (calculated)
1.	$Fe_{(s)} \rightarrow Fe^{2+} + 2e^-$	0.778 - 0.34 = 0.438
2.	$Pb_{(s)} \rightarrow Pb^{2+} + 2e^-$	0.48 - 0.34 = 0.14
3.	$Fe_{(s)} \rightarrow Fe^{2+} + 2e^-$	0.298 - 0.14 = 0.158

Effect of Temperature on Cell Potential

$-(4)(96500)(E)$

	Cell designation: E (measured)	Temperature (°C)	Temperature (K)	ΔG (calculated)
J	0.450 V	6.0	279.15	-173700 J
	0.480 V	26.0	299.15	-185280 J
	0.490 V	68.0	341.15	-189140 J
kJ	0.450 V	6.0	279.15	-173.7 kJ
	0.480 V	26.0	299.15	-185.28 kJ
	0.490 V	68.0	341.15	-189.14 kJ

ΔS determined from the slope of a plot of ΔG versus T -579 J/K, -0.579 kJ/K

$\frac{y_2-y_1}{x_2-x_1} = \frac{299.15-279.15}{-185280-(-173700)}$ flip x & y $= -11580/20 = -579$

ΔH° calculated at 298 K -357822 J/K (show calculations)

~~(ΔG+T)ΔS=H~~

$\Delta G + T(\Delta S) = -185280 + 298(-579) = -357822$ J/K

−, exo

+, endo

Is the cell reaction endothermic or exothermic? Exothermic, ΔH is negative

QUESTIONS

1. Write the net ionic equations that occur in the following cells:

   $Pb \mid Pb(NO_3)_2 \parallel AgNO_3 \mid Ag$ $Pb + 2Ag^+ \rightarrow Pb^{2+} + 2Ag$

   $Zn \mid ZnCl_2 \parallel Pb(NO_3)_2 \mid Pb$ $Zn + Pb^{2+} \rightarrow Zn^{2+} + Pb$

   $Pb \mid Pb(NO_3)_2 \parallel NiCl_2 \mid Ni$ $Pb + Ni^{2+} \rightarrow Pb^{2+} + Ni$

2. Which of the following reactions will have the larger emf under standard conditions? Why?

$$CuSO_4(aq) + Pb(s) \rightleftharpoons PbSO_4(s) + Cu(s)$$

$$Cu(NO_3)_2(aq) + Pb(s) \rightleftharpoons Pb(NO_3)_2(aq) + Cu(s)$$

This one has a larger emf because it has a lower Pb^{2+} concentration.

3. Calculate ΔG for the reaction in Example 17.3. $\Delta G = -nFE$ $n = e^-$, $F = 96{,}500$ $R = 8.3144$ $E = ?$

   $-(2)(96{,}500)(1.09\ V) = -210370\ J/K$

   or $-210.37\ kJ/K$

4. Voltages listed in textbooks and handbooks are given as *standard cell potentials* (voltages). What is meant by a standard cell? Were the cells constructed in this experiment standard cells? Why or why not?

   This means the cell has a concentration of 1 mol/L with standard temp at 25°C and a pressure of 1 atm.

5. As a standard voltaic cell runs, it "discharges" and the cell potential decreases with time. Explain.

   The oxidizer pulls electrons w/ a stronger force, which reduces the element... So, there's less of the element around to pull electrons, therefore decreasing the EMF of the cell

6. Using standard potentials given in the appendices, calculate the standard cell potentials and the equilibrium constants for the following reactions:

$$Cu(s) + 2Ag^+(aq) \rightleftharpoons Cu^{2+}(aq) + 2Ag(s)$$

$$Zn(s) + Fe^{2+}(aq) \rightleftharpoons Zn^{2+}(aq) + Fe(s)$$

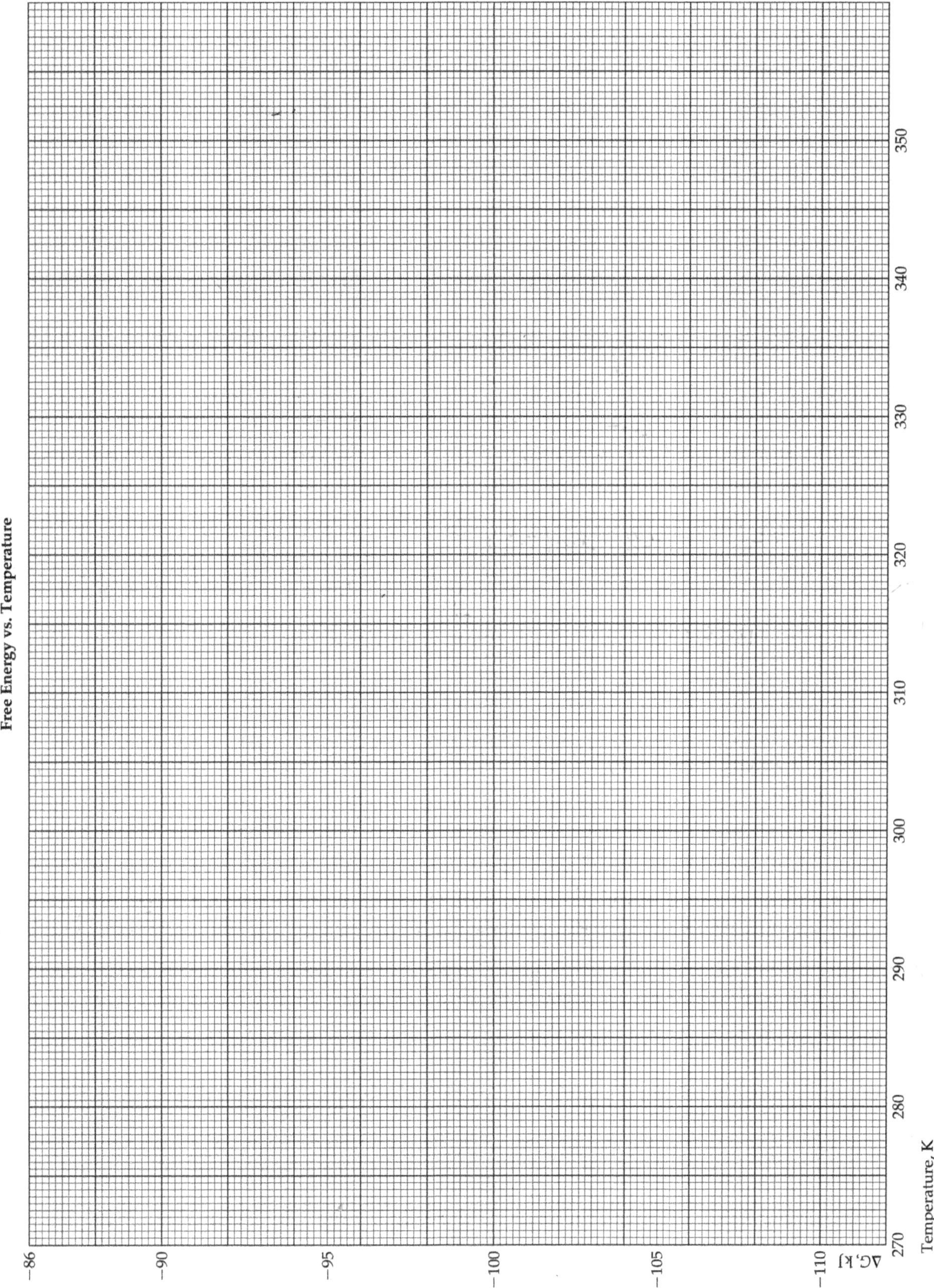

Free Energy vs. Temperature
ΔG, kJ
−86
−90
−95
−100
−105
−110
Temperature, K
270
280
290
300
310
320
330
340
350

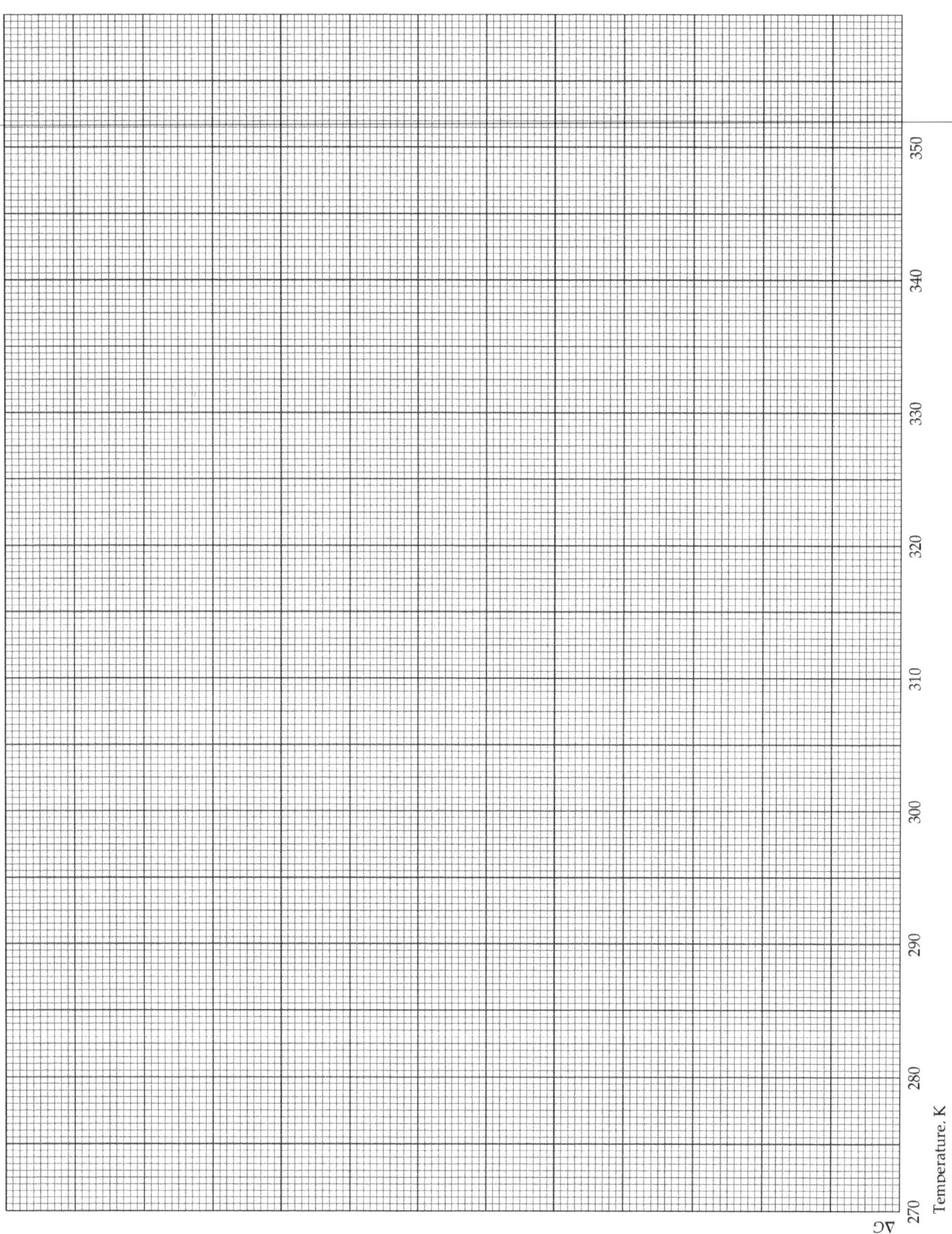
350
340
330
320
310
300
290
280
270
Temperature, K
ΔG

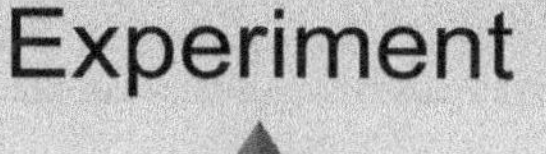

The Chemistry of Oxygen: Basic and Acidic Oxides and the Periodic Table

OBJECTIVE

To illustrate the chemistry of oxides and to become familiar with acids and bases and the concept of pH.

APPARATUS AND CHEMICALS

Apparatus

400 mL beaker	250 mL wide-mouth bottles (6)
150 mL beakers (5)	glass squares (6)
deflagration spoon	crucible and lid
pneumatic trough	pH paper
rubber tubing	Bunsen burner and hose
clay triangle	crucible tongs
no. 6 or 7 one-hole stopper	
250 mL Erlenmeyer flask	

Chemicals

3% H_2O_2*	MnO_2 (powdered)
H_3BO_3 (solid)	Na_2O_2 (solid)
ZnO (solid)	6 *M* HCl
calcium metal shavings	19 *M* NaOH
steel wool	magnesium ribbon, 10 cm
sulfur lumps	red phosphorus
Baker's yeast	charcoal pieces

WORK IN PAIRS, BUT EVALUATE YOUR DATA INDEPENDENTLY.

DISCUSSION

Most elements react with molecular oxygen under various conditions to produce oxides. These reactions are oxidation–reduction reactions and are illustrated by the examples below. The symbol "Ae" stands for any element; the formula for the oxide in general depends on the number of valence electrons of the element(Section 22.5).

$$Ae + O_2 \longrightarrow AeO_2$$

$$2Ae + O_2 \longrightarrow 2AeO$$

$$4Ae + 3O_2 \longrightarrow 2Ae_2O_3$$

Oxidation is always accompanied by a reduction reaction. In the examples above, the element Ae is oxidized while oxygen (O_2) is reduced.

*An oxygen cylinder may also be used as a source of oxygen.

Oxides of elements may be either ionic or covalent in character. In general, the more ionic oxides are formed from the elements at the left of the periodic table, whereas the elements at the right of the periodic table form covalent oxides. The ionic oxides that dissolve to any extent in water react with water to form basic solutions; for example:

$$AeO(s) + H_2O(l) \longrightarrow Ae^{2+}(aq) + 2OH^-(aq)$$

$$O^{2-}(aq) + H_2O(l) \longrightarrow 2OH^-(aq)$$

The oxide ion reacts with water to form hydroxide. If the water were evaporated from the above solution, you could obtain the base $Ae(OH)_2$. These ionic oxides are called *basic oxides* or *basic anhydrides* (base without water). For example:

$$Ba(OH)_2 - H_2O = BaO \qquad \text{(basic anhydride)}$$

The covalent oxides react with water to form acidic solutions. For example:

$$AeO(g) + H_2O(l) \longrightarrow H_2AeO_2(aq)$$

$$AeO_2(g) + H_2O(l) \longrightarrow H_2AeO_3(aq)$$

In water, these acids ionize to various degrees to furnish hydrogen ions.

$$H_2AeO_3(aq) + H_2O(l) \rightleftharpoons H_3O^+(aq) + HAeO_3^{\ -}(aq)$$

$$HAeO_3^{\ -}(aq) + H_2O(l) \rightleftharpoons H_3O^+(aq) + AeO_3^{\ 2-}(aq)$$

Thus, the covalent oxides are termed *acidic oxides* or *acid anhydrides*. You can determine the corresponding formula for the acidic or basic anhydride from the formula of a given acid or base by subtracting water from the formula to eliminate *all* hydrogen atoms. For example:

$$2HNO_2 - H_2O \longrightarrow N_2O_3$$

$$2H_3BO_3 - 3H_2O \longrightarrow B_2O_3$$

$$2NaOH - H_2O \longrightarrow Na_2O$$

To determine whether an oxide is acidic or basic, you can use litmus paper to check the water solution of the oxide. However, not all of the oxides are appreciably soluble. In those cases in which the oxides are not water soluble, you determine whether the oxides are acidic or basic by noting whether the oxide reacts with bases or acids. Acidic oxides react with bases; basic oxides, with acids.

It is not possible to classify all oxides as either acidic or basic. Some behave as both and are called *amphoteric*. Aluminum oxide is amphoteric, for it reacts with both acids and strong bases as shown by the following equations:

$$Al_2O_3(s) + 6H^+(aq) \longrightarrow 2Al^{3+}(aq) + 3H_2O(l)$$

$$Al_2O_3(s) + 3H_2O(l) + 2OH^-(aq) \longrightarrow 2Al(OH)_4^{\ -}(aq)$$

Note that the formulas $Ae(OH)_2$ and H_2AeO_2 represent the same chemical composition. By convention, you indicate bases by placing the hydroxyl group at the end of the formula. Acids are written with the acidic hydrogens at the front of the formula.

Although there are several different definitions of acids and bases, a particularly useful one for aqueous solutions was proposed in 1923 by J. N. Brønsted and T. M. Lowry (Section 16.2). In this scheme, *an acid is defined as a proton donor and a base as a proton acceptor.*

Consider the reaction of a strong acid, such as HCl, with water.

$$HCl(aq) + H_2O(l) \longrightarrow H_3O^+(aq) + Cl^-(aq) \qquad [1]$$

The HCl acts as a proton donor (acid), and the H_2O acts as a proton acceptor (base). Because HCl is a good donor of protons, the reaction essentially goes to completion in dilute solution—that is, all of the HCl is present as H_3O^+ and Cl^-. These reactions are then merely proton-transfer reactions in which the stronger base (in this case, H_2O compared with Cl^-) competes for the protons. Recall that oxidation–reduction reactions are electron–transfer reactions in which a competition for electrons is developed. The more electronegative element gains the electrons from the less electronegative element.

A weak acid, such as acetic acid, is a poor donor of protons. Such a weak acid can donate protons to water only to a limited extent, and in the resulting equilibrium, the concentrations of the undissociated reactants are much greater than the concentrations of the dissociated products. The reaction of acetic acid with water can be expressed as follows:

$$HC_2H_3O_2(aq) + H_2O(l) \rightleftharpoons H_3O^+(aq) + C_2H_3O_2^-(aq) \qquad [2]$$

This is an equilibrium, and you can write the *equilibrium constant* (K_c) expression for it (Section 15.2). Thus,

$$K_c = \frac{[H_3O^+][C_2H_3O_2^-]}{[HC_2H_3O_2][H_2O]} \qquad [3]$$

Because the concentration of molecular water is large in comparison with the concentration of all other species present, the concentration of water changes relatively little in the course of the reaction and remains essentially constant. Thus, the concentration of water can be combined with the equilibrium constant, K_c, and you obtain what is called an *acid ionization constant* or *acid dissociation constant*, K_a (Section 16.6).*

$$K_a = K_c[H_2O] = \frac{[H_3O^+][C_2H_3O_2^-]}{[HC_2H_3O_2]} \quad \text{or} \quad K_a = \frac{[H^+][C_2H_3O_2^-]}{[HC_2H_3O_3]} \qquad [4]$$

If the concentration of all of the species in Equation [4] are known, the dissociation constant, K_a, of acetic acid can be calculated.

Similarly, the weak base ammonia only partially reacts according to

$$NH_3(aq) + H_2O(l) \rightleftharpoons NH_4^+(aq) + OH^-(aq) \qquad [5]$$

for which the equilibrium constant expression is

$$K_c = \frac{[NH_4^+][OH^-]}{[NH_3][H_2O]} \qquad [6]$$

and the base dissociation constant expression is

$$K_b = K_c[H_2O] = \frac{[NH_4^+][OH^-]}{[NH_3]}$$

where K_b is called the *base dissociation constant* (Section 16.7).

---

*Note that H^+ and H_3O^+ are often used interchangeably.

The reaction of HCl, an acid, with NaOH, a base, is also an acid–base reaction. Remembering that an aqueous solution of HCl consists of H_3O^+ and Cl^- ions and that NaOH is a strong electrolyte that dissociates to produce Na^+ and OH^- ions when dissolved in water, you can write the following:

$$H_3O^+(aq) + Cl^-(aq) + Na^+(aq) + OH^-(aq) \longrightarrow 2H_2O(l) + Na^+(aq) + Cl^-(aq)$$

Here the proton is transferred from the H_3O^+ ion to the OH^- ion. Noting that the Na^+ and Cl^- ions appear as both reactants and products (that is, they are spectators and can be canceled out of the equation), you can write the following net ionic equation:

$$H_3O^+(aq) + OH^-(aq) \longrightarrow 2H_2O(l) \qquad [7]$$

Equation [7] is the net ionic equation for the reaction of any strong acid with any strong OH^--containing base. The reaction is referred to as *neutralization*.

Water is an amphoteric or amphiprotic substance—that is, it can act as either a Brønsted acid (proton donor) or a Brønsted base (proton acceptor). This is evidenced by the following reaction of water with itself:

$$H_2O(l) + H_2O(l) \rightleftharpoons H_3O^+(aq) + OH^-(aq) \qquad [8]$$

where one molecule of water donates a proton to the other. Because Equation [8] is also an equilibrium, you can write the equilibrium constant expression for this reaction as follows:

$$K_c = \frac{[H_3O^+][OH^-]}{[H_2O]^2} \qquad [9]$$

Again, because the concentration of molecular water is large in aqueous solutions when compared with the concentration of the other species present, its concentration changes very little during the course of the reaction and can be incorporated into the dissociation constant for water, known as K_w.

$$K_w = K_c[H_2O]^2 = [H_3O^+][OH^-] \text{ or } K_w = [H^+][OH^-] \qquad [10]$$

K_w is often called the ion-product constant for water. At 25 °C, $K_w = 1.0\times10^{-14}$. This relationship is essentially true for any aqueous solution at 25 °C, irrespective of the presence of other ions in solution. Hence, if you know the concentration of H_3O^+ ions in an aqueous solution, you can use Equation [10] to calculate the concentration of the OH^- ions present in the solution. In fact, because $K_w = [H_3O^+][OH^-] = 1.0\times10^{-14}$, if either $[OH^-]$ or $[H_3O^+]$ is known, the other may be calculated. As can be seen from Equation [10], because $K_w = 1.0\times10^{-14}$, the value of the H_3O^+ or OH^- concentration can be small but can *never* be equal to zero.

EXAMPLE 18.1

Calculate the $[OH^-]$ concentration in an aqueous solution where $[H_3O^+] = 3.0 \times 10^{-5}\ M$.

SOLUTION:

$$[H_3O^+][OH^-] = 1.0\times10^{-14}$$
$$[OH^-] = \frac{1.0\times10^{-14}}{[H_3O^+]} = \frac{1.0\times10^{-14}}{3.0\times10^{-5}}$$
$$= 3.3\times10^{-10}\ M$$

In a neutral solution, one where $[OH^-] = [H_3O^+]$, both concentrations are equal to 1.0×10^{-7}, or 0.00000010. The latter number and others like it are inconvenient to write. Because of this, the concept of pH was developed as a convenience in expressing the concentration of the hydronium ion in dilute solutions of acids and bases. The pH of a solution is defined as the negative logarithm of the hydronium ion concentration or hydrogen ion concentration.

$$pH = -\log[H_3O^+] \text{ or } pH = -\log[H^+] \qquad [11]$$

And consequently,

$$[H_3O^+] = 10^{-pH} \qquad [12]$$

For example, a solution containing 1.0×10^{-3} mol of HCl in 1 L of aqueous solution ($10^{-3}\ M$) contains 0.0010 M $[H_3O^+]$ and has a pH of 3.00. This pH is calculated as follows:

$$pH = -\log[H_3O^+] = -\log[1.0\times10^{-3}] = -[0 - 3.0] = 3.00$$

A review on the use of logs can be found in Appendix A of the accompanying textbook.

EXAMPLE 18.2

What is the pH of a 0.033 M HCl solution?

SOLUTION:

$$pH = -\log[H_3O^+]$$
$$= -\log[3.3\times10^{-2}]$$
$$= -(\log 3.3 + \log 10^{-2})$$
$$= -(0.52 - 2.00)$$
$$= 1.48$$

From Equation [8], you can see that in a neutral solution, the hydronium ion concentration must equal the hydroxide ion concentration. These concentrations can be calculated by letting

$$[H_3O^+] = [OH^-] = x$$

Substituting into Equation [10], you have

$$(x)(x) = x^2 = 1.0\times10^{-14}$$
$$x = 1.0\times10^{-7}\ M$$

TABLE 18.1 Important Relations in Acidic and Basic Aqueous Solutions

	$[H_3O^+]$	pH	$[OH^-]$
Acidic	$>10^{-7}$	<7	$<10^{-7}$
Neutral	10^{-7}	7	10^{-7}
Basic	$<10^{-7}$	>7	$>10^{-7}$

Then the pH of a neutral solution can be calculated as follows:

$$pH = -\log(1.0\times10^{-7}) = 7.00$$

Because an acidic solution is one in which $[H_3O^+] > [OH^-]$, $[H_3O^+] > 10^{-7}$ and $[OH^-] < 10^{-7}$.* Likewise, because a basic solution is one in which $[OH^-] > [H_3O^+]$, $[OH^-] > 10^{-7}$ and $[H_3O^+] < 10^{-7}$. These results are summarized in Table 18.1.

The hydronium ion concentration can also be calculated from the pH. For example, if a solution has a pH of 4.0 or 5.3, the hydronium ion concentration can be calculated as follows:

$$-\log[H_3O^+] = 4.0$$
$$\log[H_3O^+] = -4.0$$
$$[H_3O^+] = 1\times10^{-4}\ M$$

or

$$-\log[H_3O^+] = 5.3$$
$$\log[H_3O^+] = -5.3 = -6.0 + 0.7$$
$$[H_3O^+] = 5\times10^{-6}\ M$$

The pH of aqueous solutions may be determined two ways. One is by use of an electronic instrument having an electrode that is sensitive to hydronium ion concentration. These instruments are called pH meters. The other way is by use of organic dyes that possess two or more different pH-dependent colored forms. These dyes are called *acid–base indicators.*

Acid–base indicators derive their ability to act as pH indicators because they can exist in two forms. They are weak acids or bases themselves. The acid form possesses one color, which by loss of a proton is converted to a differently colored base form. The pertinent equilibrium for the indicator is

$$\underset{\text{(Color A)}}{HIn_A} \rightleftharpoons \underset{\text{(Color B)}}{H^+ + In_B^-} \qquad [12]$$

where HIn_A is the acid form of the indicator having color A and In_B^- is the base form of the indicator having color B. As an acid is added to a solution containing

*The symbols > and < mean "greater than" and "less than," respectively.

an indicator, Le Châtelier's principle says that the equilibrium in Equation [12] will shift to the left, the indicator will be present predominantly as HIn_A, and the solution will have color A. Likewise, if enough base is added, the indicator will be present predominantly in the form of In_B^- and the solution will have color B. Figure 18.1 shows some common indicators, with their respective colors and pH transition regions.

pH paper is paper that has been impregnated with a series of organic dyes similar to those listed in Figure 18.1. The paper changes various colors according to the pH of the solution in which it is immersed. Using pH paper, you should be able to estimate the pH of a solution to the nearest pH unit in the region pH 1 to 11.

In this experiment, you will prepare oxygen gas and a number of typical oxides of various elements. You will then dissolve these oxides in water and test the acidity or basicity of the resulting solution. You should note in particular the following key points:

1. Is there any correlation between the locations of the elements in the periodic table and the acidic and basic properties of their oxides?
2. Is there any correlation between the location of the element in the periodic table and its reactivity toward oxygen?
3. How does a catalyst affect a reaction?

PROCEDURE

A. Preparation of Oxygen

Dispose of all chemicals used in Parts A through D in the appropriate waste containers. You will prepare oxygen by the yeast-catalysed decomposition of hydrogen peroxide.

$$2H_2O_2 \xrightarrow{\text{yeast}} 2H_2O + O_2$$

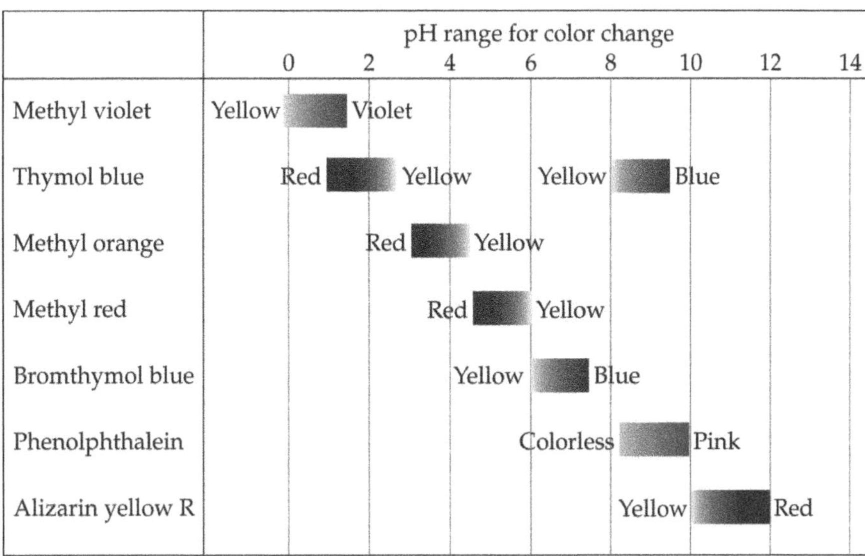

▲**FIGURE 18.1** The pH ranges for the color changes of some common acid–base indicators. Most indicators have a useful range of about 2 pH units.

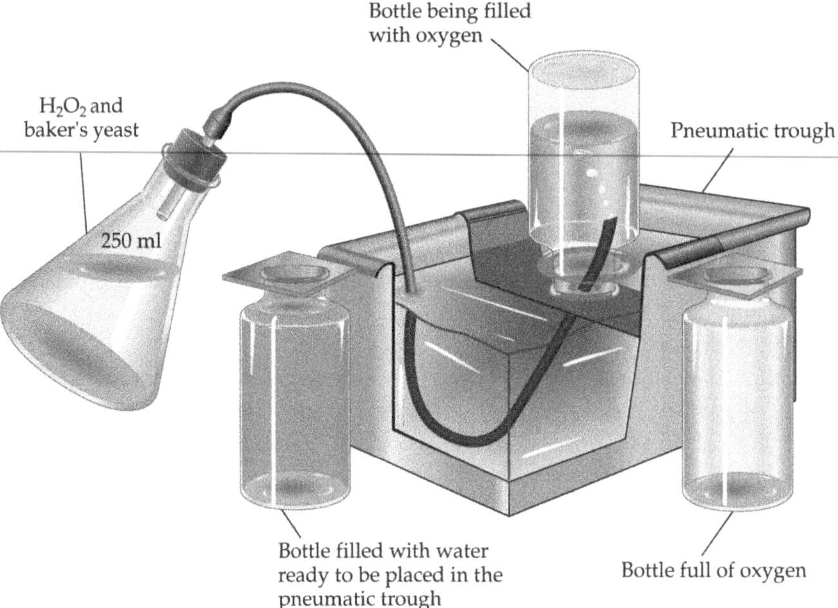

▲**FIGURE 18.2** Apparatus for oxygen production.

Completely fill six 250 mL wide-mouth bottles with tap water and cover them with glass squares. Place about 0.2 g of active dry yeast in a 250 mL Erlenmeyer flask. Assemble the apparatus as shown in Figure 18.2, but do not connect the Erlenmeyer flask to the delivery tube. Add 200 mL of 3% H_2O_2 to the Erlenmeyer flask and insert the stopper and tube into the flask. Before collecting any oxygen, allow some of the gas to escape into the atmosphere. This will sweep air from the generator and allow you to collect pure oxygen. Collect six bottles of oxygen by displacing all but about 6.35 mm of water from the bottles inverted in the pneumatic trough. Immediately cover the bottles with glass squares after they are filled with oxygen.

B. Preparation of Oxides

Prepare oxides of the elements as directed below and on the next page by burning them in the oxygen you collected. Label each bottle so that you can identify it without confusion. During the combustion, keep the bottles covered with your glass square as much as possible (Figure 18.3). Immediately after each combustion, add about 30 to 50 mL of water; then cover the bottle with the glass square, shake the bottle to dissolve the oxide, and set it aside for later tests.

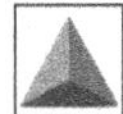

GIVE IT SOME THOUGHT

What observations indicate that an oxide formed as a product?

Magnesium *In the hood*, hold a 10 cm piece of magnesium with crucible tongs, ignite it with a Bunsen Burner, and quickly place it in a bottle of oxygen. **(CAUTION: *To avoid injury to your eyes, do not look directly at the brilliant light; always wear your safety goggles when working in the laboratory!*)**

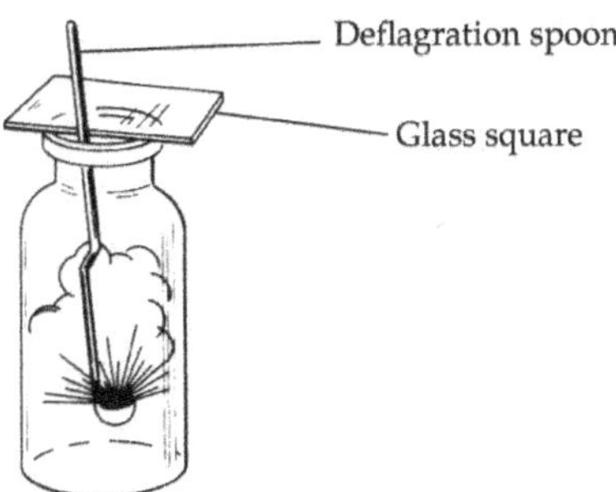

▲**FIGURE 18.3** Cover bottles with glass square as much as possible during combustion.

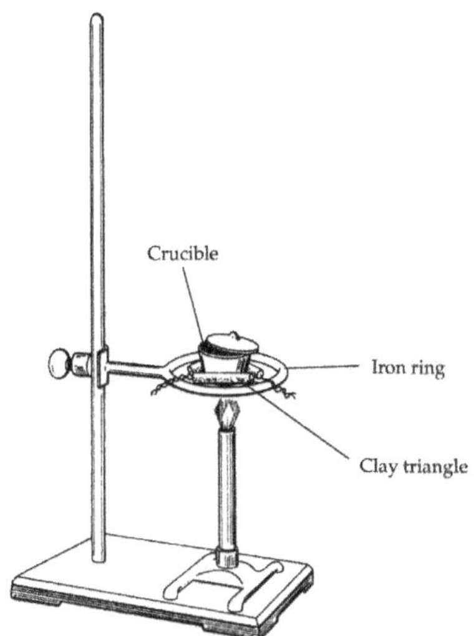

▲**FIGURE 18.4** Apparatus for ignition of calcium.

Calcium Calcium burns brilliantly but is difficult to ignite. *In the hood*, place a shaving of calcium in a crucible and ignite it in air over a Bunsen burner for 15 min using the maximum temperature possible. After the crucible has cooled, wash out the contents with 50 mL of water in a 150 mL beaker (Figure 18.4).

Iron Pour a little water into a bottle of oxygen—enough to cover the bottom. The water layer will help prevent cracking of the bottle. Ignite a small, loosely packed wad of steel wool, that has been quite spread out, by holding it with tongs in a Bunsen burner flame until it begins to glow red. Quickly thrust it into the bottle.

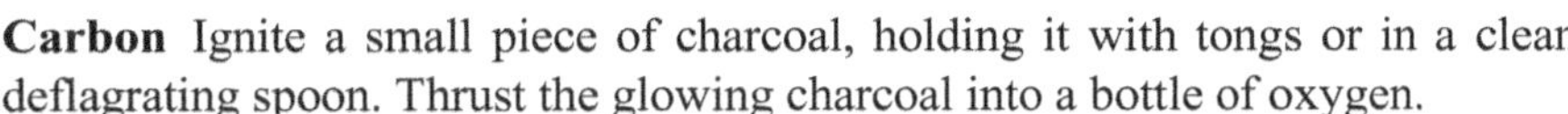

Carbon Ignite a small piece of charcoal, holding it with tongs or in a clean deflagrating spoon. Thrust the glowing charcoal into a bottle of oxygen.

Phosphorus and Sulfur DO THESE COMBUSTIONS IN THE HOOD. Clean the deflagrating spoon in the dilute HCl and rinse with water. Then heat the spoon to remove any combustible material before it is used for the combustion of each of these elements. After the spoon has cooled, add a bit (no more in size than half a pea) of sulfur or red phosphorus. **(CAUTION: *Do not touch the phosphorus with your hands; it will cause burns.*)** Ignite each with your burner and thrust the spoon into separate bottles of oxygen. After the combustion subsides, heat the spoon to burn off all remaining traces of phosphorus or sulfur; then clean the spoon by dipping it in dilute HCl.

On your report sheet, record your observations of each of the preceding combustions. Leave the spoons in the hood as directed by your instructor.

C. Reactions of Oxides with Water

Using a strip of pH paper, measure the pH of the water you used to make the preceding solutions and record your results. Then dip a 5 cm strip of pH paper into the solution in each of the bottles. Estimate the pH to the nearest unit by comparing the color of the paper to those on the pH paper container.

The oxides of elements from groups 1 and 3 of the periodic table are difficult to prepare or are extremely insoluble or unstable. Therefore, you will be provided with either the oxide or a solution of the oxide.

GIVE IT SOME THOUGHT

a. Which oxides do you expect to produce a basic solution?
b. Which oxides will produce an acidic solution?

Sodium When sodium burns in air, the peroxide, Na_2O_2, forms rather than the oxide, Na_2O. Because the peroxide is readily available and because its reaction with water is similar to that of sodium oxide, you will use it to be indicative of the group 1 elements. In a 16 cm test tube, carefully boil a very small amount (no more than the size of half a pea because larger quantities react violently) of sodium peroxide in 5 mL of distilled water for a few seconds. Be careful not to point the test tube toward yourself or another person. **(CAUTION:** ***Do not touch the*** $\mathbf{Na_2O_2}$ ***with your hands because it will cause severe burns.*****)** Cool the solution and test it with pH paper. Record your results.

Boron When the oxide of boron, B_2O_3, reacts with water, it forms a substance whose composition may be represented as either $B(OH)_3$ or H_3BO_3. Decide which of these formulas is preferred by dissolving a small amount of a substance, arbitrarily called boron hydroxide, in about 5 mL of warm distilled water. Cool the solution and test it with pH paper. Record your results. Which formula do you prefer—H_3BO_3 or $B(OH)_3$?

Neutralization Pour the sulfur oxide solution into the calcium oxide solution and test it with pH paper. How does this mixture differ from the individual solutions? Did a chemical reaction occur? If so, what was it? Now mix the phosphorus oxide and magnesium oxide solutions and test the mixture with pH paper.

Record your results on your report sheet.

D. Insoluble Oxide

Begin heating about 200 mL of water in a 400 mL beaker. Place about 0.25 g of zinc oxide, ZnO, in a 16 cm test tube and add 5 mL of distilled water. Test the pH of the mixture and record your results.

Place about 0.25 g of ZnO in another test tube and using disposable gloves, *carefully* add 5 mL of 19 *M* NaOH. **(CAUTION:** ***Do not get any of the*** **19** ***M*** **NaOH** ***on yourself. If you do, wash it off with water immediately!*****)** Place this test tube in the beaker of hot water for several minutes; then allow it to cool. Does the ZnO appear to dissolve in the NaOH solution?

Place another 0.25 g of ZnO in a test tube and add about 5 mL of 6 *M* HCl.

GIVE IT SOME THOUGHT

a. Does the ZnO react with the acid?

b. What kind of oxide is ZnO—acidic, basic, or amphoteric?

Name ______________________________ Desk ______________

Date ______________ Laboratory Instructor ______________________

The Chemistry of Oxygen: Basic and Acidic Oxides and the Periodic Table | 18 Pre-lab Questions

Before beginning this experiment in the laboratory, you should be able to answer the following questions.

1. Distinguish between ionic and covalent bonding.

2. What is the anhydride of sodium hydroxide, NaOH?

3. Define an acid and a base according to the Brønsted-Lowry definition.

4. How do K_w and K_c differ for the dissociation of water?

5. The higher the pH, the more acidic the solution. Is this true?

6. State when solutions are acidic, neutral, and basic in terms of $[OH^-]$, $[H_3O^+]$, and pH.

7. The pH of 0.1 mol/L solution of CH_3CO_2H is 2.9, whereas the pH of 0.1 mol/L solution of HCl is 1.0. What is the hydrogen ion concentration in each solution?

8. Complete and balance the following:

$$MgO(s) + H_2O(l) \longrightarrow$$

$$SO_3(g) + H_2O(l) \longrightarrow$$

9. If the pH of a solution is 5.1, what are the hydrogen and hydroxide ion concentrations?

10. What are the hydroxide and hydrogen ion concentrations of 0.012 M $Ca(OH)_2$?

11. What is the pH of a solution if the indicators methyl red and bromothymol blue impart a yellow color to the solution? Refer to Figure 18.1.

12. Which solution would have a higher pH—0.002 M H_2SO_4 or 0.2 M NaOH? Which solution is acidic? Which solution is basic?

Name ______________________________ Desk ______________

Date ______________ Laboratory Instructor ______________________

REPORT SHEET | EXPERIMENT

The Chemistry of Oxygen: Basic and Acidic Oxides and the Periodic Table

18

A. Preparation of Oxygen

1. Write the equation for the reaction by which you prepared oxygen.

B. Preparation of Oxides

Describe any changes occurring during the reaction of each element with oxygen and the properties of the products formed. Write the equation for each reaction.

	Observations	Equations	Oxidation state of element in the oxide
Mg			
Ca			
Fe			
C			
P			
S			

C. Reactions of Oxides with Water

pH of water ___________

On the line opposite its periodic group, write the formula for each oxide (or hydroxide) you studied in this experiment. Indicate the pH of its solution, the $[H_3O^+]$ and $[OH^-]$ concentrations, and the equation for the formation of the acid or base.

Group	Formula of oxide	pH	$[H_3O^+]$ M	$[OH^-]$ M	Equation for reaction
1 Na					
2 Mg					
2 Ca					
3 B					
4 C					
5 P					
6 S					
8 Fe					

Write equations for the chemical reactions that occurred between the aqueous solutions of the oxides of sulfur and calcium and between the aqueous solutions of the oxides of phosphorus and magnesium. Indicate the acids and bases in each reaction.

Acid **Base**

D. Insoluble Oxide

Compare the solubility of ZnO in H_2O, NaOH, and HCl.

What kind of oxide is ZnO?

QUESTIONS

1. Comment on the positions of the elements in the periodic table and on whether their oxides are acid or base producers.

2. Which of the following hydroxides are amphoteric—$Sn(OH)_2$, $Be(OH)_2$, $Mg(OH)_2$, and $Ca(OH)_2$?

3. Deduce and write formulas for the anhydrides of the following.

$HClO_2$	________	KOH	________	H_7SbO_6	________
H_2SO_4	________	$Ba(OH)_2$	________	HNO_3	________
H_3AsO_4	________	$Al(OH)_3$	________	H_2CO_3	________

4. If the pH of an aqueous solution is 8.23, what are the hydrogen and hydroxide ion concentrations of this solution?

5. If 3.70 g of $Ca(OH)_2$ is dissolved in sufficient water to make 100 mL of solution, what is the hydroxide ion concentration of this solution?

6. Determine the hydroxide ion concentration, the pH, and the hydrogen ion concentrations of solutions containing 15.0 g of the substances below in 1.00 L of solution. (HINT: First, write balanced equations for the reactions that occur when the oxides are dissolved in water.)

Substance	**Molarity of substance**	**$[OH^-]$**	**pH**	**$[H_3O^+]$**
CaO				
Na_2O				

7. If 50 mL of 0.50 *M* HCl solution is added to 75 mL of 0.20 *M* $Ca(OH)_2$ solution, will the solution be neutral, acidic, or basic? Write a balanced chemical equation for the reaction and justify your answer.

8. Complete and balance the following equations and tell whether you predict the oxide to be acidic, basic, or amphoteric.

$$Zn + O_2 \longrightarrow$$

$$Ga + O_2 \longrightarrow$$

$$As_4 + O_2 \longrightarrow$$

$$Li + O_2 \longrightarrow$$

Experiment 19

Freezing Point Depression

OBJECTIVE

To use the concept of freezing point depression to determine the molar mass of a compound by determination of the freezing point of a pure solvent and a solution.

APPARATUS AND CHEMICALS

Apparatus

25 × 200 mm test tube (outer)	clamp
20 × 150 mm test tube (inner)	magnetic stirring plate
two hole stopper, one side slit	top loading 0.001g balance and 0.0001 g analytical balance
thermometer, alcohol	stopwatch
wire stirring loop	spatula
800 mL beaker (×2), 250 mL beaker (×1)	weighboat
ring stand and test tube clamp	plastic funnel, 3″

Chemicals

Lauric acid	Ice
Unknown solid assigned by instructor	

WORK IN GROUPS OF TWO, DEPENDING ON EQUIPMENT AVAILABILITY, BUT ANALYZE YOUR DATA INDIVIDUALLY

DISCUSSION

Solutions are homogeneous mixtures that contain two or more substances. The major component is called the *solvent*, and the minor component is called the *solute*. Because the solution is composed primarily of solvent, physical properties of a solution resemble those of the solvent(Section 13.5). The cooling system in an automobile engine contains a mixture of primarily ethylene glycol and water. In cold climates, the low temperature freezes water into ice, yet the solution of ethylene glycol and water in the engine block and radiator doesn't freeze when exposed to the same low temperature. When a solute is dissolved into a liquid solvent, the freezing point temperature of the solution is lowered compared to that of the pure solvent due to the interactions between the solute and solvent particles. The magnitude of the freezing point depression is proportional to the number of moles, the quantity, of dissolved solute particles. It doesn't depend on the kind, identity or form of the solute particles, which could be molecules, atoms, simple monatomic ions, or polyatomic ions. The freezing point depression is a **colligative property** (Section 13.5) that depends on the collective effect of the number of solute particles (ions and/or molecules). The colligative properties include vapor pressure lowering, boiling point elevation, freezing point lowering, and osmotic pressure. Experiments have found that

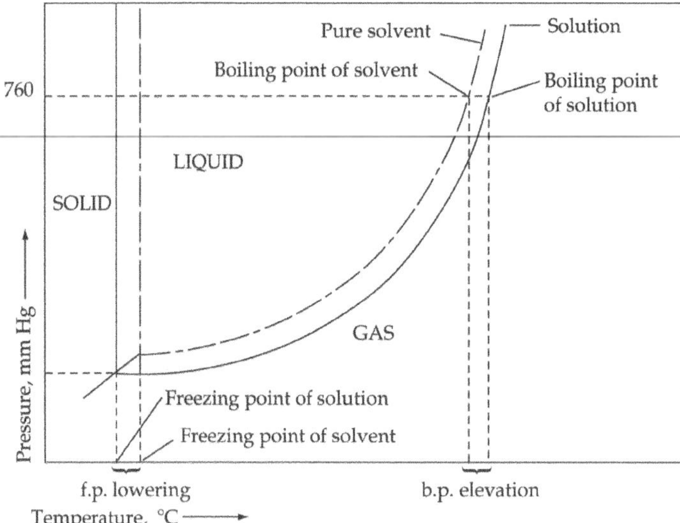

▲**FIGURE 19.1** Phase diagram for a solvent and a solution.

the dissolution of a nonvolatile solute (one with very low vapor pressure) in a solvent lowers the vapor pressure of the solvent, which in turn raises the boiling point and lowers the freezing point. This is shown graphically in the phase diagram provided in Figure 19.1.

The change in freezing point, ΔT_f, is directly proportional to the molal concentration (m) of solute in the solution, including the van't Hoff factor i, as follows

$$\Delta T_f = T_f(\text{solution}) - T_f(\text{solvent}) = -iK_f m$$

where K_f is the molal freezing point depression constant of the solvent, which has units of °C/m. Selected molal freezing point depression constants are given in Table 19.1. The molality (m) is represented by the number of moles of solute per 1 kg of solvent. The molality concentration unit is used because it is not temperature dependent. The van't Hoff factor for a nonelectrolyte is $i = 1$. For electrolytes, the value of i depends on how the solute ionizes in the solvent. A mole of sodium chloride forms two moles of ions in water, $NaCl \longrightarrow Na^+(aq) + Cl^-(aq)$, thus $i = 2$.

TABLE 19.1 Molal Freezing Point Constants

Solvent	Freezing Point (°C)	K_f (°C/m)
Acetic acid (CH_3COOH)	16.6	3.90
Acetone (C_3H_6O)	−94.9	4.04
Benzene (C_6H_6)	5.5	5.12
Chloroform ($CHCl_3$)	−63.5	4.68
Cyclohexane (C_6H_{12})	6.6	20.4
Ethyl alcohol (C_2H_5OH)	−114.6	1.99
Lauric acid ($C_{12}H_{24}O_2$)	43.2	3.90
Naphthalene ($C_{10}H_8$)	80.6	6.9
Phenol (C_6H_6O)	40.5	7.40
Water (H_2O)	0.0	1.86

GIVE IT SOME THOUGHT

What is the difference between molarity and molality?

EXAMPLE 19.1

Ethylene glyol, a nonelectrolyte with the chemical formula $C_2H_6O_2$, is a component of antifreeze. A cooling system in a vehicle holds 6.00 L of the antifreeze mixture with 25:75 volume ethylene glycol: water. The density of water is 1.00 g/cm^3 and the density of ethylene glycol is 1.11 g/cm^3. Determine the freezing point of the solution.

SOLUTION: The mixture contains 1.50 L ethylene glycol and 4.50 L water based on the given volume ratio. The molality is defined as mol solute/kg solvent. The moles of solute (ethylene glycol) are found by using the density equation to find the mass of the ethylene glycol, then find the moles of ethylene glycol.

$$\text{Mass ethylene glycol} = 1.50\ \text{L}\left(\frac{1000\ \text{mL}}{1\ \text{L}}\right)\left(\frac{1\ \text{cm}^3}{1\ \text{mL}}\right)\left(\frac{1.11\ \text{g}}{1\ \text{cm}^3}\right) = 1.67\times10^3\ \text{g}\ C_2H_6O_2$$

$$1.67\times10^3\ \text{g}\ C_2H_6O_2\left(\frac{1\ \text{mol}\ C_2H_6O_2}{62.07\ \text{g}\ C_2H_6O_2}\right) = 26.9\ \text{mol}\ C_2H_6O_2$$

The mass of the solvent (water) is found by:

$$4.50\ \text{L}\ H_2O\left(\frac{1000\ \text{mL}}{1\ \text{L}}\right)\left(\frac{1\ \text{cm}^3}{1\ \text{mL}}\right)\left(\frac{1.00\ \text{g}}{1\ \text{cm}^3}\right)\left(\frac{1\ \text{kg}}{1000\ \text{g}}\right) = 4.50\ \text{kg}\ H_2O$$

The molality of the solution is:

$$\text{molality} = \left(\frac{26.9\ \text{mol}\ C_2H_6O_2}{4.50\ \text{kg}\ H_2O}\right) = 5.98\ m$$

The freezing point depression is:

$$\Delta Tf = -iK_f m = -1(1.86\ °\text{C/m})(5.96\ \text{m}) = -11.1\ °\text{C}$$

The freezing point of pure water is 0.0 °C, thus solving $\Delta T_f = T_f(\text{solution}) - T_f(\text{solvent})$ for $T_f(\text{solution})$ gives:

$$T_f(\text{solution}) = T_f(\text{solvent}) + \Delta T_f = 0.0\ °\text{C} + (-11.1\ °\text{C}) = -11.1\ °\text{C}$$

EXAMPLE 19.2

In cold climates, salt is added to the roads to melt the ice and snow as a safety measure. A saturated solution of sodium chloride in water has a mass of 360.0 g sodium chloride per 1.00 kg of water. Determine the temperature at which the mixture will freeze.

SOLUTION: The sodium chloride is soluble in water and dissolves forming two ions in solution. The ideal van't Hoff factor (i) is 2 in this example.*

$$\text{NaCl}(s) \longrightarrow \text{Na}^+(aq) + \text{Cl}^-(aq)$$

*It should be noted that at high concentrations the van't Hoff factor deviates from the ideal integer value.

Determine the molality (mol/kg) of the solution. Find the moles of ions in solution.

$$\text{moles NaCl} = (360.0\text{ g})/(58.44\text{ g/mol}) = 6.16\text{ mol}$$

The molality of the solution is:

$$\frac{6.16\text{ mol NaCl}}{1.00\text{ kg H}_2\text{O}} = 6.16\ m$$

Similar to the previous example, the $\Delta T_f = -iK_f m = -2(1.86\ °\text{C}/m)(6.16\ m) = -22.9\ °\text{C}$

$$T_f(\text{solution}) = -22.9\ °\text{C}$$

Example 19.3

Freezing point depression experiments can be used to determine the molar mass of an unknown substance. A solution is prepared by dissolving 0.100 g of a substance in 20.0 g cyclohexane. The freezing point of the solution was lowered 1.06 °C compared to pure cyclohexane. Assuming the unknown substance is a nonelectrolyte, determine the molar mass of the solute.

SOLUTION: From $\Delta T_f = -iK_f m$, the molality of the solution may be found:

$$m = \frac{\Delta T_f}{-iK_f} = \frac{-1.06\ °\text{C}}{-1 \times 20.0\ °\text{C/m}} = 0.0530\ m$$

Find the moles of solute for the solution:

$$0.0530 \frac{\text{mol}}{\text{kg C}_6\text{H}_{12}} = \frac{\text{moles of substance}}{20.0 \times 10^{-3}\text{ kg C}_6\text{H}_{12}}$$

$$\text{moles of substance} = 0.0530 \frac{\text{mol}}{\text{kg C}_6\text{H}_{12}} \times 20.0 \times 10^{-3}\text{ kg C}_6\text{H}_{12} = 1.06 \times 10^{-3}\text{ mol}$$

Determine the molar mass of the solute:

$$\text{molar mass} = \frac{0.100\text{ g}}{1.06 \times 10^{-3}\text{ mol}} = 94.3\text{ g/mol}$$

In this experiment, you will determine the molar mass of an unknown. You will do this by determining the freezing point depression of a lauric acid solution having a known added mass of your unknown. The freezing temperature is difficult to ascertain by direct visual observation because of a phenomenon called supercooling and because solidification of solutions usually occurs over a broad temperature range. Temperature-time graphs, called cooling curves, reveal freezing temperatures rather clearly.

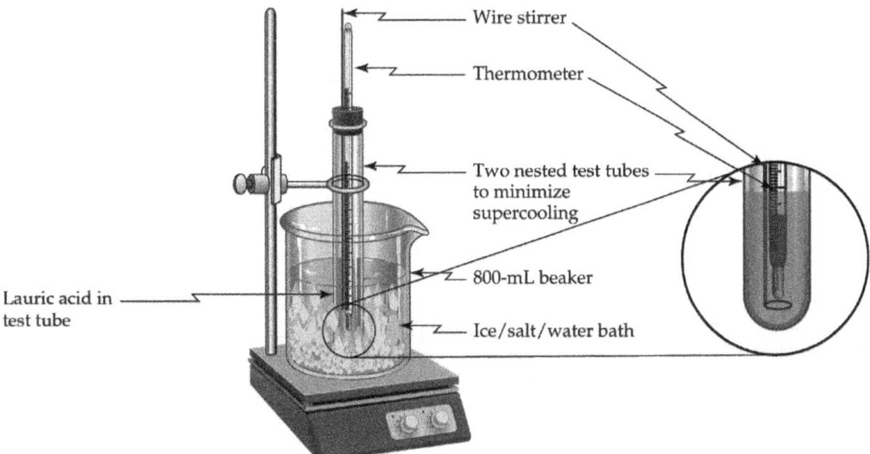

▲**FIGURE 19.2** Apparatus for determination of cooling curve.

PROCEDURE

A. Preparation of Experimental Apparatus and Freezing Point Determination of Lauric Acid

Prepare a water bath with a sufficient liquid level, about 650 mL in the 800 mL beaker, to cover the bottom one half of the test tube when it is placed into the water bath. Begin heating the beaker with the water while stirring with a stir bar. Maintain the water bath temperature near 70–80 °C and avoid boiling. Add about 300 mL of tap water and some ice cubes in a second cooling 400 mL beaker.

Construct the freezing point depression experimental apparatus similar to the one shown in Figure 19.2. Ensure the test tube, two-hole stopper, stirrer, and thermometer are clean and dry. Assemble the experimental apparatus by placing the stirrer through one hole of the stopper and the thermometer through the side slit of the stopper. Determine the mass of the freezing point depression apparatus on a top loading balance. Obtain approximately 10 g of the solid lauric acid in a weighboat, and then transfer into the inner test tube with the aid of a funnel. Insert, but don't force, the two-hole stopper with stirrer and thermometer into the test tube, and then determine the mass of the apparatus with the lauric acid.

Clamp the test tube and place it in the heated water bath, ensuring that the bottom of the test tube is not touching the beaker or stir bar. Melt all of the lauric acid and ensure that the bulb of the thermometer is below the liquid level of the lauric acid. Continue heating, with intermediate stirring, until the temperature is near 50 °C. Remove the apparatus from the water bath and clamp higher on the ring stand.

GIVE IT SOME THOUGHT

Why is it necessary to stir the solution while cooling?

Continuously stir the lauric acid with the stirring wire loop as it cools. After it cools to 50 °C, with continuous stirring, read and record the temperature to the nearest 0.2 °C every 30 seconds for the next 5 minutes. Next, lower the beaker into the cooled water and continue to stir, if the lauric acid is not completely solid, then record the temperature every 30 seconds for at least another 8 minutes. Repeat the experiment for a second trial with the same lauric acid by heating until fully liquefied, then cooling and recording the temperatures in the same manner as the first trial.

GIVE IT SOME THOUGHT

What valuable information does a temperature-time graph give you for this experiment?

B. Freezing Point Determination of a Lauric Acid Solution

Dry the cooled apparatus from part A and determine the mass. Obtain a designation of an unknown from your lab instructor. Obtain about 1 gram of the specified unknown, then transfer to the test tube. Measure the mass of the apparatus with the lauric acid and unknown. Clamp the apparatus in the water bath and melt the mixture of the two solids. Stir the completely melted solution for several minutes to ensure adequate mixing to form a homogeneous solution. In some cases, there may be small crystals remaining after stirring the liquid solution, but even if these remain for several minutes you may continue with the cooling procedure. Remove the test tube from the hot water bath and record the temperature to the nearest 0.2 °C every 30 seconds under constant stirring. Cool in air for 5 minutes, and then cool in the cold water bath for an additional 5 minutes. Add an additional ~0.5 g, precisely determine the mass, of the same unknown. Conduct the experiment again with the lauric acid and unknown by heating until fully liquefied, then cooling and recording the temperatures in the same manner as the previous trial.

C. Waste Disposal and Cleanup

At the completion of the experiment, melt the lauric acid solution and pour into a waste container. Thoroughly clean the glassware, stir wire, and thermometer by scrubbing with a brush while washing with soap and hot water in the sink.

GIVE IT SOME THOUGHT

a. What range do you expect to see for the molar mass of your unknown?

b. Would you expect the molar mass to be 1×10^{-3} g/mol? Why or why not?

c. Would you expect the molar mass to be 1×10^{9} g/mol? Why or why not?

CALCULATIONS

For each of the four freezing point determination trials of the pure solvent and solution, individually plot cooling curve graphs of the temperature (as the ordinate, vertical axis) versus time in seconds (as the abscissa, horizontal axis). The freezing point of the pure lauric acid solution may be determined by the intersection of the linear regions of the cooling portion of the liquid region and the freezing region, see Figure 19.3. It may be necessary to prepare a second graph of each with a narrower x-axis and y-axis range. Ensure the completely solid region, with a rapid temperature decrease, is not included in the freezing region that contains both liquid and solid components. Determine the average freezing point from each pure solvent experiment. Determine the freezing point from each solution trial.

Determine the ΔT_f for the solution using $T_f(\text{solution}) - T_f(\text{solvent})$. Calculate the molality of the solution using the freezing point depression constant (K_f) for lauric acid. Determine the moles of unknown solute using the freezing point depression equation and assuming the van't Hoff factor is equal to 1 for a nonelectrolyte organic compound. The mass of the solvent is the mass of the lauric acid, measured in the first trial, in units of kilograms. Calculate the experimental molar mass using the mass of the unknown organic compound added as the solute and the moles of unknown for experiments 3 and 4. Ensure correct significant figures and units are used in all calculations.

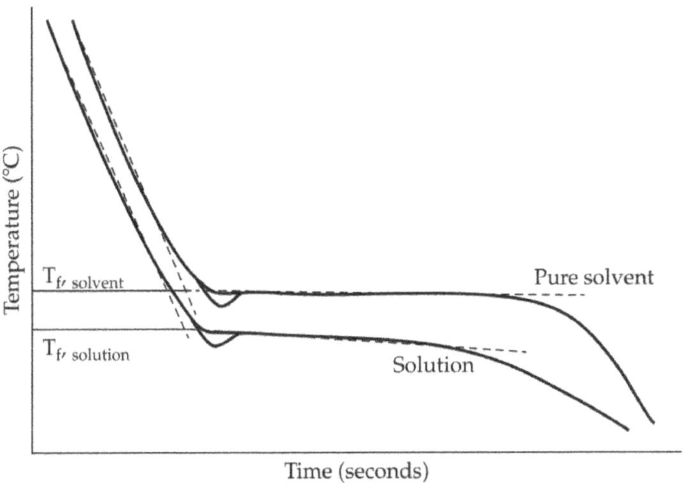

▲**FIGURE 19.3** Cooling curve for a pure solvent and solution.

NOTES AND CALCULATIONS

Name ______________________________ Desk ______________

Date ______________ Laboratory Instructor ______________________

Freezing Point Depression | 19 Pre-lab Questions

Before beginning this experiment in the laboratory, you should be able to answer the following questions.

1. What is a solution?

2. Write the expression for the freezing point depression.

3. List three colligative properties and suggest a rationale for the choice of the word colligative to describe these properties.

4. Distinguish between volatile and nonvolatile substances.

5. What effect does the presence of a nonvolatile solute have upon the freezing point of a solution?

6. Explain the difference between molarity and molality. What are the units on each?

7. What is the freezing point (°C) of a solution prepared by dissolving 15.6 g of $Al(NO_3)_3$ in 150 g of water? The molal freezing point depression constant for water is 1.86 °C/m. Assume complete dissociation of the $Al(NO_3)_3$.

---

8. What is the molality of a solution that contains 7.4 g of urea (molar mass = 60 g/mol) in 300 g of benzene (C_6H_6)?

9. Calculate the freezing point of a solution containing 1.25 g of benzene (C_6H_6) in 100 g of chloroform ($CHCl_3$).

10. A solution containing 0.050 g of an unknown nonelectrolyte in 2.50 g of cyclohexane was found to freeze at 5.1 °C. What is the molar mass of the unknown substance?

11. How many grams of $NaNO_3$ would you add to 200 g H_2O to prepare a solution that is 0.200 *m* in $NaNO_3$?

Name ______________________________ Desk ______________

Date ______________ Laboratory Instructor ______________________

REPORT SHEET | EXPERIMENT

Freezing Point Depression | 19

A. Preparation of Experimental Apparatus and Freezing Point Determination of Lauric Acid

1. Mass of freezing point depression apparatus ______________
2. Mass of freezing point depression apparatus with lauric acid ______________
3. Mass of lauric acid used ______________

B. Freezing Point Determination of a Lauric Acid Solution

4. Designation of unknown ______________
5. Mass of added unknown, Exp. 3 ______________
6. Mass of added unknown, Exp. 4 ______________

Freezing point of lauric acid from Part A, Exp. 1 ______________

Freezing point of lauric acid from Part A, Exp. 2 ______________

Average freezing point of lauric acid ______________

Freezing point of lauric acid and unknown from Part B, Exp. 3 ______________

Freezing point of lauric acid and unknown from Part B, Exp. 4 ______________

$\Delta T_f = T_{f,mixture} - T_{f,pure}$ Exp. 3 ______________ Exp. 4 ______________

Solution molality (show calculations) Exp. 3 ______________ Exp. 4 ______________

Moles of unknown (show calculation) Exp. 3 ______________ Exp. 4 ______________

Molar mass of unknown (show calculation) Exp. 3 ______________ Exp. 4 ______________

Average molar mass ______________

Show your calculations for the solution molality, mole of unknown, and molar mass of unknown.

Hand in your cooling curves with your report sheet.

Pure lauric acid

Exp. 1		Exp. 2	
Time (seconds)	**Temperature (°C)**	**Time (seconds)**	**Temperature (°C)**
0		0	
30		30	
60		60	
90		90	
120		120	
150		150	
180		180	
210		210	
240		240	
270		270	
300		300	
330		330	
360		360	
390		390	
420		420	
450		450	
480		480	
510		510	
540		540	
570		570	
600		600	
630		630	
660		660	
690		690	
720		720	
750		750	
780		780	

Solution of lauric acid and unknown (data are for camphor)

Exp. 3		Exp. 4	
Time (seconds)	**Temperature (°C)**	**Time (seconds)**	**Temperature (°C)**
0	______	0	______
30	______	30	______
60	______	60	______
90	______	90	______
120	______	120	______
150	______	150	______
180	______	180	______
210	______	210	______
240	______	240	______
270	______	270	______
300	______	300	______
330	______	330	______
360	______	360	______
390	______	390	______
420	______	420	______
450	______	450	______
480	______	480	______
510	______	510	______
540	______	540	______
570	______	570	______
600	______	600	______
630	______	630	______
660	______	660	______
690	______	690	______
720	______	720	______
750	______	750	______
780	______	780	______

QUESTIONS

1. What are the major sources of error in this experiment?

2. Suppose throughout the experiment, your thermometer consistently read a temperature 1.2 °C lower than the correct temperature. How would this have affected the molar mass you found?

3. At equal concentrations, would a nonelectrolyte (e.g. glucose) or electrolyte (e.g. NaCl) containing solution have a lower freezing point? Why?

4. Arrange the following liquids in order of increasing freezing point (lowest to highest temperature): pure H_2O, aqueous NaF (0.31 *m*), aqueous glucose (0.60 *m*), aqueous sucrose (0.50 *m*), aqueous MgI_2 (0.22 *m*)

5. If the freezing point of the solution had been incorrectly read 0.3 °C lower than the true freezing point, would the calculated molar mass of the solute have been too high or too low? Explain your answer.

6. What mass of NaCl is dissolved in 200 g of water in a 0.100 *m* solution?

7. Calculate the molalities of some commercial reagents from the following data:

	HCl(*aq*)	**NH_3(*aq*)**
Formula weight (g/mol)	36.465	17.03
Density of solution (g/mL)	1.19	0.90
Weight %	37.2	28.0
Molarity	12.1	14.8

8. A solution containing 1.00 g of an unknown nonelectrolyte liquid and 9.00 g water has a freezing point of −3.33 °C. The K_f = 1.86 °C/m for water. Calculate the molar mass of the unknown liquid in g/mol.

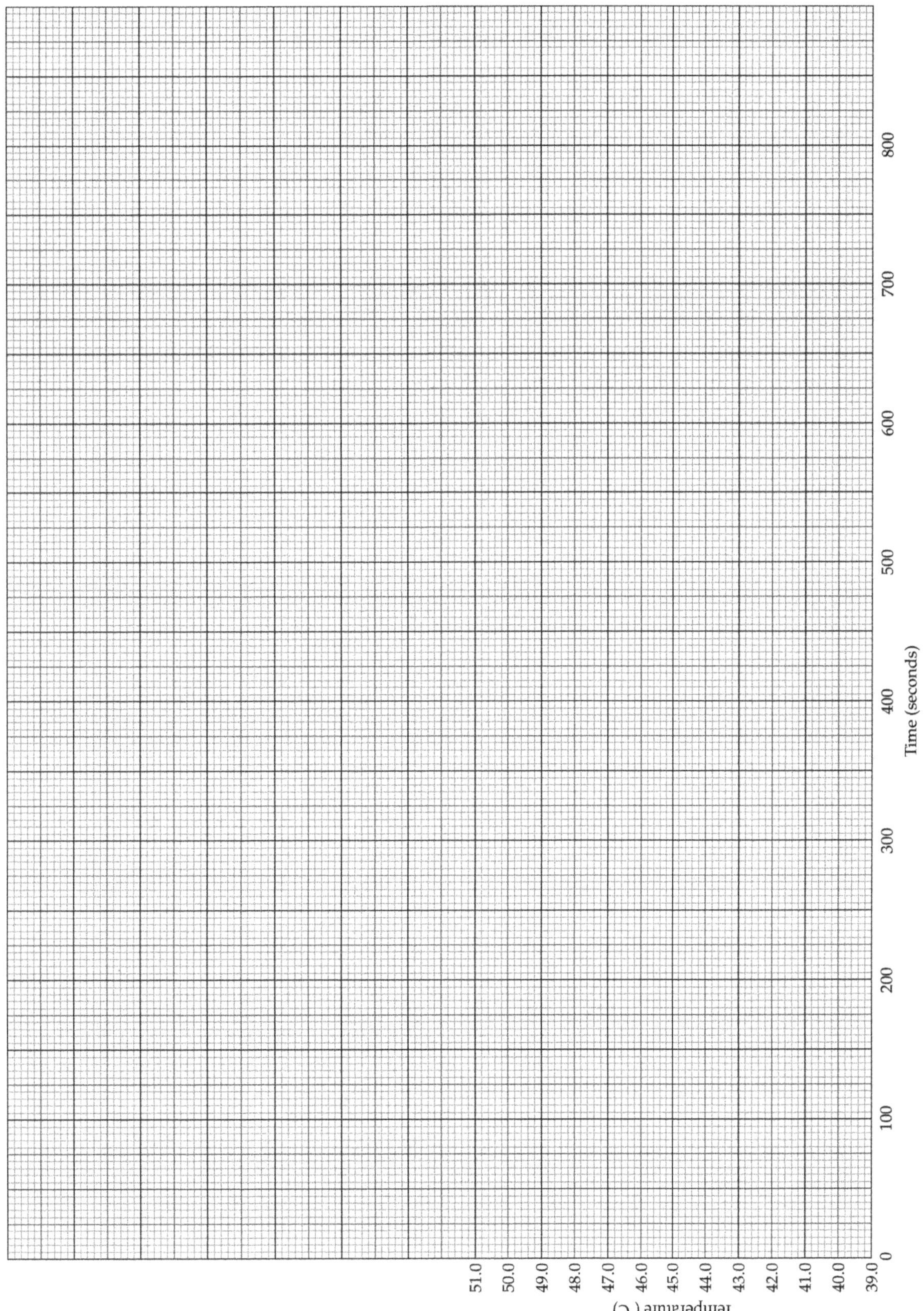
Time (seconds)
0
100
200
300
400
500
600
700
800
Temperature (°C)
39.0
40.0
41.0
42.0
43.0
44.0
45.0
46.0
47.0
48.0
49.0
50.0
51.0

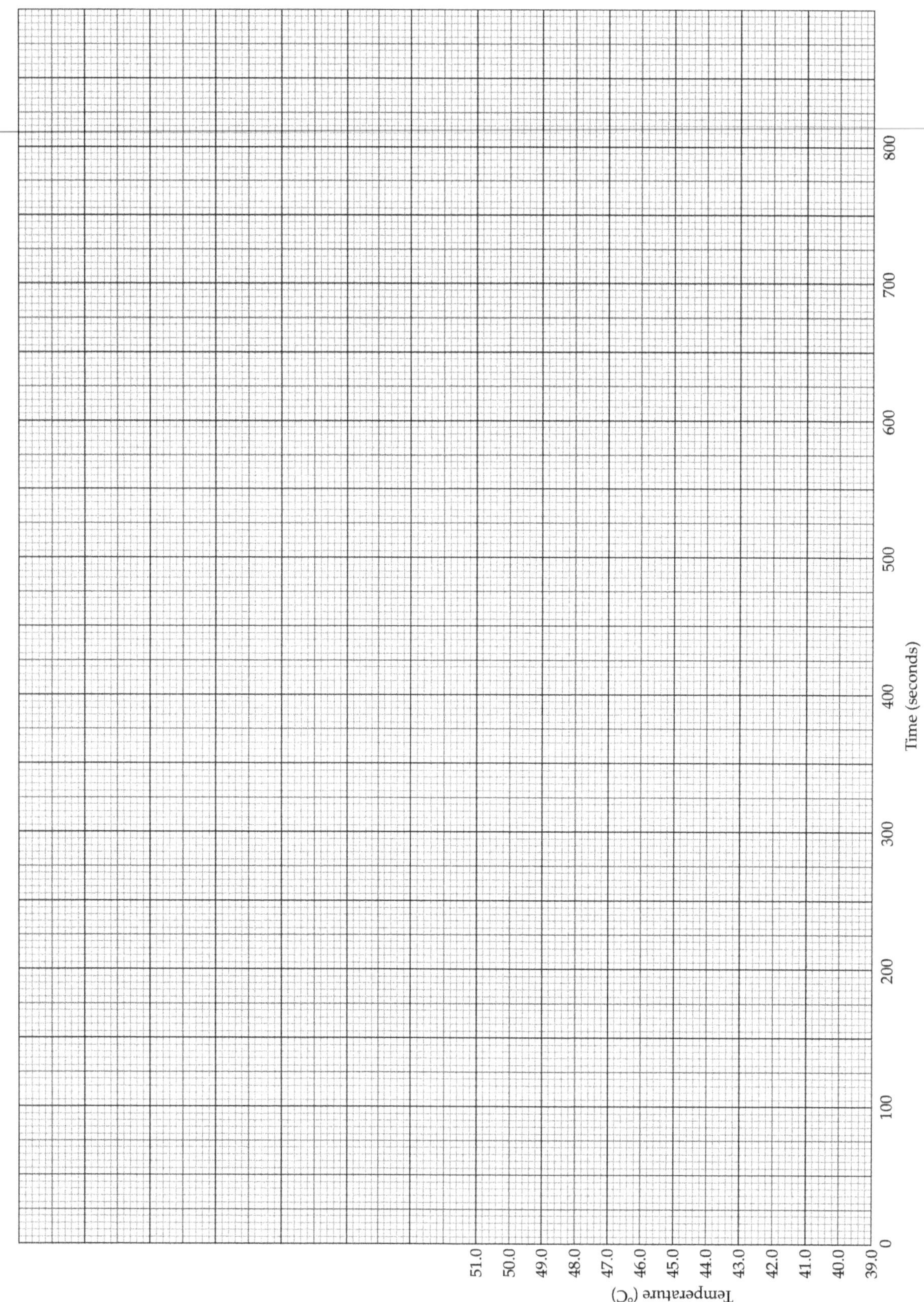

Time (seconds)
0
100
200
300
400
500
600
700
800
Temperature (°C)
39.0
40.0
41.0
42.0
43.0
44.0
45.0
46.0
47.0
48.0
49.0
50.0
51.0

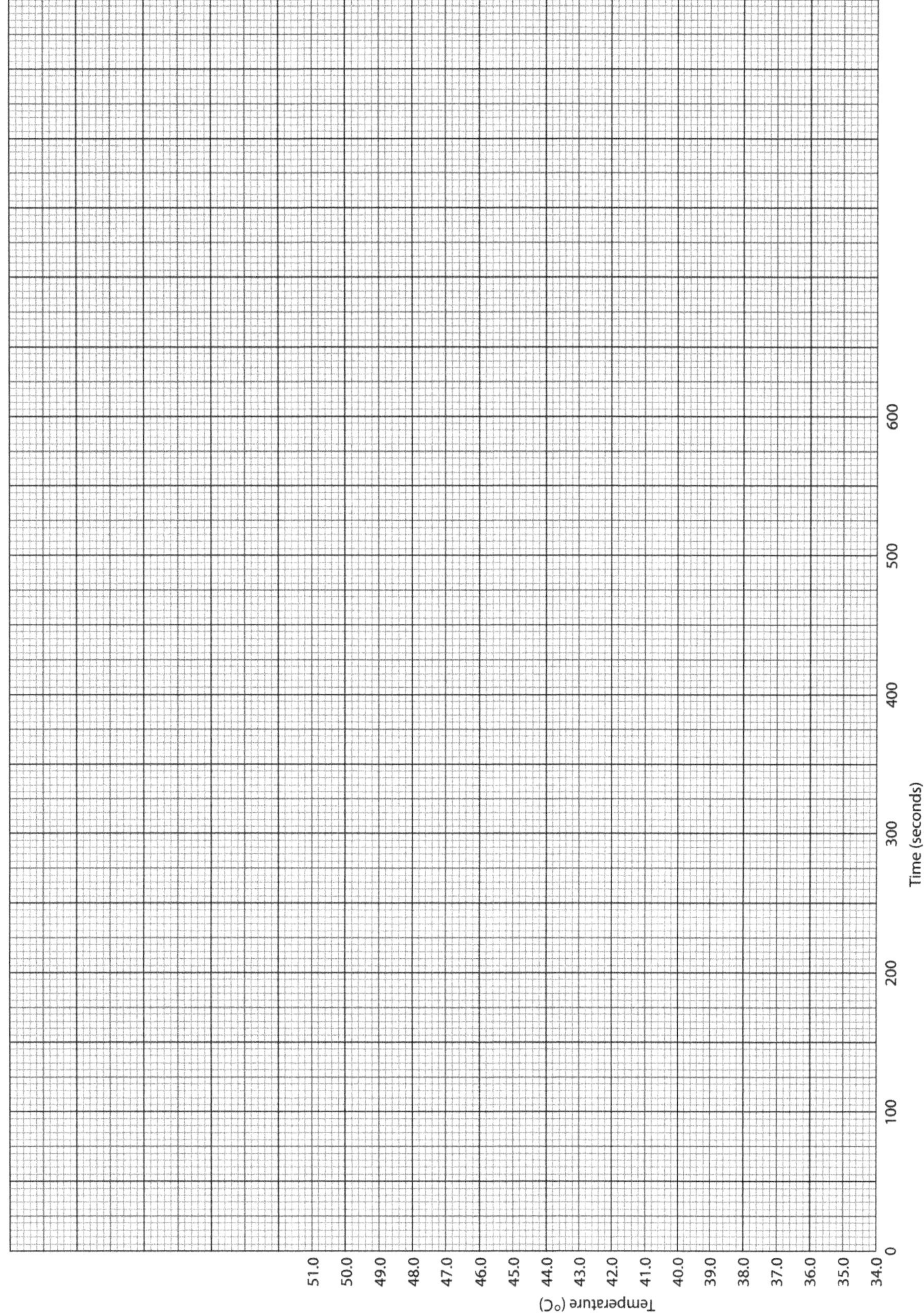
Temperature (°C)
34.0
35.0
36.0
37.0
38.0
39.0
40.0
41.0
42.0
43.0
44.0
45.0
46.0
47.0
48.0
49.0
50.0
51.0
Time (seconds)
0
100
200
300
400
500
600

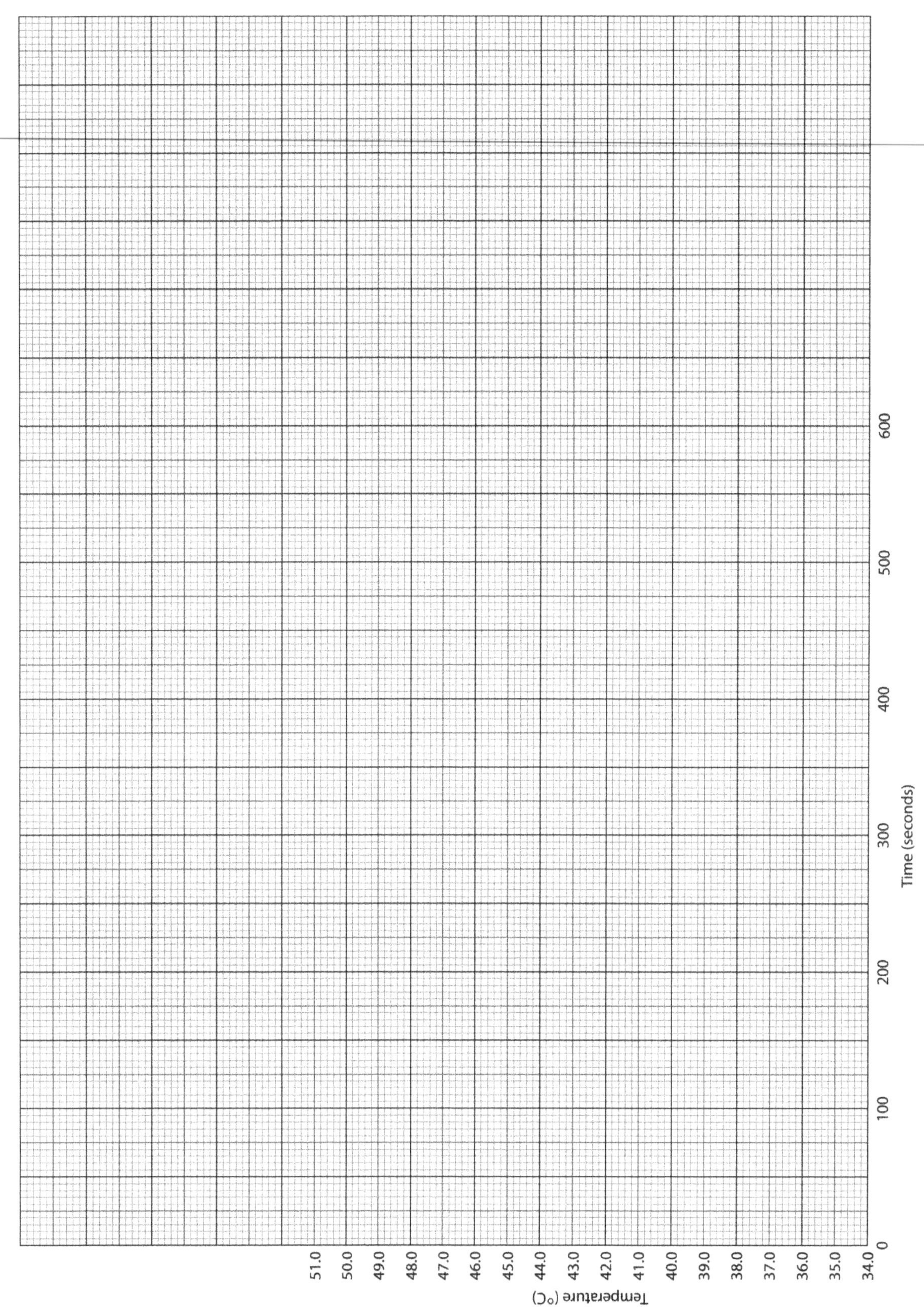

Time (seconds)
0
100
200
300
400
500
600
Temperature (°C)
34.0
35.0
36.0
37.0
38.0
39.0
40.0
41.0
42.0
43.0
44.0
45.0
46.0
47.0
48.0
49.0
50.0
51.0

Experiment

20

Titration of Acids and Bases

OBJECTIVE

To become familiar with the techniques of titration, a volumetric method of analysis; to determine the amount of acid in an unknown.

APPARATUS AND CHEMICALS

Apparatus

50 mL buret
600 mL beaker
500 mL Erlenmeyer flask
250 mL Erlenmeyer flasks (3)
weighing bottle
ring stand and ring
balance
Bunsen burner and hose
500 mL plastic bottle with plastic lid
wash bottle
buret clamp and ring stand
wire gauze

Chemicals

19 *M* NaOH
potassium hydrogen phthalate (KHP, primary standard)
phenolphthalein solution
unknown acid

DISCUSSION

One of the most common and familiar reactions in chemistry is that of an acid with a base (Section 4.3). This reaction is termed *neutralization*, and the essential feature of this process in aqueous solution is the combination of hydronium ions with hydroxide ions to form water.

$$H_3O^+(aq) + OH^-(aq) \longrightarrow 2H_2O(l)$$

In this experiment, you will use this reaction to determine accurately the concentration of a sodium hydroxide solution you have prepared. The process of determining the concentration of a solution is called *standardization*. Then you will measure the amount of acid in an unknown. To do this, using a buret, you will accurately measure, the volume of your standard base that is required to exactly neutralize the acid present in the unknown. The technique of accurately measuring the volume of a solution required to react with another reagent is termed *titration* (Section 4.6).

An indicator solution is used to determine when an acid has exactly neutralized a base and vice versa. A suitable indicator changes colors when equivalent amounts of acid and base are present. The color change is termed the *end point* of the titration. Indicators change colors at different pH values. Phenolphthalein, for example, changes from colorless to pale pink at a pH of about 9. In slightly more acidic solutions, it is colorless, whereas in more alkaline solutions, it is deep pink.

In this experiment, your solution of NaOH will be standardized by your titrating it against a very pure sample of potassium hydrogen phthalate, $KHC_8H_4O_4$, of known mass. Potassium hydrogen phthalate (often abbreviated KHP) has one

KHP

acidic hydrogen. Its structure is shown here. It is called a primary standard and can be obtained in very pure form which is stable. It is a monoprotic acid with the acidic hydrogen bonded to oxygen and has a molar mass of 204.23 g/mol.

The balanced equation for the neutralization of KHP is given in Equation [1].

$$KHC_8H_4O_4(aq) + NaOH(aq) \longrightarrow H_2O(l) + KNaC_8H_4O_4(aq) \qquad [1]$$

In the titration of the base NaOH against KHP, an equal number of moles of KHP and NaOH are present at the equivalence point. In other words, at the equivalence point

$$\text{moles NaOH} = \text{moles KHP} \qquad [2]$$

The point at which stoichiometrically equivalent quantities are brought together is known as the *equivalence point* of the titration.

GIVE IT SOME THOUGHT

a. What stoichiometrically equivalent quantities are present at the equivalence point in this experiment?

b. How do these quantities relate to concentration?

Note that the equivalence point in a titration is a theoretical point. You can estimate it by observing some physical change associated with the condition of equivalence, such as the change in color of an indicator, which as stated previously, is termed the end point. The equivalence point and end point should closely correspond.

The most common way of quantifying concentrations is molarity (symbol M), which is defined as the number of moles of solute per liter of solution, or the number of millimoles of solute per milliliter of solution.

$$M = \frac{\text{moles solute}}{\text{volume of solution in liters}} = \frac{10^{-3}\text{ mole}}{10^{-3}\text{ liter}} = \frac{\text{mmol}}{\text{mL}} \qquad [3]$$

From Equation [3], the moles of solute (or mmol solute) are related to the molarity and the volume of the solution as follows:

$$M \times \text{liters} = \text{moles solute and } M \times \text{mL} = \text{mmol solute} \qquad [4]$$

Thus, if you measure the volume of the base, NaOH, required to neutralize a known mass of KHP, you may be able to calculate the molarity of the NaOH solution.

EXAMPLE 20.1

Calculate the molarity of a solution that is made by dissolving 16.7 g of sodium sulfate, Na_2SO_4, in enough water to form 125 mL of solution.

SOLUTION:

$$\text{molarity} = \frac{\text{moles } Na_2SO_4}{\text{liters solution}}$$

Using the molar mass of Na_2SO_4, calculate the number of moles of Na_2SO_4.

$$\text{moles } Na_2SO_4 = (16.7 \text{ g } Na_2SO_4)\left(\frac{1 \text{ mol } Na_2SO_4}{142.04 \text{ g } Na_2SO_4}\right) = 0.118 \text{ mol } Na_2SO_4$$

Changing the volume of the solution to liters results in the following:

$$125 \text{ mL} \times (1 \text{ L}/1000 \text{ mL}) = 0.125 \text{ L}$$

Thus, the molarity is

$$\text{molarity} = \frac{0.118 \text{ mol } Na_2SO_4}{0.125 \text{ L}} = 0.941\ M\ Na_2SO_4$$

EXAMPLE 20.2

What is the molarity of a NaOH solution if 35.75 mL is required to neutralize 1.070 g of KHP?

SOLUTION: Recall from Equation [2] that at the equivalence point, the number of moles of NaOH equals the number of moles of KHP.

$$\text{moles KHP} = (1.070 \text{ g KHP})\left(\frac{1 \text{ mol KHP}}{204.23 \text{ g KHP}}\right) = 5.240 \times 10^{-3} \text{ mol KHP}$$

Because this is the exact number of moles of NaOH contained in 35.75 mL of solution, its molarity is

$$\text{molarity} = \frac{5.240 \times 10^{-3} \text{ mol NaOH}}{0.03575 \text{ L}} = 0.1466\ M \text{ NaOH}$$

Once the molarity of the NaOH solution is known, the base can be used to determine the amount of KHP or any other acid present in a known mass of an impure sample. The percentage of KHP in an impure sample is

$$\%\text{ KHP} = \frac{\text{g KHP}}{\text{mass of sample}} \times 100\%$$

In this experiment, an acid–base indicator, phenolphthalein, is used to signal the end point in the titration. This indicator was chosen because its color change coincides so closely with the equivalence point.

EXAMPLE 20.3

What is the percentage of KHP in an impure sample of KHP that has a mass of 2.537 g and requires 32.77 mL of 0.1466 M NaOH to neutralize it?

SOLUTION: The number of grams of KHP in the sample must be determined first. Remember that at the equivalence point, the number of millimoles of NaOH equals the number of millimoles of KHP.

$$\text{mmol NaOH} = (32.77 \text{ mL NaOH})(0.1466 \text{ mmol NaOH/mL NaOH})$$
$$= 4.804 \text{ mmol NaOH}$$

Thus, there are 4.804 mmol of KHP in the sample, which corresponds to the following number of grams of KHP in the sample:

$$\text{grams KHP} = 4.804 \text{ mmol KHP}\left(\frac{1 \text{ mol KHP}}{1000 \text{ mmol KHP}}\right)\left(\frac{204.2 \text{ g KHP}}{1 \text{ mol KHP}}\right)$$
$$= 0.9810 \text{ g KHP}$$

Therefore,

$$\% \text{ KHP} = \frac{0.9810 \text{ g}}{2.537 \text{ g}} \times 100\% = 38.67\%$$

PROCEDURE

Preparation of Approximately 0.100 *M* Sodium Hydroxide (NaOH) Heat 500 mL of distilled water to boiling in a 600 mL flask;* *after cooling under the water tap*, transfer to a 500 mL plastic bottle with a plastic lid. **(CAUTION: *Do not get any of the 19 M* NaOH *on yourself. If you do, immediately wash the area with copious amounts of water.*)** Add 3 mL of stock solution of bicarbonate-free NaOH (approximately 19 *M*) and shake vigorously for at least 1 min. The bottle should be stoppered to protect the NaOH solution from CO_2 in the air.

Preparation of a Buret for Use Clean a 50 mL buret with soap solution and a buret brush and thoroughly rinse with tap water. Then rinse with at least five 10 mL portions of distilled water. The water must run freely from the buret without leaving any drops adhering to its sides. Make sure the buret does not leak and the stopcock turns freely.

Reading a Buret All liquids, when placed in a buret, form a curved meniscus at their upper surface. In the case of water or water solutions, this meniscus is concave (Figure 20.1), and the most accurate buret readings are obtained by observing on the graduated scales the position of the lowest point on the meniscus.

To avoid parallax errors when taking readings, your eye must be level with the meniscus or use a meniscus reader. Wrap a strip of paper around the buret and hold the top edges of the strip together evenly. Adjust the strip so that the front and back edges are in line with the lowest part of the meniscus and take the reading by estimating to the nearest tenth of a marked division (0.01 mL). A simple way to do this for repeated readings on a buret is illustrated in Figure 20.1.

GIVE IT SOME THOUGHT

When you are taking a proper reading from this buret, where should your eye level be?

*The water is boiled to remove carbon dioxide (CO_2), which would react with the NaOH and change its molarity. $NaOH(aq) + CO_2(g) \longrightarrow NaHCO_3(aq)$

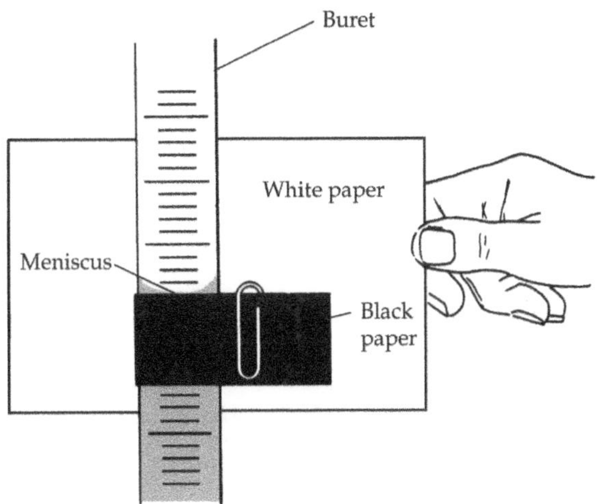

▲**FIGURE 20.1** Reading a buret.

A. Standardization of Sodium Hydroxide (NaOH) Solution

Prepare about 400 to 450 mL of CO_2-free water by boiling for about 5 min. To save time, make an additional 400 mL of CO_2-free water for Part B by boiling it now. Using a weighing bottle determine the mass of (your lab instructor will show you how to use a weighing bottle if you don't already know) triplicate samples of between 0.4 and 0.6 g each of pure potassium hydrogen phthalate (KHP) into three separate 250 mL Erlenmeyer flasks; accurately determine the mass to four significant figures.* Do not determine the mass of the flasks. Record the masses and label the three flasks so that you can distinguish among them. Add to each sample about 100 mL of distilled water that has been freed from CO_2 by boiling and warm gently with swirling until the salt is completely dissolved. Add to each flask two drops of phenolphthalein indicator solution. Cover and allow to cool before titrating the samples.

Rinse the previously cleaned buret with at least four 5 mL portions of the approximately 0.100 *M* NaOH solution that you prepared. Discard each portion into the designated receptacle. *Do not return any of the washings to the bottle.* Completely fill the buret with the solution and remove the air from the tip by dispensing some of the liquid into an empty beaker. Make sure the lower part of the meniscus is at the zero mark or slightly lower. Allow the buret to stand for at least 30 s before you read the exact position of the meniscus. Remove any hanging drop from the buret tip by touching it to the side of the beaker used for the washings. Record the initial buret reading on your report sheet.

Slowly add the NaOH solution to one of your flasks of KHP solution while gently swirling the contents of the flask, as illustrated in Figure 20.2. As the NaOH solution is added, a pink color appears where the drops of the base come in contact with the solution. This coloration disappears as you swirl the contents of the flask. As the end point is approached, the color disappears

*In cases where the mass of a sample is larger than 1 g, it is necessary to determine the mass only to the nearest milligram to obtain four significant figures. Buret readings can be read only to the nearest 0.02 mL, and for readings greater than 10 mL, this represents four significant figures.

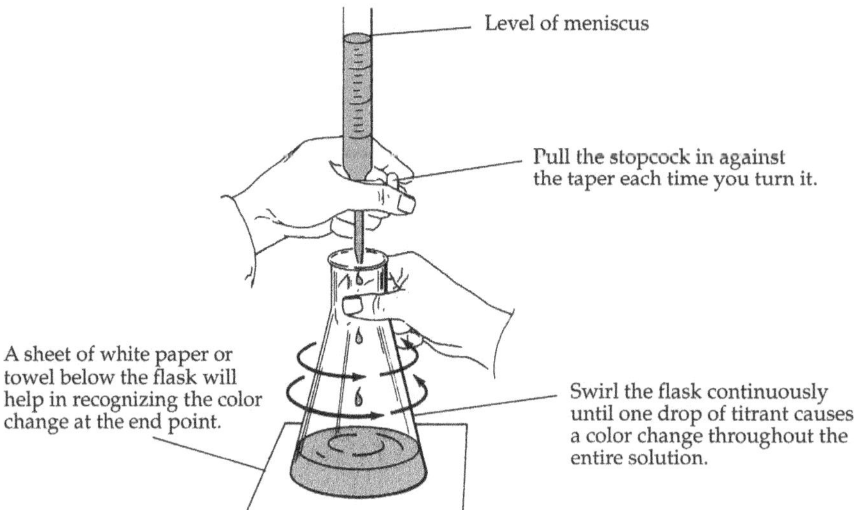
Level of meniscus
Pull the stopcock in against the taper each time you turn it.
A sheet of white paper or towel below the flask will help in recognizing the color change at the end point.
Swirl the flask continuously until one drop of titrant causes a color change throughout the entire solution.

▲**FIGURE 20.2** Titration procedure.

more slowly, at which time you should add the NaOH drop by drop. Continue swirling the flask for the entire titration. The end point is reached when one drop of the NaOH solution turns the entire solution in the flask from colorless to pink. Allow the titrated solution to stand for at least 1 min so that the buret will drain properly. Remove any hanging drop from the buret tip by touching it to the side of the flask and wash down the sides of the flask with a stream of water from the wash bottle. The solution should remain pink for about 30 s when it is swirled. Record the buret reading on your report sheet. Repeat this procedure with the other two samples. Dispose of the neutralized solutions as instructed.

From the data you obtain in the three titrations, calculate the molarity of the NaOH solution to four significant figures as in Example 20.2.

The three determinations should agree within 1.0%. If they do not, you should repeat the standardization until agreement is reached. The average of the three acceptable determinations is taken as the molarity of the NaOH. Calculate the standard deviation of your results. *Save* your standardized solution for the unknown determination.

B. Analysis of an Unknown Acid

Calculate the approximate mass of the unknown that should be taken to require about 20 mL of your standardized NaOH, assuming that your unknown sample is 75% KHP.

From a weighing bottle, determine the mass by difference of triplicate portions of the sample to four significant figures and place them in three separate 250 mL flasks. The sample size should be about the amount determined by the above computation. Dissolve the sample in 100 mL of CO_2-free distilled water (prepared by boiling and cooling) and add two drops of phenolphthalein indicator solution. Titrate with your standard NaOH solution to the faintest visible shade of pink (not red) as described previously in the standardization procedure. Calculate the percentage of KHP in the samples as in Example 20.3. For good results, the three determinations should agree within 1.0%. Your answers should have four significant figures. Compute the standard deviation of your results. Dispose of all solutions as directed.

Test your results by computing the average deviation from the mean. If one result is noticeably different from the others, perform an additional titration. If any result is more than two standard deviations from the mean, discard it and titrate another sample.

NOTES AND CALCULATIONS

Name ______________________________ Desk ______________

Date ______________ Laboratory Instructor ______________________

Titration of Acids and Bases | 20 Pre-lab Questions

Before beginning this experiment in the laboratory, you should be able to answer the following questions.

1. Define *standardization* and state how you would go about doing it.

2. Define the term *titration.*

3. Define the term *molarity.*

4. Why do you determine a mass by difference?

5. What are equivalence points and end points? How do they differ?

6. What apparatuses do you need for titration of acids and bases?

7. What is parallax? Why should you avoid it?

8. Why is it necessary to rid the distilled water of CO_2?

9. What is the molarity of a solution that contains 2.38 g of $H_2C_2O_4 \cdot 2H_2O$ in exactly 300 mL of solution?

10. If 25.21 mL of NaOH solution is required to react completely with 0.550 g KHP, what is the molarity of the NaOH solution?

11. The titration of an impure sample of KHP found that 36.00 mL of 0.100 *M* NaOH was required to react completely with 0.758 g of sample. What is the percentage of KHP in this sample?

12. How many milliliters of 0.200 *M* NaOH are required to neutralize 22.3 mL of 0.100 *M* HCl?

Name Grace Rademacher Desk ______

Date 3/9/21 Laboratory Instructor Prof. Towle

REPORT SHEET | EXPERIMENT

Titration of Acids and Bases | 20

A. Standardization of Sodium Hydroxide (NaOH) Solution

	Trial 1	Trial 2	Trial 3
~~Mass of bottle + KHP~~			
~~Mass of bottle~~			
Mass of KHP used	0.385 g	0.39 g	0.397 g
Final buret reading	9.4 mL	9.5 mL	9.6 mL
Initial buret reading	0.00 mL	0.00 mL	0.00 mL
mL of NaOH used	9.4 mL	9.5 mL	9.6 mL
Molarity of NaOH			

Average molarity 0.2014 M Standard deviation 0.0018

Show your calculations for molarity and standard deviation.

Trial 1: $0.385\text{g KHP} \times \frac{1 \text{ mol KHP}}{204.22\text{g KHP}} = .0019 \text{ mols}$

molarity: $\frac{.0019}{9.4\times10^{-3}} = 0.202\ M$

Trial 2: $0.39 \times \frac{1}{204.22} = 0.0019 \quad \frac{0.0019}{9.5\times10^{-3}} = 0.2\ M$

Trial 3: $0.397 \times \frac{1}{204.22} = 0.00194 \quad \frac{0.00194}{9.6\times10^{-3}} = 0.2021\ M$

$\left(\frac{(0.202-0.2014)^2 + (0.2-0.2014)^2 + (0.2021-0.2014)^2}{3-1}\right)^{1/2}$

$(3.225\times10^{-6})^{1/2} = 0.0018$

B. Analysis of an Unknown Acid

	Trial 1	Trial 2	Trial 3
~~Mass of bottle + unknown~~			
~~Mass of bottle~~			
Mass of unknown used	0.364 g	0.364 g	0.368 g
Final buret reading	1.0 mL	1.20 mL	1.2 mL
Initial buret reading	0.00 mL	0.00 mL	0.00 mL
mL of NaOH used	1.00 mL	1.2 mL	1.2 mL
~~Mass of KHP in unknown~~			
~~Percent of KHP in unknown~~			

Average percent of KHP 12% Standard deviation 1.218

Calculations of percent KHP and standard deviation (show using equations with units):

$\left(\frac{(11.3-12)^2+(11.3-12)^2+(13.41-12)^2}{2}\right)^{1/2}$

$(1.48405)^{1/2} = 1.218$

M from part A

(0.2014)(.001) = 2.014 x 10^-4 mol KHP

(2.014x10^-4) x 204.2 = 0.0411 g KHP

(.0411/.364) 100 = 11.3% (Sample 1+2)

(0.2014)(.0012) = 2.4168x10^-4

(2.4168x10^-4)(204.2) = 0.04934

(.04934/.368) 100 = 13.41% (sample 3)

Avg = 12%

QUESTIONS

1. Write the balanced chemical equation for the reaction of KHP with NaOH.

   KHP + NaOH → KNaP + H_2O

2. Suppose your laboratory instructor inadvertently gave you a sample of KHP contaminated with NaCl to use in standardizing your NaOH. How would this affect the molarity you calculated for your NaOH solution? Justify your answer.

   This will cause the molarity of NaOH to be higher than its true value

3. How many grams of NaOH are needed to prepare 500 mL of 0.125 *M* NaOH?

   0.125 M x 0.5 L = 0.0625 mol

   0.0625 mol x 40 g / 1 mol = 2.5 grams NaOH needed.

4. A solution of malonic acid, $H_2C_3H_2O_4$, was standardized by titration with 0.1000 *M* NaOH solution. If 20.76 mL of the NaOH solution is required to neutralize completely 12.95 mL of the malonic acid solution, what is the molarity of the malonic acid solution?

$$H_2C_3H_2O_4 + 2NaOH \longrightarrow Na_2C_3H_2O_4 + 2H_2O$$

5. Sodium carbonate is a reagent that may be used to standardize acids in the same way you used KHP in this experiment. In such a standardization, it was found that a 0.512 g sample of sodium carbonate required 26.30 mL of a sulfuric acid solution to reach the end point for the reaction.

$$Na_2CO_3(aq) + H_2SO_4(aq) \longrightarrow H_2O(l) + CO_2(g) + Na_2SO_4(aq)$$

What is the molarity of the H_2SO_4?

$\frac{0.512}{106.0} = 0.00483\text{mol}$ 1:1 ratio; $H_2SO_4 = 0.00483$ moles

$26.3\text{mL} = .0263\text{ L}$

$\frac{0.00483}{0.0263} = 0.184\text{ M } H_2SO_4$

6. A solution contains 6.30×10^{-2} g of oxalic acid, $H_2C_2O_4 \cdot 2H_2O$, in 250 mL. What is the molarity of this solution?

NOTES AND CALCULATIONS

Reactions in Aqueous Solutions: Metathesis Reactions and Net Ionic Equations

Experiment 21

OBJECTIVE

To become familiar with writing equations for metathesis reactions, including net ionic equations.

APPARATUS AND CHEMICALS

Apparatus

small test tubes (12)	100 mL beaker (2)
evaporating dish	600 mL beaker
thermometer	funnels (2)
filter paper	short-stem funnel
Bunsen burner and hose	funnel support
magnifying glass	ring stand and ring

Chemicals

sodium nitrate	0.1 *M* sodium acetate
potassium chloride	0.1 *M* lead nitrate
0.1 *M* potassium chloride	0.1 *M* copper(II) sulfate
0.1 *M* barium chloride	0.1 *M* sodium nitrate
0.1 *M* sodium phosphate	1.0 *M* sulfuric acid
0.1 *M* silver nitrate	1.0 *M* ammonium chloride
0.1 *M* nickel chloride	1.0 *M* sodium hydroxide
0.1 *M* sodium sulfide	1.0 *M* hydrochloric acid
0.1 *M* cadmium chloride	1.0 *M* sodium carbonate
ice	

ALL SOLUTIONS SHOULD BE PROVIDED IN DROPPER BOTTLES.

DISCUSSION

In molecular equations for many aqueous reactions, cations and anions appear to exchange partners. These reactions conform to the following general equation:

$$AX + BY \longrightarrow AY + BX \qquad [1]$$

Such reactions are known as *metathesis reactions*. For a metathesis reaction to lead to a net change in solution, ions must be removed from the solution. In general, three chemical processes can lead to the removal of ions from solution, thus serving as a *driving force* for metathesis to occur:

1. The formation of a precipitate.
2. The formation of a weak electrolyte or nonelectrolyte.
3. The formation of a gas that escapes from solution.

FORMATION OF A PRECIPITATE

The reaction of barium chloride with silver nitrate is a typical example of a precipitation reaction (Section 4.2).

$$BaCl_2(aq) + 2AgNO_3(aq) \longrightarrow Ba(NO_3)_2(aq) + 2AgCl(s) \qquad [2]$$

This form of the equation for this reaction is referred to as the *molecular equation*. Because the salts $BaCl_2$, $AgNO_3$, and $Ba(NO_3)_2$ are strong electrolytes and are completely dissociated in solution, you can more realistically write the equation as follows:

$$Ba^{2+}(aq) + 2Cl^-(aq) + 2Ag^+(aq) + 2NO_3^-(aq) \longrightarrow Ba^{2+}(aq) + 2NO_3^-(aq) + 2AgCl(s) \qquad [3]$$

This form, in which all ions are shown, is known as the *complete ionic equation*. Reaction [2] occurs because the insoluble substance AgCl precipitates out of solution. The other product, barium nitrate, is soluble in water and remains in solution. You see that Ba^{2+} and NO_3^- ions appear on both sides of the equation and thus do not enter into the reaction. Such ions are called *spectator ions*. If you eliminate or omit them from both sides, you obtain the *net ionic equation* as follows:

$$Ag^+(aq) + Cl^-(aq) \longrightarrow AgCl(s) \qquad [4]$$

This equation focuses your attention on the salient feature of the reaction: the formation of the precipitate AgCl. It tells you that solutions of any soluble Ag^+ salt and any soluble Cl^- salt, when mixed, will form insoluble AgCl. When you are writing net ionic equations, remember that only *strong electrolytes* are written in the ionic form. Solids, gases, nonelectrolytes, and weak electrolytes are written in the molecular form. Frequently, the symbol (*aq*) is omitted from ionic equations. The symbols (*g*) for gas, (*l*) for liquid, and (*s*) for solid should not be omitted. Thus, Equation [4] can be written as follows:

$$Ag^+ + Cl^- \longrightarrow AgCl(s) \qquad [5]$$

Consider mixing solutions of KCl and $NaNO_3$. The complete ionic equation for the reaction is

$$K^+(aq) + Cl^-(aq) + Na^+(aq) + NO_3^-(aq) \longrightarrow K^+(aq) + NO_3^-(aq) + Na^+(aq) + Cl^-(aq) \qquad [6]$$

Because all of the compounds are water-soluble and are strong electrolytes, they have been written in the ionic form. They completely dissolve in water. If you eliminate spectator ions from the equation, nothing remains. Hence, there is no reaction.

$$K^+(aq) + Cl^-(aq) + Na^+(aq) + NO_3^-(aq) \longrightarrow \text{no reaction} \qquad [7]$$

Metathesis reactions occur when a precipitate, a gas, a weak electrolyte, or a nonelectrolyte is formed. The following equations are further illustrations of such processes.

FORMATION OF A GAS

Molecular equation (Section 4.3):

$$2HCl(aq) + Na_2S(aq) \longrightarrow 2NaCl(aq) + H_2S(g)$$

Complete ionic equation:

$$2H^+(aq) + 2Cl^-(aq) + 2Na^+(aq) + S^{2-}(aq) \longrightarrow 2Na^+(aq) + 2Cl^-(aq) + H_2S(g)$$

Net ionic equation:

$$2H^+(aq) + S^{2-}(aq) \longrightarrow H_2S(g)$$

FORMATION OF A WEAK ELECTROLYTE

Molecular equation (Section 4.1); (Section 4.3):

$$HNO_3(aq) + NaOH(aq) \longrightarrow H_2O(l) + NaNO_3(aq)$$

Complete ionic equation:

$$H^+(aq) + NO_3^-(aq) + Na^+(aq) + OH^-(aq) \longrightarrow H_2O(l) + Na^+(aq) + NO_3^-(aq)$$

Net ionic equation:

$$H^+(aq) + OH^-(aq) \longrightarrow H_2O(l)$$

To decide if a reaction occurs, you need to be able to determine whether a precipitate, a gas, a nonelectrolyte, or a weak electrolyte will be formed. The following brief discussion is intended to aid you in this regard. Table 21.1 summarizes solubility rules, which you should consult while performing this experiment.

TABLE 21.1 Solubility Rules

Water-soluble salts	
Na^+, K^+, NH_4^+	All sodium, potassium, and ammonium salts are soluble.
NO_3^-, ClO_3^-, CH_3COO^-	All nitrates, chlorates, and acetates are soluble.
Cl^-	All chlorides are soluble except AgCl, Hg_2Cl_2, and $PbCl_2$*.
Br^-	All bromides are soluble except AgBr, Hg_2Br_2, $PbBr_2$,* and $HgBr_2$*.
I^-	All iodides are soluble except AgI, Hg_2I_2, PbI_2, and HgI_2.
SO_4^{2-}	All sulfates are soluble except $CaSO_4$,* $SrSO_4$, $BaSO_4$, Hg_2SO_4, $PbSO_4$, and Ag_2SO_4.
Water-insoluble salts	
CO_3^{2-}, SO_3^{2-}, PO_4^{3-}, CrO_4^{2-}	All carbonates, sulfites, phosphates, and chromates are insoluble except those of alkali metals and NH_4^+.
OH^-	All hydroxides are insoluble except those of alkali metals and $Ca(OH)_2$,* $Sr(OH)_2$,* and $Ba(OH)_2$.
S^{2-}	All sulfides are insoluble except those of alkali metals, alkaline earths, and NH_4^+.

*Slightly soluble.

The common gases are CO_2, SO_2, H_2S, and NH_3. Carbon dioxide and sulfur dioxide may be regarded as resulting from the decomposition of their corresponding weak acids, which are initially formed when carbonate and sulfite salts are treated with acid.

$$H_2CO_3(aq) \longrightarrow H_2O(l) + CO_2(g)$$

and

$$H_2SO_3(aq) \longrightarrow H_2O(l) + SO_2(g)$$

Ammonium salts form NH_3 when they are treated with strong bases.

$$NH_4^+(aq) + OH^-(aq) \longrightarrow NH_3(g) + H_2O(l)$$

Which are the weak electrolytes? The easiest way to answer this question is to identify all of the strong electrolytes, and if the substance does not fall in that category, it is a weak electrolyte. Note that water is a nonelectrolyte. Strong electrolytes are summarized in Table 21.2.

In the first part of this experiment, you will study some metathesis reactions. In some instances, it will be evident that a reaction has occurred, whereas in other instances, it will not be so apparent. In the doubtful case, use the preceding guidelines to decide whether a reaction has taken place. You will be given the names of the compounds to use, but not their formulas. This is being done deliberately to give you practice in writing formulas from names (Section 2.8).

In the second part of this experiment, you will study the effect of temperature on solubility. The effect that temperature has on solubility varies from salt to salt. As Equations [6] and [7] showed, mixing solutions of KCl and $NaNO_3$ resulted in no reaction. What would happen if you cooled such a mixture? The solution would eventually become saturated with respect to one of the salts, and crystals of that salt would begin to appear as its solubility was exceeded. Examination of Equation [6] reveals that crystals of any of the following salts could appear initially: KNO_3, KCl, $NaNO_3$, or NaCl.

Consequently, if a solution containing Na^+, K^+, Cl^-, and NO_3^- ions is evaporated at a given temperature, the solution becomes more and more concentrated and will eventually become saturated with respect to one of the four compounds. If evaporation is continued, that compound will crystallize out, removing its ions from solution. The other ions will remain in solution and increase in concentration. Before beginning this laboratory exercise, you are to plot the solubilities of the four salts given in Table 21.3 on the graph on your report sheet. From this graph, you may predict which salt will precipitate first as the temperature is lowered. In Part B, you will test your prediction.

GIVE IT SOME THOUGHT

List all of the strong acids and strong bases. How does this list compare with the table of strong electrolytes?

TABLE 21.2 Strong Electrolytes

Salts	All common **soluble** salts
Acids	$HClO_4$, HCl, HBr, HI, HNO_3, and H_2SO_4 are strong electrolytes; all others are weak.
Bases	Alkali metal hydroxides, $Ca(OH)_2$, $Sr(OH)_2$, and $Ba(OH)_2$ are strong electrolytes; all others are weak.

TABLE 21.3 Molar Solubilities of NaCl, $NaNO_3$, KCl, and KNO_3 (mol/L)

Compound	0 °C	20 °C	40 °C	60 °C	80 °C	100 °C
NaCl	5.4	5.4	5.5	5.5	5.5	5.6
$NaNO_3$	6.7	7.6	8.5	9.4	10.4	11.3
KCl	3.4	4.0	4.6	5.1	5.5	5.8
KNO_3	1.3	3.2	5.2	7.0	9.0	11.0

PROCEDURE

A. Metathesis Reactions

The report sheet lists 16 pairs of chemicals that are to be mixed. Use about 15 drops of the reagents to be combined as indicated on the report sheet. Mix the solutions in small test tubes by rapping the test tube with your finger to generate a vortex which provides efficient mixing (see appendix J) and record your observations on the report sheet. If there is no reaction, write N.R. (The reactions need not be carried out in the order listed. You can avoid congestion at the reagent shelf if everyone does not start with the reagents for reaction 1.) Dispose of the contents of your test tubes in the designated receptacles.

GIVE IT SOME THOUGHT

a. If no reaction occurs, what are the products of the complete ionic equation?
b. Do you indicate them as (*aq*) or (*s*)?
c. How is this impacted by solubility?

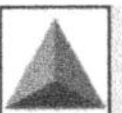

GIVE IT SOME THOUGHT

Are the results from mixing the reagents consistent with Table 21.1, Solubility Rules?

B. Solubility, Temperature, and Crystallization

Place 8.5 g of sodium nitrate and 7.5 g of potassium chloride in a 100 mL beaker and add 25 mL of water. Warm the mixture, swirling until the solids dissolve completely. Assuming a volume of 25 mL for the solution, calculate the molarity of the solution with respect to $NaNO_3$, KCl, NaCl, and KNO_3 and record these molarities (1).

Cool the solution in the 100 mL beaker to about 10 °C by dipping the beaker in ice water contained in a 600 mL beaker and stir the solution carefully with a thermometer, being careful not to break it. (SHOULD THE THERMOMETER BREAK, CONSULT YOUR INSTRUCTOR IMMEDIATELY.) When no more crystals form, at approximately 10 °C, filter the cold solution quickly and allow the filtrate to drain thoroughly into an evaporating dish. Dry the crystals between two dry pieces of filter paper or paper towels. Examine the crystals with a magnifying glass (or fill a Florence flask with water and look at the crystals through it). Describe the shape of the crystals—that is, needles, cubes, plates, rhombohedron, etc. (2). Based upon your solubility graph, which compound crystallized out of solution (3)?

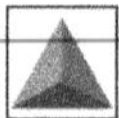

GIVE IT SOME THOUGHT

A soluble compound begins to form a precipitate at lower temperatures. What does this tell you about the temperatures at which the solubility rules were determined?

Evaporate the filtrate to about half its volume using a Bunsen burner or hot plate and ring stand. A second crop of crystals should form. Record the temperature (4) and rapidly filter the hot solution, collecting the filtrate in a clean 100 mL beaker. Dry the second batch of crystals between two pieces of filter paper and examine their shape. Compare their shape with that of the first batch of crystals (5). Based upon your solubility graph, what is the composition of this substance (6)?

GIVE IT SOME THOUGHT

a. How can you use the shape of a crystal to identity each salt?
b. Do reference materials such as the *CRC Handbook of Chemistry and Physics* contain data based on crystal shape?

Finally, cool the filtrate to 10 °C while stirring carefully with a thermometer to obtain a third crop of crystals. Carefully observe their shapes and compare them with those of the first and second batches (7). What compound is the third batch of crystals (8)? Dispose of the chemicals in the designated waste containers.

Name ______________________________ Desk ______________

Date ______________ Laboratory Instructor ______________________

Reactions in Aqueous Solutions: Metathesis Reactions and Net Ionic Equations | 21 Pre-lab Questions

Before beginning this experiment in the laboratory, you should be able to answer the following questions.

1. Write molecular, complete ionic, and net ionic equations for the reactions that occur, if any, when solutions of the following substances are mixed:

   (a) copper(II) sulphate and sodium hydroxide

   (b) silver nitrate and sodium chloride

   (c) potassium carbonate and hydrochloric acid

   (d) ammonium nitrate and potassium hydroxide

   (e) barium nitrate and sulfuric acid

2. Which of the following are not water-soluble: MgO, $NaCl$, $MgSO_4$, $BaSO_4$, NH_4NO_3, and $AgCl$?

3. Write equations for the decomposition of H_2CO_3 and H_2SO_3.

4. At what temperature (from your graph) do KNO_3 and $NaCl$ have the same molar solubility?

5. Which of the following are strong electrolytes: K_2CO_3, HCl, KCl, H_2SO_4, and $CaSO_4$?

6. Which of the following are weak electrolytes: $CuSO_4$, $NaOH$, NH_3, Na_2SO_4, and CH_3COOH?

7. For each of the following water-soluble compounds, indicate the ions present in an aqueous solution: NaI, K_2SO_4, $NaCN$, $Ba(OH)_2$, and $(NH_4)_2SO_4$.

8. Write a balanced chemical equation showing how you could prepare each of the following salts from an acid–base reaction: KNO_3, $NaBr$, and CaS.

Name Grace Rademacher Desk

Date 11/6/20 Laboratory Instructor Prof Briguglio

REPORT SHEET | EXPERIMENT 21

Reactions in Aqueous Solutions: Metathesis Reactions and Net Ionic Equations

A. Metathesis Reactions

1. Copper(II) sulfate + sodium carbonate

   Observations Pale blue precipitate forms

   Molecular equation $CuSO_4 + Na_2CO_3 \rightarrow Na_2SO_4 + CuCO_3$

   Complete ionic equation

   $2Na^+ + CO_3^{2-} + Cu^{2+} + SO_4^{2-} \rightarrow 2Na^+ + SO_4^{2-} + CuCO_3$

   Net ionic equation $CO_{3(aq)}^{2-} + Cu_{(aq)}^{2+} \rightarrow CuCO_{3(s)}$

2. Copper(II) sulfate + barium chloride

   Observations White precipitate forms

   Molecular equation $BaCl_{2(aq)} + CuSO_{4(aq)} \rightarrow Ba_2SO_{4(s)} + CuCl_{(aq)}$

   Complete ionic equation

   $Ba_{(aq)}^{2+} + Cl_{2(aq)}^{-} + Cu_{(aq)}^{2+} + SO_{4(aq)}^{2-} \rightarrow Ba_2SO_{4(s)} + Cl_{2(aq)}^{-} + Cu_{(aq)}^{2+}$

   Net ionic equation $Ba_{2(aq)}^{+} + SO_{4(aq)}^{2-} \rightarrow Ba_2SO_{4(s)}$

3. Copper(II) sulfate + sodium phosphate

   Observations Light blue precipitate forms

   Molecular equation $3CuSO_{4(aq)} + 2Na_3PO_{4(aq)} \rightarrow Cu_3(PO_4)_{2(s)} + 3Na_2SO_{4(aq)}$

   Complete ionic equation

   $3Cu_{(aq)}^{2-} + 3SO_{4(aq)} + 2Na_{3(aq)} + 2PO_{4(aq)} \rightarrow Cu_3(PO_4)_{2(s)} + 6Na_{(aq)} + 3SO_{4(aq)}$

   Net ionic equation $3Cu_{(aq)}^{2-} + 2PO_{4(aq)} \rightarrow Cu_3(PO_4)_{2(s)}$

4. Sodium carbonate + sulfuric acid

   Observations Gas bubbles form in a colorless solution

   Molecular equation $H_2SO_4 + Na_2CO_3 \rightarrow Na_2SO_4 + H_2O + CO_2$

   Complete ionic equation

   $2H_{(aq)}^{+} + SO_{4(aq)}^{2-} + 2Na_{(aq)}^{2+} + C_{(aq)}^{-} + 3O_{(aq)}^{-} \rightarrow Na_2SO_{4(s)} + 2H_{(aq)}^{+} + 3O_{(aq)}^{-} + C_{(aq)}^{+}$

   Net ionic equation $2Na_{(aq)}^{2+} + SO_{4(aq)}^{2-} \rightarrow Na_2SO_{4(s)}$

5. Sodium carbonate + hydrochloric acid

   Observations Gas bubbles form in a colorless solution

   Molecular equation $Na_2CO_3 + 2HCl \rightarrow 2NaCl + H_2CO_3$

   Complete ionic equation

   $2Na_{(aq)}^{2+} + C_{(aq)}^{+} + 3O_{(aq)}^{-} + 2H_{(aq)}^{+} + 2Cl_{(aq)}^{-} \rightarrow 2NaCl_{(s)} + 2H_{(s)}^{-} + C_{(aq)}^{+} + 3O_{(aq)}^{-}$

   Net ionic equation $2Na_{(aq)}^{2+} + 2Cl_{(aq)}^{-} \rightarrow 2NaCl_{(s)}$

6. Cadmium chloride + sodium sulfide

   Observations Yellow precipitate forms

Molecular equation $Na_2S_{(aq)} + CdCl_{2(aq)} \rightarrow 2NaCl_{(aq)} + CdS_{(s)}$

Complete ionic equation $2Na^{+}_{(aq)} + S^{2-}_{(aq)} + Cd^{2+}_{(aq)} + 2Cl^{-}_{(aq)} \rightarrow 2Na^{+}_{(aq)} + 2Cl^{-}_{(aq)} + CdS_{(s)}$

Net ionic equation $Cd^{2+}_{(aq)} + S^{2-}_{(aq)} \rightarrow CdS_{(s)}$

7. Cadmium chloride + sodium hydroxide

Observations White precipitate

Molecular equation $CdCl_2 + 2NaOH \rightarrow 2NaCl + Cd(OH)_2$

Complete ionic equation

$Cd^{2+}_{(aq)} + 2Cl^{-}_{(aq)} + 2Na^{+}_{(aq)} + 2OH^{-}_{(aq)} \rightarrow 2Na^{+}_{(aq)} + 2Cl^{-}_{(aq)} + Cd(OH)_{2(s)}$

Net ionic equation $Cd^{+2}_{(aq)} + 2OH^{-}_{(aq)} \rightarrow Cd(OH)_{2(s)}$

8. Nickel chloride + silver nitrate

Observations pink color / thick white precipitate

Molecular equation $NiCl_2 + 2AgNO_3 \rightarrow Ni(NO_3)_2 + 2NaCl$

Complete ionic equation

$Ni^{2+}_{(aq)} + 2Cl^{-}_{(aq)} + 2Ag^{+}_{(aq)} + 2NO^{-}_{3(aq)} \rightarrow AgCl_{(s)} + Ni^{2+}_{(aq)} + 2NO^{-}_{3(aq)}$

Net ionic equation $Cl^{-}_{(aq)} + Ag^{+}_{(aq)} \rightarrow AgCl_{(s)}$

9. Nickel chloride + sodium carbonate

Observations purple cloudy Precipitate

Molecular equation $NiCl_2 + Na_2CO_3 \rightarrow NiCO_3 + 2NaCl$

Complete ionic equation

$Ni^{2+}_{(aq)} + 2Cl^{-}_{(aq)} + 2Na^{+}_{(aq)} + C^{+}_{(aq)} + 3O^{-}_{(aq)} \rightarrow NiCO_{3(s)} + 2Na^{+}_{(aq)} + 2Cl^{-}_{(aq)}$

Net ionic equation $Ni^{2+}_{(aq)} + CO^{2-}_{3(aq)} \rightarrow NiCO_{3(s)}$

10. Hydrochloric acid + sodium hydroxide

Observations Colorless solution formed that gets warm

Molecular equation $HCl + NaOH \rightarrow H_2O + NaCl$

Complete ionic equation $H^{+}_{(aq)} + Cl^{-}_{(aq)} + Na^{+}_{(aq)} + OH^{-}_{(aq)} \rightarrow NaCl_{(s)} + 2H^{+}_{(aq)} + O^{2-}_{(aq)}$

Net ionic equation $H^{+}_{(aq)} + OH^{-}_{(aq)} \rightarrow H_2O_{(l)}$

11. Ammonium chloride + sodium hydroxide

Observations Odor of NH_3 (ammonia gas) known

Molecular equation $NH_4Cl + NaOH \rightarrow NaCl + NH_3 + H_2O$

Complete ionic equation

$NH^{+}_{4(aq)} + Cl^{-}_{(aq)} + Na^{+}_{(aq)} + OH^{-}_{(aq)} \rightarrow NH_{3(g)} + H_2O_{(l)} + Na^{+}_{(aq)} + Cl^{-}_{(aq)}$

Net ionic equation $NH^{+}_{4(aq)} + OH^{-}_{(aq)} \rightarrow NH_{3(g)} + H_2O_{(l)}$

12. Sodium acetate + hydrochloric acid

Observations Odor of vinegar (HC_2O_2)

Molecular equation $HCl + NaOH \rightarrow NaCl + H_2O$

Complete ionic equation

$H^{+}_{(aq)} + Cl^{-}_{(aq)} + Na^{+}_{(aq)} + OH^{-}_{(aq)} \rightarrow Na^{+}_{(aq)} + Cl^{-}_{(aq)} + H_2O_{(l)}$

Net ionic equation $H^{+}_{(aq)} + OH^{-}_{(aq)} \rightarrow H_2O_{(l)}$

13. Sodium sulfide + hydrochloric acid

Observations Gas w/ rotten egg odor

Molecular equation $Na_2S + 2HCl \rightarrow 2NaCl + H_2S$

Complete ionic equation $2Na^{+}_{(aq)} + S^{2-}_{(aq)} + 2H^{+}_{(aq)} + 2Cl^{-}_{(aq)} \rightarrow 2Na^{+}_{(aq)} + 2Cl^{-}_{(aq)} + H_2S_{(g)}$

Net ionic equation $2H^{+}_{(aq)} + S^{2-}_{(aq)} \rightarrow H_2S_{(g)}$

Tuesday 3-4 office hrs. Bruigilio

14. Lead nitrate + sodium sulfide
    Observations ______
    Molecular equation ______
    Complete ionic equation ______
    Net ionic equation ______
15. Lead nitrate + sulfuric acid
    Observations ______
    Molecular equation
    Complete ionic equation
    ______
    Net ionic equation ______
16. Potassium chloride + sodium nitrate
    Observations No reactions
    Molecular equation $KCl(aq) + NaNO_3 \rightarrow$ N.R.
    Complete ionic equation
    $K^+(aq) + Cl^-(aq) + Na^+(aq) + NO_3^-(aq) \rightarrow$ N.R.
    Net ionic equation N.A.

B. Solubility, Temperature, and Crystallization

Don't need to Complete pt B

Solubilities as a function of temperature

Solubility (mol/L): 0, 2, 4, 6, 8, 10, 12

Temperature, °C: 0, 20, 40, 60, 80, 100

1. Molarities
   ______ M $NaNO_3$, ______ M KCl, ______ M NaCl, ______ M KNO_3

2. Crystal shape ______
3. Identity of crystals ______
4. Temperature of filtrate ______
5. Crystal shape of second batch ______
6. Identity of second batch of crystals ______
7. Crystal shape of third batch ______
8. Identity of third batch of crystals ______

QUESTIONS

1. Which of the following reactions are metathesis reactions?
   (a) $2KClO_3 \longrightarrow 2KCl + 3O_2$
   (b) $Cu(NO_3)_2 + Zn \longrightarrow Cu + Zn(NO_3)_2$
   (c) $BaCO_3 + 2\ HCl \longrightarrow BaCl_2 + H_2O + CO_2$
   (d) $Na_2CO_3 + CuSO_4 \longrightarrow Na_2SO_4 + CuCO_3$

2. (a) How many grams of each of the following substances will dissolve in 100 mL of cold water? Consult a handbook or the Internet and cite your source.

   $Ce(IO_3)_4$ $RaSO_4$ $Pb(NO_3)_2$ $(NH_4)_2SeO_4$

2. (b) Which of these substances is least soluble on a gram-per-100 mL basis?

3. Suppose you have a solution that might contain any or all of the following cations: Cu^{2+}, Ag^{+}, Ba^{2+}, and Mn^{2+}. The addition of HBr causes a precipitate to form. After the precipitate is separated by filtration, H_2SO_4 is added to the supernatant liquid, and another precipitate forms. This precipitate is separated by filtration, and a solution of NaOH is added to the supernatant liquid until it is strongly alkaline. No precipitate is formed. Which ions are present in each of the precipitates? Which cations are not present in the original solution?

Bromides Cu^{2+}, Ag^{+}, Ba^{2+}, Mn^{2+} are soluble, and won't form insoluble precipitate, However, Ba^{2+} ions as barium sulfate is an insoluble precipitate and is still present.

4. Write balanced net ionic equations for the reactions, if any, that occur between (a) $Fe_2S_3(s)$ and $HBr(aq)$, (b) $K_2CO_3(aq)$ and $Cu(NO_3)_2(aq)$, (c) $Fe(NO_3)_2(aq)$ and $HCl(aq)$, and (d) $Bi(OH)_3(s)$ and $HNO_3(aq)$.

a) $Fe_2S_{3(s)} + 6H^+_{(aq)} \rightarrow 6HBr_{(g)} + Fe^{3+}$

b) $Cu^{2+}_{(aq)} + CO^{2-}_{3(aq)} \rightarrow CuCO_{3(s)}$

c) No reaction - no precipitate

d) $Bi(OH)_{3(s)} + 3H^+_{(aq)} \rightarrow Bi^{3+}_{(aq)} + 3H_2O$

Colorimetric Determination of an Equilibrium Constant in Aqueous Solution

Experiment

OBJECTIVE

To become familiar with the concept of equilibrium by determination of an equilibrium constant for a reaction in solution.

APPARATUS AND CHEMICALS

Apparatus

50 mL volumetric flasks (6)	10 mL graduated pipet
spectrophotometer	10 mL volumetric pipet
cuvettes (2)	600 mL beaker
thermometer	18×150 mm test tubes (6)

Chemicals

0.200 *M* $Fe(NO_3)_3$	0.10 *M* HNO_3
2.00×10^{-3} *M* NaSCN	2.00×10^{-3} *M* $Fe(NO_3)_3$

WORK IN GROUPS OF THREE OR FOUR, BUT ANALYZE THE DATA INDIVIDUALLY.

DISCUSSION

If you have ever seen the beautiful stalactites and stalagmites that form in limestone caves, you may have wondered how they were formed. The key aspect of the formation of these natural wonders made up primarily of calcium carbonate is the reversibility of chemical reactions. Calcium carbonate occurs in underground deposits as a remnant from the shellfish that inhabited ancient oceans. When water that contains dissolved CO_2 seeps through these deposits, they slowly dissolve as a result of the following reaction:

$$CaCO_3(s) + CO_2(g) + H_2O(l) \longrightarrow Ca^{2+}(aq) + 2HCO_3^-(aq) \qquad [1]$$

When the water saturated with $Ca(HCO_3)_2$ reaches a cave, the reverse reaction occurs, liberating gaseous CO_2 and very slowly depositing solid $CaCO_3$ as stalactites and stalagmites. The expression for the *reversible* equilibrium reaction is written as Equation [2], with the double arrow indicating that the reaction proceeds in both directions simultaneously (Section 15.1).

$$CaCO_3(s) + CO_2(g) + H_2O(l) \rightleftharpoons Ca^{2+}(aq) + 2HCO_3^-(aq) \qquad [2]$$

All chemical equilibria are dynamic and are constantly proceeding in both directions. The system reaches a state of *equilibrium* when the rate of the reaction in the forward direction is equal to the rate of the reaction in the reverse direction. The *equilibrium expression* for the equilibrium represented by Equation [2] with *equilibrium constant K* is as follows (Section 15.2):

$$K = \frac{[Ca^{2+}][HCO_3^-]^2}{[CaCO_3][CO_2][H_2O]} \qquad [3]$$

For the general equilibrium Equation [4], where A, B, C, and D are the chemical species involved and a, b, c, and d are their coefficients in the balanced chemical equation, the equilibrium constant expression is

$$aA + bB \rightleftharpoons cC + dD \qquad [4]$$

$$K_{eq} = \frac{[C]^c[D]^d}{[A]^a[B]^b} \qquad [5]$$

GIVE IT SOME THOUGHT

If a chemical equation is not balanced properly, how will this change the associated K_{eq}?

The numerator in Equation [5] is the product of the molar concentrations of the products raised to the powers of their respective coefficients. The denominator is the product of the molar concentrations of reactants raised to the powers of their respective coefficients. To determine the value of an *equilibrium constant*, K_c, you need to know the *equilibrium concentrations* of all species involved in the equilibrium.

In this experiment, you will determine the value of K_{eq} for the following equilibrium:

$$\underset{\text{pale yellow}}{Fe^{3+}(aq)} + \underset{\text{colorless}}{SCN^-(aq)} \rightleftharpoons \underset{\text{blood red}}{FeNCS^{2+}(aq)} \qquad [6]$$

The equilibrium in Equation [6] has been written in a simplified form. The pale yellow $Fe^{3+}(aq)$ is, in reality, $[Fe(H_2O)_6]^{3+}$. Actually, the $FeNCS^{2+}$, which contains the *ambidentate** thiocyanate ion bound to iron through nitrogen rather than sulfur, is, in reality, $[Fe(NCS)(H_2O)_5]^{2+}$. The waters, both the coordinated and the free, have been omitted for simplicity from this equilibrium reaction.

If you determine the equilibrium $FeNCS^{2+}$ concentration in solutions of known initial concentrations of Fe^{3+} and SCN^- by simple stoichiometry, you can calculate the final equilibrium concentrations of all three species and hence the value of the equilibrium constant, K_{eq}. This is illustrated in Example 22.1.

EXAMPLE 22.1

Assuming that the equilibrium $FeNCS^{2+}$ concentration is 6.08×10^{-5} *M* in a solution that initially was 1.00×10^{-3} *M* in Fe^{3+} and 2.00×10^{-4} *M* in SCN^-, calculate the equilibrium concentrations of Fe^{3+} and SCN^- and the value of K_{eq}.

SOLUTION: Equation [6] shows that 1 mol of Fe^{3+} reacts with 1 mol of SCN^- to form 1 mol of $FeNCS^{2+}$. Because the volume does not change in this equilibrium reaction, you can calculate the concentrations of all species by simple differences.

*An ambidentate ion is capable of binding to a metal through two different sites.

equilibrium $[Fe^{3+}]$ = initial $[Fe^{3+}]$ − equilibrium $[FeNCS^{2+}]$; therefore,

equilibrium $[Fe^{3+}] = 1.00 \times 10^{-3}\ M - 6.08 \times 10^{-5}\ M = 9.39 \times 10^{-4}\ M$

Similarly, equilibrium $[SCN^-] = 2.00 \times 10^{-4}\ M - 6.08 \times 10^{-5}\ M = 1.39 \times 10^{-4}\ M$.

Inserting these values into the equilibrium constant expression, you can calculate its value as follows (Section 15.5):

$$K_{eq} = \frac{[FeNCS^{2+}]}{[Fe^{3+}][SCN^-]} = \frac{[6.08 \times 10^{-5}]}{[9.39 \times 10^{-4}][1.39 \times 10^{-4}]} = 466$$

GIVE IT SOME THOUGHT

a. What does a K_{eq} of 466 imply?

b. Will the equilibrium reaction expression favor the products or the reactants?

You can determine the $FeNCS^{2+}$ concentration spectrophotometrically because it is highly colored. You will use a solution of iron(III) nitrate, $Fe(NO_3)_3$, as your source of Fe^{3+} ions and a solution of sodium thiocyanate, NaSCN, as your source of SCN^- ions. Equilibrium constants are somewhat dependent upon the total ionic concentration of the reaction mixture. Therefore, you will prepare all solutions using 0.10 M nitric acid rather than water to ensure that all mixtures have comparable ionic concentrations. This also keeps the solution sufficiently acidic to prevent any significant formation of iron(III) hydroxide. Because equilibrium constants are also temperature-dependent, you will record the temperature at which your measurements are made.

Although the eye can discern differences in color intensity with reasonable accuracy, an instrument known as a *spectrophotometer*, which eliminates "human" limitations, is commonly used for this purpose. Basically, it is an instrument that measures the ratio I/I_0 where I is the intensity of light transmitted by the sample and I_0 is the intensity of the incident beam. A schematic representation of a spectrophotometer is shown in Figure 22.1. The instrument has these five fundamental components:

- A light source that produces light with a wavelength range from about 375 to 650 nm for a visible range instrument only.

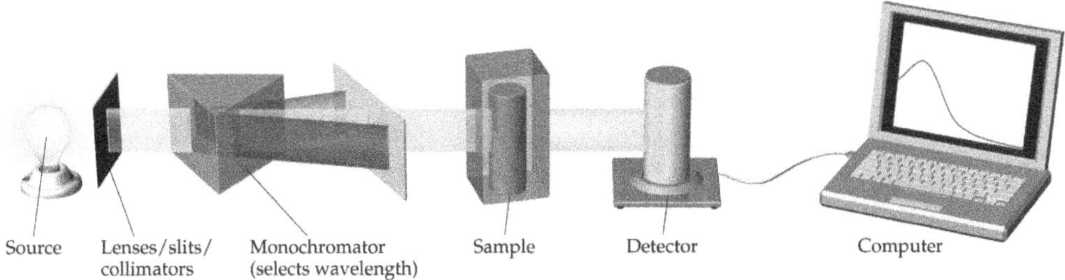

▲**FIGURE 22.1** Schematic representation of a spectrophotometer.

- A monochromator, which *selects* a particular wavelength of light and sends it to the sample cell with an intensity of I_0.
- The sample cell, which contains the solution being analyzed.
- A detector that measures the intensity, I, of the light transmitted from the sample cell; if the intensity of the incident light is I_0 and the solution absorbs light, the intensity of the transmitted light, I, is less than I_0.
- A display that indicates the intensity of the transmitted light.

For a given substance, the amount of light absorbed depends on the following:

- Concentration and identity of the absorbing species
- Cell or path length
- Wavelength of light
- Solvent

Plots of the amount of light absorbed versus wavelength are called *absorption spectra*. There are two common ways of expressing the amount of light absorbed. One is in terms of *percent transmittance*, *%T*, which is defined as

$$\%T = \frac{I}{I_0} \times 100\% \qquad [7]$$

As the term implies, percent transmittance corresponds to the percentage of light transmitted. When the sample in the cell is a solution, I is the intensity of light transmitted by the solution and I_0 is the intensity of light transmitted when the cell contains only solvent. Another method of expressing the amount of light absorbed is in terms of *absorbance*, A, which is defined by

$$A = \log\frac{I_0}{I} \qquad [8]$$

GIVE IT SOME THOUGHT

a. What happens to the absorbance when the concentration is increased?

b. What happens to the absorbance when the concentration is decreased?

If a sample absorbs no light at a given wavelength, the percent transmittance is 100 and the absorbance is 0. On the other hand, if the sample absorbs all of the light, $\%T = 0$ and $A = \infty$. Absorbance is more useful in quantitative work since it varies linearly with concentration.

Absorbance is related to concentration by the Beer-Lambert law

$$A = abc$$

where A is absorbance, b is solution path length, c is concentration in moles per liter, and a is molar absorptivity or molar extinction coefficient. There is a linear relationship between absorbance and concentration when the

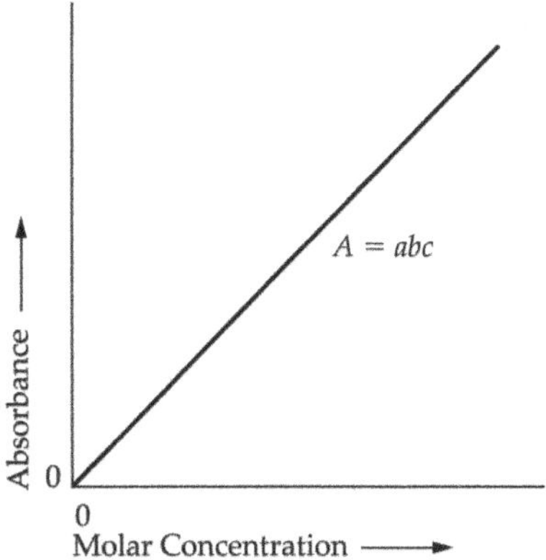

▲**FIGURE 22.2** Relationship between absorbance and molar concentration according to the Beer-Lambert law.

Beer-Lambert law is obeyed, as illustrated in Figure 22.2. However, because deviations from this law occasionally occur, it is wise to construct a calibration curve of absorbance versus molar concentration.

A. Preparation of the Calibration Curve

You will measure the absorbance of a series of solutions with known $FeNCS^{2+}$ ion concentrations to prepare a calibration curve. The experimental approach to this involves an interesting dilemma. The reaction that you will be using to prepare these solutions (equation [6]) is reversible and does not proceed to completion. How then can solutions with known $FeNCS^{2+}$ concentrations be prepared? You will take advantage of Le Châtelier's principle to shift the equilibrium of Equation [6] to the right (Section 15.7). When you make an equilibrium mixture in which the Fe^{3+} ion concentration is relatively high, the equilibrium expression (see Example 22.1) indicates that the SCN^- concentration for this mixture is relatively low. Consequently, by preparing a solution in which the initial Fe^{3+} ion concentration is much larger than that of the SCN^- ion, you can force the reaction to proceed nearly to completion. Under these conditions, you will assume within the limits of the experiment that all of the SCN^- is in the form of $FeNCS^{2+}$.

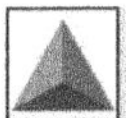

GIVE IT SOME THOUGHT

When a high concentration of Fe^{3+} is used, which way will the equilibrium shift?

Label six 50.0 mL volumetric flasks 1 through 6. Pipet 10.00 mL of 2.00×10^{-1} M $Fe(NO_3)_3$ solution into each volumetric flask. Then pipet 1.00, 2.00, 3.00, 4.00, and 5.00 mL of 2.00×10^{-3} M NaSCN solution into flasks 2 through 6, respectively.* **(CAUTION:** ***Nitric acid and solutions containing it can cause burns and skin discoloration. If you come into contact with it,***

*Instructor: Students can save time in this experiment if you set up several burets filled with $Fe(NO_3)_3$, NaSCN, and HNO_3 solutions on side tables in the laboratory. Students may obtain these solutions from the burets and refill the burets from bottles of stock solutions also located there.

immediately wash the area with copious amounts of water.) Add sufficient 0.10 *M* nitric acid to each flask to make the total volume 50.00 mL. Calculate the final $Fe(NO_3)_3$ concentrations and record them on your report sheet. Stopper the flasks. While holding the stopper firmly in the flask, invert the flask several times to thoroughly mix the solution. Use the information in Table 22.1 to calculate the $FeNCS^{2+}$ concentration in each flask, assuming that all of the SCN^- is present as $FeNCS^{2+}$. Enter the results on your report sheet.

Obtain two cuvettes. Rinse one cuvette with solution 1, discarding the rinse into a 600 mL beaker labeled "Waste." Fill the rinsed cuvette with solution 1. Insert the cuvette into the spectrophotometer; set the wavelength to 447 nm, and adjust the light control knob until the meter reads zero absorbance (100 %T) at 447 nm. Save this reference cuvette for periodic calibration checks.

Rinse the second cuvette with solution 2, discarding the rinse into the waste beaker; then measure the absorbance and transmittance of solution 2 at 447 nm. Repeat this process with solutions 3 through 6. Prepare your calibration curve by plotting absorbance versus concentration as in Figure 22.2. You should submit this curve with your report sheet.

TABLE 22.1 Standard Solutions for $FeNCS^{2+}$ Ion Beer-Lambert Plot

Solution	mL of 2.00×10^{-1} *M* $Fe(NO_3)_3$ in 0.10 *M* HNO_3	mL of 2.00×10^{-3} *M* NaSCN in 0.10 *M* HNO_3	Total volume (mL)
1	10.00	0.00	50.00
2	10.00	1.00	50.00
3	10.00	2.00	50.00
4	10.00	3.00	50.00
5	10.00	4.00	50.00
6	10.00	5.00	50.00

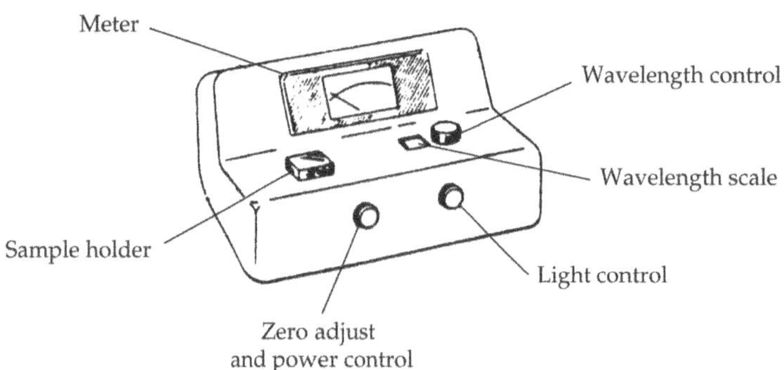

▲**FIGURE 22.3** Typical spectrophotometer controls. Those on the spectrometer you use may differ from the ones illustrated.

Operating Instructions for Spectronic 20*

1. Turn the wavelength control knob (Figure 22.3) to the desired wavelength (447 nm).
2. Turn on the instrument by rotating the power control clockwise and allow the instrument to warm up about 5 min. With no sample in the holder but with the cover closed, turn the zero adjust to bring the meter needle to zero on the percent transmittance scale.
3. Fill the cuvette about halfway with distilled water (or solvent blank) and insert it in the sample holder, aligning the line on the cuvette with that of the sample holder. Close the cover and rotate the light control knob until the meter reads 100% transmittance.
4. Remove the blank from the sample holder and replace it with the cuvette containing the sample whose absorbance is to be measured. Align the lines on the cuvette with the holder and close the cover. Read the percent transmittance or absorbance from the meter.

B. Determination of the Equilibrium Constant

Label six clean, dry, 18 mm × 150 mm test tubes 1 through 6. Pipet 5.00 mL of 2.00×10^{-3} *M* $Fe(NO_3)_3$ solution into each of these test tubes. Add 1.00, 2.00, 3.00, 4.00, and 5.00 mL of 2.00×10^{-3} *M* NaSCN solution to test tubes 2 through 6, respectively. Add 5.00, 4.00, 3.00, 2.00, and 1.00 mL of 0.10 *M* HNO_3 to test tubes 1 through 6, respectively. The total volume in each test tube should now be 10 mL. On your report sheet, record the volume of each of the component solutions in each of the test tubes. Measure and record the absorbances and transmittances of these solutions at 447 nm as described above for determining the calibration curve. From your calibration curve, determine the equilibrium concentration of $FeNCS^{2+}$ in each of the test tubes.

GIVE IT SOME THOUGHT

How is the absorbance related to the equilibrium concentration on the calibration curves?

Waste Disposal Instructions

The solutions used in this experiment are acidic and mildly oxidizing. Dispose of them in an appropriate waste container.

*Follow the instructions for the spectrometer you are using.

NOTES AND CALCULATIONS

Name ______________________ Desk ______________

Date ______________ Laboratory Instructor ______________________

Colorimetric Determination of an Equilibrium Constant in Aqueous Solution | 22 Pre-lab Questions

Before beginning this experiment in the laboratory, you should be able to answer the following questions.

1. Write the reaction for the formation of $FeNCS^{2+}$.

2. Write the equilibrium constant expression for the formation of $FeNCS^{2+}$.

3. What $FeNCS^{2+}$ concentrations in solutions 1 through 6 are used for the calibration curve?

4. Calculate the value of K_{eq} from the following equilibrium concentrations: $[FeNCS^{2+}] = 1.71 \times 10^{-4}\ M$, $[Fe^{3+}] = 8.28 \times 10^{-4}\ M$, and $[SCN^{-}] = 4.28 \times 10^{-4}\ M$.

5. What does *abc* in the Beer-Lambert law stand for?

6. What is the unit of absorbance in the Beer-Lambert law? Does it increase or decrease as the solution concentration of the absorbing substance decreases?

7. Why is a calibration curve constructed? How?

8. Briefly explain the meanings of the following terms as they relate to this experiment (a) *reversible reactions*, (b) *state of dynamic equilibrium*, (c) *equilibrium constant expression*, and (d) *equilibrium constant*.

9. How are percent transmittance and absorbance related algebraically?

Name ______________________ Desk ______________________

Date ______________________ Laboratory Instructor ______________________

Unknown no. ______________________

REPORT SHEET | EXPERIMENT

Colorimetric Determination of an Equilibrium Constant in Aqueous Solution

22

A. Preparation of the Calibration Curve

Concentration of $Fe(NO_3)_3$ in 0.10 *M* HNO_3 solution ______________

Concentration of NaSCN in 0.10 *M* HNO_3 solution ______________

	Flask Number					
	1	2	3	4	5	6
Volume of NaSCN, mL Solution						
Initial $[SCN^-]$, *M*						
Equil. $[FeNCS^{2+}]$, *M*						
Percent T						
Absorbance						

B. Determination of the Equilibrium Constant

Concentration of $Fe(NO_3)_3$ in 0.10 *M* HNO_3 solution ______________

Concentration of NaSCN in 0.10 *M* HNO_3 solution ______________

	Test Tube Number					
	1	2	3	4	5	6
Solution Temperature						
Volume of $Fe(NO_3)_3$ Solution, mL						
Volume of NaSCN Solution, mL						
Initial $[Fe^{3+}]$, *M*						
Initial $[SCN^-]$, *M*						

Absorbance					
Equil. $[FeNCS^{2+}]$, M					
Equil. $[Fe^{3+}]$, M					
Equil. $[SCN^-]$, M					
K_{eq} (show calculations)					

Average K_{eq} (show calculations) __________

QUESTIONS

1. How would the accuracy of your determined K_{eq} change if all of your volume measurements were made with graduated cylinders rather than pipets?

2. If all of the SCN^- was not converted completely to $FeNCS^{2+}$ when the calibration curve was prepared, would this raise or lower the value of K_{eq}? Explain.

3. Use the mean value you determined for K_{eq} to calculate the SCN^- concentration in a solution whose initial Fe^{3+} concentration was 4.00×10^{-2} M and initial SCN^- concentration was 1.00×10^{-3} M. Is all of the SCN^- in the form of $FeNCS^{2+}$?

4. Does the result of the calculation in question 3 justify your original assumption that all of the SCN^- is in the form of $FeNCS^{2+}$?

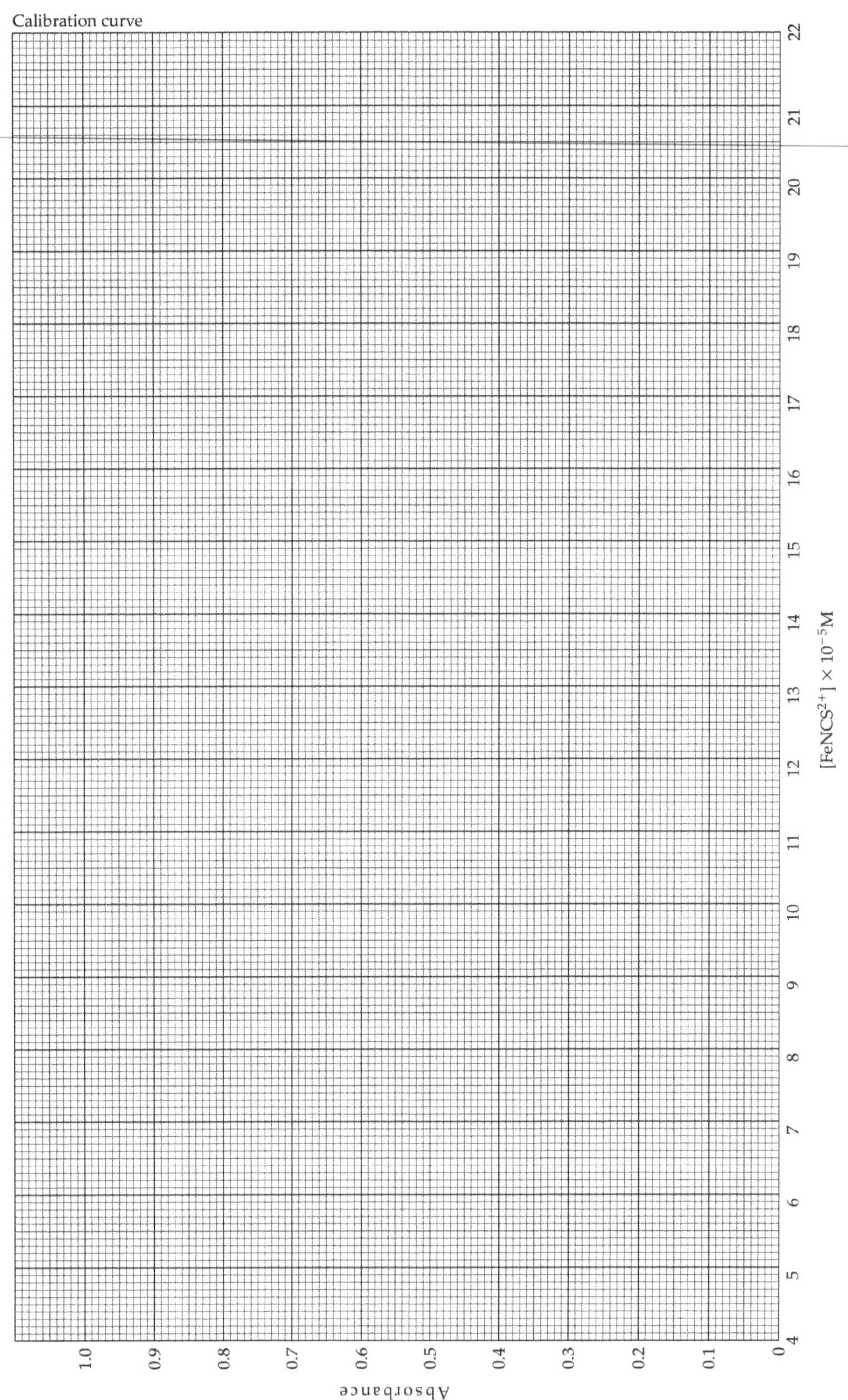
Calibration curve
Absorbance
0
0.1
0.2
0.3
0.4
0.5
0.6
0.7
0.8
0.9
1.0
$[FeNCS^{2+}] \times 10^{-5}M$
4
5
6
7
8
9
10
11
12
13
14
15
16
17
18
19
20
21
22

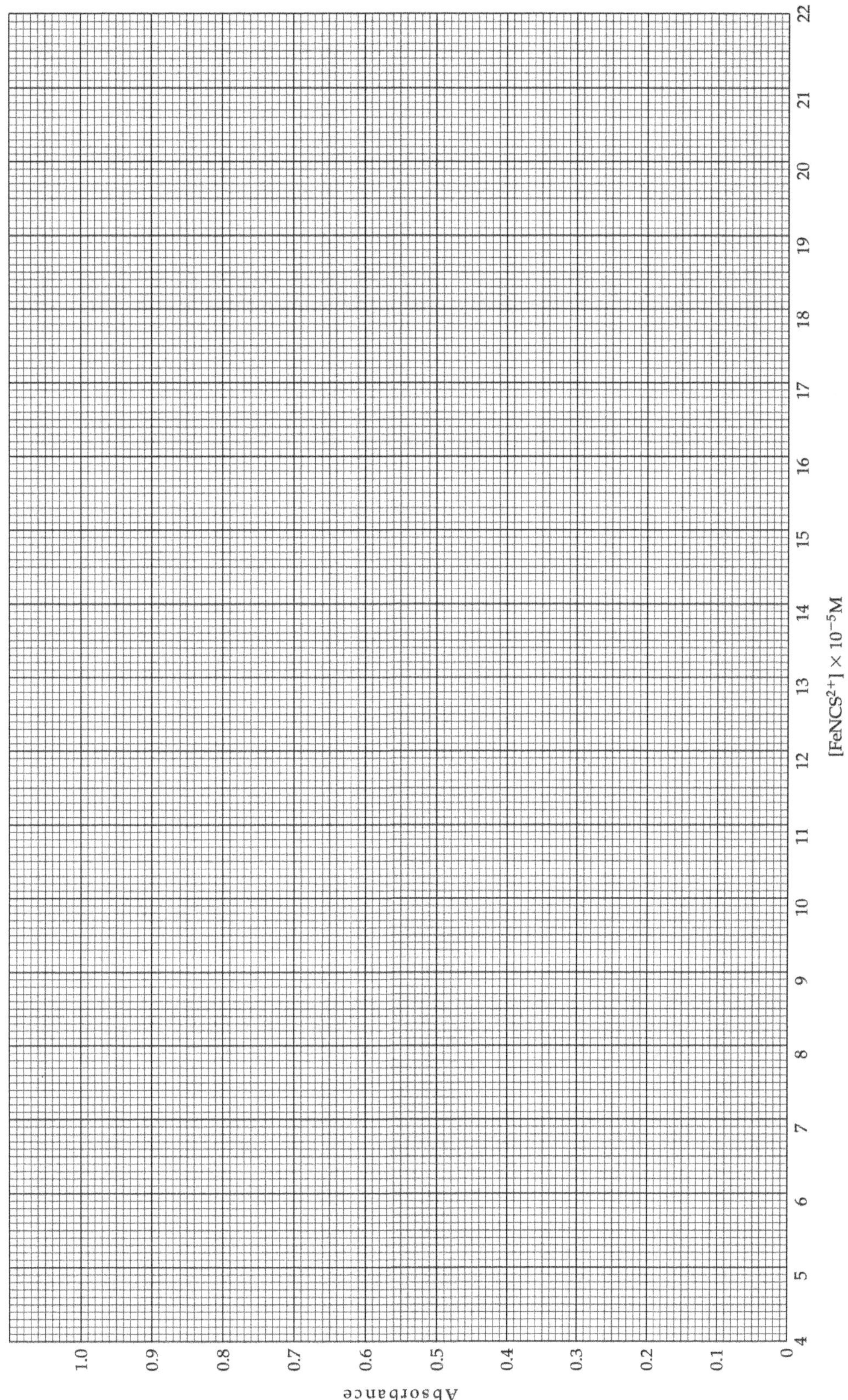
Absorbance
0
0.1
0.2
0.3
0.4
0.5
0.6
0.7
0.8
0.9
1.0
[FeNCS^{2+}] × 10^{-5}M
4
5
6
7
8
9
10
11
12
13
14
15
16
17
18
19
20
21
22

NOTES AND CALCULATIONS

Experiment

Chemical Equilibrium: Le Châtelier's Principle

OBJECTIVE

To study the effects of concentration and temperature on equilibrium positions.

APPARATUS AND CHEMICALS

Apparatus

medicine droppers (4)	large test tube (1)
250 mL beaker	ring stand, iron ring, and wire gauze
100 mL graduated cylinder	Bunsen burner and hose
small test tubes (3)	

Chemicals

0.1 *M* $CuSO_4$	0.01 *M* $AgNO_3$
0.1 *M* $NiCl_2$	6 *M* HNO_3
1.0 *M* $CoCl_2$	0.1 *M* HCl
0.1 *M* KI	1 *M* HCl
0.1 *M* Na_2CO_3	15 *M* NH_3

DISCUSSION

Many chemical reactions do not go to completion, that is, do not produce 100% yield of products. After a certain amount of time, many of these reactions appear to "stop"—colors stop changing, gases stop evolving, etc. In several of these instances, the process apparently stops before the reaction is complete, leading to a mixture of reactants and products.

For example, consider the interconversion of gaseous nitrogen oxides in a sealed tube.

$$\underset{\text{colorless}}{N_2O_4(g)} \rightleftharpoons \underset{\text{brown}}{2NO_2(g)} \qquad [1]$$

When frozen N_2O_4 is warmed above its boiling point (21.2 °C), the gas in a sealed tube progressively turns darker as colorless N_2O_4 dissociates into brown NO_2. The color change eventually stops even though N_2O_4 is still present in the tube.

The condition in which the concentrations of all reactants and products in a closed system cease to change with time is called *chemical equilibrium* (Section 15.1). Chemical equilibrium occurs when the rate at which products are formed from reactants equals the rate at which reactants are formed from products. For equilibrium to occur, neither reactants nor products can escape from the system.

If the concentration of any one of the reactants or products involved in a chemical equilibrium is changed or if the temperature is changed, the position of the equilibrium shifts to *minimize the change*. For example, assuming that the reaction represented by Equation [1] is at equilibrium, if more NO_2 is added, the probability of it reacting with other NO_2 molecules is increased, and the concentration of NO_2 decreases until a new state of equilibrium is attained. The equilibrium reaction is said to *shift to the left* (Section 15.7). *Le Châtelier's principle* states that if a system at equilibrium is disturbed (by altering the concentration of reactants or products, the temperature, or the pressure), the equilibrium will shift to minimize the disturbing influence. By this principle, if a reactant or product is added to a system at equilibrium, the equilibrium will shift to consume the added substance. Conversely, if reactant or product is removed, the equilibrium will shift to replenish the substance that was removed. The enthalpy change for a reaction indicates how a change in temperature affects the equilibrium. For an endothermic reaction in the forward direction, an increase in temperature shifts the equilibrium to the right to absorb the added heat; for an exothermic reaction in the forward direction, an increase in temperature shifts the equilibrium to the left. The equilibrium of Equation [1] is endothermic, $\Delta H = +58$ kJ. Increasing the temperature will shift this equilibrium in the direction that absorbs the heat; so the equilibrium shifts to the right.

GIVE IT SOME THOUGHT

If this reaction happened to have a negative ΔH, which way would the equilibrium shift?

It is important to remember that changes in concentrations, while causing shifts in the equilibrium positions, do not cause a change in the value of the equilibrium constant. Only changes in temperature affect the values of equilibrium constants.

In this experiment, you will observe two ways that a chemical equilibrium can be disturbed: (1) by adding or removing a reactant or product and (2) by changing the temperature. You will interpret your observations and conclusions using Le Châtelier's principle.

Part I: Changes in Reactant or Product Concentrations

A. Copper and Nickel Ions

Aqueous solutions of copper(II) and nickel(II) appear blue and green, respectively. However, when aqueous ammonia, NH_3, is added to these solutions, their colors change to dark blue and pale violet, respectively (Section 15.7).

$$\underset{\text{blue}}{[Cu(H_2O)_4]^{2+}(aq)} + 4NH_3(aq) \rightleftharpoons \underset{\text{dark blue}}{[Cu(NH_3)_4]^{2+}(aq)} + 4H_2O(l)$$

$$\underset{\text{green}}{[Ni(H_2O)_6]^{2+}(aq)} + 6NH_3(aq) \rightleftharpoons \underset{\text{pale violet}}{[Ni(NH_3)_6]^{2+}(aq)} + 6H_2O(l)$$

Ammonia substitutes for water in these two reactions because the metal–ammonia bond is stronger than the metal–water bond, and the equilibria shift to the right, accounting for the color changes.

If a strong acid such as HCl is added to these ammoniacal solutions, their colors revert back to the original colors of blue and green. The equilibria shift left because the reactant ammonia, NH_3, is removed from the equilibria. It reacts with the acid to form ammonium ion according to reaction [2].

$$H^+(aq) + NH_3(aq) \rightleftharpoons NH_4^+(aq) \qquad [2]$$

Place about 1 mL of 0.1 *M* $CuSO_4$ in a small, clean test tube. Record the color of the solution on your report sheet (1). **(CAUTION: *Concentrated NH_3 has a strong irritating odor; do not inhale. If you come in contact with it, immediately wash the area with copious amounts of water.*)** Add 15 *M* NH_3 dropwise until a color change occurs and the solution is clear, not colorless. Some solid may initially form because $Cu(OH)_2$ is sparingly soluble. It will dissolve as more $NH_3(aq)$ is added. Record your observation on your report sheet (2). Mix the solution in the test tube by "tickling" the test tube with your fingers as you add the NH_3 (see Appendix J, Figure J.1). Add 1 *M* HCl dropwise while carefully mixing the solution in the test tube until the solution becomes clear and the color changes. Note the color (3).

GIVE IT SOME THOUGHT

a. When concentrated NH_3 is added to $[Ni(H_2O)_6]^{2+}$, which way will the equilibrium shift?

b. Which way will it shift when HCl is added?

Repeat the same procedure using 0.1 *M* $NiCl_2$ in place of the $CuSO_4$ and record your corresponding observations on your report sheet (4–6).

Dispose of the solutions in the test tubes in the designated waste containers.

B. Cobalt Ions

Cobalt(II) ions in aqueous solution appear pale pink. In the presence of a large concentration of chloride ions, the solution changes color to blue and the following equilibrium is established:

$$\underset{\text{pale pink}}{[Co(H_2O)_6]^{2+}(aq)} + 4Cl^-(aq) \rightleftharpoons \underset{\text{deep blue}}{[CoCl_4]^{2-}(aq)} + 6H_2O(l) \qquad [3]$$

GIVE IT SOME THOUGHT

a. When you add 12 *M* HCl, which concentrations are affected in Equation [3]?

b. How will this shift the equilibrium?

Place about 0.5 mL (10 drops) of 1 *M* $CoCl_2$ in a small, clean test tube and note the color (8). **(CAUTION: *Avoid inhalation and contact with concentrated HCl. If you come in contact with it, immediately wash the area with copious amounts of water.*)** Add dropwise 12 *M* HCl to the test tube until a distinct color change occurs. Record the color on your report sheet (9). Slowly add water to the test tube while mixing. Record the color change on your report sheet (10).

Dispose of the solution in the test tube in the designated waste container.

Part II. Equilibria Involving Sparingly Soluble Salts

Silver carbonate, silver chloride, and silver iodide salts are only very slightly soluble in water. They can be precipitated from silver nitrate solutions by the addition of sodium salts containing the corresponding anions (Section 17.4). For example, silver carbonate will precipitate by the mixing of solutions of $AgNO_3$ and Na_2CO_3:

$$2AgNO_3(aq) + Na_2CO_3(aq) \rightleftharpoons Ag_2CO_3(s) + 2NaNO_3(aq)$$

for which the net ionic equation is:

$$2Ag^+(aq) + CO_3^{2-}(aq) \rightleftharpoons Ag_2CO_3(s) \qquad [4]$$

There is a dynamic equilibrium in the saturated solution of silver carbonate between the solid silver carbonate and its constituent silver and carbonate ions as shown in Equation [4]. In all saturated solutions, a dynamic equilibrium exists between the solid and the ions in solution.

The silver carbonate precipitate can be dissolved by the addition of nitric acid. Protons, H^+, from the HNO_3 react with the carbonate ions, CO_3^{2-}, to form unstable carbonic acid.

$$2H^+(aq) + CO_3^{2-}(aq) \rightleftharpoons H_2CO_3(aq);\ H_2CO_3(aq) \longrightarrow CO_2(g) + H_2O(l)$$

Removal of carbonate ions results in the dissolution of silver carbonate by a shift to the left of the equilibrium represented by Equation [4].

To 0.5 mL (10 drops) of 0.1 *M* Na_2CO_3 in a large, clean test tube, add 10 drops of 0.01 *M* $AgNO_3$. Record your observations on your report sheet (11). **(CAUTION:** ***Avoid contact with nitric acid,*** **HNO_3*. If you come in contact with it, immediately wash the area with copious amounts of water. Avoid contact with the*** **$AgNO_3$** ***solution as it will stain your skin purple. This stain is harmless and will eventually wear away.*)** Cautiously add 6 *M* HNO_3 dropwise to the test tube until you observe a change in appearance of the contents of the test tube (12). Save the contents for the next steps.

The above solution contains silver ions and nitrate ions because the Ag_2CO_3 dissolved in the nitric acid. Addition of chloride ions to this solution from HCl results in the precipitation of AgCl. The precipitated AgCl is in dynamic equilibrium with Ag^+ and Cl^- ions.

$$Ag^+(aq) + Cl^-(aq) \rightleftharpoons AgCl(s) \qquad [5]$$

This dynamic equilibrium can be disturbed by removing the Ag^+ ions, thereby forcing the equilibrium to shift to the left; as a result, the AgCl dissolves. Silver ions can be removed by the addition of NH_3 because they react with NH_3 to form $[Ag(NH_3)_2]^+$.

$$Ag^+(aq) + 2NH_3(aq) \rightleftharpoons [Ag(NH_3)_2]^+(aq) \qquad [6]$$

Because the equilibrium of Equation [6] lies much farther to the right than that of Equation [5], the AgCl will dissolve.

Adding acid to this ammoniacal solution will remove the NH_3 by forming NH_4^+ (see Equation [2]). This causes equilibrium 6 to shift to the left. The released Ag^+ will combine again with the Cl^- present to precipitate AgCl, as shown in Equation [5]. The reprecipitated AgCl can be redissolved by the addition of excess NH_3 for the same reason given previously (see Equation [6]).

To the solution you just saved, add 0.1 *M* HCl dropwise until you observe a change in the appearance of the contents of the test tube. Record your observations on your report sheet (13). **(CAUTION: *Concentrated* NH_3 *has a strong irritating odor; do not inhale. Do not get it on your skin. If you come in contact with it, immediately wash the area with copious amounts of water.*)** While mixing the contents of the test tube, add 15 *M* NH_3 dropwise until evidence of a chemical change occurs (14). Acidify the solution by the dropwise addition of 6 *M* HNO_3 until there is evidence of a chemical change. Record your observations on your report sheet (15). Again, while mixing, add 15 *M* NH_3 dropwise until there is no longer a change in the appearance of the contents of the test tube. Record your observations on your report sheet (16). Save the solution for the next step.

The equilibrium of Equation [6] can be disturbed by the addition of I^- from KI. Silver iodide will precipitate, removing Ag^+ and causing the equilibrium to shift to the left. The reason AgI will precipitate is because the equilibrium of Equation [7] lies much farther to the right than does the equilibrium of Equation [6].

$$I^-(aq) + Ag^+(aq) \rightleftharpoons AgI(s) \qquad [7]$$

To the solution from above, continue to add 0.1 *M* KI dropwise until you see evidence of a chemical reaction. Record your observations on your report sheet (17).

Dispose of the silver salt solution in the designated waste container.

Part III. Effect of Temperature on Equilibria

Heat about 75 mL of water to boiling in a 250 mL beaker on a ring stand. Place about 1 mL of 1.0 *M* $CoCl_2$ in a small test tube and place the test tube in the boiling water without spilling its contents. Compare the color of the cool cobalt solution to that of the hot solution (18).

GIVE IT SOME THOUGHT

a. When you heat the solution, what color does it turn?

b. Based on Equation [3], would heat be a product or a reactant in the equilibrium expression?

c. Based on these results, is the reaction endothermic or exothermic?

Dispose of the solution in the designated waste container.

NOTES AND CALCULATIONS

Name ____________________ Desk ____________________

Date ____________________ Laboratory Instructor ____________________

Chemical Equilibrium: Le Châtelier's Principle | 23 Pre-lab Questions

Before beginning this experiment in the laboratory, you should be able to answer the following questions.

1. Briefly state Le Châtelier's principle.

2. Consider the following equilibrium:

$$2CrO_4^{2-}(aq) + 2H^+(aq) \rightleftharpoons Cr_2O_7^{2-}(aq) + H_2O(l)$$

   a. Describe the change in appearance of the solution when the equilibrium shifts to the right.

   b. What happens when more acid is added?

   c. What happens when more alkali is added?

3. Consider the following reaction:

$$CO(g) + H_2(g) \rightleftharpoons C(s) + H_2O(g) \quad \Delta H = -131 \text{ kJ mol}^{-1}$$

Explain what would happen to the equilibrium position and the concentration of H_2 if

a. more C is added?

b. the concentration of CO is increased?

4. Complete and balance the following equations; then write balanced net ionic equations.

a. $AgNO_3(aq) + HCl(aq) \rightleftharpoons$

b. $NH_3(aq) + HCl(aq) \rightleftharpoons$

c. $Na_2CO_3(aq) + HNO_3(aq) \rightleftharpoons$

5. On the basis of Le Châtelier's principle, explain why Ag_2CO_3 dissolves when HNO_3 is added.

Name Grace Rademacher Desk

Date 2/23/21 Laboratory Instructor Prof. Towle

REPORT SHEET | EXPERIMENT

Chemical Equilibrium: Le Châtelier's Principle

23

Part I. Changes in Reactant or Product Concentrations

A. Copper and Nickel Ions

Colors:

1. $CuSO_4(aq)$ light blue
2. $[Cu(NH_3)_4]^{2+}(aq)$ dark blue
3. After HCl addition sky blue
4. $NiCl_2(aq)$ light blue-green
5. $[Ni(NH_3)_6]^{2+}(aq)$ mid-blue, purpleish
6. After HCl addition light blue-green

Explain the effects of $NH_3(aq)$ and $HCl(aq)$ on the $CuSO_4$ solution in terms of Le Châtelier's principle.

When adding NH_3, the equilibrium will shift to the right in order to maintain equilibrium due to being reactant-heavy.

(V↑ P↓) $CuSO_4 + 4NH_3 \rightleftharpoons [Cu(NH_3)_4]SO_4$ (V↓ P↑)

When adding HCl, the reaction will shift left because the color went back to the original. This means more product was produced as a result of HCl's addition.

7. What initially forms as pale blue and pale green precipitates when $NH_3(aq)$ is added to $[Cu(H_2O)_6]^{2+}$ and $[Ni(H_2O)_6]^{2+}$ solutions, respectively?

Pale blue is formed w/ the formation of $[Cu(NH_3)_6]^2$

Pale green is formed w/ the formation of $[Ni(NH_3)_6]^2$

B. Cobalt Ions

8. Color of $CoCl_2(aq)$ red-pink
9. Color after the addition of $HCl(aq)$ dark blue
10. Color after the addition of H_2O pink

Account for the changes you observed for the cobalt solutions in terms of Le Châtelier's principle.

Due to the color returning to its original hue, the equilibrium will shift to the left to balance out its product-heaviness.

back to original; shift left
Changes color; shift right

Part II. Equilibria Involving Sparingly Soluble Salts

11. Light pink $NO_2CO_3 + AgHO_3$
12. Clear $+HNO_3$

Account for your observations.

White solid of Ag_2CO_3 is formed, and then dissolves along CO_2

11) $NO_2CO_{3(aq)} + 2AgNO_{3(aq)} \rightarrow Ag_2CO_{3(s)} + 2NaNO_{3(aq)}$ white solid

12) $Ag_2CO_{3(s)} + 2HNO_{3(aq)} \rightarrow 2AgNO_{3(aq)} + H_2O_{(l)} + CO_{2(g)}$ Clear sol

13. Foggy white $+HCl$

Account for your observations.

$AgNO_{3(aq)} + HCl_{(aq)} \rightarrow AgCl_{(s)} + HNO_{3(aq)}$

White precipitate (silver chloride) forms when silver nitrate is added to the HCl

14. Did the precipitated AgCl dissolve? Explain.

Yes. AgCl has a white precipitate which dissolves to form a clear solution.

15. What effect did the addition of HNO_3 have on the contents of the test tube? Explain.

$Ag^+_{(aq)} + 2NH_{3(aq)} \rightarrow [Ag(NH_3)_2]^+$

HNO_3 being added will result in the creation of a silver ion, causing the reaction to turn into

$2HNO_3 + Ag(NH_3)_2Cl \rightarrow AgCl_{(s)} + 2NH_4^+ + 2NO_3^-$

16. What effect did the addition of NH_3 have on the contents of the test tube? Explain.

Caused the contents to turn cloudy white since AgCl precipitate reappears when nitric acid is added.

$[Ag(NH_3)_2]Cl_{(aq)} + 2HNO_{3(aq)} \rightarrow AgCl_{(s)} + 2NH_4NO_{3(aq)}$

★

17. Explain the effect of the addition of KI. - Catalyst

$AgNO_3 + KI \rightarrow AgI + KNO_3$

The addition of KI causes a deep yellow precipitate to form (AgI)

Part III. Effect of Temperature on Equilibria

18. Color of cool $CoCl_2$ Pale pink

Color of hot $[CoCl_4]^{2-}$ Blue

Is the reaction exothermic? No Explain.

Heating changes the color from pink to blue-ish, causing an equilibrium shift to the right. Temperature increases favor endothermic reactions, so its not exothermic.

NOTES AND CALCULATIONS

Experiment

Hydrolysis of Salts and pH of Buffer Solutions

OBJECTIVE

To learn about the concept of hydrolysis and to gain familiarity with acid–base indicators and the behavior of buffer solutions.

APPARATUS AND CHEMICALS

Apparatus

500 mL Erlenmeyer flask
150 mL beakers (2)
10 and 100 mL graduated cylinders
test tubes (6)
test-tube rack
wire gauze
stirring rods (2)
pH meter
balance
1 mL pipet
Bunsen burner and hose
ring stand and iron ring
plastic wash bottle

Chemicals

$NaC_2H_3O_2 \cdot 3H_2O$
0.1 *M* $ZnCl_2$
0.1 *M* NH_4Cl
0.1 *M* $KAl(SO_4)_2$
0.1 *M* Na_2CO_3
0.1 *M* NaCl
0.1 *M* $NaC_2H_3O_2$
6.0 *M* HCl
3.0 *M* $HC_2H_3O_2$
6.0 *M* NaOH
dropper bottles of
methyl orange
methyl red
bromothymol blue
phenolphthalein
alizarin yellow R
phenol red
standard buffer solution (pH 4.0)

DISCUSSION

You expect solutions of substances such as HCl and HNO_2 to be acidic and solutions of NaOH and NH_3 to be basic. However, you may be somewhat surprised to discover that aqueous solutions of some salts (for example, sodium nitrite, $NaNO_2$, and potassium acetate, $KC_2H_3O_2$) are basic, whereas others (for example, NH_4Cl and $FeCl_3$) are acidic. Recall that salts are the products formed in neutralization reactions of acids and bases (Section 4.3). For example, when NaOH and HNO_2 (nitrous acid) react, the salt $NaNO_2$ is formed.

$$NaOH(aq) + HNO_2(aq) \longrightarrow NaNO_2(aq) + H_2O(l)$$

Nearly all soluble salts are strong electrolytes and exist as ions in aqueous solutions. Many ions react with water to produce acidic or basic solutions. The reactions of ions with water are frequently called *hydrolysis reactions* (Section 16.9). You will see that anions such as CN^- and $C_2H_3O_2^-$ that are the conjugate bases of the weak acids HCN and $HC_2H_3O_2$, respectively, react with water to form OH^- ions. Cations such as NH_4^+ and Fe^{3+} come from weak bases and react with water to form H^+ ions.

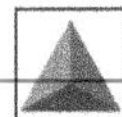

GIVE IT SOME THOUGHT

How do you determine the acid/base strength of these cations?

A. Hydrolysis of Anions: Basic Salts

Consider the behavior of anions first. Anions of weak acids react with proton sources. When placed in water, these anions react to some extent with water to accept protons and generate OH^- ions and thus cause the solution pH values to be greater than 7. Recall that proton acceptors are Brønsted-Lowry bases (Section 16.2). Thus, the anions of weak acids are basic in two senses: They are proton acceptors, and their aqueous solutions have pH values greater than 7. The nitrite ion, for example, reacts with water to increase the concentration of OH^- ions.

$$NO_2^-(aq) + H_2O(l) \rightleftharpoons HNO_2(aq) + OH^-(aq)$$

This reaction of the nitrite ion is similar to that of weak bases such as NH_3 with water.

$$NH_3(aq) + H_2O(l) \rightleftharpoons NH_4^+(aq) + OH^-(aq)$$

Thus, both NH_3 and NO_2^- are bases and as such have a base dissociation constant, K_b, (Section 16.7) associated with their corresponding equilibria.

According to the Brønsted-Lowry theory, the nitrite ion is the conjugate base of nitrous acid. Now consider the conjugate acid–base pair HNO_2 (Section 16.8) and NO_2^- and their behavior in water.

$$HNO_2(aq) \rightleftharpoons H^+(aq) + NO_2^-(aq) \qquad K_a = \frac{[H^+][NO_2^-]}{[HNO_2]}$$

$$NO_2^-(aq) + H_2O(l) \rightleftharpoons HNO_2(aq) + OH^-(aq) \qquad K_b = \frac{[HNO_2][OH^-]}{[NO_2^-]}$$

Multiplication of these dissociation constants yields

$$K_a \times K_b = \left(\frac{[H^+][\cancel{NO_2^-}]}{[\cancel{HNO_2}]}\right)\left(\frac{[\cancel{HNO_2}][OH^-]}{[\cancel{NO_2^-}]}\right) = [H^+][OH^-] = K_w$$

where K_w is the ion-product constant of water (Section 16.3).

Thus, the product of the acid-dissociation constant for an acid and the base-dissociation constant for its conjugate base is the ion-product constant for water.

$$K_a \times K_b = K_w = 1.0 \times 10^{-14} \text{ at } 25\ °C \qquad [1]$$

Knowing the K_a for a weak acid, you can easily find the K_b for the anion of the acid.

$$K_b = \frac{K_w}{K_a} \qquad [2]$$

By consulting a table of acid-dissociation constants, such as Appendix F, you can find that K_a for nitrous acid is 4.5×10^{-4}. Using this value, you can readily determine K_b for NO_2^-.

$$K_b = \frac{1.0\times10^{-14}}{4.5\times10^{-4}} = 2.2\times10^{-11}$$

Note that the stronger the acid the larger its K_a and that the weaker its conjugate base, the smaller its K_b. Likewise, the weaker the acid (the smaller the K_a), the stronger the conjugate base (the larger the K_b).

Anions derived from *strong acids*, such as Cl^- from HCl, do not react with water to affect the pH. Nor do Br^-, I^-, NO_3^-, SO_4^{2-}, and ClO_4^- affect the pH, for the same reason. They are spectator ions in the acid–base sense and can be described as neutral ions. Similarly, cations from strong bases, such as Na^+ from NaOH and K^+ from KOH, do not react with water to affect the pH. Hydrolysis of an ion occurs only when it can form a molecule or anion that is a weak electrolyte in the reaction with water. Strong acids and bases do not exist as molecules in dilute water solutions.

GIVE IT SOME THOUGHT

a. Do you write an equilibrium expression for a reaction involving strong acids or strong bases?

b. How does this come into play in describing these neutral cations and anions?

EXAMPLE 24.1

What is the pH of a 0.10 *M* NaClO solution if K_a for HClO is 3.0×10^{-8}?

SOLUTION: The salt NaClO exists as Na^+ and ClO^-. The Na^+ ions are spectator ions, but ClO^- ions undergo hydrolysis to form the weak acid HClO. Let x equal the equilibrium concentration of HClO (and OH^-).

$$\begin{array}{lcl} ClO^-(aq) + H_2O(l) & \rightleftharpoons & HClO(aq) + OH^-(aq) \\ (0.10 - x)\,M & & x\,M \qquad\qquad x\,M \end{array}$$

The value of K_b for the reaction is $(1.0\times10^{-14})/(3.0\times10^{-8}) = 3.3\times10^{-7}$. Because K_b is so small, you can neglect x in comparison with 0.10; thus, $0.10 - x \simeq 0.10$.

$$\frac{[HClO][OH^-]}{[ClO^-]} = K_b$$

$$\frac{x^2}{0.10} = 3.3\times10^{-7}$$

$$x^2 = 3.3\times10^{-8}$$

$$x = 1.8\times10^{-4}\ M$$

$$pOH = 3.74$$

$$\text{and } pH = 14 - 3.74 = 10.26$$

Anions with ionizable protons such as HCO_3^-, $H_2PO_4^-$, and HPO_4^{2-} may be acidic or basic depending upon the relative values of K_a and K_b for the ion. You will not consider such ions in this experiment.

Hydrolysis of Cations: Acidic Salts

Cations that are derived from weak bases react with water to increase the hydrogen-ion concentration; they form acidic solutions. The ammonium ion is derived from the weak base NH_3 and reacts with water as follows:

$$NH_4^+(aq) + H_2O(l) \rightleftharpoons H_3O^+(aq) + NH_3(aq)$$

This reaction is completely analogous to the dissociation of any other weak acid, such as acetic acid or nitrous acid. You can represent this acid dissociation of NH_4^+ more simply.

$$NH_4^+(aq) \rightleftharpoons NH_3(aq) + H^+(aq)$$

Here too the acid dissociation constant is related to the K_b of NH_3, which is the conjugate base of NH_4^+.

$$NH_3(aq) + H_2O(l) \rightleftharpoons NH_4^+(aq) + OH^-(aq)$$

Knowing the value of K_b for NH_3, you can readily calculate the acid dissociation constant from Equation [3].

$$K_a = \frac{K_w}{K_b} \qquad [3]$$

Cations of the alkali metals (group 1A) and the larger alkaline earth ions, Ca^{2+}, Sr^{2+}, and Ba^{2+}, do not react with water because they come from strong bases. Thus, these ions have no influence on the pH of aqueous solutions. They are merely spectator ions in acid–base reactions. Consequently, they are described as being neutral in the acid–base sense. The cations of most other metals do hydrolyze to produce acidic solutions. Metal cations are coordinated with water molecules, and it is the hydrated ion that serves as the proton donor. The following equations illustrate this behavior for the hexaaquairon (III) ion:

$$[Fe(H_2O)_6]^{3+}(aq) + H_2O(l) \rightleftharpoons [Fe(OH)(H_2O)_5]^{2+}(aq) + H_3O^+(aq) \qquad [4]$$

The coordinated water molecules are frequently omitted from such equations. For example, Equation [4] may be written as follows:

$$Fe^{3+}(aq) + H_2O(l) \rightleftharpoons Fe(OH)^{2+}(aq) + H^+(aq) \qquad [5]$$

Additional hydrolysis reactions can occur to form $Fe(OH)_2^+$ and may even lead to the precipitation of $Fe(OH)_3$. The equilibria for such cations are often complex, and not all species have been identified. However, equations such as [4] and [5] serve to illustrate the acidic character of dipositive and tripositive ions and account for most of the H^+ in these solutions.

Summary of Hydrolysis Behavior

Whether a solution of a salt will be acidic, neutral, or basic can be predicted on the basis of the strengths of the acid and base from which the salt was formed.

1. *Salt of a strong acid and a strong base*: Examples: NaCl, KBr, and $Ba(NO_3)_2$. Neither the cation nor anion hydrolyzes, and the solution has a pH of 7. It is neutral.
2. *Salt of a strong acid and a weak base*: Examples: NH_4Br, $ZnCl_2$, and $Al(NO_3)_3$. The cation hydrolyzes, forming H^+ ions, and the solution has a pH less than 7. It is acidic.
3. *Salt of a weak acid and a strong base*: Examples: $NaNO_2$, $KC_2H_3O_2$, and $Ca(OCl)_2$. The anion hydrolyzes, forming OH^- ions, and the solution has a pH greater than 7. It is basic.
4. *Salt of a weak acid and a weak base*: Examples: NH_4F, $NH_4C_2H_3O_2$, and $Zn(NO_2)_2$. Both ions hydrolyze. The pH of the solution is determined by the relative extent to which each ion hydrolyzes.

In this experiment, you will test the pH of water and of several aqueous salt solutions to determine whether these solutions are acidic, basic, or neutral. In each case, the salt solution will be 0.1 *M*. Knowing the concentration of the salt solution and the measured pH of each solution allows you to calculate K_a or K_b for the ion that hydrolyzes. Example 24.2 illustrates such calculations.

EXAMPLE 24.2

Calculate K_b for OBr^- assuming that a 0.10 *M* solution of NaOBr has a pH of 10.85.

SOLUTION: The spectator ion is Na^+. Alkali metal ions do not react with water and have no influence on pH. The ion OBr^- is the anion of a weak acid and thus reacts with water to produce OH^- ions.

$$OBr^- + H_2O \rightleftharpoons HOBr + OH^-$$

The corresponding expression for the base dissociation constant is

$$K_b = \frac{[HOBr][OH^-]}{[OBr^-]} \qquad [6]$$

If the pH is 10.85,

$$pOH = 14.00 - 10.85 = 3.15$$

and

$$[OH^-] = \text{antilog}\,(-3.15) = 7.1\times10^{-4}\ M$$

The concentration of HOBr that is formed along with OH^- must also be 7.1×10^{-4} *M*. The concentration of OBr^- that has not hydrolyzed is

$$[OBr^-] = 0.10\ M - 0.00071\ M \simeq 0.10\ M$$

Substituting these values into Equation [6] for K_b yields

$$K_b = \frac{[7.1\times10^{-4}][7.1\times10^{-4}]}{[0.10]}$$
$$= 5.0\times10^{-6}$$

You will use a set of indicators to determine the pH of various salt solutions. The dark areas in Figure 24.1 denote the transition ranges for the indicators you will use.

GIVE IT SOME THOUGHT

What is the function of an indicator?

Generally, you will find the solutions tested to be more acidic than you would have predicted them to be. A major reason for this increased acidity is the occurrence of CO_2 dissolved in the solutions. CO_2 reacts with water to generate H^+.

$$CO_2(g) + H_2O(l) \rightleftharpoons H_2CO_3(aq) \rightleftharpoons H^+(aq) + HCO_3^-(aq)$$

The solubility of CO_2 is greatest in basic solutions, intermediate in neutral solutions, and least in acidic solutions. Therefore, even distilled water will be somewhat acidic unless it is boiled to remove the dissolved CO_2.

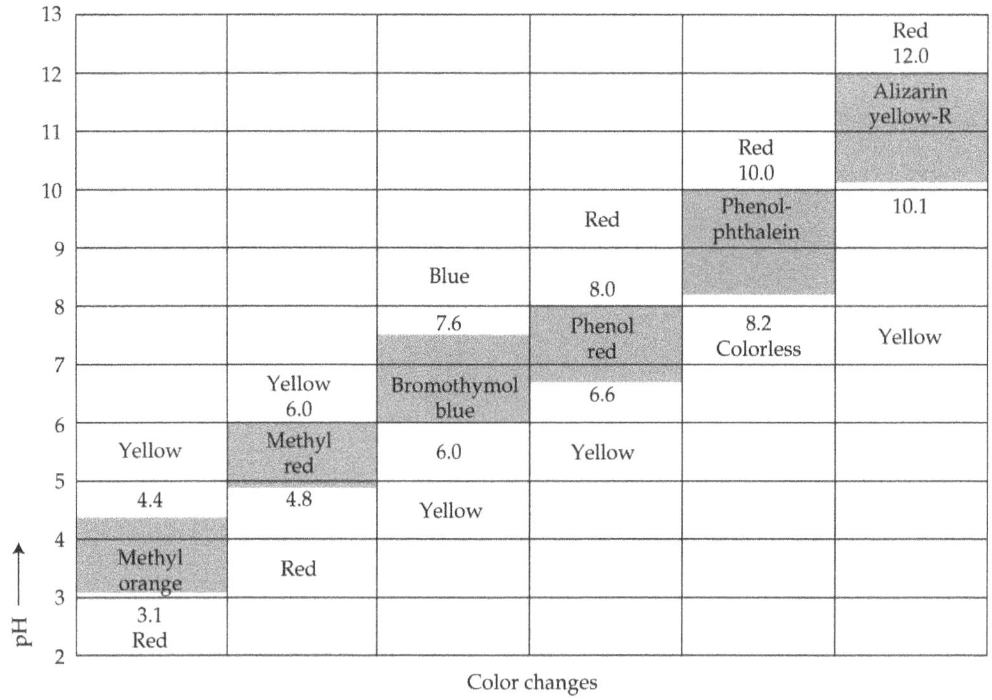

pH →						
13–12						Red 12.0
12–11						Alizarin yellow-R
11–10					Red 10.0	
10–9				Red	Phenol-phthalein	10.1
9–8			Blue	8.0		
8–7			7.6	Phenol red	8.2 Colorless	Yellow
7–6		Yellow 6.0	Bromothymol blue	6.6		
6–5	Yellow	Methyl red	6.0	Yellow		
5–4	4.4	4.8	Yellow			
4–3	Methyl orange	Red				
3–2	3.1 Red					

Color changes

▲**FIGURE 24.1** The color behavior of indicators.

B. Buffer Solutions

Chemists, biologists, and environmental scientists frequently need to control the pH of aqueous solutions. The effects on pH caused by the addition of a small amount of a strong acid or base to water are dramatic. The addition of a mere 0.001 mole of HCl to 1 L of water causes the pH to drop instantly from 7.0 to 3.0 as the hydronium ion concentration increases from 1×10^{-7} to 1×10^{-3} mol/L. On the other hand, the addition of 0.001 mole of NaOH to 1 L of water causes the pH to increase from 7.0 to 11.0. That life could not exist without some mechanism for controlling or absorbing excess acid or base is indicated by the narrow normal range of blood pH—7.35 to 7.45.

GIVE IT SOME THOUGHT

What is the function of a buffer?

The control of pH is often accomplished by use of *buffer solutions* (often simply called *buffers*) (Section 17.2). A buffer solution has the important property of resisting large changes in pH upon the addition of small amounts of strong acids or bases. A buffer solution must have two components—one that will react with H^+ and the other that will react with OH^-. The two components of a buffer solution are usually a weak acid and its conjugate base, such as $HC_2H_3O_2$–$C_2H_3O_2^-$ or NH_4^+–NH_3. Thus, buffers are often prepared by mixing a weak acid or a weak base with a salt of that acid or base. For example, the $HC_2H_3O_2$–$C_2H_3O_2^-$ buffer can be prepared by adding $NaC_2H_3O_2$ to a solution of $HC_2H_3O_2$; the NH_4^+–NH_3 buffer can be prepared by adding NH_4Cl to a solution of NH_3. By the appropriate choice of components and their concentrations, buffer solutions of virtually any pH can be made.

Now examine how a buffer works. Consider a buffer composed of a hypothetical weak acid HX and one of its salts MX, where M^+ could be Na^+, K^+, or another cation. The acid dissociation equilibrium in this buffer involves both the acid, HX, and its conjugate base X^-.

$$HX(aq) \rightleftharpoons H^+(aq) + X^-(aq) \qquad [1]$$

The corresponding acid dissociation constant expression is

$$K_a = \frac{[H^+][X^-]}{[HX]} \qquad [2]$$

Solving this expression for $[H^+]$, you have

$$[H^+] = K_a \frac{[HX]}{[X^-]} \qquad [3]$$

You can see from this expression that the hydrogen ion concentration (and therefore the pH) is determined by two factors: the value of K_a for the weak acid component of the buffer and the ratio of the concentrations of the weak acid and its conjugate base, $[HX]/[X^-]$.

When OH^- ions are added to the buffered solution, they react with the acid component of the buffer.

$$OH^-(aq) + HX(aq) \longrightarrow H_2O(l) + X^-(aq) \qquad [4]$$

This reaction results in a slight decrease in the [HX] and a slight increase in the $[X^-]$ as long as the amounts of HX and X^- in the buffer are large compared to the amount of the added OH^-. In that case, the ratio $[HX]/[X^-]$ doesn't change much; thus, the change in the pH is small.

When H^+ ions are added to the buffered solution, they react with the base component of the buffer.

$$H^+(aq) + X^-(aq) \longrightarrow HX(aq) \qquad [5]$$

This reaction causes a slight decrease in the $[X^-]$ and a slight increase in the [HX]. Once again, as long as the change in the ratio $[HX]/[X^-]$ is small, the change in the pH will be small.

Buffers resist changes in pH most effectively when the concentrations of the conjugate acid–base pair, HX and X^-, are about the same. From examining Equation [3], you can see that under these conditions, their ratio is close to one; thus, the $[H^+]$ is approximately equal to K_a. For this reason, you try to select a buffer whose acid form has a pK_a close to the desired pH.

With regard to pH, you take the negative logarithm of both sides of Equation [3] and obtain

$$-\log[H^+] = -\log K_a - \log\frac{[HX]}{[X^-]}$$

Because $-\log[H^+] = \text{pH}$ and $-\log[K_a] = \text{p}K_a$, you have

$$\text{pH} = \text{p}K_a - \log\frac{[HX]}{[X^-]}$$

And making use of the properties of logarithms (see Appendix B), you have

$$\text{pH} = \text{p}K_a + \log\frac{[X^-]}{[HX]} \qquad [6]$$

In general,

$$\text{pH} = \text{p}K_a + \log\frac{[\text{conjugate base}]}{[\text{weak acid}]} \qquad [7]$$

This relationship is known as the *Henderson-Hasselbalch equation* (Section 17.2). Biochemists, biologists, and others who frequently work with buffers often use this equation to calculate the pH of buffers. What makes this equation particularly convenient is that you can usually neglect the amounts of the acid and base of the buffer that ionize because they are comparatively small. Therefore, you can use the *initial concentrations* of the acid and conjugate base components of the buffer directly in Equation [7].

EXAMPLE 24.3

What is the pH of a buffer that is 0.120 *M* in lactic acid, $HC_3H_5O_3$, and 0.100 *M* in sodium lactate, $NaC_3H_5O_3$? For lactic acid, $K_a = 1.4 \times 10^{-4}$.

SOLUTION: Because lactic acid is a weak acid, you may assume that its initial concentration is 0.120 *M* and that none of it has dissociated. You may also assume that the lactate ion concentration is that of the salt, sodium lactate, 0.100 *M*. Let *x* represent the concentration in mol/L of the lactic acid that dissociates. The initial and equilibrium concentrations involved in this equilibrium are as follows:

$$HC_3H_5O_3(aq) \rightleftharpoons H^+(aq) + C_3H_5O_3^-(aq)$$

Initial	0.120 *M*	0	0.100 *M*
Change	$-x\ M$	$+x\ M$	$+x\ M$
Equilibrium	$(0.120 - x)\ M$	$x\ M$	$(0.100 + x)\ M$

The equilibrium concentrations are governed by the equilibrium expression

$$K_a = 1.4 \times 10^{-4} = \frac{[H^+][C_3H_5O_3^-]}{[HC_3H_5O_3]} = \frac{x(0.100 + x)}{0.120 - x}$$

Because K_a is small and because of the presence of a common ion, you can expect *x* to be small relative to 0.120 and 0.100 *M*, i.e. that [initial] approximately equals [initial-*x*]. Thus, the equation can be simplified to give

$$1.4 \times 10^{-4} = \frac{x(0.100)}{0.120}$$

Solving for *x* gives a value that justifies your neglecting it.

$$x = [H^+] = \left(\frac{0.120}{0.100}\right)(1.4 \times 10^{-4}) = 1.7 \times 10^{-4}\ M$$
$$pH = -\log(1.7 \times 10^{-4}) = 3.77$$

Alternatively, you could have used the Henderson-Hasselbalch equation to calculate the pH directly.

$$pH = pK_a + \log\left(\frac{[\text{conjugate base}]}{[\text{weak acid}]}\right) = 3.85 + \log\left(\frac{0.100}{0.120}\right)$$
$$= 3.85 + (-0.08) = 3.77$$

Addition of Strong Acids or Bases to Buffers

Now consider in a quantitative way the manner in which a buffer solution responds to the addition of a strong acid or base. Consider a buffer that consists of the weak acid HX and its conjugate base X^- (from the salt NaX). When a strong acid is added to this buffer, the H^+ is consumed by the X^- to produce HX; thus, [HX] increases and $[X^-]$ decreases. Whereas when a strong base is added to the buffer, the OH^- is consumed by HX to produce X^-; in this case, [HX] decreases and $[X^-]$ increases. Basically, two steps are involved in calculating how the pH of the buffer responds to the addition of a strong acid or base. First, consider the acid–base neutralization reaction and determine its effect on [HX] and [X^-]. Second, after performing this stoichiometric calculation, use K_a and the new concentrations of [HX] and [X^-] to calculate the [H^+]. This second step in the calculation is a standard equilibrium calculation. This procedure is illustrated in Example 24.4.

EXAMPLE 24.4

A buffer is made by adding 0.120 mol of $HC_3H_5O_3$ and 0.100 mol of $NaC_3H_5O_3$ to enough water to make 1.00 L of solution. The pH of the buffer is 3.77 (see Example 24.3). Calculate the pH of the solution after 0.001 mol of NaOH is added. Assume no volume change.

SOLUTION: Solving this problem involves two steps.

Stoichiometric calculation: The OH^- provided by the NaOH reacts with the $HC_3H_5O_3$, the weak acid component of the buffer. The following table summarizes the concentrations before and after the neutralization reaction:

$$HC_3H_5O_3(aq) + OH^-(aq) \longrightarrow H_2O(l) + C_3H_5O_3^-(aq)$$

Before reaction	0.120 *M*	0.001 *M*	–	0.100 *M*
Change	−0.001 *M*	−0.001 *M*	–	+ 0.001 *M*
After reaction	0.119 *M*	0.0 *M*	–	0.101 *M*

Equilibrium calculation: After neutralization, the solution contains different concentrations for the weak acid–conjugate base pair. Now consider the following proton–transfer equilibrium to determine the pH of the solution:

$$HC_3H_5C_3(aq) + H_2O(aq) \rightleftharpoons H_3O^+(aq) + C_3H_5O_3^-(aq)$$

Before reaction	0.119 *M*	–	0	0.101 *M*
Change	− *x M*	–	+ *x M*	+ *x M*
After reaction	(0.119 − *x*) *M*	–	*x M*	(0.101 +*x*) *M*

$$K_a = \frac{[H_3O^+][C_3H_5O_3^-]}{[HC_3H_5O_3]} = \frac{(x)(0.101+x)}{0.119-x} \approx \frac{(x)(0.101)}{0.119} \cong 1.4\times10^{-4}$$

$$x = [H_3O^+] = \frac{(0.119)(1.4\times10^{-4})}{(0.101)} = 1.65\times10^{-4}\ M$$

$$pH = -\log(1.65\times10^{-4}) = 3.78$$

Note how the buffer resists a change in its pH. The addition of 0.001 mol of NaOH to a liter of this buffer results in a change in the pH of only 0.01 units whereas the addition of the same amount of NaOH to a liter of water results in a change of pH from 7.0 to 11.0, a change of 4.0 pH units!

PROCEDURE

A. Hydrolysis of Salts

Boil approximately 450 mL of distilled water for about 10 min to expel dissolved carbon dioxide. Cover and allow the water to cool to room temperature. While the water is boiling and subsequently cooling, add about 5 mL of unboiled distilled water to each of six test tubes. Add three drops of a different indicator to each of the test tubes (one indicator per tube) and record the colors on the report sheet. From these colors and the data given in Figure 24.1, determine the pH of the unboiled water to the nearest pH unit. (Remember that its pH is likely to be below 7 because of dissolved CO_2.) Empty the contents of the test tubes and rinse the test tubes three times with about 3 mL of boiled distilled water. Then pour about 5 mL of the boiled distilled water into each of the test tubes and add three drops of each indicator (one indicator per tube) to each tube. Record the colors and determine the pH. Empty the contents of the test tubes and rinse each tube three times with about 3 mL of boiled distilled water.

Repeat the same procedure to determine the pH of each of the following solutions that are 0.1 *M*: NaCl, $NaC_2H_3O_2$, NH_4Cl, $ZnCl_2$, $KAl(SO_4)_2$, and Na_2CO_3. Use 5 mL of each of these solutions per test tube. Do not forget to rinse the test tubes with boiled distilled water when you go from one solution to the next.

GIVE IT SOME THOUGHT

Before you perform the experiment, would you predict the pH of each of these salts to be greater than 7, equal to 7, or less than 7. Why?

From the pH values you determined, calculate the hydrogen and hydroxide ion concentrations for each solution. Complete the tables on the report sheets and calculate the K_a or K_b as appropriate.

Dispose of chemicals in designated receptacles as directed.

B. pH of Buffer Solutions

1. Preparation of Acetic Acid–Sodium Acetate Buffer

Determine the mass of about 3.5 g of $NaC_2H_3O_2 \cdot 3H_2O$ to the nearest 0.01 g, record its mass, and add it to a 150 mL beaker. Using a 10 mL graduated cylinder, measure 8.8 mL of 3.0 *M* acetic acid and add it to the beaker containing the sodium acetate. Using a graduated cylinder, measure 55.6 mL of distilled water and add it to the solution of acetic acid and sodium acetate. Assume that the volumes are additive. Stir the solution until all of the sodium acetate is dissolved. You will measure the pH of this solution using a pH meter. The operation and calibration of the pH meter are described in the next section. Calibrate the pH meter using a standard buffer with a pH of 4.0. After you have calibrated the pH meter, measure the pH of the buffer solution you have prepared and record the value. Save this buffer solution for Part 2.

GIVE IT SOME THOUGHT

Why is a solution of acetic acid and sodium acetate a candidate for a buffer solution?

Operation and Calibration of the pH Meter

1. Obtain a buffer solution of known pH.
2. Plug in the pH meter to the line current and allow at least 10 min for it to warm up. You should leave it plugged in until you are finished using it. *This does not apply to battery-operated meters.*

3. Turn the function knob on the pH meter to the standby position as directed.
4. *Prepare the electrodes*. Make certain the solution in the reference electrode extends well above the internal electrode. If it does not, ask your instructor to fill it with saturated KCl solution. Remove the rubber tip and slide down the rubber collar on the reference electrode. Thoroughly rinse the outside of the electrodes with distilled water as directed.
5. *Standardize the pH meter*. Carefully immerse the electrodes in the buffer solution contained in a small beaker. *Remember that the glass electrode is very fragile; it breaks easily!* Do not touch the bottom of the beaker with the electrodes! Turn the function knob to "read" or "pH." Turn the standardize knob until the pH meter indicates the exact pH of the buffer solution. Wait 5 s to make certain the reading remains constant. *Once you have standardized the pH meter, do not readjust the standardize knob.* Turn the function knob to standby. Carefully lift the electrodes from the buffer and rinse them with distilled water. The pH meter is now ready to use to measure pH.

2. Effect of Acid and Base on the Buffer pH

Pour half (32 mL) of the buffer solution you prepared earlier into another 150 mL beaker. Label the two beakers 1 and 2. **(CAUTION: *Concentrated* HCl *can cause severe burns. Avoid contact with it. If you come in contact with it, immediately wash the area with copious amounts of water.*)** Pipet 1.0 mL of 6.0 *M* HCl into beaker 1, mix, and then measure the pH of the resultant solution and record the pH. Remember to rinse the electrodes between pH measurements. **(CAUTION: *Sodium hydroxide can cause severe burns. Avoid contact with it. If you come in contact with it, immediately wash the area with copious amounts of water.*)** Similarly, pipet 1.0 mL of 6.0 *M* NaOH into beaker 2, mix, and then measure and record the pH of the resultant solution. Calculate the pH values of the original buffer solution and the values after the additions of the HCl and NaOH. How do the measured and calculated values compare? Dispose of the chemicals in the designated receptacles.

Name ______________________________ Desk ____________________

Date ____________________ Laboratory Instructor ______________________________

Hydrolysis of Salts and pH of Buffer Solutions | 24 Pre-lab Questions

Before beginning this experiment in the laboratory, you should be able to answer the following questions.

1. Define Brønsted-Lowry acids and bases.

2. Which of the following ions will react with water in a hydrolysis reaction: K^+, Ba^{2+}, Cu^{2+}, Zn^{2+}, F^-, SO_3^{2-}, and Cl^-?

3. For those ions in question 2 that undergo hydrolysis, write net ionic equations for the hydrolysis reaction.

4. The K_a for HCN is 4.9×10^{-10}. What is the value of K_b for CN^-?

5. What are the conjugate base and conjugate acid of HCO_3^-?

6. From what acid and what base were the following salts made: NaCl, $CaSO_4$, and K_3PO_4?

7. Define the term *salt.*

8. Tell whether 0.1 *M* solutions of the following salts would be acidic, neutral, or basic: $BaCl_2$, $CuSO_4$, $(NH_4)_2SO_4$, $ZnCl_2$, and NaCN.

9. If the pH of a solution is 6, what are the hydrogen and hydroxide ion concentrations?

10. Phenylboronic acid, $C_6H_5B(OH)_2$, is a monoprotic acid with a pK_a value of 8.86. Calculate the pH of 0.0100 mol/L phenylboronic acid.

11. What is the pH of a solution that is 0.20 *M* $HC_2H_3O_2$ and 0.40 *M* $NaC_2H_3O_2$? K_a for acetic acid is 1.8×10^{-5}.

Name Grace Rademacher Desk

Date 3/23/21 Laboratory Instructor Prof. Towle

+ perlite

REPORT SHEET | EXPERIMENT

Hydrolysis of Salts and pH of Buffer Solutions | 24

A. Hydrolysis of Salts

Solution	Ion expected to hydrolyze (if any)	Spectator ion(s) (if any)
0.1 *M* NaCl	–	Na^+, Cl^-
0.1 *M* Na_2CO_3	CO_3^{2-}	Na^+
0.1 *M* $NaC_2H_3O_2$	$C_2H_3O_2^-$	Na^+
0.1 *M* NH_4Cl	NH_4^+	Cl^-
0.1 *M* $ZnCl_2$	Zn^{2+}	Cl^-
0.1 *M* $KAl(SO_4)_2$	$Al(SO_4)_2^-$	K^+

Solution	Indicator Color*								
	Methyl orange	Methyl red	Bromo-thymol blue	Phenol red	Phenol-phtha-lein	Alizarin yellow R	pH	$[H^+]$ *M*	$[OH^-]$ *M*
H_2O (unboiled)							6.22	6.03×10^{-7}	1.66×10^{-8}
H_2O (boiled)							6.55	2.82×10^{-7}	3.55×10^{-8}
NaCl							7.36	4.37×10^{-8}	2.29×10^{-7}
$NaC_2H_3O_2$							7.56	2.75×10^{-8}	3.63×10^{-7}
NH_4Cl							6.01	9.77×10^{-7}	1.02×10^{-8}
$ZnCl_2$							5.6	2.51×10^{-6}	3.98×10^{-9}
$KAl(SO_4)_2$							3.27	5.37×10^{-4}	1.86×10^{-11}
Na_2CO_3							11.45	3.55×10^{-12}	2.82×10^{-3}

*Color key: org = orange; — = colorless; yell = yellow.

10^{-pH}

$14 - pH = pOH$

10^{-pOH}

CALCULATIONS

Solution	Net-ionic equation for hydrolysis	Expression for equilibrium constant (K_a or K_b)	Value of (K_a or K_b)
$NaC_2H_3O_2$	$C_2H_3O_2^- + H_2O \rightleftharpoons HC_2H_3O_2 + OH^-$	K_b	1.32×10^{-12}
Na_2CO_3	$CO_3^{2-} + H_2O \rightleftharpoons HCO_3^- + OH^-$	K_b	7.95×10^{-5}
NH_4Cl	$NH_4^+ + H_2O \rightleftharpoons H_3O^+ + NH_3$	K_a	1.04×10^{-15}
$ZnCl_2$	$Zn^{2+}_{(aq)} + H_2O_{(l)} \rightleftharpoons ZnOH^+_{(aq)} + H^+_{(aq)}$	K_a	1.58×10^{-16}
$KAl(SO_4)_2$	$KAl(SO_4)_2 + 4H_2O \rightleftharpoons 2H_2SO_4 + KOH + Al(OH)_3$	K_a	3.5×10^{-21}

K_a: pH < 7

Soln:

$NaC_2H_3O_2$: $K_b = \frac{[HC_2H_3O_2][OH^-]}{[C_2H_3O_2]}$ $\frac{(3.63 \times 10^{-7})^2}{0.1} = 1.32 \times 10^{-12}$

Na_2CO_3: $K_b = \frac{[HCO_3^-][OH^-]}{[CO_3]}$ $\frac{(2.82 \times 10^{-3})^2}{0.1} = 7.95 \times 10^{-5}$

NH_4Cl: $K_a = \frac{[H_3O^+][NH_3]}{[NH_4^+]}$ $\frac{(1.02 \times 10^{-8})^2}{0.1} = 1.04 \times 10^{-15}$

$ZnCl_2$: $K_a = \frac{[ZnOH^+][H^+]}{[Zn^{2+}]}$ $\frac{(3.98 \times 10^{-9})^2}{0.1} = 1.58 \times 10^{-16}$

$KAl(SO_4)_2$: $\frac{(1.86 \times 10^{-11})}{0.1} = 3.5 \times 10^{-21}$

QUESTIONS

X 1. Using the K_a's for $HC_2H_3O_2$ and HCO_3^- (from Appendix F), calculate the K_b's for the $C_2H_3O_2^-$ and CO_3^{2-} ions. Compare these values with those calculated from your measured pH's.

2. Using K_b for NH_3 (from Appendix G), calculate K_a for the NH_4^+ ion. Compare this value with that calculated from your measured pH's. $K_b(NH_3) = 1.8 \times 10^{-5}$ $K_w = 1.0 \times 10^{-14}$

$K_a = \frac{K_w}{K_b} = \frac{1.0 \times 10^{-14}}{1.8 \times 10^{-5}} = 5.56 \times 10^{-10}$

$K_a(NH_4^+) = 5.56 \times 10^{-10}$ is larger than my calculated value of 1.04×10^{-15}.

3. How should the pH of a 0.1 *M* solution of $NaC_2H_3O_2$ compare with that of a 0.1 *M* solution of $KC_2H_3O_2$? Explain briefly.

Since the concentration of CH_3COO^- is the same for each solution, their pH will be the same.

4. What is the greatest source of error in this experiment? How could you minimize this source of error?

Calibration errors were a big error source here - this was minimized by re-calibrating

B. pH of Buffer Solutions

Mass of $NaC_2H_3O_2 \cdot 3H_2O$ (FW = 136 g/mol) __________

pH of original buffer solution __________

pH of buffer + HCl __________

pH buffer + NaOH __________

Calculate pH of original buffer (show calculations) __________

Calculate pH of buffer + HCl (show calculations) ____________________

Calculate pH of buffer + NaOH (show calculations) ____________________

NOTES AND CALCULATIONS

Determination of the Dissociation Constant of a Weak Acid

OBJECTIVE

To become familiar with the operation of a pH meter and use pH to determine the magnitude of the equilibrium constant of a weak acid.

APPARATUS AND CHEMICALS

Apparatus

pH meter with electrodes	600 mL Erlenmeyer flask
balance	250 mL Erlenmeyer flasks (3)
150 mL beakers (4)	25 mL pipet and pipet bulb
buret	buret clamp and ring stand
500 mL plastic bottle with plastic lid	weighing bottle
Bunsen burner and hose	ring stand and ring
wire gauze	wash bottle

Chemicals

potassium hydrogen phthalate (KHP)	0.1 *M* NaOH or 19 *M* NaOH
standard buffer solution	phenolphthalein indicator solution
unknown solution of a weak acid (~ 0.1 *M*)	Buffer solution pH = 4.0

DISCUSSION

Acid–Base Equilibria

According to the Brønsted-Lowry acid–base theory, the strength of an acid is related to its ability to donate protons (Section 16.2). All acid–base reactions are then competitions between bases of various strengths for these protons. For example, the strong acid HCl reacts with water according to Equation [1].

$$HCl(aq) + H_2O(l) \longrightarrow H_3O^+(aq) + Cl^-(aq) \qquad [1]$$

This acid is a strong acid and is completely dissociated—in other words, 100% dissociated—in dilute aqueous solution. Consequently, the $[H_3O^+]$ concentration of 0.1 *M* HCl is 0.1 *M*.

By contrast, acetic acid, CH_3COOH (abbreviated HOAc), is a weak acid and is only slightly dissociated to form hydronium and acetate (abbreviated OAc^-) ions, as shown in Equation [2],

$$H_2O(l) + HOAc(aq) \rightleftharpoons H_3O^+(aq) + OAc^-(aq) \qquad [2]$$

Therefore, its acid dissociation constant, as shown by Equation [3], is small.

$$K_a = \frac{[H_3O^+][OAc^-]}{[HOAc]} = 1.8\times10^{-5} \qquad [3]$$

Acetic acid only partially dissociates in aqueous solution, and an appreciable quantity of undissociated acetic acid remains in solution.

For the general weak acid HA in aqueous solution, the dissociation reaction and dissociation constant expression are

$$HA(aq) + H_2O(l) \rightleftharpoons H_3O^+(aq) + A^-(aq) \qquad [4]$$

$$K_a = \frac{[H_3O^+][A^-]}{[HA]} \qquad [5]$$

Recall that pH is defined as (Section 16.4)

$$-\log[H_3O^+] = pH \qquad [6]$$

Solving Equation [5] for $[H_3O^+]$ and substituting this quantity into Equation [6] yields

$$[H_3O^+] = K_a \frac{[HA]}{[A^-]} \qquad [7]$$

$$-\log[H_3O^+] = -\log K_a - \log\frac{[HA]}{[A^-]} \qquad [8]$$

$$pH = pK_a - \log\frac{[HA]}{[A^-]} \qquad [9]$$

$$\text{where } pK_a = -\log K_a$$

If you titrate the weak acid HA with a base, there will be a point in the titration at which the number of moles of base added is half the number of moles of acid initially present. This is the point at which 50% of the acid has been titrated to produce A^- and 50% remains as HA. At this point, $[HA] = [A^-]$, the ratio $[HA]/[A^-] = 1$, and $\log [HA]/[A^-] = 0$. Hence, at this point in a titration (that is, at half the equivalence point) Equation [9] becomes

$$pH = pK_a \qquad [10]$$

By titrating a weak acid with a strong base and recording the pH versus the volume of base added, you can determine the ionization constant of the weak acid. This is accomplished by using the titration curve based on the titration data, as explained in the following paragraph.

From the titration curve (Figure 25.1), you can see that at the point denoted as half the equivalence point, where $[HA] = [A^-]$, the pH is 4.3. Thus, from Equation [10], at this point, $pH = pK_a$, or

$$pK_a = 4.3$$

$$-\log K_a = 4.3$$

$$\log K_a = -4.3$$

$$K_a = 5 \times 10^{-5}$$

(Your instructor will show you a graphical method for locating the equivalence point on your titration curves.)

Operation of the pH Meter

To measure the pH during the course of the titration, you will use an electronic instrument called a pH meter. This device consists of a meter and an electrode assembly, as illustrated in Figure 25.2. You may have a digital pH meter rather than the analog one illustrated here.

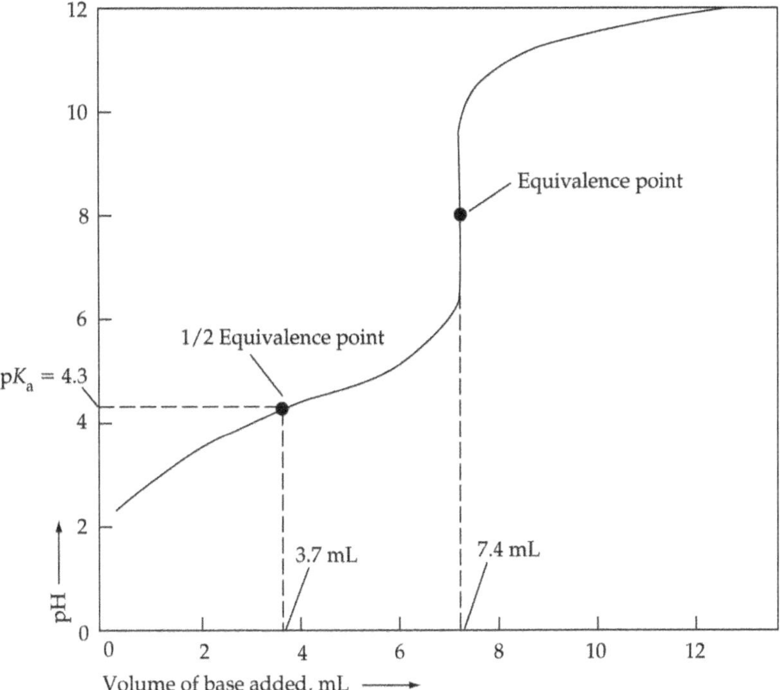

▲FIGURE 25.1 Exemplary titration curve for the titration of a weak acid HA with a strong base.

The main differences among pH meters involve the positions of the control knobs and the types of electrodes and electrode-mounting devices. The measurement of pH requires two electrodes: a sensing electrode that is sensitive to H_3O^+ concentrations and a reference electrode or a combination electrode. The two electrodes are necessary because the pH meter is really just a voltmeter that measures the electrical potential of a solution. Typical sensing and reference electrodes are illustrated in Figure 25.3.

The reference electrode is an electrode that develops a known potential that is essentially independent of the contents of the solution into which it is

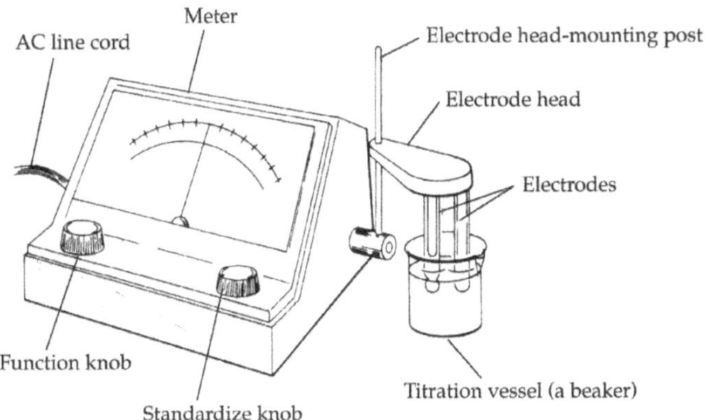

▲FIGURE 25.2 An analog pH meter.

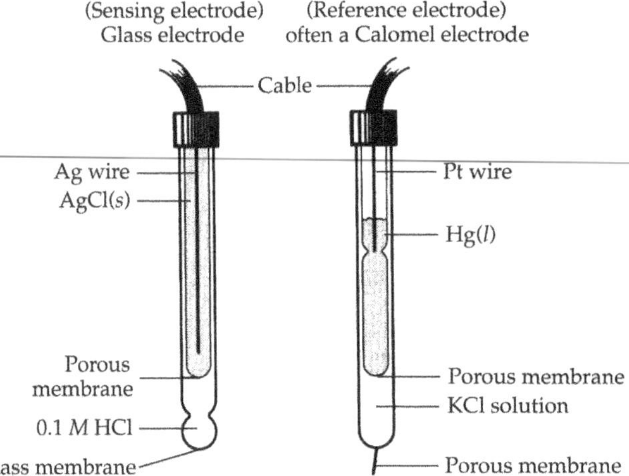

▲FIGURE 25.3 A typical sensing and reference electrode.

placed. The glass electrode is sensitive to the H_3O^+ concentration of the solution into which it is placed; its potential is a function of $[H_3O^+]$. It operates by transport of H_3O^+ ions through the glass membrane. This can be described more precisely, but for purposes here, it is sufficient for you to understand that two electrodes are required. These two electrodes are sometimes combined into an electrode called a combination electrode which appears to be a single electrode. However, the combination electrode does contain both a reference and a sensing electrode.

Preliminary Operations with the pH Meter

1. Obtain a buffer solution of known pH.

GIVE IT SOME THOUGHT

What is the function of a buffer solution?

2. Plug in the pH meter to line current and allow at least 10 min for warm-up. You should leave the meter plugged in until you are finished with it. *This does not apply to battery-operated meters.*
3. Turn the function knob on the pH meter to the standby position.
4. *Prepare the electrodes.* Make certain the solution in the reference electrode extends well above the internal electrode. If it does not, ask your instructor to fill it with saturated KCl solution. Remove the rubber tip and slide down the rubber collar on the reference electrode. Rinse the outside of the electrodes thoroughly with distilled water.
5. *Standardize the pH meter.* Carefully immerse the electrodes in the buffer solution contained in a small beaker. *Remember that the glass electrode is very fragile; it breaks easily! Do not* touch the bottom of the beaker with the electrodes! Turn the function knob to "read" or "pH." Turn the standardize knob until the pH meter indicates the exact pH of the buffer solution. Wait 5 s to make sure the reading remains constant. *Once you*

have standardized the pH meter, don't readjust the standardize knob. Turn the function knob to standby. Carefully lift the electrodes from the buffer and rinse them with distilled water. The pH meter is now ready to use to measure pH.

RECORD ALL DATA DIRECTLY ON THE REPORT SHEETS.

PROCEDURE

A. Preparation of Approximately 0.100 *M* Sodium Hydroxide (NaOH)

Heat 500 mL of distilled water to boiling in a 600 mL flask*; *after cooling under the water tap*, transfer it to a 500 mL plastic bottle with a plastic lid. **(CAUTION: *Concentrated* NaOH *can cause severe burns. If you come in contact with it, immediately wash the area with copious amounts of water.*)** Add 3 mL of stock solution of carbonate-free NaOH (approximately 19 *M*) and shake vigorously for at least 1 min.

Preparation of a Buret for Use Clean a 50 mL buret with soap solution and a buret brush and thoroughly rinse with tap water. Then rinse with at least five 10 mL portions of distilled water. The water must run freely from the buret without any drops adhering to the sides. Make sure the buret does not leak and the stopcock turns freely.

Reading a Buret All liquids, when placed in a buret, form a curved meniscus at their upper surface. In the case of water and water solutions, this meniscus is concave (Figure 25.4), and the most accurate buret readings are obtained by observing the position of the lowest point on the meniscus on the graduated scales.

GIVE IT SOME THOUGHT

a. What is a parallax error?
b. Where should your eye level be when you are reading the buret?

To avoid parallax errors when taking readings, your eye must be on a level with the meniscus or use a meniscus reader. Wrap a strip of paper around the buret and hold the top edges of the strip together, making sure they are even. Adjust the strip so that the front and back edges are in line with the lowest part of the meniscus and take the reading by estimating to the nearest tenth of a marked division (0.01 mL). A simple way to do this for repeated readings on a buret is illustrated in Figure 25.4.

*The water is boiled to remove carbon dioxide (CO_2), which would react with the NaOH and change its molarity.

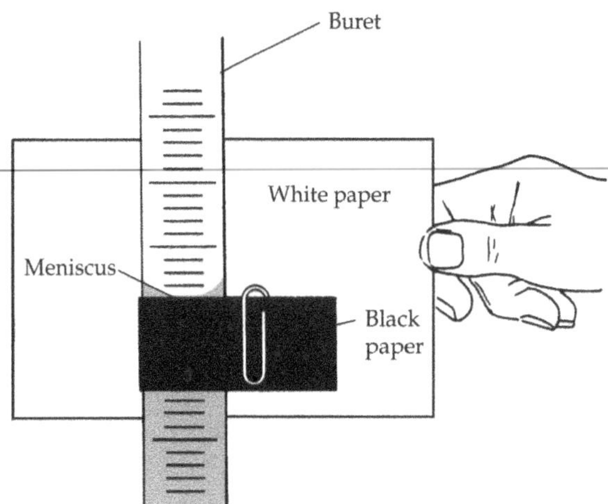

▲**FIGURE 25.4** Reading a buret.

B. Standardization of Sodium Hydroxide (NaOH) Solution

Prepare about 400 to 450 mL of CO_2-free water by boiling for about 5 min. Determine the mass of a weighing bottle (your lab instructor will show you how to use a weighing bottle if you don't already know) of triplicate samples of between 0.4 and 0.6 g each of pure potassium hydrogen phthalate (KHP, $KHC_8H_4O_4$, molar mass 204.22 g/mol) in three separate 250 mL Erlenmeyer flasks; accurately determine the mass to four significant figures.* Do not determine the mass of the flasks. Record the masses and label the three flasks so that you can distinguish among them. Add to each sample about 100 mL of distilled water that has been freed from CO_2 by boiling and warm gently by swirling until the salt is completely dissolved. Cover and allow to cool before titrating the samples. Add to each flask two drops of phenolphthalein indicator solution.

Rinse the previously cleaned buret with at least four 5 mL portions of the approximately 0.100 M NaOH solution that you prepared. Discard each portion into the designated receptacle. *Do not return any of the washings to the bottle.* Completely fill the buret with the solution and remove the air from the tip by dispensing some of the liquid into an empty beaker. Make sure the lower part of the meniscus is at the zero mark or slightly lower. Remove any hanging drop from the buret tip by touching it to the side of the beaker used for the washings. Allow the buret to stand for at least 30 s before you read the exact position of the meniscus. Record the initial buret reading on your report sheet.

Slowly add the NaOH solution to one of your flasks of KHP solution while gently swirling the contents of the flask, as illustrated in Figure 25.5. As the NaOH solution is added, a pink color appears where the drops of the base come in contact with the solution. This coloration disappears as you swirl the contents of the flask. As the end point is approached, the color disappears more slowly, at which time you should add the NaOH drop by drop. Continue

---

*In cases where the mass of a sample is larger than 1 g, it is necessary to determine the mass only to the nearest milligram to obtain four significant figures. Buret readings can be read only to the nearest 0.02 mL, and for readings greater than 10 mL, this represents four significant figures.

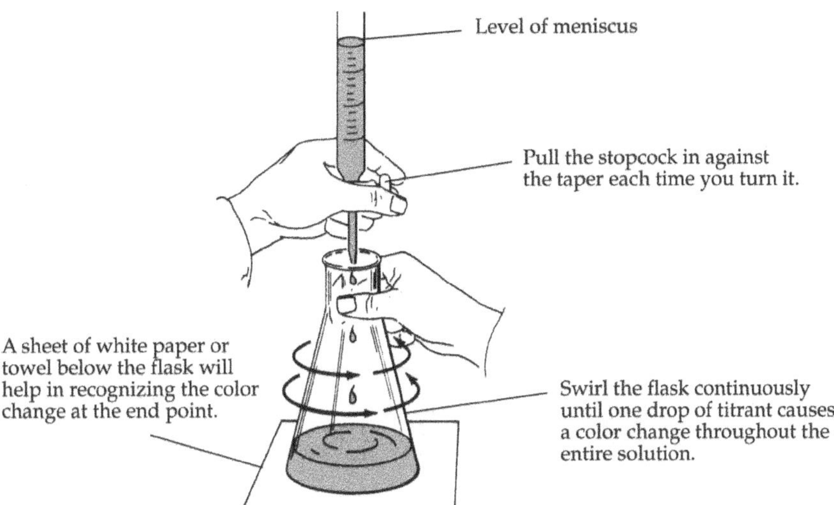
Level of meniscus
Pull the stopcock in against the taper each time you turn it.
A sheet of white paper or towel below the flask will help in recognizing the color change at the end point.
Swirl the flask continuously until one drop of titrant causes a color change throughout the entire solution.

▲**FIGURE 25.5** Titration procedure.

swirling the flask for the entire titration. The end point is reached when one drop of the NaOH solution turns the entire solution in the flask from colorless to pink that persists for at least 30 s. The solution should remain pink when it is swirled. Remove any hanging drop from the buret tip by touching it to the side of the flask and wash down the sides of the flask with a stream of water from the wash bottle. Allow the titrated solution to stand for at least 1 min so that the buret will drain properly. Record the buret reading on your report sheet. Repeat this procedure with the other two samples.

GIVE IT SOME THOUGHT

What shade of pink indicates the end point?

From the data you obtain in the three titrations, calculate the molarity of the NaOH solution to four significant figures.

The three determinations should agree within 1.0%. If they do not, repeat the standardization until agreement is reached. The average of the three acceptable determinations is taken as the molarity of the NaOH. Calculate the standard deviation of your results. *Save* your standardized solution for determining the pK_a of the unknown acid.

C. Determination of pK_a of Unknown Acid

With the aid of a pipet bulb, pipet a 25 mL aliquot of your unknown acid solution into a 250 mL beaker and carefully immerse the previously rinsed electrodes in this solution. Measure the pH of this solution by turning the function knob to "read" or "pH." Record the pH on the report sheet. Begin your titration, under constant mixing, by adding 1 mL of your standardized base from a buret and record the volume of titrant and pH. Repeat with successive additions of 1 mL of base until you approach the end point; then add 0.1 mL increments of base and record the pH and milliliters of NaOH added. After the rapid rise in pH is completed, the base can be added again in 1 mL increments. When the pH no longer changes upon addition of NaOH, your titration is complete. From these data, plot a titration curve of pH versus mL titrant added. Repeat the titration with two more 25 mL aliquots of your unknown acid and plot the titration curves. From these curves calculate the acid-dissociation constant. You may be able to save time if the first titration is run rapidly with larger-volume increments of the titrant to locate an approximate equivalence point; then the second and third titrations may be run with the small increments indicated previously. Turn the function knob to standby; rinse the electrodes with distilled water; and wipe them with a clean, dry tissue. Return them to the appropriate storage solution.

GIVE IT SOME THOUGHT

a. At what point in the procedure do you have to be extremely careful?

b. What can you do to save time?

D. Concentration of Unknown Acid

Using the volume of base at the equivalence point, its molarity, and the fact that you used 25.00 mL of acid, calculate the concentration of the unknown acid and record this on the report sheet.

GIVE IT SOME THOUGHT

Before you perform any calculations, think about the following: Based on the volume and the concentration of the standardized base, what do you expect the concentration of your unknown acid to be?

Name ______________________ Desk ______________

Date ______________ Laboratory Instructor ______________________

Determination of the Dissociation Constant of a Weak Acid | 25 Pre-lab Questions

Before beginning this experiment in the laboratory, you should be able to answer the following questions.

1. Define Brønsted-Lowry acids and bases.

2. At one-half equivalence point, the pH of an acid in an acid base titration was found to be 4.66. What is the value of K_a for this unknown acid?

3. Why isn't the pH at the equivalence point always equal to 7 in a neutralization titration? When is it 7?

4. What is the pK_a of an acid whose K_a is 4.2×10^{-5}?

5. Why must two electrodes be used to make an electrical measurement such as pH?

6. What is a buffer solution?

7. The pH at one-half the equivalence point in an acid–base titration was found to be 5.67. What is the value of K_a for this unknown acid?

8. If 30.15 mL of 0.0995 *M* NaOH is required to neutralize 0.279 g of an unknown acid, HA, what is the molar mass of the unknown acid?

9. Assuming that K_a is 1.85×10^{-5} for acetic acid, calculate the pH at one-half the equivalence point and at the equivalence point for a titration of 50 mL of 0.100 *M* acetic acid with 0.100 *M* NaOH.

Name Grace Rademacher Desk ________

Date 3/30/21 Laboratory Instructor Prof. Towle

REPORT SHEET | EXPERIMENT

Determination of the Dissociation Constant of a Weak Acid

25

B. Standardization of Sodium Hydroxide (NaOH) Solution

	Trial 1	*Trial 2*	*Trial 3*
Mass of bottle + KHP			
Mass of bottle			
Mass of KHP used			
Final buret reading			
Initial buret reading			
mL of NaOH used			
Molarity of NaOH			

Average molarity (show calculations and standard deviation) ________

Standard deviation (see Experiment 8) ________

x^3 type graph

C. Determination of pK_a of Unknown Acid

Don't graph

x-axis | y-axis

First determination		*Second determination*		*Third determination*	
mL NaOH	*pH*	*mL NaOH*	*pH*	*mL NaOH*	*pH*
0	2.62	0	2.62	0	2.68
2	3.32	2	3.34	2	3.34
4	3.68	4	3.71	4	3.75
6	3.95	6	3.96	6	4.02
8	4.15	8	4.18	8	4.24
10	4.33	10	4.35	10	4.4
12	4.50	12	4.51	12	4.56
14	4.65	14	4.67	14	4.71
16	4.82	16	4.82	16	4.87
18	4.99	18	5.0	18	5.03
20	5.19	20	5.21	20	5.23
22	5.46	22	5.44	22	5.47
24	5.92	24	5.84	24	5.92
26	9.94	24.5	6.07	24.5	6.08
28	11.49	25	6.33	25	6.38
30	11.75	25.5	6.94	25.5	6.8
32	11.89	26	9.85	26	9.69
34	11.99	26.5	10.71	26.5	10.72
36	12.05	27	11.09	27	11.11
		29	11.62	29	11.64
		31	11.82	31	11.83
		33	11.94	33	11.95
		35	12.02	35	12.03

Volume at equivalence point ~~_____~~ _____ _____

Volume at one-half equivalence point ~~_____~~ _____ _____

where pH = pK_a

26, 13

pK_a ~~_____~~ pK_a 4.64 pK_a 4.59

K_a ~~_____~~ K_a 2.29×10^{-5} K_a 2.57×10^{-5}

Average K_a (show calculations) 2.43×10^{-5} ~~Standard deviation of K_a~~ _____

This lab & hydrolysis for quiz next week

D. Concentration of Unknown Acid

	Trial 1	*Trial 2*	*Trial 3*
Volume of unknown acid	25mL	25mL	25mL
Average molarity of NaOH from above	0.09814	0.09814	0.09814
mL of NaOH at equivalence point	26	25.75	26.0
Molarity of unknown acid	0.100	0.1	0.1

Average molarity (show calculations) 0.103M ~~Standard deviation~~ ______

2) 0.09814M × 25.75
= 2.527
= 0.1032M

3) 0.09814 M × 26.0 mL
= 2.55
= 0.1036

Avg: $\frac{0.1032 + 0.1036}{2} = 0.1034$

QUESTIONS

1. What are the largest sources of error in this experiment?

The largest source of error here would be human error - determining amounts based on graph alone will skew numbers.

2. What is the pH of the solution obtained by mixing 30.00 mL of 0.250 *M* HCl and 30.00 mL of 0.125 *M* NaOH? We assume additive volumes.

3. Calculate the pH of 0.10 mol/L solution of methylamine, given the K_b of methylamine is 4.37×10^{-4} mol/L.

$4.37 \times 10^{-4} = \frac{[CH_3NH_3^+][OH^-]}{[CH_3NH_2]}$ $0.1(4.37 \times 10^{-4}) = \left(\frac{xx}{0.1}\right)0.1$

	$CH_3NH_{2(aq)}$ + $H_2O_{(l)}$ →	$CH_3NH^+_{3(aq)}$ +	$OH^-_{(aq)}$
I	0.1	0	0
C	-x	+x	+x
E	(0.1-x)	x	x

$\sqrt{x^2} = \sqrt{4.37 \times 10^{-5}}$

$x = 6.61 \times 10^{-3}$

$[OH^-] = 6.61 \times 10^{-3}$

$pOH = -\log[OH^-]$

$= -\log(6.61 \times 10^{-3})$

$pOH = 2.1797$

$pH = 14 - 2.1797$

$pH = 11.8203$

4. Calculate the pH of a solution prepared by mixing 15.0 mL of 0.10 *M* NaOH and 30.0 mL of 0.10 *M* benzoic acid solution. (Benzoic acid is monoprotic; its dissociation constant is 6.3×10^{-5}.) Assume additive volumes.

5. K_a for hypochlorous acid, HClO, is 3.0×10^{-8}. Calculate the pH after 10.0, 20.0, 30.0, and 40.0 mL of 0.100 *M* NaOH have been added to 40.0 mL of 0.100 *M* HClO.

Titration curve

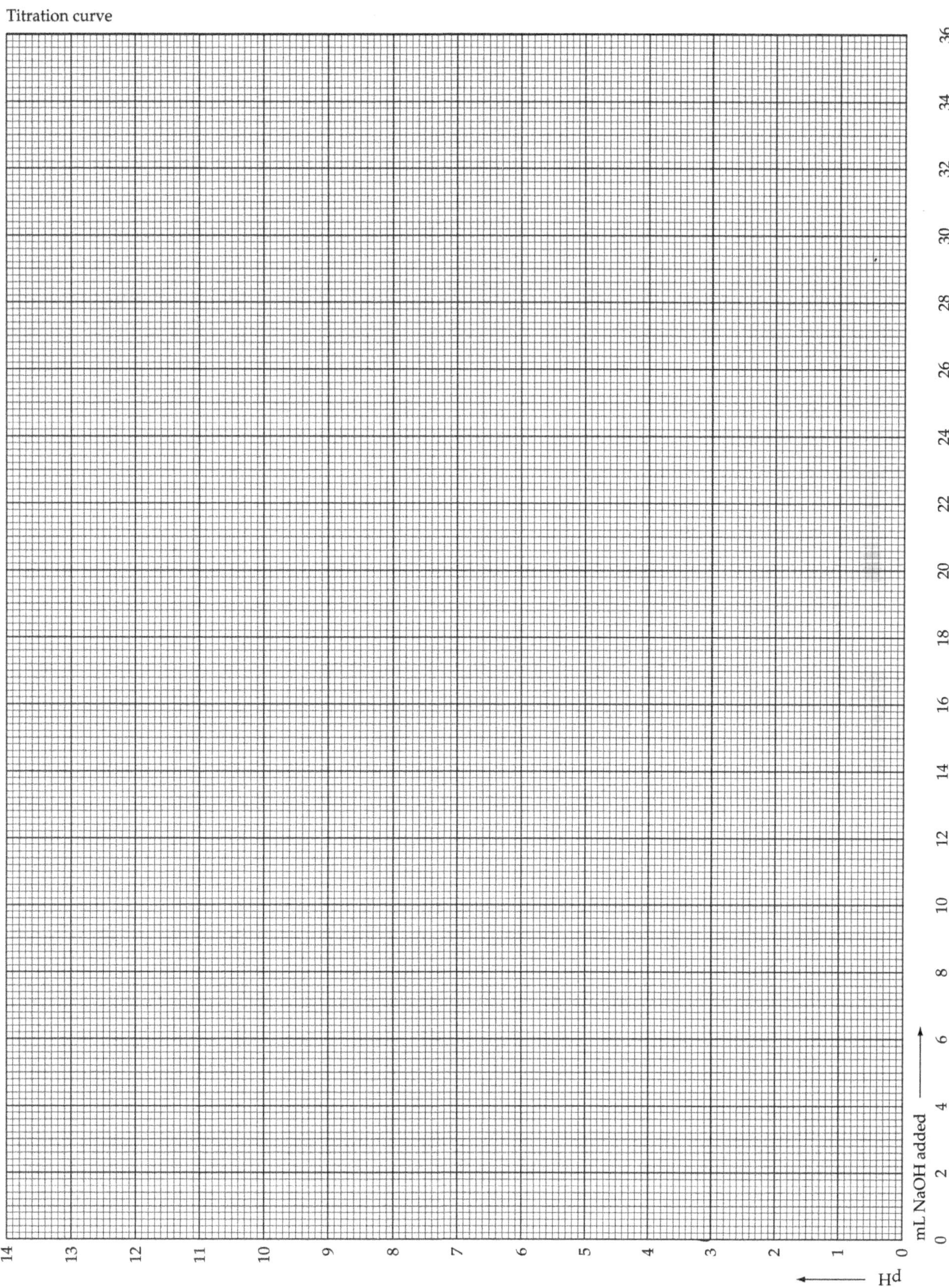

Titration curve

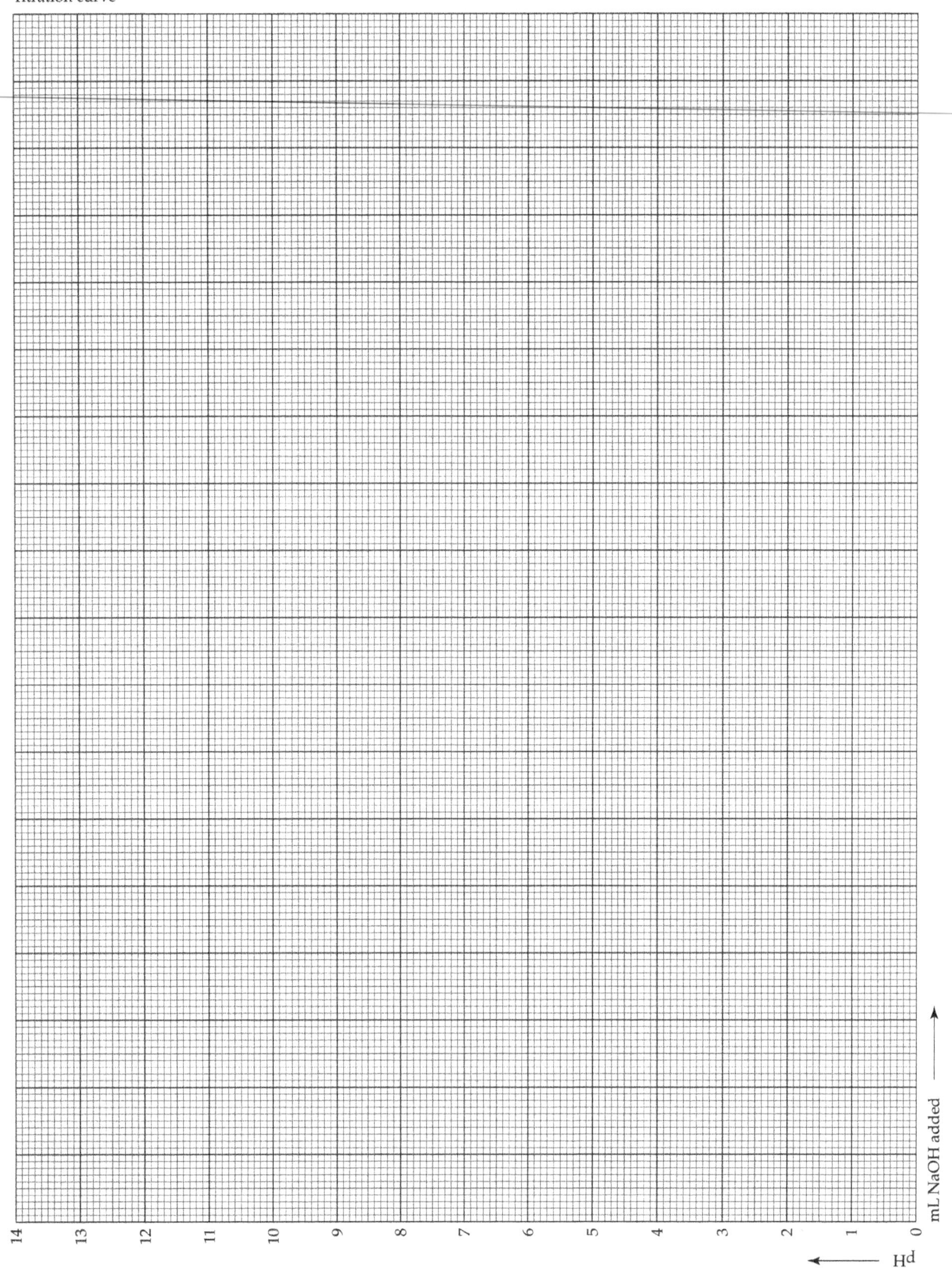

Titration curve

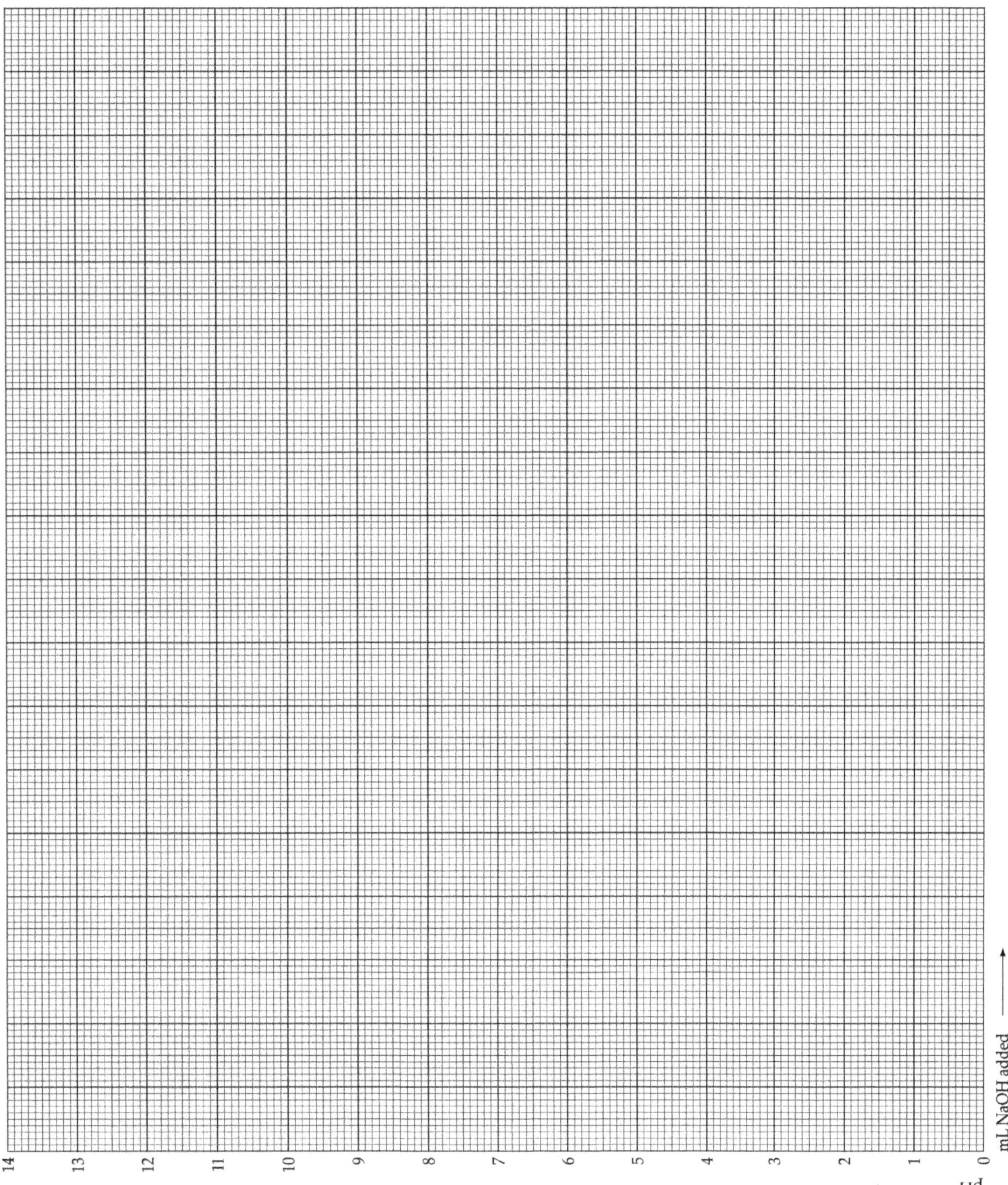

NOTES AND CALCULATIONS

Experiment

Titration Curves of Polyprotic Acids

OBJECTIVE

To become familiar with consecutive equilibria, acid-dissociation constants, and molarity.

APPARATUS AND CHEMICALS

Apparatus

pH meter with electrodes	150 mL beaker
100 mL beakers (3)	500 mL plastic bottle with a plastic lid.
600 mL beaker	buret, buret clamp, and ring stand
25 mL pipet	weighing bottle
iron ring and wire gauze	wash bottle
Bunsen burner	ring stand and ring
balance	

Chemicals

sodium hydroxide solution (0.1 *M* or 19 *M* stock solution)	potassium hydrogen phthalate
unknown solution of a polyprotic acid (approximately 0.1 *M* solution)	phenolphthalein indicator solution
	standard buffer solution pH = 4.0

DISCUSSION

In this experiment, you will perform an acid–base titration to determine the dissociation constants of an unknown polyprotic acid, H_nA. Consider the triprotic acid H_3PO_4. It undergoes the following dissociations in aqueous solution:

$$H_3PO_4 + H_2O \rightleftharpoons H_2PO_4^- + H_3O^+ \qquad K_{a1} = \frac{[H_2PO_4^-][H_3O^+]}{[H_3PO_4]} \qquad [1]$$

$$H_2PO_4^- + H_2O \rightleftharpoons HPO_4^{2-} + H_3O^+ \qquad K_{a2} = \frac{[HPO_4^{2-}][H_3O^+]}{[H_2PO_4^-]} \qquad [2]$$

$$HPO_4^{2-} + H_2O \rightleftharpoons PO_4^{3-} + H_3O^+ \qquad K_{a3} = \frac{[PO_4^{3-}][H_3O^+]}{[HPO_4^{2-}]} \qquad [3]$$

The acid H_3PO_4 possesses three ionizable protons, and for this reason, it is termed a triprotic acid (Section 16.6). If you were to perform a titration of H_3PO_4 with NaOH, the following reactions would occur in turn:

$$H_3PO_4 + NaOH \rightleftharpoons NaH_2PO_4 + H_2O \qquad [4]$$

$$NaH_2PO_4 + NaOH \rightleftharpoons Na_2HPO_4 + H_2O \qquad [5]$$

$$Na_2HPO_4 + NaOH \rightleftharpoons Na_3PO_4 + H_2O \qquad [6]$$

The resultant titration curve, when plotted as pH versus milliliters of NaOH added, would be similar to that shown in Figure 26.1 (Section 17.3). At the point at which half the protons in the first dissociation step of H_3PO_4 have been titrated with NaOH (that is halfway to the first equivalence point), the H_3PO_4 concentration is equal to the $H_2PO_4^-$ concentration. Substituting $[H_3PO_4] = [H_2PO_4^-]$ into the expression for K_{a1} (Equation [1]) yields $K_{a1} = [H_3O^+]$, or $pH = pK_{a1}$ at this point.

Similarly, at one-half the second equivalence point, one-half the $H_2PO_4^-$ has been neutralized and $[H_2PO_4^-] = [HPO_4^{2-}]$. Substituting this into the expression for K_{a2} (Equation [2]) yields $K_{a2} = [H_3O^+]$, or $pK_{a2} = pH$ at this point.

In the same manner, at one-half the third equivalence point, $[HPO_4^{2-}] = [PO_4^{3-}]$. Substituting this into the expression for K_{a3} (Equation [3]) results in the expression $K_{a3} = [H_3O^+]$, or $pK_{a3} = pH$ (see Figure 26.1).

The same type of result is obtained for any polyprotic acid. If a titration of the acid is performed with a pH meter, the dissociation constants may be obtained from titration curves as long as the dissociation constants exceed the ion product of water, which you should recall is 1.00×10^{-14} for the reaction

$$2H_2O \rightleftharpoons H_3O^+ + OH^-$$

In practice, if the acidity of the acid being studied approaches that of water, as in the case for the titration of the third proton of H_3PO_4 for which K_{a3} is 4.2×10^{-13}, it is difficult to determine the dissociation constant. Thus, for H_3PO_4, both K_{a1} and K_{a2} are readily obtained this way, but K_{a3} is not.

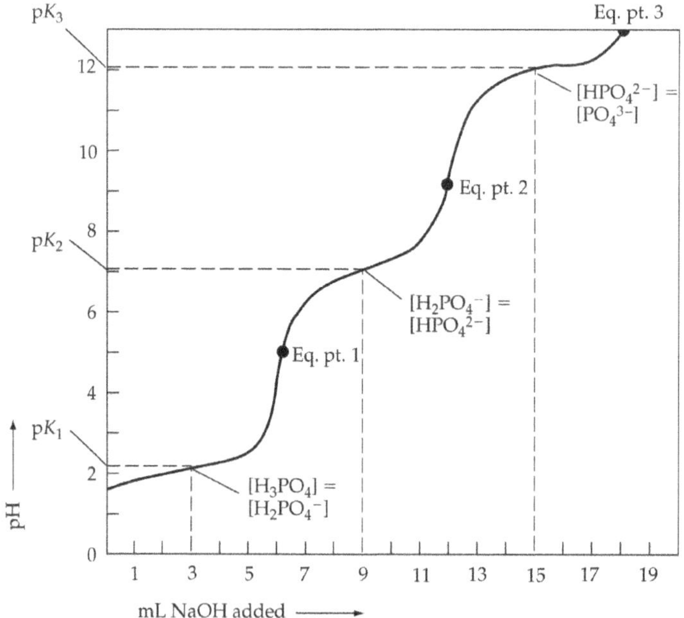

FIGURE 26.1 Titration curve for titration of H_3PO_4 with NaOH.

In this experiment, you will determine the dissociation constants K_{a1} and K_{a2} of a diprotic acid and its molarity (Section 4.5). Recall that the definition for molarity is

$$\text{molarity } (M) = \frac{\text{number of moles solute}}{\text{liter solution}} = \frac{\text{mmoles solute}}{\text{mL solution}}$$

Operation of the pH Meter

To measure the pH during the course of the titration, you will use an electronic instrument called a pH meter. This device consists of a meter and two electrodes, as illustrated in Figure 26.2. You may use a digital pH meter rather than the analog one illustrated here.

The main differences among pH meters involve the positions of the control knobs and the types of electrodes and electrode-mounting devices. The measurement of pH requires two electrodes: a sensing electrode that is sensitive to H_3O^+ concentrations and a reference electrode or a combination electrode. The two electrodes are necessary because the pH meter is really just a voltmeter that measures the electrical potential of a solution. Typical sensing and reference electrodes are illustrated in Figure 26.3.

The reference electrode is an electrode that develops a known potential that is essentially independent of the contents of the solution into which it is placed. The glass electrode is sensitive to the H_3O^+ concentration of the solution into which it is placed; its potential is a function of $[H_3O^+]$. It operates by transport of H_3O^+ ions through the glass membrane. This can be described more precisely, but for purposes here, it is sufficient for you to understand that two electrodes are required. These two electrodes are sometimes combined into an electrode called a combination electrode, which appears to be a single electrode. However, the combination electrode does contain both a reference and a sensing electrode.

Preliminary Operations with the pH Meter

1. Obtain a buffer solution of known pH.
2. Plug in the pH meter to line current and allow at least 10 min for warm-up. You should leave the meter plugged in until you are finished with it. *This does not apply to battery-operated meters.*

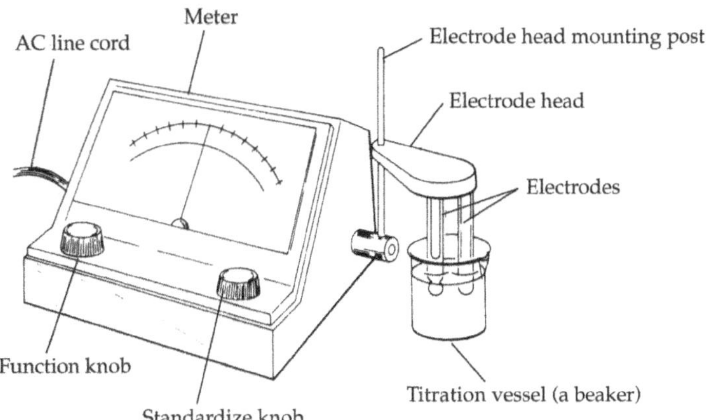

▲FIGURE 26.2 An analog pH meter.

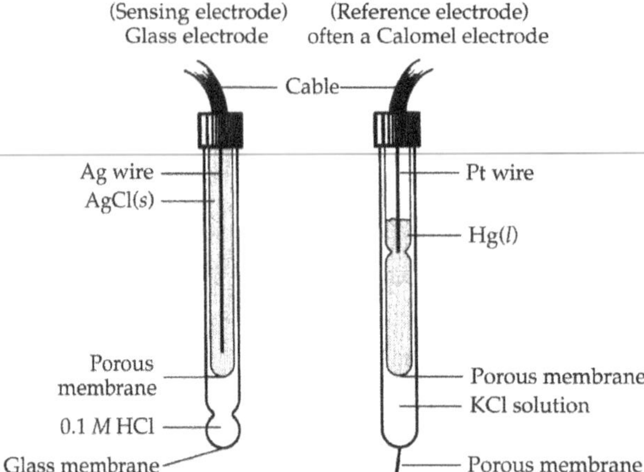

▲**FIGURE 26.3** A typical sensing and reference electrode.

3. Turn the function knob on the pH meter to the standby position.
4. *Prepare the electrodes.* Make certain the solution in the reference electrode extends well above the internal electrode. If it does not, ask your instructor to fill it with saturated KCl solution. Remove the rubber tip and slide down the rubber collar on the reference electrode. Rinse the outside of the electrodes thoroughly with distilled water.
5. *Standardize the pH meter.* Carefully immerse the electrodes in the buffer solution contained in a small beaker. *Remember that the glass electrode is very fragile; it breaks easily! Do not* touch the bottom of the beaker with the electrodes! Turn the function knob to "read" or "pH." Turn the standardize knob until the pH meter indicates the exact pH of the buffer solution. Wait 5 s to be sure the reading remains constant. *Once you have standardized the* pH *meter, don't readjust the standardize knob.* Turn the function knob to standby. Carefully lift the electrodes from the buffer and rinse them with distilled water. The pH meter is now ready to use to measure pH.

RECORD ALL DATA DIRECTLY ON THE REPORT SHEETS.

PROCEDURE

A. Preparation of Approximately 0.100 *M* Sodium Hydroxide (NaOH)*

Heat 500 mL of distilled water to boiling in a 600 mL flask†; *after cooling under the water tap*, transfer to a 500 mL plastic bottle with a plastic lid.‡ **(CAUTION: *Concentrated* NaOH *can cause severe burns. Avoid contact with it. If you come in contact with it, immediately wash the area with copious amounts of water.*)** Add 3 mL of stock solution of carbonate-free NaOH (approximately 19 *M*) and shake vigorously for at least 1 min.

*Students may save time in this experiment if you provide an approximately 0.1 *M* NaOH solution, or if they use the NaOH solution they standardized in Experiment 20.

†The water is boiled to remove carbon dioxide, which would react with the NaOH and change its molarity. (Recall that $H_2O + CO_2 \rightarrow ?$)

‡A rubber stopper should be used for a bottle containing NaOH solution. A strongly alkaline solution tends to cement a glass stopper so firmly that it is difficult to remove.

Preparation of a Buret for Use Clean a 50 mL buret with soap solution and a buret brush and thoroughly rinse with tap water. Then rinse with at least five 10 mL portions of distilled water. The water must run freely from the buret without any drops adhering to the sides. Make sure the buret does not leak and the stopcock turns freely.

Reading a Buret All liquids, when placed in a buret, form a curved meniscus at their upper surface. In the case of water and water solutions, this meniscus is concave (Figure 26.4), and the most accurate buret readings are obtained by observing the position of the lowest point on the meniscus on the graduated scales.

To avoid parallax errors when taking readings, your eye must be on a level with the meniscus. Wrap a strip of paper around the buret and hold the top edges of the strip together, making sure they are even, or use a meniscus reader. Adjust the strip so that the front and back edges are in line with the lowest part of the meniscus and take the reading by estimating to the nearest tenth of a marked division (0.01 mL). A simple way to do this for repeated readings on a buret is illustrated in Figure 26.4.

GIVE IT SOME THOUGHT

When you read the buret, where should your eye level be?

B. Standardization of Sodium Hydroxide (NaOH) Solution

Prepare about 400 to 450 mL of CO_2-free water by boiling for about 5 min. Using a weighing bottle (your lab instructor will show you how to use a weighing bottle if you don't already know) accurately determine the mass of triplicate samples of between 0.4 and 0.6 g each of pure potassium hydrogen phthalate (KHP, $KHC_8H_4O_4$, molar mass 204.22 g/mol) in three separate 250 mL Erlenmeyer flasks; accurately determine mass to four significant figures. Do not determine the mass of the flasks. Record the masses and label the three flasks so that you can distinguish among them. Add to each sample about 100 mL of distilled water that has been freed from CO_2 by boiling and

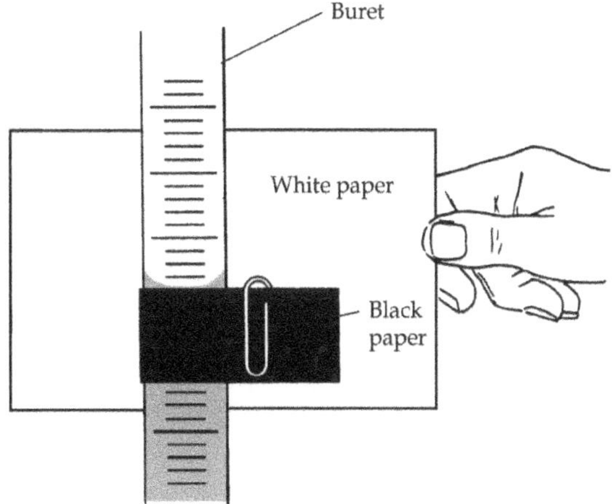

▲**FIGURE 26.4** Reading a buret.

warm gently by swirling until the salt is completely dissolved. Cover and allow to cool before titrating samples. Add to each flask two drops of phenolphthalein indicator solution.

Rinse the previously cleaned buret with at least four 5 mL portions of the approximately 0.100 *M* NaOH solution that you prepared. Discard each portion. *Do not return any of the washings to the bottle.* Completely fill the buret with the room temperature solution and remove the air from the tip by dispensing some of the liquid into an empty beaker. Make sure the lower part of the meniscus is at the zero mark or slightly lower. Allow the buret to stand for at least 30 s before you read the exact position of the meniscus. Remove any hanging drop from the buret tip by touching it to the side of the beaker used for the washings. Record the initial buret reading.

Slowly add the NaOH solution to one of your flasks of KHP solution while gently swirling the contents of the flask, as illustrated in Figure 26.5. As the NaOH solution is added, a pink color appears where the drops of the base come in contact with the solution. This coloration disappears as you swirl the contents of the flask. As the end point is approached, the color disappears more slowly, at which time you should add the NaOH drop by drop. Continue swirling the flask for the entire titration. The end point is reached when one drop of the NaOH solution turns the entire solution in the flask from colorless to pink that persists for at least 30 s. The solution should remain pink when it is swirled. Remove any hanging drop from the buret tip by touching it to the side of the flask and wash down the sides of the flask with a stream of water from the wash bottle. Allow the titrated solution to stand for at least 1 min so that the buret will drain properly. Record the buret reading. Repeat this procedure with the other two samples.

GIVE IT SOME THOUGHT

What shade of pink indicates that the end point has been reached?

From the data you obtain in the three titrations, calculate the molarity of the NaOH solution to four significant figures.

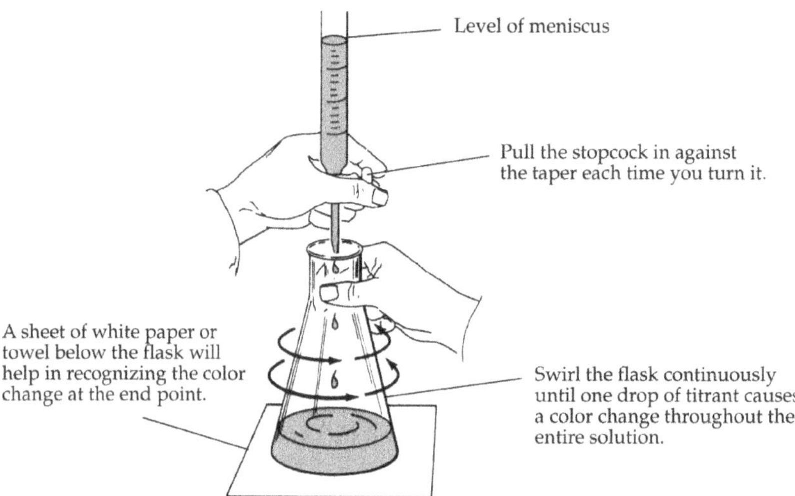
Level of meniscus
Pull the stopcock in against the taper each time you turn it.
A sheet of white paper or towel below the flask will help in recognizing the color change at the end point.
Swirl the flask continuously until one drop of titrant causes a color change throughout the entire solution.

▲**FIGURE 26.5** Titration procedure.

GIVE IT SOME THOUGHT

Why is it necessary to standardize the NaOH solution?

The three determinations should agree within 1.0%. If they do not, repeat the standardization until agreement is reached. The average of the three acceptable determinations is taken as the molarity of the NaOH. Calculate the standard deviation of your results. *Save* your standardized solution for the determination of the acid dissociation constants.

C. Determination of the Acid Dissociation Constants and the Molarity of the Unknown Acid

Using a 25.00 mL pipet accurately deliver three separate 25.00 mL aliquots of the unknown acid into three separate 100 mL beakers. Titrate these aliquots, under constant stirring, using a pH meter as described in experiment 25. Plot the titration curves as shown in Figure 26.1. From the volumes at the two equivalence points, calculate the volumes halfway to each equivalence point and determine the pH's at these two points. These two pH's are equal to the two pK_a's for the unknown acid. Record values of these pK_a's and calculate the values of the corresponding K_a's. Calculate the molarity of your unknown acid from the first equivalence point volume. For this calculation, recall that at the first equivalence point, moles acid equal moles base so that (mL acid) (M acid) = (mL base) (M base) at this point.

GIVE IT SOME THOUGHT

a. What is the significance of the volume halfway to each equivalence point?

b. What is the relationship between pH and pK_a at this point?

(HINT: You can save time if you do your first titration rapidly so that you know the approximate volumes of the equivalence points; then you can do the next two titrations with large-volume increments away from the equivalence points and small-volume increments near the equivalence points.)

NOTES AND CALCULATIONS

Name ______________________ Desk ______________

Date ______________ Laboratory Instructor ______________________

Titration Curves of Polyprotic Acids | 26 Pre-lab Questions

Before beginning this experiment in the laboratory, you should be able to answer the following questions.

1. What is a polyprotic acid?

2. If 9.74 mL of 0.100 *M* NaOH is required to reach the first equivalence point of a solution of citric acid ($H_3C_6H_5O_7$), how many millilitres of NaOH are required to completely neutralize this solution?

3. How many millimoles of NaOH will react completely with 25 mL of 1.0 *M* $H_3C_6H_5O_7$?

4. How many moles of H_3O^+ are present in 100 mL of a 0.40 *M* solution of H_2SO_4?

5. Why is it necessary to standardize a pH meter?

6. If the pH at one-half the first and second equivalence points of a dibasic acid is 4.20 and 7.34, respectively, what are the values for pK_{a1} and pK_{a2}? From pK_{a1} and pK_{a2}, calculate the K_{a1} and K_{a2}.

---

7. Derive the relationship between pH and pK_a at one-half the equivalence point for the titration of a weak acid with a strong base.

8. Could K_b for a weak base be determined the same way that K_a for a weak acid is determined in this experiment?

9. What is the relationship of the successive equivalence point volumes in the titration of a polyprotic acid?

10. If the pK_{a1} of a diprotic acid is 2.90, what is the pH of a 0.10 M solution of this acid that has been one-quarter neutralized?

Name ______________________________ Desk ______________

Date ______________ Laboratory Instructor ______________________

REPORT SHEET | EXPERIMENT

Titration Curves of Polyprotic Acids | 26

B. Standardization of Sodium Hydroxide (NaOH) Solution

	Trial 1	*Trial 2*	*Trial 3*
Mass of bottle + KHP	________	________	________
Mass of bottle	________	________	________
Mass of KHP used	________	________	________
Final buret reading	________	________	________
Initial buret reading	________	________	________
mL of NaOH used	________	________	________
Molarity of NaOH	________	________	________

Average molarity (show calculations) ________

Standard deviation (show calculations) ________

C. Determination of the Acid Dissociation Constants and the Molarity of the Unknown Acid

	Trial 1	*Trial 2*	*Trial 3*
Volume of unknown acid	________	________	________
Molarity of NaOH from above	________	________	________
mL of NaOH at eq. point 1	________	________	________
mL of NaOH at eq. point 2	________	________	________
Molarity of unknown acid	________	________	________

Average molarity ________________

Standard deviation (show calculations) ____________

Determination of pK_a Values of Unknown Acid

First determination		*Second determination*		*Third determination*	
mL NaOH	*pH*	*mL NaOH*	*pH*	*mL NaOH*	*pH*

First determination		*Second determination*		*Third determination*	
mL NaOH	*pH*	*mL NaOH*	*pH*	*mL NaOH*	*pH*

pK_{a1} __________ pK_{a1} __________ pK_{a1} __________

K_{a1} __________ K_{a1} __________ K_{a1} __________

Average K_{a1} __________ Standard deviation __________

pK_{a2} __________ pK_{a2} __________ pK_{a2} __________

K_{a2} __________ K_{a2} __________ K_{a2} __________

Average K_{a2} __________ Standard deviation __________

pH of Unknown Acid

Using the measured K_{a1} and the concentration, calculate the pH of your original unknown acid solution and compare it with that measured. The use of the quadratic equation is required because of the magnitude of K_{a1} and acid concentration.

Calculated pH ________ Calculated pH ________ Calculated pH ________
Measured pH ________ Measured pH ________ Measured pH ________

Average: Calculated pH ________ Measured pH ________

Identity of Unknown Acid

Consult a reference book such as the *CRC Handbook of Chemistry and Physics* or the Internet and using tables of acid dissociation constants contained therein, identify your unknown acid. ____________________________

Internet Cite source: *CRC Handbook of Chemistry and Physics*

Titration curve

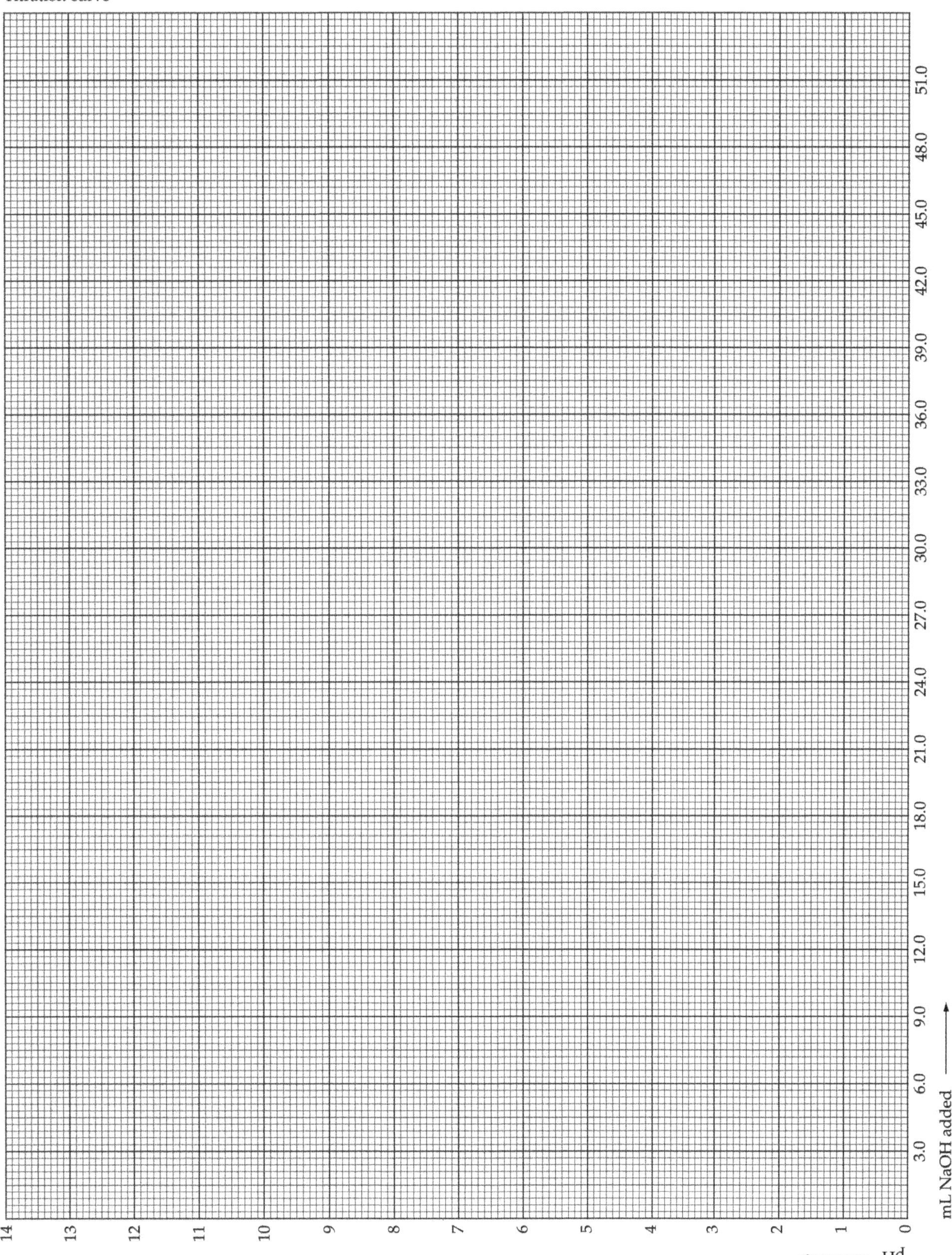

Titration curve

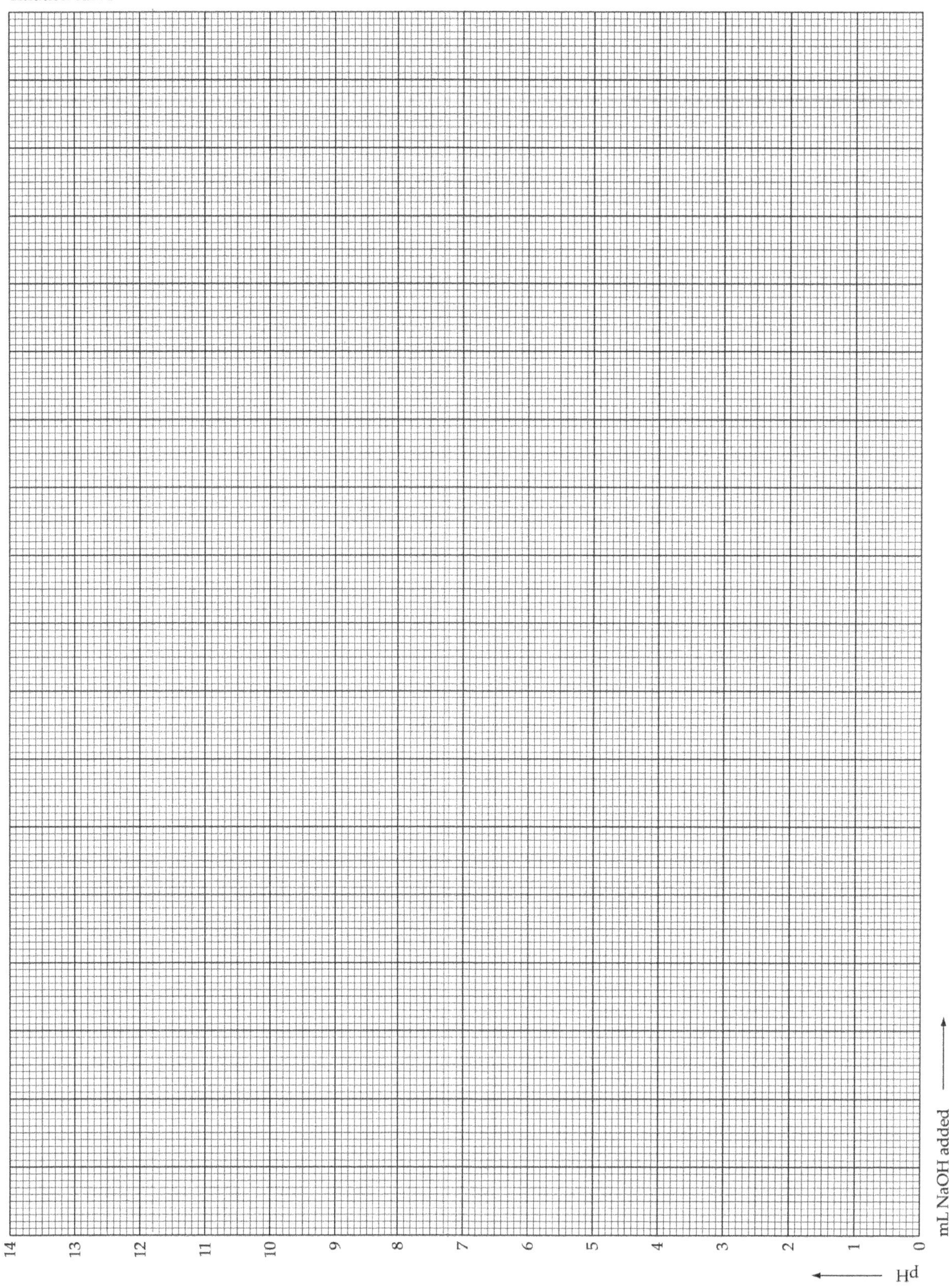

Titration curve

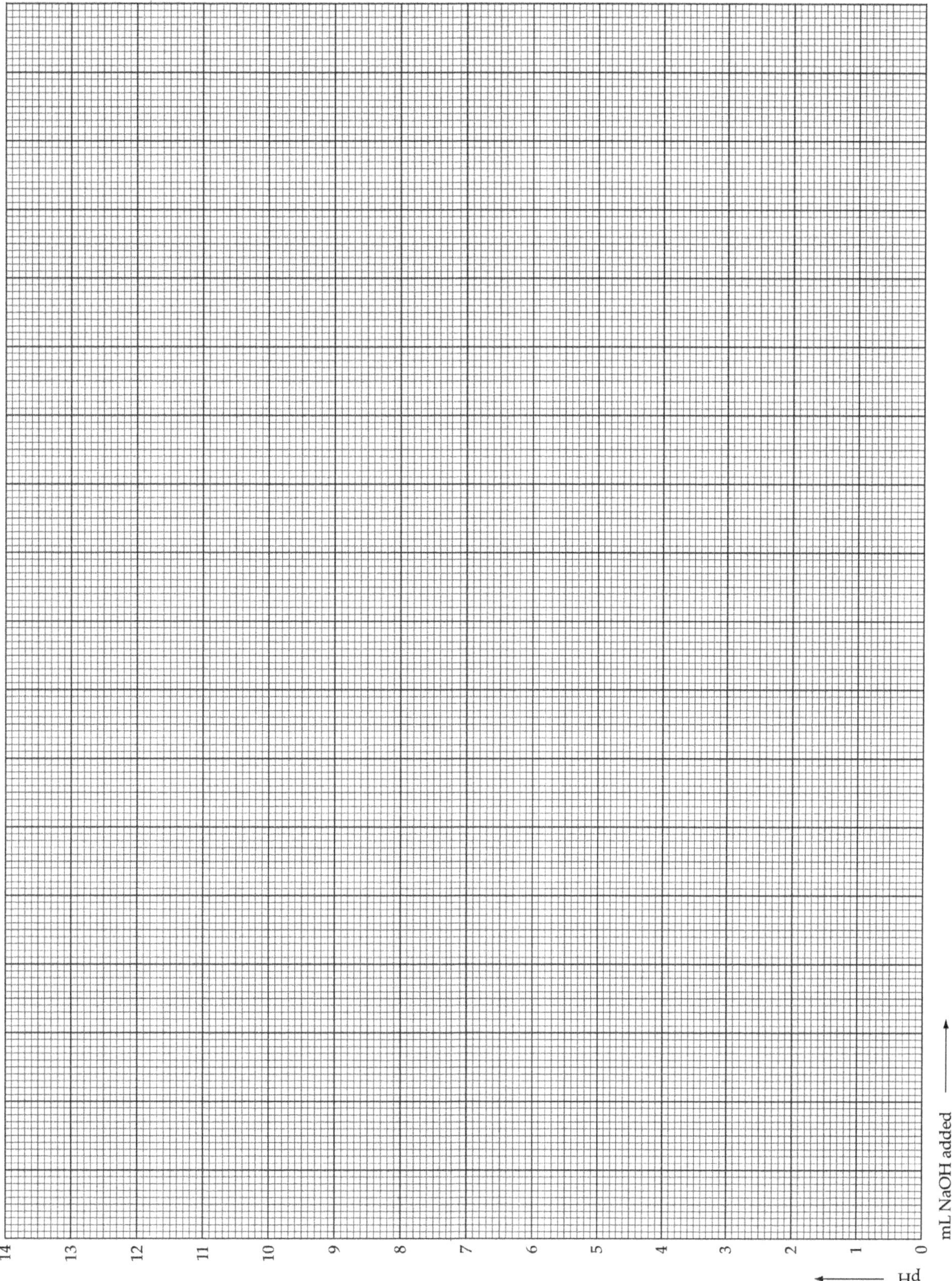

NOTES AND CALCULATIONS

Determination of the Solubility-Product Constant for a Sparingly Soluble Salt

Experiment

OBJECTIVE

To become familiar with equilibria involving sparingly soluble substances by determining the value of the solubility-product constant for a sparingly soluble salt.

APPARATUS AND CHEMICALS

Apparatus

spectrophotometer and cuvettes	100 mL volumetric flasks (4)
5 mL pipets (2)	buret
75 mm test tubes (3)	centrifuge
150 mm test tubes (3)	ring stand and buret clamp
No. 1 corks (3)	

Chemicals

0.0024 M K_2CrO_4	0.0040 M $AgNO_3$
0.25 M $NaNO_3$	disposable gloves

DISCUSSION

Inorganic substances may be broadly classified into three different categories: acids, bases, and salts. According to the Brønsted-Lowry theory, acids are proton donors and bases are proton acceptors (Section 16.2). When an acid reacts with a base in aqueous solution, the products are a salt and water, as illustrated by the following reaction of H_2SO_4 and $Ba(OH)_2$:

$$H_2SO_4\,(aq) + Ba(OH)_2\,(aq) \longrightarrow BaSO_4\,(s) + 2H_2O\,(l) \qquad [1]$$

With but a few exceptions, nearly all common salts are strong electrolytes. The solubilities of salts span a broad spectrum, ranging from slightly or sparingly soluble to very soluble. This experiment is concerned with heterogeneous equilibria of slightly soluble salts. For a true equilibrium to exist between a solid and a solution, the solution must be saturated. Barium sulfate is a slightly soluble salt, and in a saturated solution, this equilibrium may be represented as follows:

$$BaSO_4(s) \rightleftharpoons Ba^{2+}(aq) + SO_4^{\ 2-}(aq) \qquad [2]$$

The equilibrium constant expression for Equation [2] is

$$K_c = \frac{[Ba^{2+}][SO_4^{\ 2-}]}{[BaSO_4]} \qquad [3]$$

The terms in the numerator refer to the molar concentration of ions in solution. The term in the denominator refers to the "concentration" of solid $BaSO_4$.

Because the concentration of a pure solid is a constant, $[BaSO_4]$ can be combined with K_c to give a new equilibrium constant, K_{sp}, which is called the solubility-product constant (Section 17.4).

$$K_{sp} = K_c[BaSO_4] = [Ba^{2+}][SO_4^{2-}]$$

At a given temperature, the value of K_{sp} is a constant. The solubility-product constant for a sparingly soluble salt can be easily calculated by determining the solubility of the substance in water. Suppose, for example, that you determined that 2.42×10^{-4} g of $BaSO_4$ dissolves in 100 mL of water. The molar solubility of this solution (that is, the molarity of the solution) is

$$\left(\frac{2.42\times10^{-4}\text{ g BaSO}_4}{100\text{ mL}}\right)\left(\frac{1000\text{ mL}}{\text{liter}}\right)\left(\frac{1\text{ mol BaSO}_4}{233.4\text{ g BaSO}_4}\right) = 1.04\times10^{-5}\ M$$

You can see from Equation [2] that for each mole of $BaSO_4$ that dissolves, one mole of Ba^{2+} and one mole of SO_4^{2-} are formed. It follows, therefore, that

$$\begin{aligned}\text{solubility of BaSO}_4\text{ in moles/liter} &= [Ba^{2+}]\\ &= [SO_4^{2-}]\\ &= 1.04\times10^{-5}\ M\end{aligned}$$

and

$$\begin{aligned}K_{sp} &= [Ba^{2+}][SO_4^{2-}]\\ &= [1.04\times10^{-5}][1.04\times10^{-5}]\\ &= 1.08\times10^{-10}\end{aligned}$$

In a saturated solution, the product of the molar concentrations of Ba^{2+} and SO_4^{2-} cannot exceed 1.08×10^{-10}. If the ion product $[Ba^{2+}][SO_4^{2-}]$ exceeds 1.08×10^{-10}, precipitation of $BaSO_4$ would occur until this product is reduced to the value of K_{sp}. Or if a solution of Na_2SO_4 is added to a solution of $Ba(NO_3)_2$, $BaSO_4$ would precipitate if the ion product $[Ba^{2+}][SO_4^{2-}]$ is greater than K_{sp}.

Similarly, if you determine that the solubility of Ag_2CO_3 is 3.49×10^{-3} g/100 mL, you could calculate the solubility-product constant for Ag_2CO_3 as follows. The solubility equilibrium involved is

$$Ag_2CO_3(s) \rightleftharpoons 2Ag^+(aq) + CO_3^{2-}(aq) \qquad [4]$$

and the corresponding solubility-product expression is

$$K_{sp} = [Ag^+]^2[CO_3^{2-}]$$

The rule for writing the solubility-product expression states that K_{sp} is equal to the product of the concentration of the ions involved in the equilibrium, each raised to the power of its coefficient in the equilibrium equation.

The solubility of Ag_2CO_3 in moles per liter is

$$\left(\frac{3.49\times10^{-3}\text{ g }Ag_2CO_3}{100\text{ mL}}\right)\left(\frac{1000\text{ mL}}{\text{liter}}\right)\left(\frac{1\text{ mol }Ag_2CO_3}{275.8\text{ g }Ag_2CO_3}\right)=1.27\times10^{-4}\ M$$

so that

$$[CO_3^{2-}]=1.27\times10^{-4}\ M \quad \text{(from Equation [4])}$$

and

$$\begin{aligned}[Ag^+]&=2(1.27\times10^{-4})\\&=2.54\times10^{-4}\ M \quad \text{(based on the stoichiometry of Equation [4])}\\K_{sp}&=[Ag^+]^2[CO_3^{2-}]\\&=[2.54\times10^{-4}]^2[1.27\times10^{-4}]\\&=8.19\times10^{-12}\end{aligned}$$

To determine the solubility of Ag_2CrO_4, you will first prepare it by the reaction of $AgNO_3$ with K_2CrO_4:

$$2AgNO_3(aq)+K_2CrO_4(aq)\rightleftharpoons Ag_2CrO_4(s)+2KNO_3(aq)$$

If a solution of $AgNO_3$ is added to a solution of K_2CrO_4, precipitation will occur when the ion product $[Ag^+]^2[CrO_4^{2-}]$ numerically exceeds the value of K_{sp}; if it doesn't exceed that value, no precipitation will occur (Section 17.6).

EXAMPLE 27.1

If the K_{sp} for PbI_2 is 7.1×10^{-9}, will precipitation of PbI_2 occur when 10 mL of $1.0\times10^{-4}\ M$ $Pb(NO_3)_2$ is mixed with 10 mL of $1.0\times10^{-3}\ M$ KI? Assume additive volumes.

SOLUTION:

$$PbI_2(s)\rightleftharpoons Pb^{2+}(aq)+2I^-(aq)$$

$$K_{sp}=[Pb^{2+}][I^-]^2=7.1\times10^{-9}$$

Precipitation will occur if the reaction quotient Q is greater than K_{sp} (that is if $Q=[Pb^{2+}][I^-]^2>7.1\times10^{-9}$).

$$\begin{aligned}[Pb^{2+}]&=\left(\frac{10\text{ mL}}{20\text{ mL}}\right)(1.0\times10^{-4}\ M)\\&=5.0\times10^{-5}\ M\\[I^-]&=\left(\frac{10\text{ mL}}{20\text{ mL}}\right)(1.0\times10^{-3}\ M)\\&=5.0\times10^{-4}\ M\\[Pb^{2+}][I^-]^2&=[5.0\times10^{-5}][5.0\times10^{-4}]^2\\&=125\times10^{-13}\\&=1.3\times10^{-11}\end{aligned}$$

Because $1.3\times10^{-11} < 7.1\times10^{-9}$, no precipitation will occur. However, if 10 mL of $1.0\times10^{-2}\ M$ $Pb(NO_3)_2$ is added to 10 mL of $2.0\times10^{-2}\ M$ KI,

$$
\begin{aligned}
[Pb^{2+}] &= \left(\frac{10\ \text{mL}}{20\ \text{mL}}\right)(1.0\times10^{-3}\ M)\\
&= 5.0\times10^{-3}\ M\\
[I^{-}] &= \left(\frac{10\ \text{mL}}{20\ \text{mL}}\right)(2.0\times10^{-2})\\
&= 1.0\times10^{-2}\ M
\end{aligned}
$$

and

$$[Pb^{2+}][I^{-}]^2 = [5.0\times10^{-3}][1.0\times10^{-2}]^2 = 5.0\times10^{-7}$$

Because $5.0\times10^{-7} > 7.1\times10^{-9}$, precipitation of PbI_2 will occur in this solution.

To determine the solubility-product constant for a sparingly soluble substance, you need to determine the concentration of only one of the ions because the concentration of the other ion is related to the first ion's concentration by a simple stoichiometric relationship. Any method that accurately determines the concentration would be suitable. In this experiment, you will determine the solubility-product constant for Ag_2CrO_4. This substance contains the yellow chromate ion, CrO_4^{2-}. You will determine the concentration of the chromate ion spectrophotometrically at 375 nm.

Although the eye can discern differences in color intensity with reasonable accuracy, an instrument known as a *spectrophotometer*, which eliminates "human" limitations, is commonly used for this purpose. Basically, it is an instrument that measures the ratio I/I_0 where I is the intensity of light transmitted by a sample and I is the intensity of the incident beam. A schematic representation of a spectrophotometer is shown in Figure 27.1. The instrument has these five fundamental components:

- A light source that produces light with a wavelength range from about 375 to 650 nm if a visible range instrument only.
- A monochromator, which *selects* a particular wavelength of light and sends it to the sample cell with an intensity of I_0.
- The sample cell, which contains the solution being analyzed.

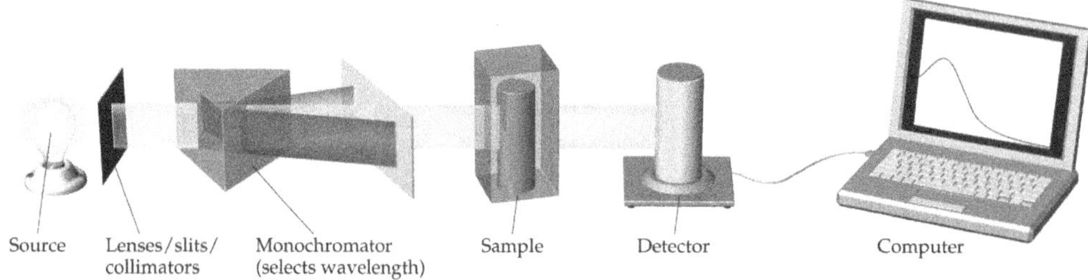

▲**FIGURE 27.1** Schematic representation of a spectrophotometer.

- A detector that measures the intensity, I, of the light transmitted from the sample cell; if the intensity of the incident light is I_0 and the solution absorbs light, the intensity of the transmitted light, I, is less than I_0.
- A display that indicates the intensity of the transmitted light.

For a given substance, the amount of light absorbed depends on the following:

- Concentration and identity of the absorbing species
- Cell or path length
- Wavelength of light
- Solvent

Plots of the amount of light absorbed versus wavelength are called *absorption spectra*. There are two common ways of expressing the amount of light absorbed. One is in terms of *percent transmittance*, $\%T$, which is defined as

$$\%T = \frac{I}{I_0} \times 100\% \qquad [5]$$

As the term implies, percent transmittance corresponds to the percentage of light transmitted. When the sample in the cell is a solution, I is the intensity of light transmitted by the solution and I_0 is the intensity of light transmitted when the cell contains only solvent. Another method of expressing the amount of light absorbed is in terms of *absorbance*, A, which is defined by

$$A = \log\frac{I_0}{I} \qquad [6]$$

If a sample absorbs no light at a given wavelength, the percent transmittance is 100 and the absorbance is 0. On the other hand, if the sample absorbs all of the light, $\%T = 0$ and $A = \infty$. A is more useful in quantitative work since it varies linearly with concentration.

Absorbance is related to concentration by the Beer-Lambert law

$$A = abc$$

where A is absorbance, b is solution path length, c is concentration in moles per liter, and a is molar absorptivity or molar extinction coefficient. There is a linear relationship between absorbance and concentration when the Beer-Lambert law

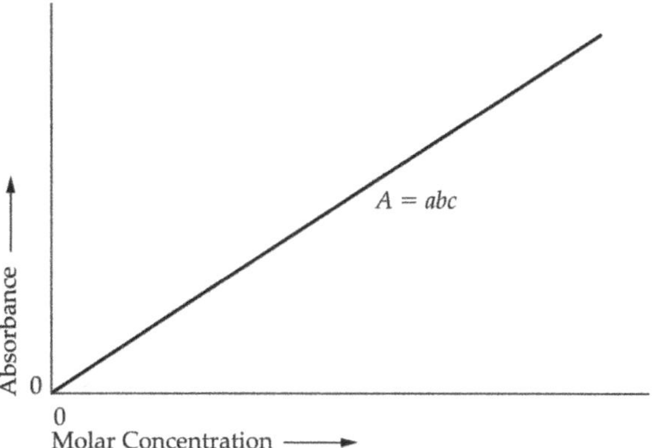

▲**FIGURE 27.2** Relationship between absorbance and molar concentration according to the Beer-Lambert law.

is obeyed, as illustrated in Figure 27.2. However, because deviations from this law occasionally occur, it is wise to construct a calibration curve of absorbance versus concentration.

PROCEDURE | A. Preparation of a Calibration Curve

WORK IN GROUPS OF FOUR TO OBTAIN YOUR CALIBRATION CURVE, BUT EVALUATE YOUR DATA INDIVIDUALLY. Using a buret, add 1, 5, 10, and 15 mL of standardized 0.0024 *M* K_2CrO_4 to each of four clean 100 mL volumetric flasks and dilute to the 100 mL mark with 0.25 *M* $NaNO_3$. Calculate the CrO_4^{2-} concentration in each solution. Measure the absorbance of these solutions at 375 nm and plot the absorbance versus concentration to construct your calibration curve as shown in Figure 27.2.

Operating Instructions for Spectronic 20

You may use a different spectrophotometer. If so, your instructor will explain how to use it.

1. Turn the wavelength control knob (Figure 27.3) to the desired wavelength.
2. Turn on the instrument by rotating the power control clockwise and allow the instrument to warm up about 5 min. With no sample in the holder but with the cover closed, turn the zero adjust to bring the meter needle to zero on the percent transmittance scale.
3. Fill the cuvette about halfway with distilled water (or solvent blank) and insert it in the sample holder, aligning the line on the cuvette with that of the sample holder. Close the cover and rotate the light control knob until the meter reads 100% transmittance.
4. Remove the blank from the sample holder and replace it with the cuvette containing the sample whose absorbance is to be measured. Align the lines on the cuvette with the holder and close the cover. Read the percent transmittance or optical density from the meter.

PROCEDURE | B. Determination of the Solubility-Product Constant

Accurately prepare three solutions in separate 150 mm test tubes by adding 5.00 mL of 0.0040 *M* $AgNO_3$ to 5.00 mL of 0.0024 *M* K_2CrO_4.

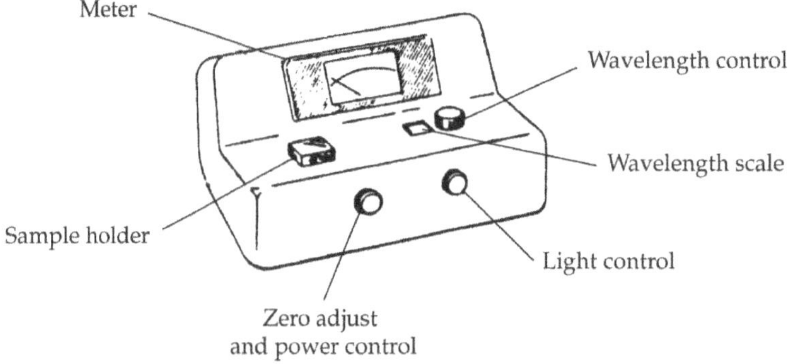

▲**FIGURE 27.3** Spectrophotometer controls.

Waste Disposal Instructions Because chromates are hazardous, you must treat all chromate solutions with care. Avoid spilling or touching these solutions. All excess K_2CrO_4 solution from Part A should be returned to a specially marked waste container, not to the original stock solution. All chromate solutions from Part B of the experiment should be placed in the same waste container. Likewise, all of the Ag_2CrO_4 samples should be disposed of in the second specially marked waste container. Silver nitrate ($AgNO_3$) solution is also hazardous. Any $AgNO_3$ solution that is spilled on the skin will cause discoloration after a few minutes, which can be avoided by immediately and thoroughly washing the affected area. All excess $AgNO_3$ solution should be returned to a third specially marked container. The other solution used in this experiment, $NaNO_3$, should be disposed of as indicated by your instructor.

GIVE IT SOME THOUGHT

a. Calculate the reaction quotient, Q, when these two solutions are mixed.

b. If a solid forms, what does that tell you about the value of Q versus K_{sp}?

c. If a solid forms, what happens to the concentration of CrO_4^{2-}?

Mix by rapping the solutions with your fingers to generate a vortex as in Appendix J. Perform this at periodic intervals for about 15 min to establish equilibrium between the solid phase and the ions in solution. Transfer approximately 3 mL of each solution along with most of the insoluble Ag_2CrO_4 to 75 mm test tubes and centrifuge. Discard the supernatant liquid and retain the precipitate as directed. To each of the test tubes add 2 mL of 0.25 *M* $NaNO_3$. Mix each test tube thoroughly and centrifuge again. Discard the supernatant liquid; then add 2 mL of 0.25 *M* $NaNO_3$ to each of the test tubes, mixing them vigorously and periodically for about 15 min to establish an equilibrium between the solid and the solution. Some solid Ag_2CrO_4 must remain in the test tubes. If none remains, start over. After mixing the test tubes for about 15 min, centrifuge the mixtures. Transfer the clear, pale yellow supernatant liquid from each of the three test tubes to a clean, dry cuvette. Measure and record the absorbance of the three solutions. Using your calibration curve, calculate the molar concentration of CrO_4^{2-} in each solution.

GIVE IT SOME THOUGHT

a. Why is $NaNO_3$ an appropriate choice for this solution?

b. Why can't you use NaCl?

c. Why is it necessary to have solid Ag_2CrO_4 present in the test tubes?

Note on Calculations

You are determining the K_{sp} of Ag_2CrO_4 in this experiment. The equilibrium reaction for the dissolution of Ag_2CrO_4 is

$$Ag_2CrO_4(s) \rightleftharpoons 2Ag^+(aq) + CrO_4^{2-}(aq)$$

for which $K_{sp} = [Ag^+]^2[CrO_4^{2-}]$.

You should note that at equilibrium, $[Ag^+] = 2[CrO_4^{2-}]$; hence, having determined the concentration of chromate ions, you know the silver ion concentration.

Name ______________________ Desk ______________

Date ______________ Laboratory Instructor ______________

Determination of the Solubility-Product Constant for a Sparingly Soluble Salt | 27 Pre-lab Questions

Before beginning this experiment in the laboratory, you should be able to answer the following questions.

1. Write the solubility equilibrium and the solubility-product constant expression for the slightly soluble hydroxide $Ba(OH)_2$.

2. Calculate the number of moles of Ag^+ in 5.00 mL of 0.0040 M $AgNO_3$ and the number of moles of CrO_4^{2-} in 5.00 mL of 0.0024 M K_2CrO_4.

3. If 5 mL of 0.0040 M $AgNO_3$ is added to 5 mL of 0.0024 M K_2CrO_4, is either Ag^+ or CrO_4^{2-} in stoichiometric excess? If so, which one?

4. A 20.0 mL sample of saturated and aqueous solution of calcium hydroxide required 18.2 mL of 0.050 mol/L hydrochloric acid for neutralisation. Calculate the hydroxide ion concentration and the pH of the original solution of calcium hydroxide.

5. The K_{sp} for $BaCO_3$ is 5.1×10^{-9}. How many grams of $BaCO_3$ will dissolve in 1000 mL of water?

6. Distinguish between the equilibrium constant expression and K_{sp} for the dissolution of a sparingly soluble salt.

7. List as many experimental techniques as you can that may be used to determine K_{sp} for a sparingly soluble salt.

8. Why must some solid remain in contact with a solution of a sparingly soluble salt in order to ensure equilibrium?

9. In general, when will a sparingly soluble salt precipitate from solution?

$$\frac{1.75}{1.32\times10^{4}}$$

Name Grace Rademacher Desk

Date 04/06/21 Laboratory Instructor Prof. Towle

REPORT SHEET | EXPERIMENT 27

Determination of the Solubility-Product Constant for a Sparingly Soluble Salt

A. Preparation of a Calibration Curve

$\frac{.0024}{100}$

Initial $[CrO_4^{2-}]$ 0.0024 M

Volume of 0.0024 M K_2CrO_4	*Total volume*	*$[CrO_4^{2-}]$*	*Absorbance*
1. 1.0 mL	100 mL	2.4×10^{-5}	0.091
2. 5.0 mL	100 mL	1.2×10^{-4}	0.474
3. 10.0 mL	100 mL	2.4×10^{-4}	0.986
4. 15.0 mL	100 mL	3.6×10^{-4}	1.515

2 ions = x^2

Molar extinction coefficient for $[CrO_4^{2-}]$

1. 3.79×10^{3} $\frac{0.091}{2.4\times10^{-5}}$ 2. 3.95×10^{3} $\frac{0.474}{1.2\times10^{-4}}$ 3. 4.1×10^{4} $\frac{0.986}{2.4\times10^{-5}}$ 4. 4.2×10^{3} $\frac{1.515}{3.6\times10^{-4}}$

Average molar extinction coefficient 1.32×10^{4}

$$\frac{(3.79\times10^{-7}+3.95\times10^{-5}+4.12\times10^{-6}+4.2\times10^{-5})}{4} = 1.32\times10^{4}$$

~~Standard deviation (show calculations)~~ ______

$$Ag_2CrO_{4(s)} \rightleftharpoons 2Ag^+_{(aq)} + CrO^{2-}_{4(aq)}$$

B. Determination of the Solubility-Product Constant

Absorbance	*$[CrO_4^{2-}]$*	*$[Ag^+]$*	*K_{sp} of Ag_2CrO_4*
1. 1.750	1.33×10^{-4}	2.66×10^{-4}	9.41×10^{-12}
2. 1.194	9.05×10^{-5}	1.81×10^{-4}	2.96×10^{-12}
3. 1.277	9.67×10^{-5}	1.93×10^{-4}	3.6×10^{-12}

$\frac{1.75}{1.32\times10^{4}}$

$K_{sp} = [Ag^+]^2[CrO_4^{2-}]$

Average K_{sp} (show calculations) 5.32×10^{-12}

~~Standard deviation (show calculations)~~

$$\frac{9.41\times10^{-12} + 2.96\times10^{-12} + 3.6\times10^{-12}}{3} = 5.32\times10^{-12}$$

QUESTIONS

1. If the standard solutions had unknowingly been made up to be 0.0024 M $AgNO_3$ and 0.0040 M K_2CrO_4, would this have affected your results? If so, how?

2. If your cuvette had been dirty, how would this have affected the calculated value of K_{sp}?
There'd be a higher absorbance, increasing the ion concentration, therefore a higher K_{sp} value too.

3. The solubility of lead(II) iodide in water at 15 °C is 0.46 g/L. For a saturated solution of lead(II) iodide at 15 °C, calculate the concentration in mol/L of lead(II) ions, the concentration in mol/L of iodide ions, and the K_{sp} of lead(II) iodide.

Solu. of lead iodide $\frac{0.46}{461.01} = 0.0009978 = 9.978\times10^{-4}$ mol/L

PbI_2 ions

$[I^-] = 2(PbI_2) = 2(9.978\times10^{-4}) = 1.99\times10^{-3}$

$K_{sp} = (9.978\times10^{-4})(1.99\times10^{-3})^2$
$= (9.978\times10^{-4})(3.96\times10^{-6}) = 3.95\times10^{-9}$ mol/L

4. The experimental procedure for this experiment has you add 5 mL of 0.0040 M $AgNO_3$ to 5 mL of 0.0024 M K_2CrO_4. Is either of these reagents in excess? If so, which one?

5. Use your experimentally determined value of K_{sp} and show, by calculation, that Ag_2CrO_4 should precipitate when 5 mL of 0.0040 *M* $AgNO_3$ are added to 5 mL of 0.0024 *M* K_2CrO_4.

6. In the back of your textbook, look up the accepted value of K_{sp} for Ag_2CrO_4. Calculate the percentage error in your experimentally determined value for K_{sp}.

7. Although Ag_2CrO_4 is insoluble in water, it is soluble in dilute HNO_3. Explain using chemical equations.

8. What is the greatest source of error in this experiment?

9. Reviewing the procedure, indicate where careless work could contribute to the error source you identified in question 8.

Calibration curve

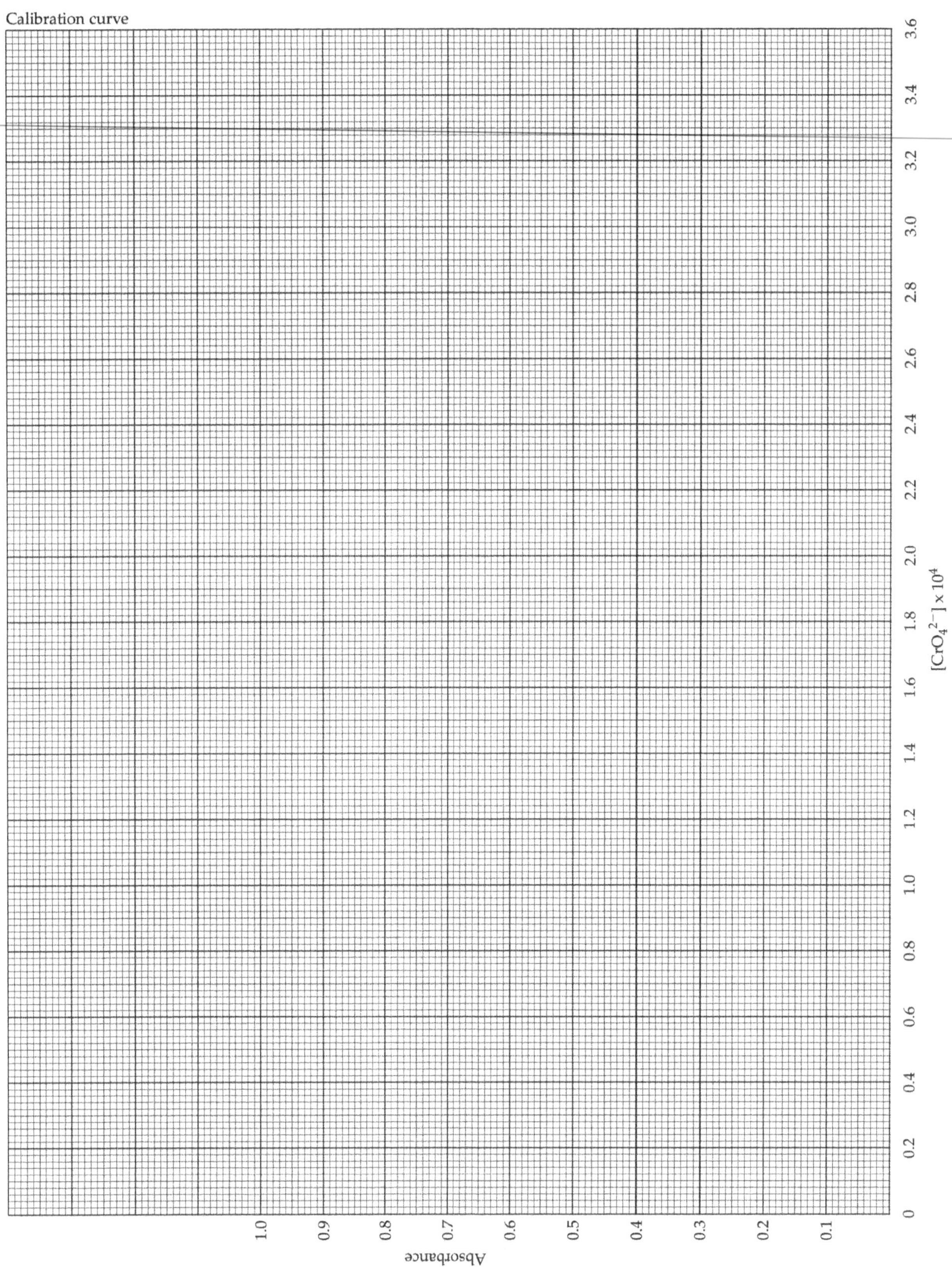

Experiment

Heat of Neutralization

OBJECTIVE

To measure, using a calorimeter, the enthalpy changes accompanying neutralization reactions.

APPARATUS AND CHEMICALS

Apparatus

Bunsen burner and hose
Styrofoam cups (2)
cardboard square with hole in center
400 mL beaker
wire gauze
thermometers (2)
50 mL graduated cylinder
split one-hole rubber stopper
250 mL beaker
ring stand and ring

Chemicals

1 *M* HCl
1 *M* acetic acid (CH_3COOH)
1 *M* NaOH

DISCUSSION

WORK IN PAIRS, BUT EVALUATE YOUR DATA INDIVIDUALLY.

Every chemical change is accompanied by a change in energy, usually in the form of heat. The energy change of a reaction that occurs at constant pressure is termed the *heat of reaction* or the *enthalpy change* (Section 5.3). The symbol ΔH (the symbol Δ means "change in") is used to denote the enthalpy change. If heat is evolved, the reaction is *exothermic* ($\Delta H < 0$); if heat is absorbed, the reaction is *endothermic* ($\Delta H > 0$). In this experiment, you will measure the heat of neutralization (or the enthalpy of neutralization) when an acid and a base react to form water.

This quantity of heat is measured experimentally by allowing the reaction to take place in a thermally insulated vessel called a *calorimeter* (Section 5.5). The heat liberated in the neutralization will cause an increase in the temperature of the solution and of the calorimeter. If the calorimeter were perfect, no heat would be radiated to the laboratory. The calorimeter you will use in this experiment is shown in Figure 28.1.

Because the heat of the reaction and some heat is absorbed by the calorimeter itself, you must know the amount of heat absorbed by the calorimeter. This requires that you determine the heat capacity of the calorimeter. *Heat capacity of the calorimeter* means the amount of heat (that is, the number of joules) required to raise its temperature 1 kelvin, which is the same as 1 °C. In this experiment, the temperature of the calorimeter and its contents is measured

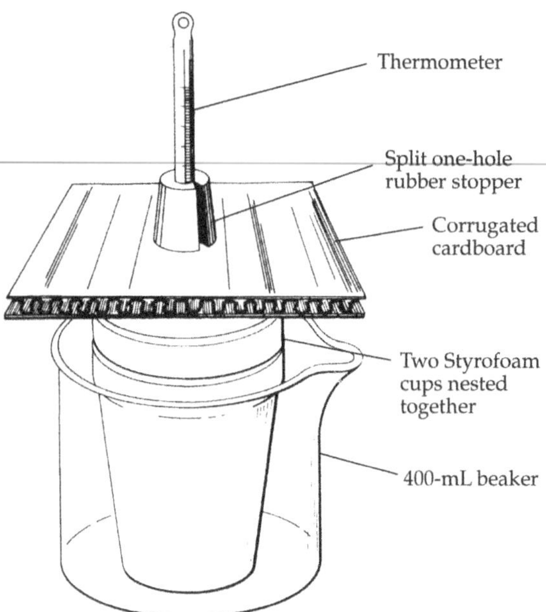

▲**FIGURE 28.1** A simple coffee cup calorimeter (Figure 5.17).

before and after the reaction. The change in the enthalpy, ΔH, is equal to the negative product of the temperature change, ΔT, times the heat capacity of the calorimeter and its contents.

$$\Delta H = -\Delta T \text{ (heat capacity of calorimeter + heat capacity of contents)} \qquad [1]$$

Note that the *numerical difference* on the Celsius scale is the same as the *numerical difference* on the Kelvin scale, where ΔT is the difference between the final and initial temperatures: $\Delta T = T_f - T_i$. Because ΔH is negative for an exothermic reaction whereas ΔT is positive, a negative sign is required in Equation [1].

The heat capacity of the calorimeter is determined by measuring the temperature change that occurs when a known amount of hot water is added to a known amount of cold water in the calorimeter. The heat lost by the warm water is equal to the heat gained by the cold water and the calorimeter. (Assume that no heat is lost to the laboratory.) For example, if T_1 equals the temperature of a calorimeter and 50 mL of cooler water, if T_2 equals the temperature of 50 mL of warmer water added to it, and if T_f equals the temperature after mixing, the heat lost by the warmer water is

$$\text{heat lost by warmer water} = (T_2 - T_f) \times 50\ \text{g} \times 4.184\ \text{J/K-g} \qquad [2]$$

The specific heat of water is 4.184 J/K-g, and the density of water is 1.00 g/mL. The heat gained by the cooler water is

$$\text{heat gained by cooler water} = (T_f - T_1) \times 50\ \text{g} \times 4.184\ \text{J/K-g} \qquad [3]$$

The heat lost to the calorimeter is the difference between heat lost by the warmer water and that gained by the cooler water.

$$\text{(heat lost by warmer water)} - \text{(heat gained by cooler water)}$$
$$= \text{heat gained by the calorimeter}$$

Substituting Equations [2] and [3], you have

$$[(T_2 - T_f) \times 50 \text{ g} \times 4.184 \text{ J/K-g}] - [(T_f - T_1) \times 50 \text{ g} \times 4.184 \text{ J/K-g}]$$
$$= (T_f - T_1) \times \text{heat capacity of calorimeter} \quad [4]$$

Note that the heat lost to the calorimeter equals its temperature change times its heat capacity. Thus, by measuring T_1, T_2, and T_f, the heat capacity of the calorimeter can be calculated from Equation [4]. This is illustrated in Example 28.1.

EXAMPLE 28.1

Given the following data, calculate the heat lost by the warmer water, the heat gained by the cooler water, the heat lost to the calorimeter, and the heat capacity of the calorimeter.

Temperature of 50.0 mL warmer water: $37.92\ °\text{C} = T_2$
Temperature of 50.0 mL cooler water: $20.91\ °\text{C} = T_1$
Temperature after mixing: $29.11\ °\text{C} = T_f$

SOLUTION: The heat lost by the warmer water, where $\Delta T = 37.92\ °\text{C} - 29.11\ °\text{C}$, is

$$8.81 \text{ K} \times 50.0 \text{ g} \times 4.184 \text{ J/K-g} = 1840 \text{ J}$$

The heat gained by the cooler water, where $\Delta T = 29.11\ °\text{C} - 20.91\ °\text{C}$, is

$$8.20 \text{ K} \times 50.0 \text{ g} \times 4.184 \text{ J/K-g} = 1710 \text{ J}$$

The heat gained by the calorimeter is

$$1840 \text{ J} - 1710 \text{ J} = 130 \text{ J}$$

Therefore, the heat capacity of the calorimeter, using the temperature change of the cold water since it is located within the calorimeter initially, is

$$130 \text{ J}/8.20 \text{ K} = 15.9 \text{ J/K}$$

Once the heat capacity of the calorimeter is determined, Equation [1] can be used to determine the ΔH for the neutralization reaction. Example 28.2 illustrates such a calculation.

EXAMPLE 28.2

Given the following data, calculate the heat gained by the solution, the heat gained by the calorimeter, and the heat of reaction.

Temperature of 50.0 mL of acid before mixing: 21.02 °C
Temperature of 50.0 mL of base before mixing: 21.02 °C
Temperature of 100.0 mL of solution after mixing: 27.53 °C

Assume that the density of these solutions is 1.00 g/mL and that the volumes are additive.

SOLUTION: The heat gained by the solution, where $\Delta T = 27.53\ °\text{C} - 21.02\ °\text{C}$, is

$$6.51 \text{ K} \times 100 \text{ g} \times 4.184 \text{ J/K-g} = 2720 \text{ J}$$

The heat gained by the calorimeter, where $\Delta T = 27.53\ °\text{C} - 21.02\ °\text{C}$, is

$$6.51 \text{ K} \times 15.9 \text{ J/K} = 104 \text{ J}$$

Therefore, the heat of reaction is

$$2720 \text{ J} + 104 \text{ J} = 2824 \text{ J}$$

or

$$2.82 \text{ kJ}$$

PROCEDURE

A. Heat Capacity of Calorimeter

Construct a calorimeter similar to the one shown in Figure 28.1 by nesting two Styrofoam cups together. Use a cork borer to make a hole in the lid just big enough to admit the thermometer; then slip the thermometer into a split one-hole rubber stopper, which prevents you from placing the thermometer too deeply in the calorimeter. The thermometer should not touch the bottom of the cup. Rest the entire apparatus in a 400 mL beaker to provide stability.

Place exactly 50.0 mL of distilled water in the calorimeter cup and replace the cover and thermometer. Allow 5 to 10 min for the system to reach thermal equilibrium; then record the temperature to the nearest 0.1 °C on the report sheet (1).

Place exactly 50.0 mL of distilled water in a clean, dry 250 mL beaker and heat the water with a low flame until the temperature is approximately 15° to 20 °C above room temperature. Do not heat to boiling; otherwise, appreciable water will be lost, leading to an erroneous result. Allow the hot water to stand for a minute or two; quickly record its temperature to the nearest 0.1 °C on the report sheet (2) and pour as much as possible into the calorimeter. Replace the lid with the thermometer and carefully stir the water with the thermometer. Observe the temperature for the next 3 min and record the temperature every 15 s on the temperature versus time data sheet at the end of the report pages. Plot the temperature as a function of time, as shown in Figure 28.2. Determine ΔT from your curve; then do the calculations indicated on the report sheet.

B. Heat of Neutralization of HCl–NaOH

Dry the calorimeter and the thermometer with a towel. Carefully measure 50.0 mL of 1.0 *M* NaOH and add it to the calorimeter. Place the lid on the calorimeter, but leave the thermometer out. Measure exactly 50.0 mL of 1.0 *M* HCl into a dry beaker. Allow it to stand near the calorimeter for 3 to 4 min. Measure the temperature of the acid, rinse the thermometer with distilled water, and wipe dry. Insert the thermometer into the calorimeter and measure the temperature of the NaOH solution.

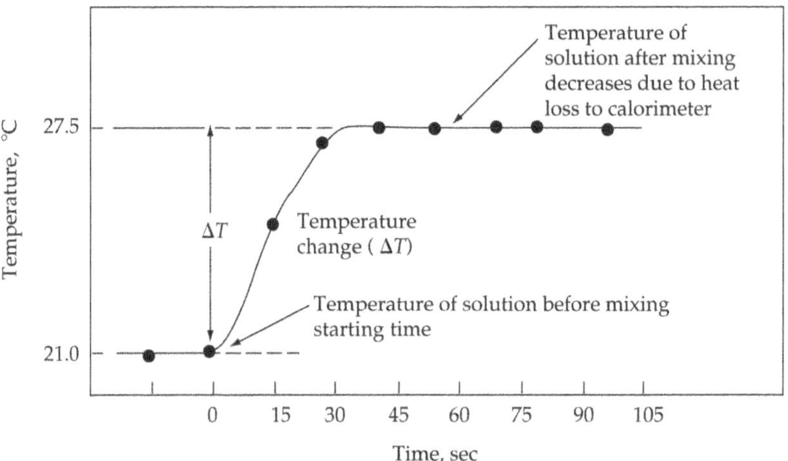

▲**FIGURE 28.2** Temperature as a function of time.

The temperatures of the NaOH and the HCl should not differ by more than 0.5 °C. If the difference is greater than that, adjust the temperature of the HCl by *either* warming it by holding the beaker in your hands or cooling the outside of the beaker with tap water until the temperature of the HCl is within 0.5 °C of that of the NaOH; record the temperature on the report sheet 1.

GIVE IT SOME THOUGHT

Here you observe a temperature increase ($\Delta T > 0$). Does this indicate that the reaction is exothermic or endothermic?

Record the temperature of the NaOH solution. Lift the lid and carefully add the 1.0 *M* HCl all at once. Be careful not to splash any on the upper sides of the cup. Stir the solution gently with the thermometer and record the temperature as a function of time every 15 s for the next 3 min. Construct a temperature-versus-time curve and determine ΔT. Calculate the heat of neutralization per mole of water formed. You may assume that the NaCl solution has the same density and specific heat as water.

GIVE IT SOME THOUGHT

a. Based solely on the ΔT, would you predict the reaction to be endothermic or exothermic?

b. After you perform your calculations, does the sign of ΔH agree with your prediction?

C. Heat of Neutralization of CH_3COOH–NaOH

Follow the same procedure as in Part B, but substitute 1.0 *M* CH_3COOH for 1.0 *M* HCl. Calculate the heat of neutralization per mole of water formed.

GIVE IT SOME THOUGHT

a. Do you predict ΔH to be the same for the neutralizations of acetic acid and hydrochloric acid?

b. Why?

Waste Disposal Instructions Handle the stock solutions carefully. You may use a wet sponge or paper towel to clean up any spills. The reaction mixtures produced in the Styrofoam cups contain salts. They should be disposed of in the designated waste containers.

NOTES AND CALCULATIONS

Name ______________________________ Desk ______________

Date ______________ Laboratory Instructor ______________________

Heat of Neutralization | 28 Pre-lab Questions

Before beginning this experiment in the laboratory, you should be able to answer the following questions.

1. Define endothermic and exothermic reactions in terms of the sign of ΔH.

2. A 500 mL sample of water was cooled from 10 °C to 50 °C. How much heat was lost?

3. Define the term *heat capacity*.

4. How many joules are required to change the temperature of 20.0 g of water from 33.2 °C to 51.5 °C?

5. Define the term *endothermic*.

6. Calculate the final temperature when 50 mL of water at 20 °C are added to 50 mL of water at 60 °C.

7. Describe how you could determine the specific heat of a metal by using the apparatus and techniques from this experiment.

8. A piece of metal weighing 5.10 g at a temperature of 48.6 °C was placed in a calorimeter in 20.00 mL of water at 22.1 °C. The final equilibrium temperature was found to be 29.2 °C. What is the specific heat of the metal?

9. If the specific heat of methanol is 2.44 J/K-g, how many joules are necessary to raise the temperature of 40 g of methanol from 50 °C to 60 °C?

10. When a 3.25 g sample of solid sodium hydroxide was dissolved in a calorimeter in 100.0 g of water, the temperature rose from 23.9 °C to 32.0 °C. Calculate ΔH (in kJ/mol NaOH) for the following solution process:

$$NaOH(s) \longrightarrow Na^{+}(aq) + OH^{-}(aq)$$

Assume that it's a perfect calorimeter and that the specific heat of the solution is the same as that of pure water.

Name Grace Rademacher Desk

Date 04/13/21 Laboratory Instructor Prof. Towle

REPORT SHEET | EXPERIMENT

Heat of Neutralization 28

A. Heat Capacity of Calorimeter

1. Temp. of calorimeter and water before mixing 25.0 °C - initial
2. Temp. of warm water 45.0 °C - dump in
3. Maximum temp. determined from your curve 35.0 °C - final
4. Heat lost by warm water (temp decrease × 50.0 g × 4.184 J/K-g) = -2029 J

   35-45 = -10

   (-10)(50)(4.184) = -2092
5. Heat gained by cooler water (temp. increase × 50.0 g × 4.184 J/K-g) = 2029 J

   35-25 = 10

   (10)(50)(4.184) = 2029
6. Heat gained by the calorimeter [(4) – (5)] = 0 J
7. Heat capacity of calorimeter:

   $$\frac{\text{heat gained by the calorimeter}}{\text{temperature increase}}$$ 0 J/C

B. Heat of Neutralization of HCl–NaOH

1. Temp. of calorimeter and NaOH 25.5 °C

   Temp. of HCl 25.5
2. ΔT determined from your curve after adding HCl to the NaOH 9.5 °C

   35-25.5 = 9.5°C +273 = 282.5 K
3. Heat gained by solution (temperature increase × 100 g × 4.184 J/K-g) = 3974.8 J

   (9.5)(100)(4.184) = 3974.8
4. Heat gained by calorimeter (temperature increase × heat capacity of calorimeter) = 0 J

   (9.5)(0) = 0
5. Total joules released by reaction [(3) + (4)] = 3974.8 J

6. Complete: HCl + NaOH ⟶ NaCl + H_2O

7. The number of moles of HCl in 50 mL of 1.0 *M* HCl (show calculations) M×L
1.0M × 0.05L = 0.05 mol
0.05 mol

8. The number of moles of H_2O produced in reaction of 50 mL 1.0 *M* HCl and 50 mL 1.0 *M* NaOH (show calculations) 1 HCl + 1 mol = 1 mol H_2O
1.0×0.05 = 0.05 mol NaOH
HCl: 1.0×0.05 = 0.05 mol HCl
Therefore, 0.05 mol H_2O present
0.05 mol

9. Joules released per mole of water formed:

$$\frac{\text{total joules released (5)}}{\text{number of moles water produced (8)}} = \frac{3974.8}{0.05}$$

79496 kJ/mol

Convert to kJ $\frac{79{,}496}{1000} = 79.496$

C. Heat of Neutralization of CH_3COOH–NaOH

1. Temperature of calorimeter and NaOH 25.5 °C
2. ΔT determined from cooling curve after adding CH_3COOH to NaOH 25.5 °C
3. Heat gained by solution (temp. increase × 100 g × 4.184 J/K-g) = (25.5)(100)(4.184) 10669.2 J
4. Heat gained by calorimeter (temp. increase × heat capacity of calorimeter) = (25.5)(74.826) 1908.06 J
5. Total joules released by reaction [(3) + (4)] = 12577.26 J
6. Complete: CH_3COOH + NaOH ⟶ $CH_3COONa + H_2O$
7. The number of moles of H_2O produced in reaction of 50 mL 1.0 *M* CH_3COOH and 50 mL 1.0 *M* NaOH (show calculations) For both, 1.0M × 0.05L = 0.05
0.05 mol
8. Joules released per mole of water formed:

$$\frac{\text{total joules released (5)}}{\text{number of moles water produced (7)}} =$$

251.55 kJ/mol

$$\frac{12577.26}{0.05}$$

Temperature versus Time Data

A. Heat Capacity of Calorimeter		B. Heat of Neutralization of NaOH–HCl		C. Heat of Neutralization of NaOH–CH_3COOH	
Time (s)	Temp (°C)	Time (s)	Temp (°C)	Time (s)	Temp (°C)
0	30	0	28	0	28
15	35	15	31	15	30.5
30	35	30	31	30	30.5
45	35	45	31	45	30.5
60	35	60	31	60	30.5
75	35	75	31	75	30.5
90	35	90	31	90	30.5
105	35	105	31	105	30.5
120	35	120	31	120	30.5
135	35	135	31	135	30.5
150	35	150	31	150	30.5
165	35	165	31	165	30.5
180	35	180	31	180	30.5

QUESTIONS

50mL = 50 g's (for this exp)

1. What is the largest source of error in the experiment?

If measuring equipment is uncalibrated, like a graduated cylinder, or that leftover material was in these tools and resulting in an excess/defecit of material.

2. How should the two heats of reaction for the neutralization of NaOH and the two acids compare? Why?

They should be similar because the same # of moles of NaOH were used for both acids, so they generated the same amount of moles of H_2O, hence making these reactions similar and comparable.

3. The experimental procedure has you wash your thermometer and dry it after you measure the temperature of the NaOH solution and before you measure the temperature of the HCl solution. Why?

Because if not, some NaOH could stick to the thermometer and would react with the HCl, messing up temp calculations.

4. A 50.0 mL sample of a 1.00 *M* solution of $CuSO_4$ is mixed with 50.0 mL of 2.00 *M* KOH in a calorimeter. The temperature of both solutions was 20.2 °C before mixing and 26.3 °C after mixing. The heat capacity of the calorimeter is 12.1 J/K. From these data, calculate ΔH for the process

$$CuSO_4(1\ M) + 2KOH(2\ M) \longrightarrow Cu(OH)_2(s) + K_2SO_4(0.5\ M)$$

Assume that the specific heat and density of the solution after mixing are the same as those of pure water and that the volumes are additive. Total volume: 100 mL, 100 g^3 Density

Specific heat is the same as water's → 4.184 J/g

Temp change: 26.3 − 20.2 = 6.1 °C

Heat absorbed: (100)(4.184)(6.1) = 2549.8 J

Heat absorbed by calori: (12.1 J/K)(279.25 K) = 3378.93

ΔH = (2549.8)(3378.93) = 8615595.71 / 1000 = 8616.6 KJ

Time (s)

0 15 30 45 60 75 90 105 120 135 150 165 180

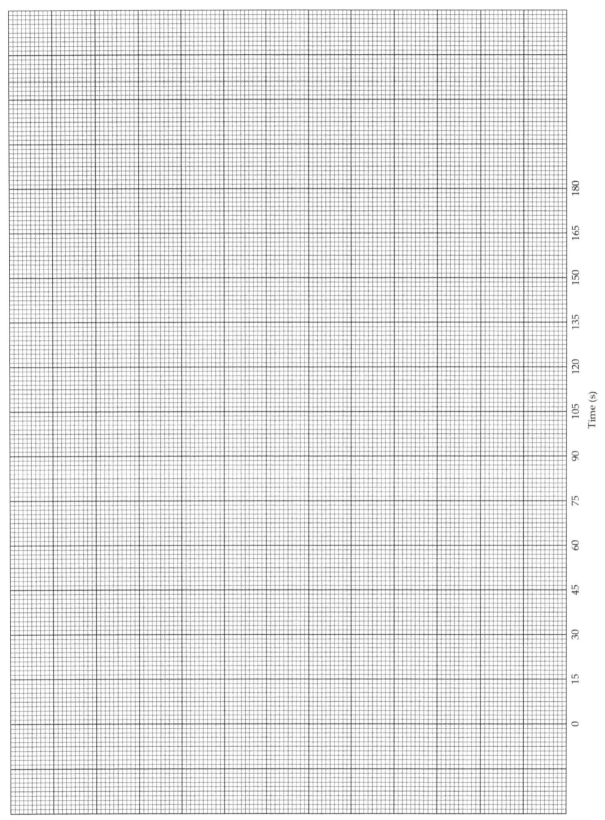

Temperature (°C)

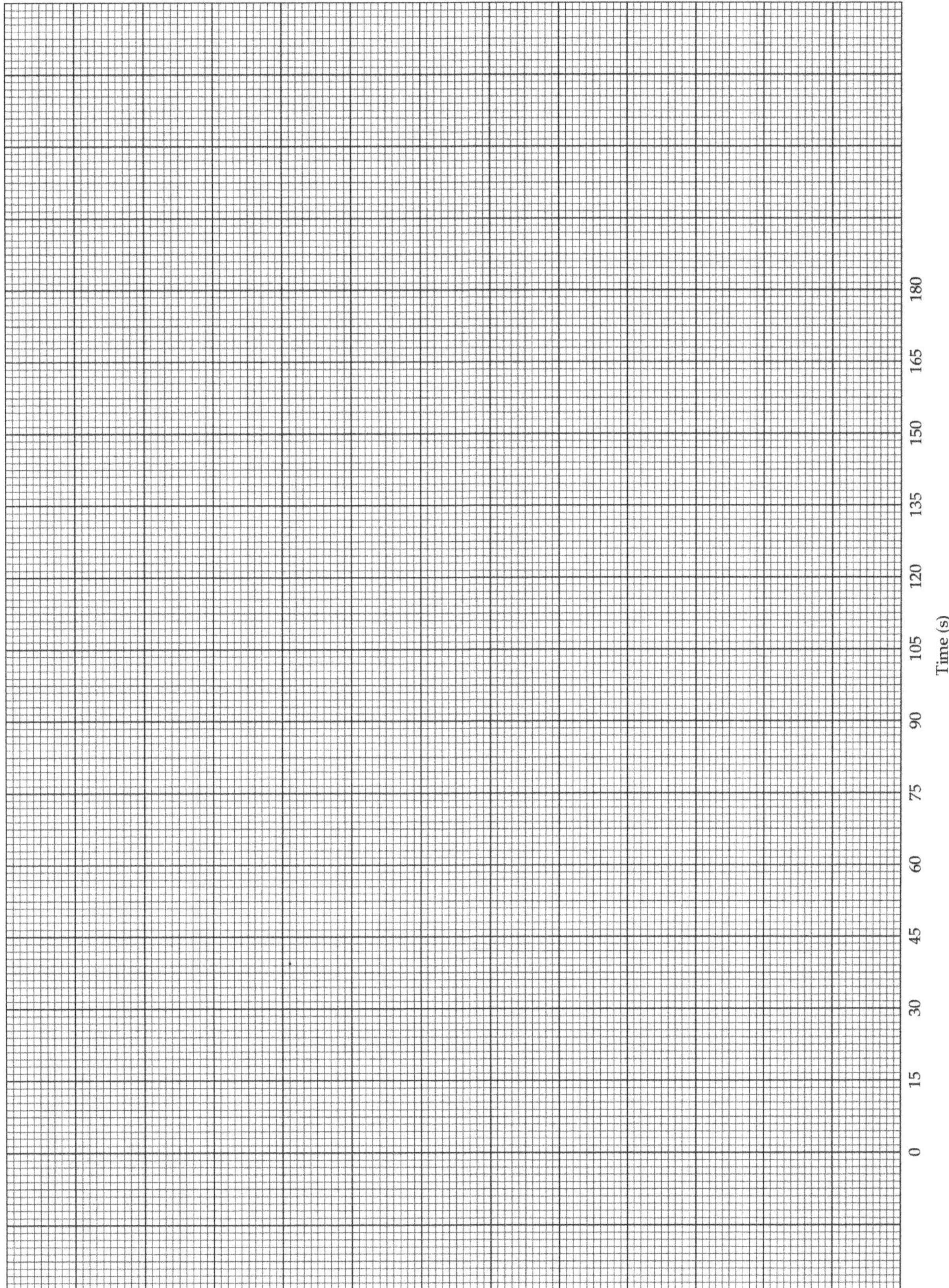

0
15
30
45
60
75
90
105
120
135
150
165
180
Time (s)
Temperature (°C)

Time (s)

0 15 30 45 60 75 90 105 120 135 150 165 180

Temperature (°C)

Rates of Chemical Reactions I: A Clock Reaction

OBJECTIVE

To measure the effect of concentration upon the rate of the reaction of peroxydisulfate ion with iodide ion; to determine the order of the reaction with respect to the reactant concentrations; and to obtain the rate law for the chemical reaction.

APPARATUS AND CHEMICALS

Apparatus

burets (2)	25 mL volumetric pipet
1 mL volumetric pipets (2)	50 mL volumetric pipet
clock or watch with second hand	test tubes (8)
pipet bulb	250 mL Erlenmeyer flasks (4)
buret clamp	100 mL beakers (4)
ring stand	

Chemicals

0.200 *M* KI	0.400 *M* $Na_2S_2O_3$ (freshly prepared)
1% starch solution, boiled	0.1 *M* solution of Na_2H_2EDTA
0.200 *M* $(NH_4)_2S_2O_8$ (freshly prepared)	0.200 *M* KNO_3

WORK IN PAIRS, BUT EVALUATE YOUR DATA INDIVIDUALLY.

DISCUSSION

Factors Affecting Rates of Reactions

On the basis of the experiments you've performed, you may have noticed that reactions occur at varying rates. There is an entire spectrum of rates of reactions, ranging from very slow to extremely fast. For example, the rusting of iron is reasonably slow, whereas the decomposition of TNT is extremely fast. The branch of chemistry concerned with the rates of reactions is called *chemical kinetics* (Section 14.1). Experiments show that rates of homogeneous reactions in solution depend upon the following:

1. The nature of the reactants
2. The concentration of the reactants
3. The temperature
4. The presence of a catalyst

Before a reaction can occur, the reactants must come in direct contact via collisions of the reacting particles (Section 14.5). However, even then, the reacting particles (ions or molecules) must collide with sufficient energy to result in a reaction. If they do not, their collisions are ineffective and analogous to collisions of billiard balls. Keeping these considerations in mind, you can qualitatively explain how the various factors influence the rates of reactions.

Concentration Changing the concentration of a solution alters the number of particles per unit volume (Section 14.3). The more particles present in a given volume, the greater the probability of their colliding. Hence, increasing the concentration of a solution increases the number of collisions per unit time and therefore may increase the rate of reaction.

Temperature Because temperature is a measure of the average kinetic energy, an increase in temperature increases the kinetic energy of the particles (Section 14.5). An increase in kinetic energy increases the velocity of the particles and therefore the number of collisions between them in a *given period of time.* Thus, the rate of reaction increases. Also, an increase in kinetic energy results in a greater proportion of the collisions having the required activation energy for the reaction. As a rule of thumb, for each 10 °C increase in temperature, the rate of reaction doubles.

Catalyst Catalysts, in some cases, are believed to increase reaction rates by bringing particles into close juxtaposition in the correct geometrical arrangement for reaction to occur (Section 14.7). In other instances, catalysts offer an alternative route to the reaction, one that requires less energetic collisions between reactant particles. If less energy is required for a successful collision, a larger percentage of the collisions will have the requisite energy, and the reaction will occur faster. Actually, the catalyst may take an active part in the reaction, but at the end of the reaction, the catalyst can be recovered chemically unchanged.

Order of Reaction Defined

Now examine precisely what is meant by the expression *rate of reaction* (Section 14.2). Consider the hypothetical reaction

$$\mathrm{A + B \longrightarrow C + D} \qquad [1]$$

You can measure the rate of this reaction by observing the rate of disappearance of either of the reactants A and B or the rate of appearance of either of the products C and D. In practice, then, you measure the change of concentration with time of A, B, C, or D. Which species you choose to observe is a matter of convenience. For example, if A, B, and D are colorless and C is colored, you could conveniently measure the rate of appearance of C by observing an increase in the intensity of the color of the solution as a function of time. Mathematically, the rate of reaction may be expressed as follows:

$$\text{rate of disappearance of A} = \frac{\text{change in concentration of A}}{\text{time required for change}} = \frac{-\Delta[\mathrm{A}]}{\Delta t}$$

$$\text{rate of appearance of C} = \frac{\text{change in concentration of C}}{\text{time required for change}} = \frac{\Delta[\mathrm{C}]}{\Delta t}$$

In general, the rate of the reaction will depend upon the concentrations of the reactants. Thus, the rate of the hypothetical reaction may be expressed as

$$\text{rate} = k[\mathrm{A}]^x[\mathrm{B}]^y \qquad [2]$$

where [A] and [B] are the molar concentrations of A and B, x and y are the powers to which the respective concentrations must be raised to describe the rate, and k is the *specific rate constant.* One of the objectives of chemical kinetics is to determine the rate law. Stated slightly differently, one goal of measuring the rate of the reaction is to determine the numerical values of x and y in Equation [2]. Suppose you found that $x = 2$ and $y = 1$ for this reaction. Then

$$\text{rate} = k[\text{A}]^2[\text{B}] \qquad [3]$$

would be the rate law. It should be evident from Equation [3] that doubling the concentration of B (keeping [A] the same) would cause the reaction rate to double. On the other hand, doubling the concentration of A (keeping [B] the same) would cause the rate to increase by a factor of 4 because the rate of the reaction is proportional to the *square* of the concentration of A. The powers to which the concentrations in the rate law are raised are termed the *order of the reaction* (Section 14.3). In this case, the reaction is said to be second order in A and first order in B. The *overall order* of the reaction is the sum of the exponents, $2 + 1 = 3$, or a third-order reaction (Section 14.3). It is possible to determine the order of the reaction by noting the effects of changing reagent concentrations on the rate of the reaction. Note that the order of a reaction may be (and frequently is) different from the stoichiometry of the reaction.

Keep in mind that k, the specific rate constant, has a definite value that is independent of the concentration. It is characteristic of a given reaction and depends upon temperature only. Once that rate law and the rate are known, the value of k can be calculated.

Reaction of Peroxydisulfate Ion with Iodide Ion

In this experiment, you will measure the rate of the reaction

$$S_2O_8^{2-}(aq) + 2I^-(aq) \longrightarrow I_2(aq) + 2SO_4^{2-}(aq) \qquad [4]$$

and you will determine the rate law by measuring the amount of peroxydisulfate, $S_2O_8^{2-}$, that reacts as a function of time. The rate law to be determined is of the form

$$\text{rate of disappearance of } S_2O_8^{2-} = k[S_2O_8^{2-}]^x[I^-]^y \qquad [5]$$

or

$$\frac{\Delta[S_2O_8^{2-}]}{\Delta t} = k[S_2O_8^{2-}]^x[I^-]^y$$

Your goal will be to determine the values of x and y as well as the specific rate constant, k.

You will add to the solution a small amount of another reagent (sodium thiosulfate, $Na_2S_2O_3$), which will cause a change in the color of the solution. The amount is such that the color change will occur when 2×10^{-4} mol of $S_2O_8^{2-}$ has reacted. For reasons to be explained shortly, the solution will turn blue-black when 2×10^{-4} mol of $S_2O_8^{2-}$ has reacted. You will quickly add another portion of $Na_2S_2O_3$ after the appearance of the color, and the blue-black color will disappear. When the blue-black color reappears the second time, *another* 2×10^{-4} mol of $S_2O_8^{2-}$ has reacted, making a total of $2(2\times10^{-4})$ mol of $S_2O_8^{2-}$ that has reacted. You will repeat this procedure several times, keeping *careful* note of the time for the appearance of the blue-black colors.

By graphing the amount of $S_2O_8^{2-}$ consumed versus time, you will be able to determine the rate of the reaction. By changing the initial concentrations of $S_2O_8^{2-}$ and I^- and observing the effects upon the rate of the reaction, you will determine the order of the reaction with respect to $S_2O_8^{2-}$ and I^-.

The blue-black color that appears in the reaction is due to the presence of a starch–iodine complex that is formed from iodine, I_2, and starch in the solution. Therefore, the color will not appear until a detectable amount of I_2 is formed according to Equation [4]. The thiosulfate that is added to the solution reacts *extremely rapidly* with the iodine, as follows:

$$I_2(aq) + 2S_2O_3^{2-}(aq) \longrightarrow 2I^-(aq) + S_4O_6^{2-}(aq) \qquad [6]$$

Consequently, until all of the $S_2O_3^{2-}$ that is added is consumed, there will not be a sufficient amount of I_2 in the solution to yield the blue-black color. You will add 4×10^{-4} mol of $S_2O_3^{2-}$ each time (these equal portions are termed *aliquots*). From the stoichiometry of Equations [4] and [6], you can verify that when this quantity of $S_2O_3^{2-}$ has reacted, 2×10^{-4} mol of $S_2O_8^{2-}$ has reacted. Note also that although iodide, I^-, is consumed according to Equation [4], it is rapidly regenerated according to Equation [6]; therefore, its concentration does not change during a given experiment.

Graphical Determination of Rate

The more rapidly the 2×10^{-4} mol of $S_2O_8^{2-}$ is consumed, the faster the reaction. To determine the rate of the reaction, a plot of moles of $S_2O_8^{2-}$ that have reacted versus the time required for the reaction is made, as shown in Figure 29.1. The best straight line passing through the origin is drawn, and the slope is determined. The slope, $\Delta S_2O_8^{2-}/\Delta t$, corresponds to the moles of $S_2O_8^{2-}$ that have been consumed per second and is proportional to the rate. Because the rate corresponds to the change in the concentration of $S_2O_8^{2-}$ per second, dividing the slope by the volume of the solution yields the rate of disappearance of $S_2O_8^{2-}$ (that is, $\Delta[S_2O_8^{2-}]/\Delta t$). If the total volume of the solution in this example was 75 mL, the rate would be as follows:

$$\frac{4.5\times10^{-5}\text{ mol/s}}{0.075\text{ L}} = 6.0\times10^{-4}\text{ mol/L-s}$$

If you obtain a rate of 6.0×10^{-4} mol/L-s when $[S_2O_8^{2-}] = 2.0\ M$ and $[I^-] = 2.0\ M$ and a rate of 3.0×10^{-4} mol/L-s when $[S_2O_8^{2-}] = 1.0\ M$ and $[I^-] = 2.0\ M$, you know that doubling the concentration of $S_2O_8^{2-}$ doubles the rate of the reaction and the reaction is first order in $S_2O_8^{2-}$. By varying the initial concentrations of $S_2O_8^{2-}$ and I^-, you can, via the above type of analysis, determine the order of the reaction with respect to both species.

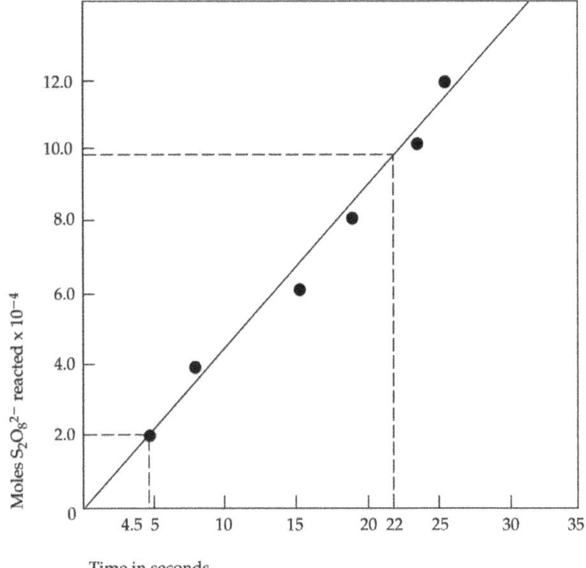

$$\text{Slope} = \frac{\Delta S_2O_8^{2-}}{\Delta t}$$

$$= \frac{(9.8 - 2.0) \times 10^{-4}\,\text{mol}}{(22.0 - 4.5)\,\text{s}}$$

$$= \frac{7.8 \times 10^{-4}\,\text{mol}}{17.5\,\text{s}}$$

$$= 4.5 \times 10^{-5}\,\text{mol/s}$$

or use a linear regression analysis to determine the slope (see Appendix C)

▲**FIGURE 29.1** Graphical determination of rate.

GIVE IT SOME THOUGHT

What does the steepness of the slope tell you about the rate of the reaction?

Helpful Comments

1. According to the procedure of this experiment, the solution will turn blue-black when exactly 2×10^{-4} mol of $S_2O_8^{2-}$ has reacted.
2. The purpose of the KNO_3 solution in this reaction is to keep the *reaction medium* the same in each run in terms of the concentration of ions; it does not enter into the reaction in any way.
3. The reaction studied in this experiment is catalyzed by metal ions. The purpose of the drop of the EDTA solution is to minimize the effects of trace quantities of metal ion impurities that would cause spurious effects on the reaction.
4. You will perform a few preliminary experiments to become acquainted with the observations in this experiment so that you will know what to expect in the reactions.
5. The initial concentrations of the reactants have been provided on the report sheet.

A. Preliminary Experiments

PROCEDURE

1. Dilute 5.0 mL of 0.2 *M* KI solution with 10.0 mL of distilled water in a test tube, add three drops of starch solution and mix thoroughly, and then add 5.0 mL of 0.2 *M* $(NH_4)_2S_2O_8$ solution. Mix. Wait awhile and observe color changes.

2. Repeat the procedure in (1), but when the solution changes color, add four drops of 0.4 *M* $Na_2S_2O_3$, mix the solution, and note the effect that the addition of $Na_2S_2O_3$ has on the color.

B. Kinetics Experiment

Equipment Setup Set up two burets held by a clamp on a ring stand as shown in Figure 29.2. Use these burets to accurately measure the volumes of the KI and KNO_3 solutions. Use two separate 1 mL pipets for measuring the volumes of the $Na_2S_2O_3$ and starch solutions and use 25 mL and 50 mL pipets to measure the volumes of the $(NH_4)_2S_2O_8$ solutions.

Solution Preparation Prepare four reaction solutions as follows (prepare the next solution only when you have completely finished with the previous one):

Each solution must be freshly prepared before you to begin the rate study—that is, *prepare solutions 1, 2, 3, and 4 one at a time as you make your measurements.*

Solution 1: 25.0 mL KI solution
1.0 mL starch solution
1.0 mL $Na_2S_2O_3$ solution
48.0 mL KNO_3 solution
1 drop EDTA solution
Total volume = 75.0 mL

Solution 2: 25.0 mL KI solution
1.0 mL starch solution
1.0 mL $Na_2S_2O_3$ solution
23.0 mL KNO_3 solution
1 drop EDTA solution
Total volume = 50.0 mL

Solution 3: 50.0 mL KI solution
1.0 mL starch solution
1.0 mL $Na_2S_2O_3$ solution
23.0 mL KNO_3 solution
1 drop EDTA solution
Total volume = 75.0 mL

Solution 4: 12.5 mL KI solution
1.0 mL starch solution
1.0 mL $Na_2S_2O_3$ solution
35.5 mL KNO_3 solution
1 drop EDTA solution
Total volume = 50.0 mL

GIVE IT SOME THOUGHT

Why is a drop of EDTA added to each solution?

Rate Measurements Prepare solution 1 in a 250 mL Erlenmeyer flask that has been scrupulously cleaned and dried. Pipet 25.0 mL of $(NH_4)_2S_2O_8$ solution into a clean, dry 100 mL beaker. *Be ready* to begin timing the reaction when the solutions are mixed (READ AHEAD). The reaction starts the moment the solutions are mixed. BE PREPARED! ZERO TIME! Quickly pour the 25.0 mL of $(NH_4)_2S_2O_8$ solution into solution 1 and swirl vigorously; note to the nearest second the time you begin mixing. At the instant the blue-black color appears, 2×10^{-4} mol of $S_2O_8{}^{2-}$ has reacted. *Immediately* (be prepared!) add a 1 mL aliquot of $Na_2S_2O_3$ solution from the pipet and swirl the solution; the color will disappear. By filling each of seven clean, dry test tubes with 1 mL of $Na_2S_2O_3$ solution, to avoid losing time, you can add these aliquots to your reactions when the blue color appears.

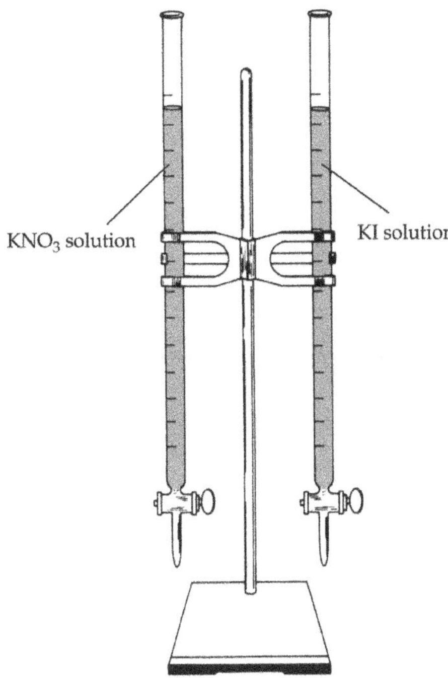

▲FIGURE 29.2 Experimental apparatus with two burets clamped to a ring stand.

Record the time for the reappearance of the blue-black color. Add another 1 mL aliquot of $Na_2S_2O_3$ solution and note the time for the reappearance of the color. The time interval being measured is that between appearances of the blue-black color. For good results, these aliquots of $Na_2S_2O_3$ must be measured as quickly, accurately, and reproducibly as possible. Continue this procedure until you have added seven aliquots to solution 1.

GIVE IT SOME THOUGHT

What does the blue-black color indicate?

You are finished with solution 1 when you have recorded all of your times on the report sheet. *(The time intervals are cumulative.)*

Solutions 2, 3, and 4 should be treated in the same manner except that 50.0 mL portions of $(NH_4)_2S_2O_8$ solutions should be added to solutions 2 and 4 and 25 mL of $(NH_4)_2S_2O_8$ solution should be added to solution 3. **(CAUTION: *Be on guard—solution 2 will react more rapidly than solution 1.*)** In each of these reactions, the final total solution volume is exactly 100 mL.

Calculations

Use the data sheet to tabulate the following for each aliquot of $Na_2S_2O_3$ added to each of the four solutions:

1. The time interval from the start of the reaction (addition of $S_2O_8^{2-}$) to the appearance of color for the first aliquot of $S_2O_3^{2-}$ and the time interval from the preceding color appearance for each succeeding aliquot (column 2).

2. The cumulative time from the start of the reaction to each appearance of color (column 3).
3. The corresponding numbers of moles $S_2O_8^{2-}$ consumed (column 4).

For each solution, use the graph paper provided to plot the moles of $S_2O_8^{2-}$ consumed (as the ordinate, vertical axis) versus time in seconds (as the abscissa, horizontal axis), using the data in columns 3 and 4 on the report sheet. Calculate the slope of each plot and from these calculations, answer the questions on your report sheet.

GIVE IT SOME THOUGHT

a. Do you expect your graph to be linear or exponential?
b. What does this tell you about the overall order?

Waste-Disposal Instructions No wastes from this experiment should be flushed down the sink. Locate the special containers placed in the laboratory for the disposal of excess iodide and peroxydisulfate solutions as well as for the reaction mixtures from the test tubes or flasks. All wastes should be disposed of in these containers.

Name ______________________________ Desk ______________________________

Date ______________ Laboratory Instructor ______________________________

Rates of Chemical Reactions I: A Clock Reaction | 29 Pre-lab Questions

Before beginning this experiment in the laboratory, you should be able to answer the following questions.

1. Besides temperature and the presence of a catalyst, what are the other factors that affect the rate of a chemical reaction?

2. What is the general form of a rate law?

3. What is the order of reaction with respect to A and B for a reaction that obeys the rate law: rate = $k[A]^3[B]^2$?

4. Write the chemical equations involved in this experiment and show that the rate of disappearance of $[S_2O_8^{2-}]$ is proportional to the rate of appearance of the blue-black color of the starch–iodine complex.

5. It is found for the reaction $2A + 3B \longrightarrow C$ that doubling the concentration of either A or B doubles the rate of reaction. Write the rate law for this reaction.

6. If 2.0×10^{-4} moles of $S_2O_8^{2-}$ in 50 mL of solution is consumed in 188 seconds, what is the rate of consumption of $S_2O_8^{2-}$?

---

7. Why are chemists concerned with the rates of chemical reactions? What possible practical value does this type of information have?

8. Suppose you were to dissolve a sugar cube in hot water. How would the temperature of the hot water affect the rate of its dissolution?

9. Assuming that a chemical reaction doubles in rate for each 10 °C temperature increase, by what factor would the rate increase if the temperature increased by 50 °C?

10. A reaction between substances A and B has been found to give the following data:

$$3A + 2B \longrightarrow 2C + D$$

[A] (*mol/L*)	*[B]* (*mol/L*)	*Rate of appearance of C* (*mol/L-hr*)
1.0×10^{-2}	1.0	0.300×10^{-6}
1.0×10^{-2}	3.0	8.10×10^{-6}
2.0×10^{-2}	3.0	3.24×10^{-5}
2.0×10^{-2}	1.0	1.20×10^{-6}
3.0×10^{-2}	3.0	7.30×10^{-5}

Using the above data, determine the order of the reaction with respect to A and B, the rate law, and calculate the specific rate constant.

B=1

$\frac{.0222}{.0111} = \frac{[0.17]^m[0.5]^n}{[0.17]^m[0.25]^n}$

$2 = 2^n$

$n = 1$

$\frac{.0222}{.0111} = \frac{[0.25]^n[0.34]^m}{[0.25]^n[0.17]^m}$

$2 = 2^m$

$.0111 = k[0.17]^{m\,1}[0.25]^{n\,1}$

Name Grace Rademacher Desk ____________

Date 02/09/21 Laboratory Instructor Professor Tolle

REPORT SHEET

Rates of Chemical Reactions I: A Clock Reaction

EXPERIMENT 29

A. Preliminary Experiments

1. What are the colors of the solutions containing the following ions? K^+ Brown ; I^- Brown

2. The color of the starch·I_2 complex is ____________.

B. Kinetics Experiment

Divide slope of whole graph by 0.05

Solution 1. Initial $[S_2O_8^{2-}] = 0.050\ M$; initial $[I^-] = 0.050\ M$. Time experiment started ____________

rate of slope (mols per sec) Y axis

Aliquot no.	Time (s) between appearances of color	X axis Cumulative times (s)	Total moles of $S_2O_8^{2-}$ consumed
1	1.06	66	2.0×10^{-4}
2	1.56	116	4.0×10^{-4}
3	2.4	160	6.0×10^{-4}
4	3.19	199	8.0×10^{-4}
5	4.16	256	$10.\times10^{-4}$
6	5.12	312	12×10^{-4}
7	5.58	358	14×10^{-4}

Divide by 0.05 L, change to molar in order to plot slope (mol/s)

Solution 2. Initial $[S_2O_8^{2-}] = 0.10\ M$; initial $[I^-] = 0.050\ M$. Time experiment started ____________

Aliquot no.	Time (s) between appearances of color	Cumulative times (s)	Total moles of $S_2O_8^{2-}$ consumed
1	265	185	2.0×10^{-4}
2	1:01	61	4.0×10^{-4}
3	132	92	6.0×10^{-4}
4	150	110	8.0×10^{-4}
5	235	155	$10.\times10^{-4}$
6	322	202	12×10^{-4}
7	355	235	14×10^{-4}

Solution 3. Initial $[S_2O_8^{2-}] = 0.050\ M$; initial $[I^-] = 0.10\ M$. Time experiment started ______ 120

Aliquot no.	*Time (s) between appearances of color*	*Cumulative times (s)*	*Total moles of $S_2O_8^{2-}$ consumed*
1	37	37	2.0×10^{-4}
2	1.13	73	4.0×10^{-4}
3	1.47	107	6.0×10^{-4}
4	2.34	154	8.0×10^{-4}
5	3.06	186	$10.\times10^{-4}$
6	3.50	230	12×10^{-4}
7	4.28	268	14×10^{-4}

147

306

→ .05 L to get the Rate!

Solution 4. Initial $[S_2O_8^{2-}] = 0.10\ M$; initial $[I^-] = 0.025\ M$. Time experiment started ______

Aliquot no.	*Time (s) between appearances of color*	*Cumulative times (s)*	*Total moles of $S_2O_8^{2-}$ consumed*
1			2.0×10^{-4}
2			4.0×10^{-4}
3			6.0×10^{-4}
4			8.0×10^{-4}
5			$10.\times10^{-4}$
6			12×10^{-4}
7			14×10^{-4}

Calculations

1. Rate of reaction, $\Delta[S_2O_8^{2-}]/\Delta t$, as calculated from graphs (that is, from slopes of lines):

   Solution 1 (8E-5)x + 0.0012 Solution 3 0.0001x + 0.0004

   Solution 2 (8E-5)x + 0.004 ~~Solution 4~~

2. What effect does doubling the concentration of I^- have on the rate of this reaction?

   It increases the Cumulative time of the reaction.

3. What effect does changing the $[S_2O_8^{2-}]$ have on the reaction?

   Heavily decreases the Cumulative time of the reaction.

4. Write the rate law for this reaction that is consistent with your data.

   $\frac{4}{2} = \frac{0.05}{0.1}$ $\frac{4.0}{2.0} = \frac{0.05}{0.1}$

   $2^x = 0.5$ $2^x = 0.5$

   $x = -1$ $x = -1$

5. From your knowledge of x and y in the equation (as well as the rate in a given experiment from your graph), calculate k from your data. Rate = $k[S_2O_8^{2-}]^x[I^-]^y$.

$X = 1 \quad y = 1$

$2.0 \times 10^{-4} = k[0.05]^1[0.1]^1$

$\frac{2.0 \times 10^{-4}}{0.005} = \frac{k[0.005]}{0.005}$

$k = 0.04$

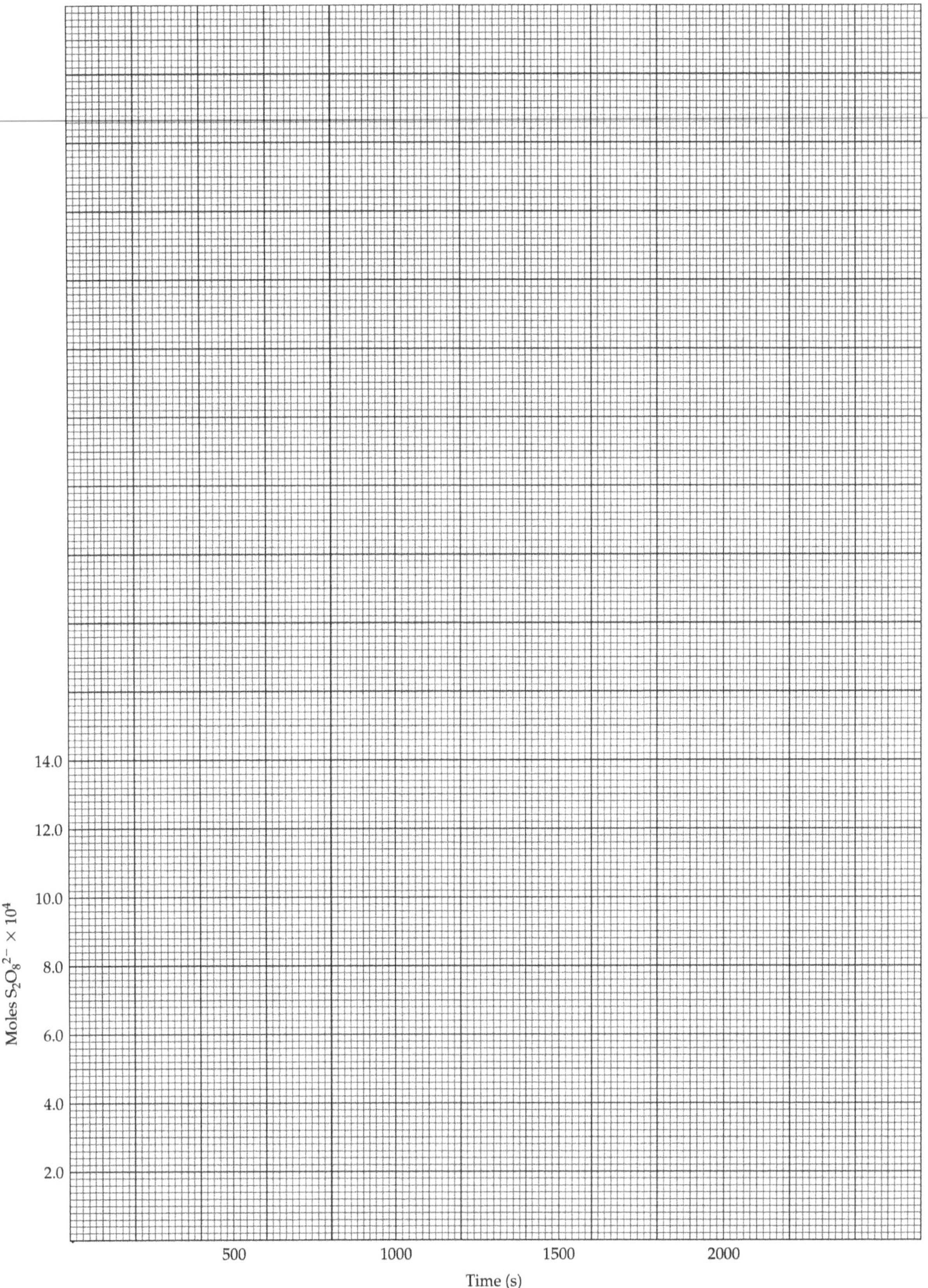
14.0
12.0
10.0
8.0
6.0
4.0
2.0
Moles $S_2O_8^{2-} \times 10^4$
500
1000
1500
2000
Time (s)

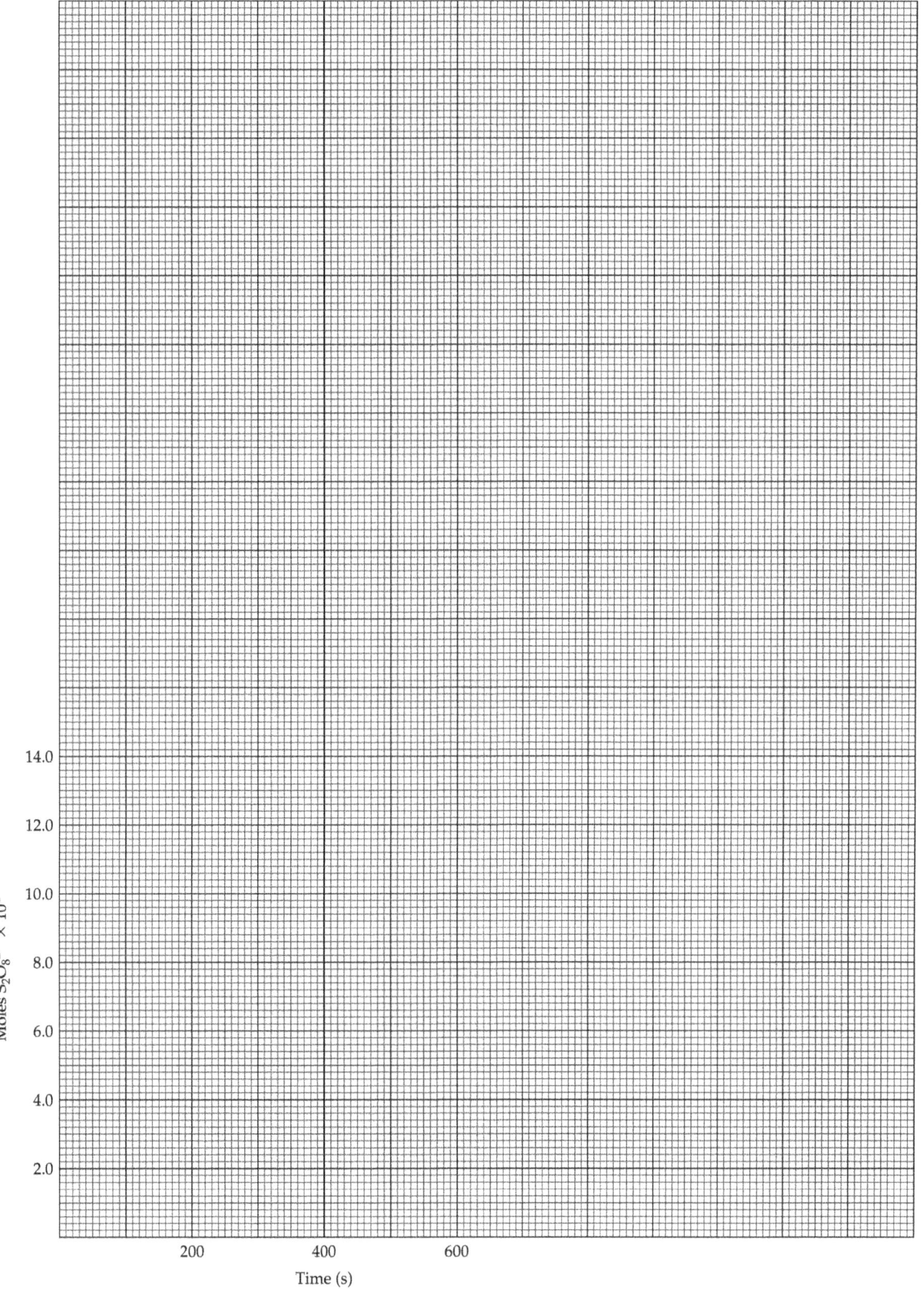
14.0
12.0
10.0
8.0
6.0
4.0
2.0
Moles $S_2O_8^{2-} \times 10^4$
200
400
600
Time (s)

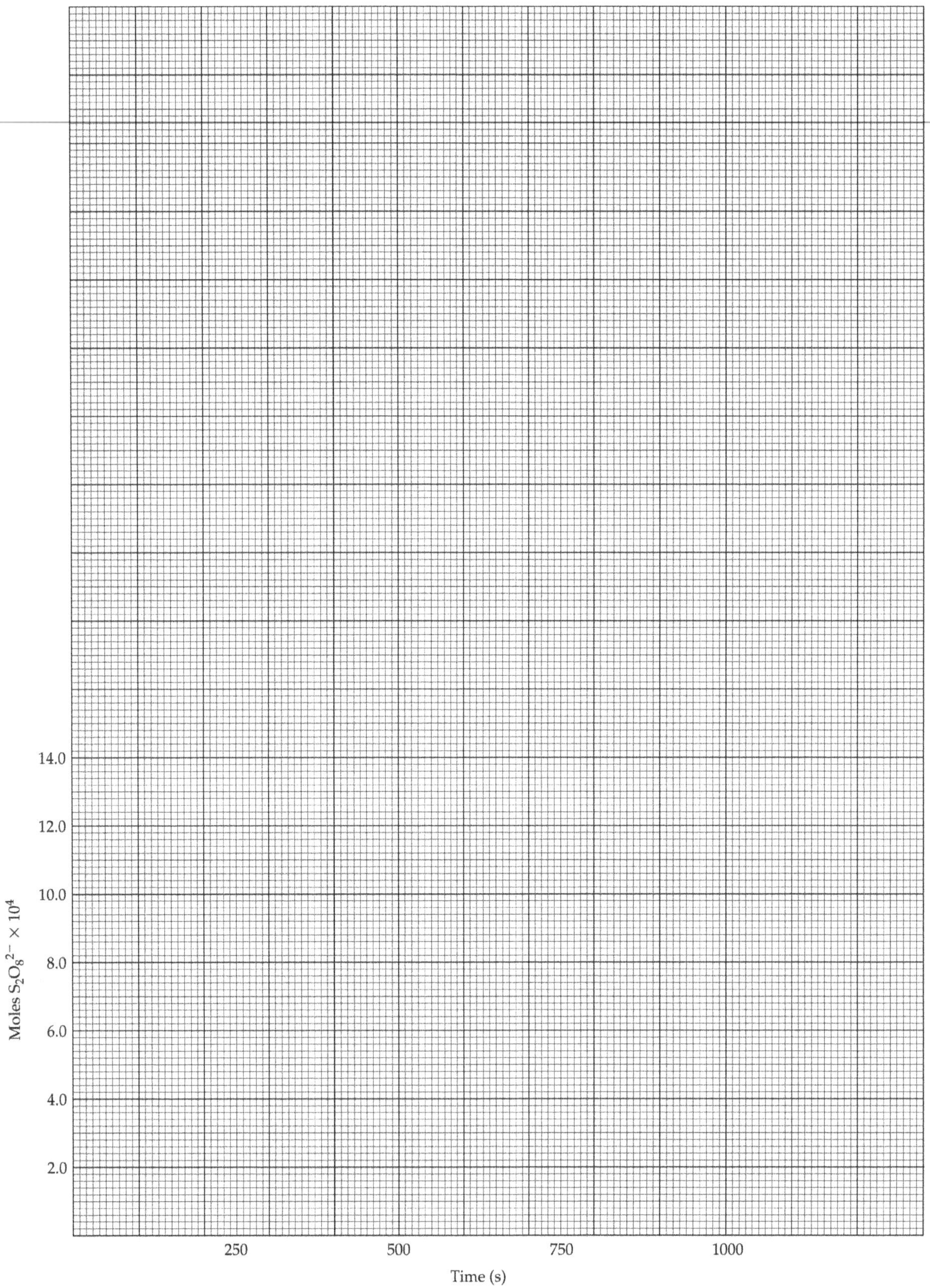
14.0
12.0
10.0
8.0
6.0
4.0
2.0
Moles $S_2O_8^{2-} \times 10^4$
250
500
750
1000
Time (s)

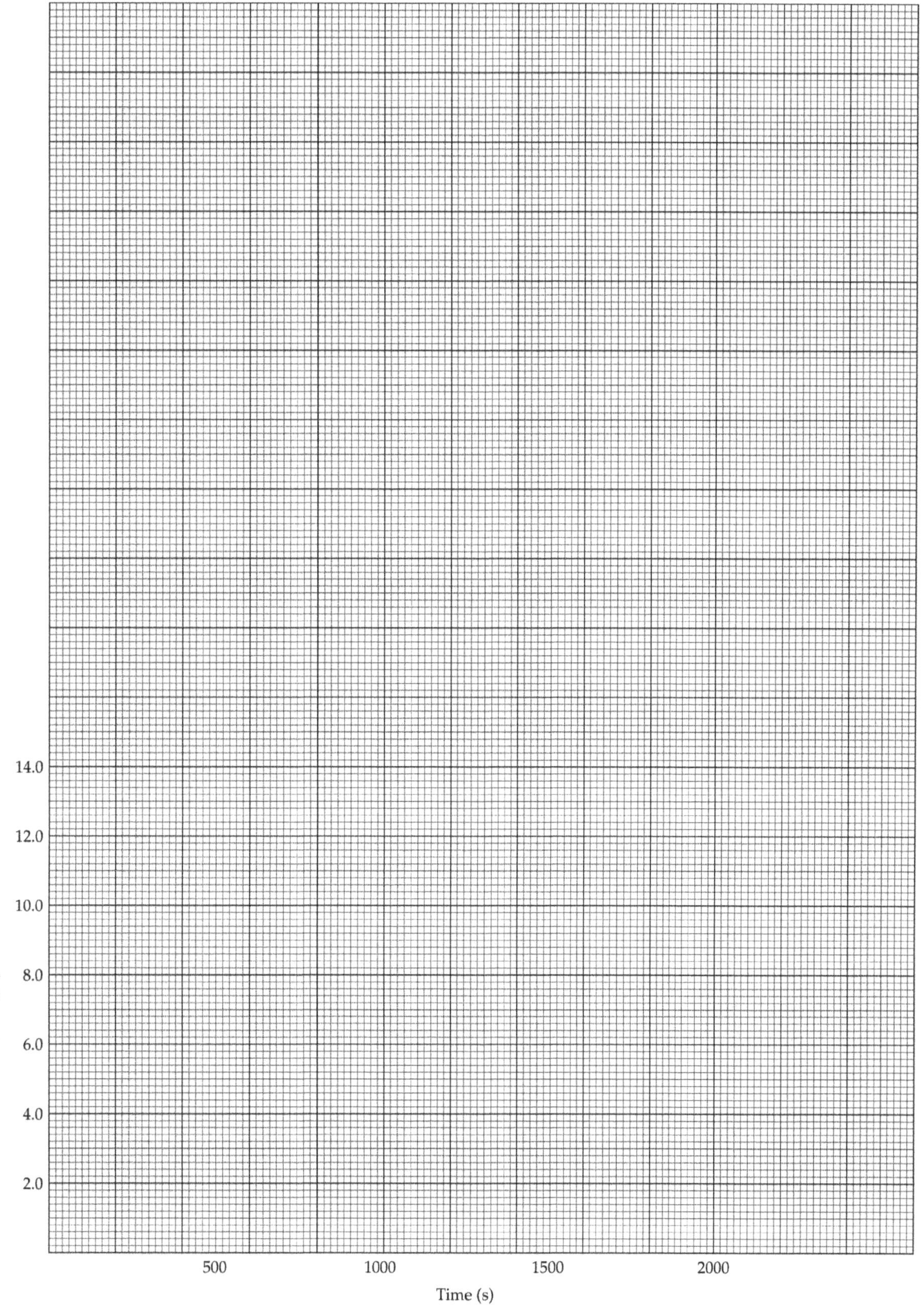
14.0
12.0
10.0
8.0
6.0
4.0
2.0
Moles $S_2O_8^{2-} \times 10^4$
500
1000
1500
2000
Time (s)

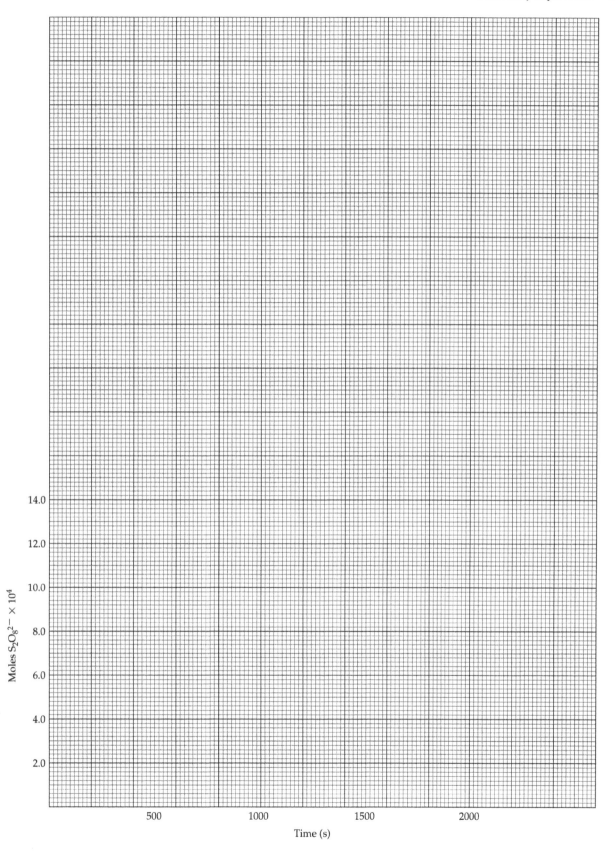

NOTES AND CALCULATIONS

Rates of Chemical Reactions II: Rate and Order of H_2O_2 Decomposition

Experiment

30

OBJECTIVE

To determine the rate and order of reaction for the decomposition of hydrogen peroxide.

APPARATUS AND CHEMICALS

Apparatus

Mohr buret or standard buret	barometer
leveling bulb or funnel	ring stand and iron ring
125 mL Erlenmeyer flask	wristwatch or timer
buret clamp	thermometer
graduated cylinder	pneumatic trough
rubber tubing	No. 00 and No. 6 one-hole stoppers
test tube	
medicine droppers (2)	

Chemicals

3% H_2O_2	0.10 M KI
0.10 M $Pb(NO_3)_2$	

WORK IN PAIRS, BUT EVALUATE YOUR DATA INDEPENDENTLY.

DISCUSSION

You have learned that much information may be obtained from a chemical equation—for example, the identity of the products formed from the specified reactants and the quantitative mass relationships between the substances involved in the reaction. The equation, however, does not provide anything about the conditions required for the reaction to occur, nor does it give any information about the speed or rate at which the reaction takes place. The area of chemistry concerned with rates of chemical reactions is called *chemical kinetics* (Section 14.1).

Experiments show that rates of homogeneous reactions in solution depend upon the following:

1. The nature of the reactants
2. The concentration of the reactants
3. The temperature
4. The presence of a catalyst

Before a reaction can occur, the reactants must come in direct contact via collisions of the reacting particles. However, even then, the reacting particles (ions or molecules) must collide with sufficient energy to result in a reaction. If they do not, their collisions are ineffective and analogous to collisions of billiard balls. With these considerations in mind, you can qualitatively explain how the various factors influence the rates of reactions.

Concentration Changing the concentration of a solution alters the number of particles per unit volume (Section 14.3). The more particles present in a given volume, the greater the probability of their colliding. Hence, increasing the concentration of a solution increases the number of collisions per unit time and therefore the rate of reaction may increase.

Temperature Because temperature is a measure of the average kinetic energy, an increase in temperature increases the kinetic energy of the particles (Section 14.5). An increase in kinetic energy increases the velocity of the particles and therefore the number of collisions between them in a *given period of time*. Thus, the rate of reaction increases. Also, an increase in kinetic energy results in a greater proportion of the collisions having the required activation energy for the reaction. As a rule of thumb, for each 10 °C increase in temperature, the rate of reaction doubles.

Catalyst In some cases, catalysts are believed to increase reaction rates by bringing particles into close juxtaposition in the correct geometrical arrangement for reaction to occur (Section 14.7). In other instances, catalysts offer an alternative route to the reaction, one that requires less energetic collisions between reactant particles. If less energy is required for a successful collision, a larger percentage of the collisions will have the requisite energy and the reaction will occur faster. Actually, the catalyst may take an active part in the reaction, but at the end of the reaction, the catalyst can be recovered chemically unchanged in any case.

Order of Reaction Defined

Now examine precisely what is meant by the expression *rate of reaction*. Consider the hypothetical reaction (Section 14.2)

$$A + B \longrightarrow C + D \qquad [1]$$

The rate of this reaction may be measured by observing the rate of disappearance of either of the reactants A and B or the rate of appearance of either of the products C and D. In practice then, you measure the change of concentration with time of A, B, C, or D. Which species you choose to observe is a matter of convenience. For example, if A, B, and D are colorless and C is colored, you could conveniently measure the rate of appearance of C by observing an increase in the intensity of the color of the solution as a function of time. Mathematically, the rate of reaction may be expressed as follows:

$$\text{rate of disappearance of A} = \frac{\text{change in concentration of A}}{\text{time required for change}} = \frac{-\Delta[A]}{\Delta t}$$

$$\text{rate of appearance of C} = \frac{\text{change in concentration of C}}{\text{time required for change}} = \frac{\Delta[C]}{\Delta t}$$

In general, the rate of the reaction will depend upon the concentrations of the reactants. Thus, the rate of the hypothetical reaction may be expressed as

$$\text{rate} = k[A]^x[B]^y \qquad [2]$$

where [A] and [B] are the molar concentrations of A and B, x and y are the powers to which the respective concentrations must be raised to describe the rate, and k is the *specific rate constant*. One of the objectives of chemical kinetics is to determine the rate law. Stated slightly differently, one goal of measuring the rate of the reaction is to determine the numerical values of x and y in Equation 2. Suppose you found that $x = 2$ and $y = 1$ for this reaction. Then

$$\text{rate} = k[\text{A}]^2[\text{B}] \qquad [3]$$

would be the rate law. It should be evident from Equation [3] that doubling the concentration of B (keeping [A] the same) would cause the reaction rate to double. On the other hand, doubling the concentration of A (keeping [B] the same) would cause the rate to increase by a factor of 4 because the rate of the reaction is proportional to the *square* of the concentration of A. The powers to which the concentrations in the rate law are raised are termed the *order of the reaction* (Section 14.4). In this case, the reaction is said to be second order in A and first order in B. The *overall order* of the reaction is the sum of the exponents, $2+1=3$, or a third-order reaction (Section 14.3). It is possible to determine the order of the reaction by noting the effects of changing reagent concentrations on the rate of the reaction. Note that the order of a reaction may be (and frequently is) different from the stoichiometry of the reaction.

It should be emphasized that k, the specific rate constant, has a definite value that is independent of the concentration (Section 14.3). It is characteristic of a given reaction and depends only on temperature. Once that rate law and the rate are known, the value of k can be calculated.

The term *rate of reaction* refers to the speed at which reactants are being consumed or products are being formed. This must be determined experimentally by measuring the rate of change in the concentration of one of the reactants or one of the products or some physical property (such as the volume of a gas or the color intensity of a solution) that is directly proportional to one of these concentrations. The rate may be expressed, for example, as moles per liter of product being formed per minute, milliliters of gas being produced per minute, or moles per liter of reactant being consumed per second.

In this experiment, you will study the rate of the iodide-catalyzed decomposition of hydrogen peroxide to form water and oxygen.

$$2H_2O_2(aq) \xrightarrow{I^-} O_2(g) + 2H_2O(l) \qquad [4]$$

The standard enthalpy change, $\Delta H°$, and standard free-energy change, $\Delta G°$, for the reaction in Equation [4] are −196.1 kJ and −233.5 kJ, respectively. Even though $\Delta G°$ is negative and you can conclude that the reaction will proceed spontaneously at room temperature, you cannot predict anything about the *rate* at which H_2O_2 decomposes. Recall the relation between free energy and the equilibrium constant. Reactions generally involve an energy barrier (Figure 30.1) that must be overcome in going from reactants to products. This energy barrier is called the *activation energy* and is commonly designated by the term E_a. It corresponds to the minimum energy that molecules must have for the reaction to occur. The magnitude of E_a (the activation energy) determines the rate of the reaction, whereas the magnitude of the free-energy change (ΔG) determines the extent of the reaction. By observing the volume of oxygen formed as a function of time, you will determine how the rate of reaction [4] is affected by different initial concentrations of hydrogen peroxide and iodide ion (which is introduced to the reaction in the form of the strong electrolyte KI). Although iodide does not appear in the chemical equation, it does have a pronounced effect on the rate of the reaction, for it serves as a catalyst.

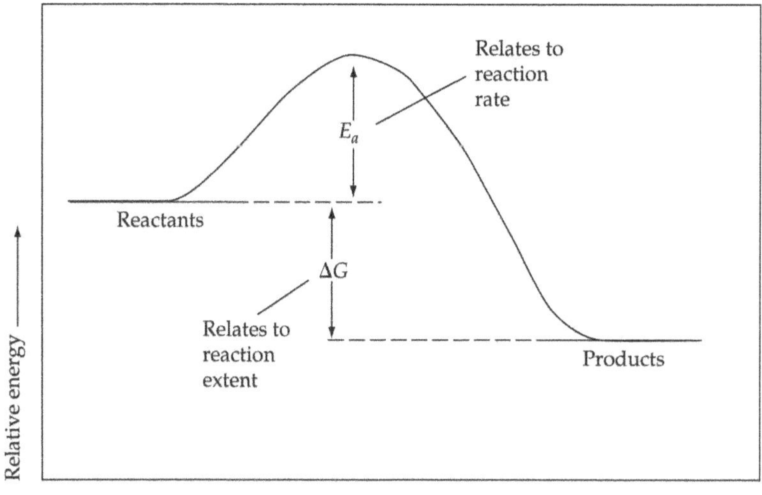

▲FIGURE 30.1 Reaction profile for an exothermic reaction illustrating ΔG and E_a.

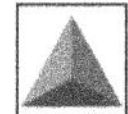

GIVE IT SOME THOUGHT

a. What do thermodynamic quantities tell you about the speed of a reaction?
b. What does a catalyst do to a reaction?
c. What portion of this graph is dictated by kinetics?
d. Which portion is dictated by thermodynamics?

Your ultimate goal in this experiment is to deduce a rate law for the reaction, showing the dependence of the rate on the concentrations of H_2O_2 and I^-. Your rate law will be of the form

$$\text{rate of oxygen production} = k[H_2O_2]^x[I^-]^y$$

where k is the specific rate constant and depends only on temperature. Thus, your objective is to determine the numerical values of the exponents (x and y) and k.

GIVE IT SOME THOUGHT

How would you express the rate of the overall reaction in terms of x and y?

You will also make a qualitative study of the effect that temperature has on the rate of decomposition of H_2O_2. At higher temperatures, the average kinetic energy of molecules is greater. Hence, the fraction of molecules with kinetic energy sufficient for reaction is larger, and the rate of reaction is correspondingly larger. Arrhenius noted a nonlinear relation between rate and temperature and found that most reaction-rate data obeyed the equation

$$\log k = \log A - \frac{E_a}{2.30RT}$$

where k is the specific rate constant, E_a is the activation energy, R is the gas constant, and T is the absolute temperature. A is a constant, or nearly so, as temperature is varied. It is called the frequency factor and is related to the frequency of collisions and the probability that the molecules are suitably oriented for reaction.

PROCEDURE

A. Order of Reaction

Assemble an apparatus for the collection of oxygen similar to that shown in Figure 30.2. (*The rubber tube connections must be snug.*) Fill the trough with water at room temperature and record the temperature. It may be necessary to add some hot water to achieve room temperature. Add room temperature water to the assembly until the height of water in the buret is about 10 mL from the top when the water in the leveling bulb is at the same level (see Figure 30.2). Check for leaks in the apparatus by lowering the bulb with the system closed. In the absence of leaks, only small changes in the water level in the buret due to pressure changes will occur as the leveling bulb is lowered.

Solution 1 Add 10.0 mL of 0.10 *M* KI and 15.0 mL of distilled water to a clean 125 mL Erlenmeyer flask. Carefully swirl the flask for a few minutes so that the solution attains the bath temperature in the trough. Add 5.0 mL of 3% H_2O_2 and quickly stopper the flask. One student should keep swirling the flask in the bath as vigorously as possible during the remainder of the experiment; the other student should observe the volume of oxygen produced during the reaction at various times. You should begin recording volume and time after approximately 2 mL of gas have been evolved. Because of the importance of measuring the volume of oxygen evolved at constant pressure, you must measure the volume with the level of water in the leveling bulb that is the same as

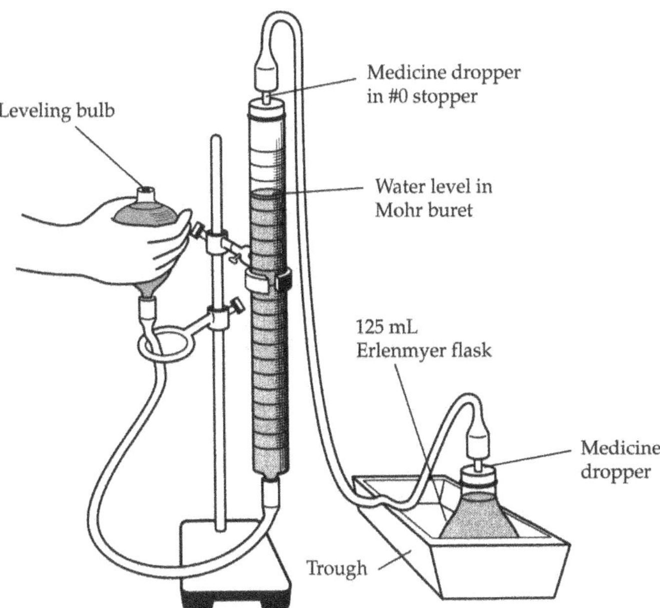

▲**FIGURE 30.2** Experimental apparatus.

that in the buret. One student should match the water levels in the buret and leveling bulb (by lowering the leveling bulb) and read the volume. The other student should continue to swirl the flask and record on the report sheet the total elapsed time at the instant the gas volume is read. You should take time and volume readings at approximately 2 mL intervals until a total of about 14 mL of oxygen has been evolved. Although it is not necessary, you will find it convenient to use the stopwatch function of a digital wristwatch or a cell phone for noting the time intervals.

Solution 2 Rinse the Erlenmeyer flask and graduated cylinder thoroughly with distilled water and allow them to drain completely. Make certain the bath temperature is the same as that used for solution 1. You may need to adjust it by adding a little warm water. Repeat the experiment, this time adding 10.0 mL of 0.10 *M* KI and 10.0 mL of distilled water, swirling, and then adding 10.0 mL of 3% H_2O_2. Quickly stopper the flask and take volume-time readings as before.

Solution 3 Rinse the flask and graduated cylinder. Check bath temperature and adjust if necessary. Repeat as follows: Add 20.0 mL of 0.10 *M* KI and 5.0 mL of distilled water, swirl, and add 5.0 mL of H_2O_2. Take volume-time readings as before.

B. Effect of Temperature

Adjust the bath temperature so that it is approximately 10 °C to 12 °C higher than what it was for Part A. Then repeat the experiment, following the directions given in Part A for solution 1. After you have collected 14 mL of O_2, do not take any more volume-time readings, but allow the reaction to continue to completion (that is, until no more oxygen is evolved). Save the solution for Part C. On your report sheet, record the total volume of O_2 collected and the barometric pressure.

GIVE IT SOME THOUGHT

How do you think increasing the temperature will influence the reaction rate?

C. Identification of the Catalyst

You can show that iodide has been a catalyst in the decomposition of H_2O_2 by noting the fact that it has not been consumed during the reaction even though it appears in the rate law. You can identify the iodide ion by precipitating it as PbI_2. Pour about 1 mL of the reaction mixture you saved from Part B into a small test tube. Add 10 drops of 0.1 *M* $Pb(NO_3)_2$ to the test tube, mix, and record your observations. **(CAUTION:** ***Lead nitrate is toxic, and you should not get any on yourself. Should you come in contact with the*** **$Pb(NO_3)_2$** ***solution, immediately wash it off with copious amounts of water.*****)** Dilute about 0.2 mL of 0.10 *M* KI with about 5 mL of distilled water. Add two drops of this solution to 10 drops of 0.1 *M* $Pb(NO_3)_2$ and record your observations.

Waste Disposal Instructions Any of the lead-containing solutions should be disposed of in a designated waste container. The other solutions should be disposed of according to your instructor's directions.

Calculations

After you have completed Parts A, B, and C, graph your data, showing for each run the volume of oxygen on the ordinate (vertical axis) versus the elapsed time along the abscissa (horizontal axis). Zero time for each run corresponds to the time when the first 2 mL of O_2 were evolved. Draw the best straight line through your points for each run. Your line need not pass through all of the points. Label each line. Calculate the slopes for each of your four lines. The slopes of the lines correspond to the initial rates of the reaction.

Decide how formation of the initial rate of oxygen is affected by doubling the concentrations of H_2O_2 and I^-. How do the concentrations of H_2O_2 and I^- change from solution 1 to solution 2 and then to solution 3? Write a rate law for the reaction using the values of x and y you have determined by rounding them off to whole numbers.

GIVE IT SOME THOUGHT

If the reaction rate doubles when the concentration doubles, what can you conclude about the order of the reaction?

NOTES AND CALCULATIONS

Name ______________________ Desk ______________

Date ______________ Laboratory Instructor ______________________

Rates of Chemical Reactions II: Rate and Order of H_2O_2 Decomposition | 30 Pre-lab Questions

Before beginning this experiment in the laboratory, you should be able to answer the following questions.

1. What four factors may influence the rate of a reaction?

   Conc, temp, pressure, volume

2. Assume that the rate law for a reaction is rate = $k[A][B]$.

   1/3

   (a) What is the overall order of the reaction?

   overall: 2

   (b) If the concentration of both A and B are tripled, how will this affect the rate of reaction?

   (c) What name is given to k? Concentration constant

3. Define the term *catalyst.*

   Speeds up a rxn by offering a different metabolic pathway

4. Write the chemical equation for the iodide-catalyzed decomposition of H_2O_2.

   $H_2O_{2(aq)} + I^-_{(aq)} \rightleftharpoons H_2O_{(l)} + OI^-_{(aq)}$

5. It is found for the reaction $A + 3B \longrightarrow C$ that tripling the concentration of either A or B triples the reaction rate. Write a rate law for the reaction.

6. If the concentration of a substance is tripled, what effect does it have on the rate if the order with respect to that reactant is (a) 0, (b) 1, (c) 2, (d) 3, and (e) $\frac{1}{2}$?

7. The following data were collected for the volume of O_2 produced in the decomposition of H_2O_2.

Time (s)	*mL O_2*
0.0	0.0
45.0	2.0
88.0	3.9
131.0	5.8

Calculate the average rate of reaction for each time interval.

Name Grace Rademacher Desk ______

Date 2/16/2021 Laboratory Instructor Prof. Towle

$2H_2O_{2(aq)} + I^- \rightarrow O_2 + 2H_2O$

REPORT SHEET | EXPERIMENT

Rates of Chemical Reactions II: Rate and Order of H_2O_2 Decomposition

30

Rate = k ↓ only impacted by temp

A. Order of Reaction

$k = [H_2O_2]^x [I^-]^y$ x = 1

BATH TEMPERATURE

Solution 1 (24 °C) Buret reading (mL)	Vol. of O_2 (mL) (y)	Time (x)	Solution 2 (24 °C) Buret reading (mL)	Vol. of O_2 (mL)	Time	Solution 3 (24 °C) Buret reading (mL)	Vol. of O_2 (mL)	Time
23	0	0s	23	0	0s	23	0	0s
25	2	40s	25	2	19s	25	2	25s
27	4	75s	27	4	37s	27	4	43s
29	6	115s	29	6	55s	29	6	60s
31	8	156s	31	8	72s	31	8	81s
33	10	200s	33	10	89s	33	10	104s
35	12	236s	35	12	107s	35	12	125s
37	14	279s	37	14	124s	37	14	148s

(comes from graph)

Slope 0.05x   Slope 0.1134x   Slope 0.0964x

$\frac{0.05}{0.0964} = \frac{k[0.01]^n[0.02]^m}{k[0.01]^n[0.04]^m}$ $0.5186 = 0.5^m$, m = 1

$k = [H_2O_2]^1[I^-]^1$

Rate law: Since doubling either $[H_2O_2]$ or $[I^-]$ doubles the reaction rate the rate law is:

2:1 ratio

$k = [H_2O_2]^x [I^-]^y$

$\frac{0.05}{0.1134} = \frac{k[0.01]^n[0.02]^m}{k[0.02]^n[0.02]^m}$

$0.4409 = 0.5^n$

$\log(0.4409) = n\log(0.5)$

$\frac{-0.3557}{-0.3103} = \frac{n(-0.30103)}{-0.30103}$ n = 1.146, n = 1.15 → 1

Rate Data

Solution	mL H_2O_2	mL KI	Rate (mL O_2/s)
1	~~5.0~~ 10	~~10.0~~ 20	0.05 mL/s
2	~~10.0~~ 20	~~10.0~~ 20	0.1134 mL/s
3	~~5.0~~ 10	~~20.0~~ 40	0.0964 mL/s

B. Effect of Temperature

BATH TEMPERATURE 34 °C

Buret reading (mL)	*Vol. of O_2 (mL)*	*Time*
23	0	0s
25	2	15s
27	4	28s
29	6	44s
31	8	58s
33	10	74s
35	12	90s
37	14	107s

Total volume of O_2 collected 14.0 mL

Barometric pressure 757 mm Hg

Vapor pressure of water at bath temperature (see Appendix L) 39.9 mm Hg

Slope 0.1315x $\frac{x(0.05) = 0.1315}{0.05}$ $\frac{}{0.05} = 2.63$ $PO_2 = 757 - 39.9$ divide by 760 to get atm (pressure)

Compared with the rate found for solution 1, the rate is 2.63 times faster.

Using the ideal gas law, calculate the moles of O_2 collected 0.005 mol O_2

(show calculations) PV=nRT - solve for n

$n = \frac{PV}{RT} = \frac{(0.944\,atm)(0.014\,L)}{(0.0821)(307\,K)} = \frac{0.13216}{25.2047} = 0.005$ mol O_2

vol = 14 convert to L
.0821
34 + 273

Based on the moles of O_2 evolved, calculate the molar concentration of the *original* 3% H_2O_2 solution (show calculations).

$\frac{\text{moles } O_2}{L\ O_2} = \frac{0.005}{0.014} = 0.357$ M O_2

Molar concentration: 3.57×10^{-1} M

C. Identification of the Catalyst

Observation when $Pb(NO_3)_2$ is added to the following:

(a) Solution in which reaction has gone to completion

•Yellow/ brown precipitate → formation of PbI_2

(b) Original KI solution

•Yellow/ brown precipitate, formation of PbI_2

(c) Conclusion regarding nature of KI

If a yellow/brown precipitate forms, it shows that KI is present after the rxn is complete. KI as catalyst doesn't get changed chemically

QUESTIONS

1. How does the rate of formation of O_2 compare with (a) the rate of formation of H_2O and (b) the rate of disappearance of H_2O_2 for reaction [4]?

a) H_2O forms twice as fast as O_2 (2 H_2O molecules form for every O_2 molecule)

b) Rate of disappearance of H_2O_2 is twice as fast as the rate of formation for O_2. (2 H_2O molecules needed to form 1 O_2 molecule)

2. Why should the levels of water in the leveling bulb and buret be the same?

Leaks can be created from increasing pressure which causes gas to escape.

3. Why were you instructed to keep swirling the Erlenmeyer flask?

To use as much of the reactant as possible; assuring everything's being used

4. If you use 0.20 *M* KI instead of 0.10 *M* KI, how would this affect (a) the slopes of your curves, (b) the rate of the reactions, and (c) the numerical value of *k* for the reaction?

All of a, b, and c will remain the same. More catalyst will not increase the volume of O_2, nor will it speed up the rxn. K remains the same because it is only impacted by temperature changes.

5. How did an increase in temperature affect the reaction rate?

Speeds up the reaction.

6. Could you determine the activation energy for this from the data you have obtained?

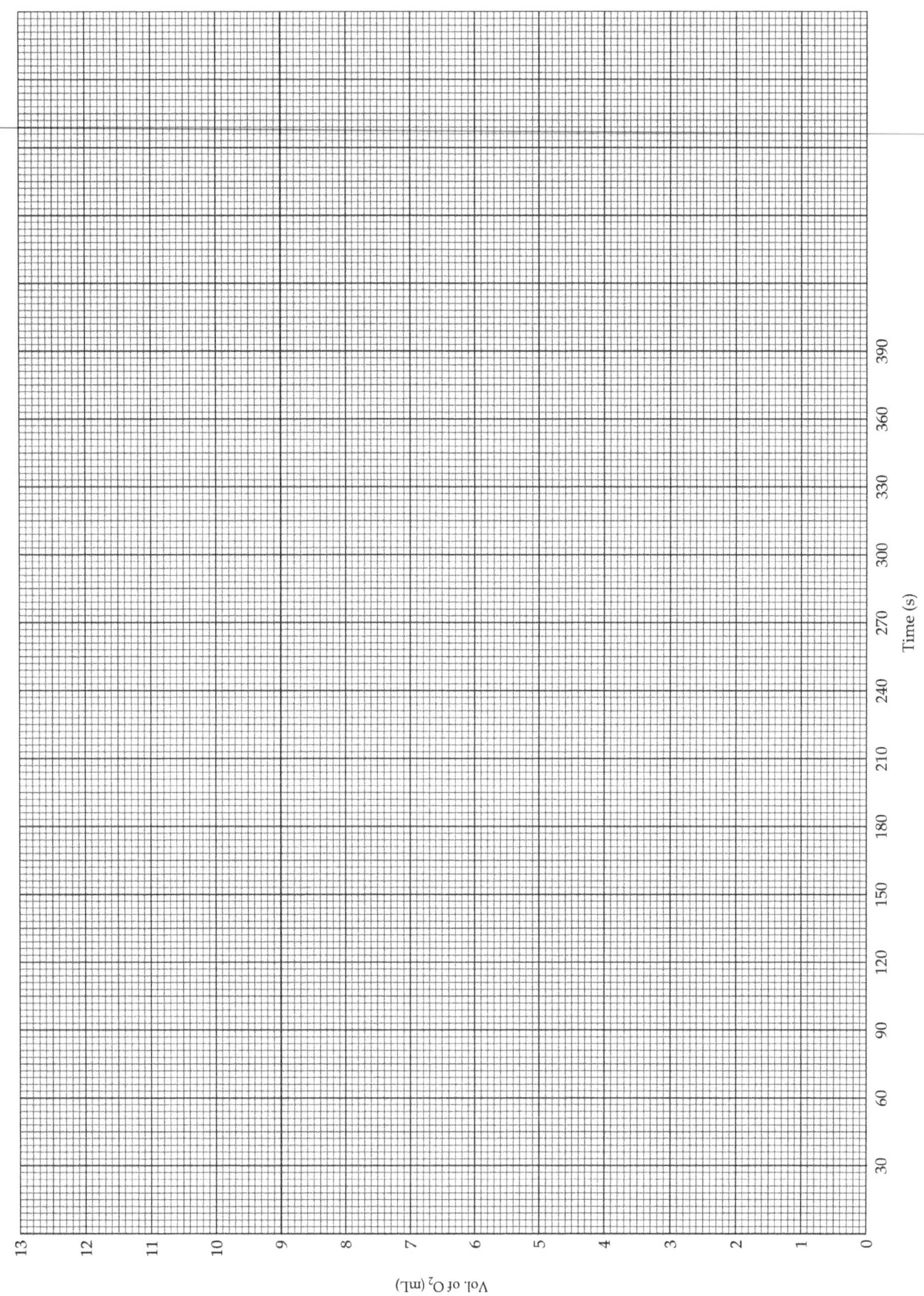

Time (s)
30
60
90
120
150
180
210
240
270
300
330
360
390
Vol. of O2 (mL)
0
1
2
3
4
5
6
7
8
9
10
11
12
13

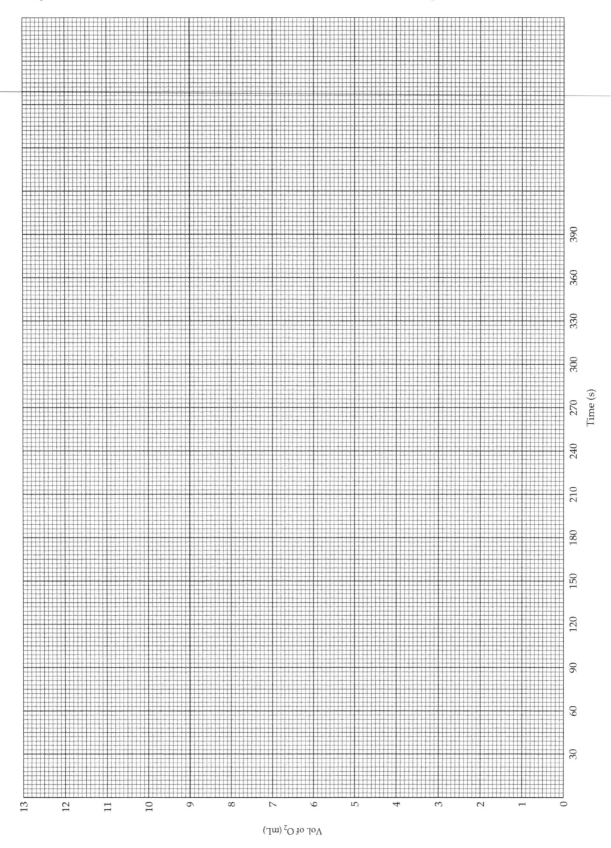

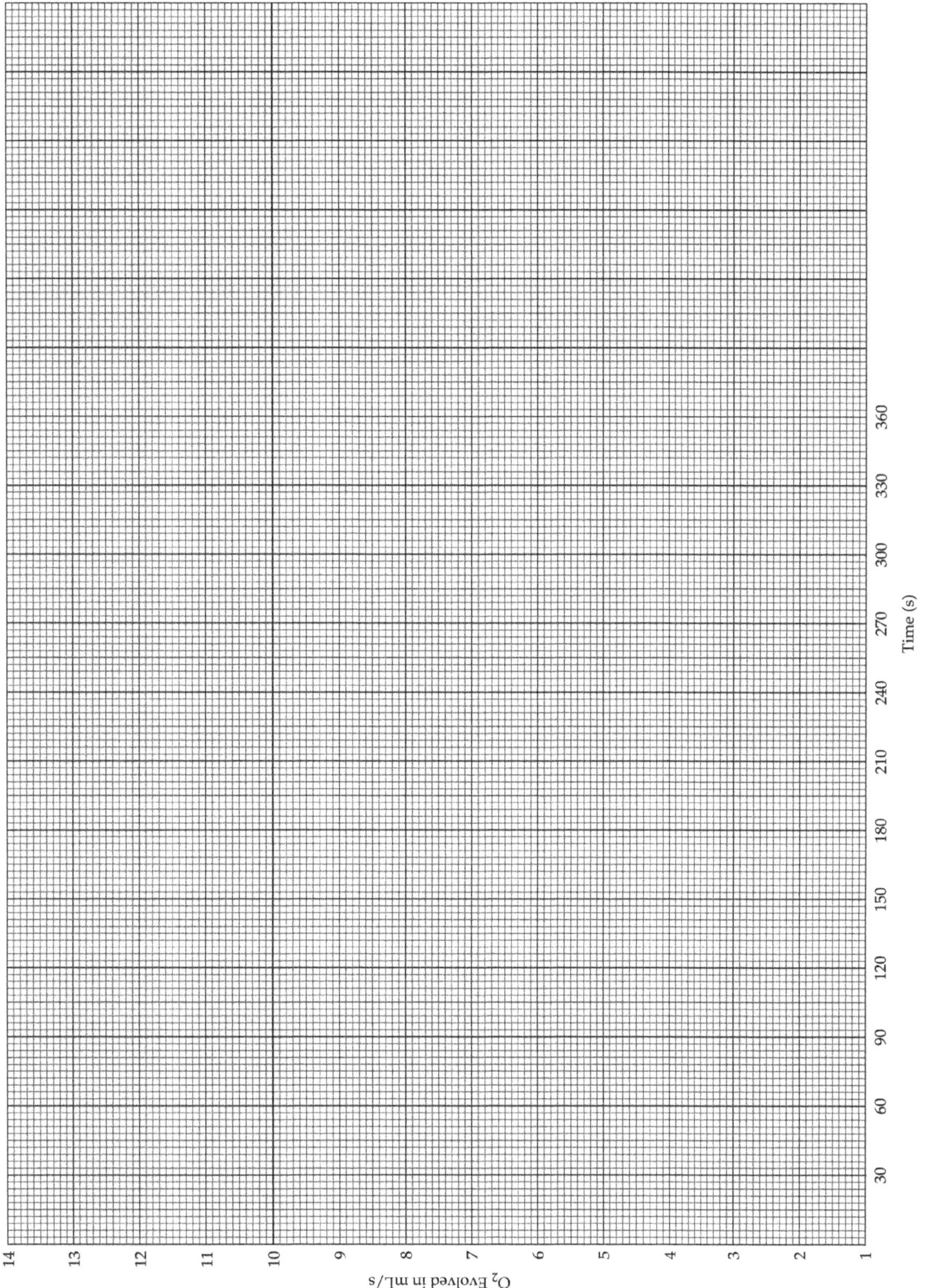
Time (s)
30
60
90
120
150
180
210
240
270
300
330
360
O_2 Evolved in mL/s
1
2
3
4
5
6
7
8
9
10
11
12
13
14

NOTES AND CALCULATIONS

Experiment

Introduction to Qualitative Analysis

CATIONS: Na^+, NH_4^+, Ag^+, Fe^{3+}, Al^{3+}, Cr^{3+}, Ca^{2+}, Mg^{2+}, Ni^{2+}, Zn^{2+}

ANIONS: SO_4^{2-}, NO_3^-, CO_3^{2-}, Cl^-, Br^-, I^-

OBJECTIVE

To become acquainted with the chemistry of several elements and the principles of qualitative analysis.

APPARATUS AND CHEMICALS

Apparatus

small test tubes (12)
centrifuge
evaporating dish or crucible
red litmus
droppers (6)
Bunsen burner and hose
10 cm Nichrome wire with loop at end

Chemicals

6 M H_2SO_4
18 M H_2SO_4
3 M NaOH
6 M HCl
12 M HCl
6 M NH_3
15 M NH_3
6 M HNO_3
15 M HNO_3
5 M NH_4Cl
3% H_2O_2
0.2 M $K_4Fe(CN)_6$
0.1% Aluminon reagent*
1% dimethylglyoxime†
0.2 M $BaCl_2$
0.2 M $(NH_4)_2C_2O_4$
Magnesium reagent‡
0.2 M $FeSO_4$ (stabilized with iron wire and 0.01 M H_2SO_4)
0.2 M NaCl
Cl_2 water
mineral oil
0.1 M $Ba(OH)_2$
Solids: Na_2SO_4, $NaNO_3$, Na_2CO_3, NaCl, NaBr, NaI
0.1 M NH_4NO_3
0.1 M $AgNO_3$
0.1 M $Fe(NO_3)_3$
0.1 M $Cr(NO_3)_3$
0.1 M $Al(NO_3)_3$
0.1 M $Ca(NO_3)_2$
0.1 M $NaNO_3$
0.1 M $Zn(NO_3)_2$
0.1 M $Ni(NO_3)_2$
0.1 M $Mg(NO_3)_2$
unknown cation solution
unknown anion salt, solid

---

*One gram of ammonium aurintricarboxylic acid in 1 L H_2O.
†In 95% ethyl alcohol solution.
‡A solution of 0.1% (0.1 g/L) *p*-nitrobenzene-azo-resorcinol in 0.025 M NaOH.

ALL SOLUTIONS SHOULD BE PROVIDED IN DROPPER BOTTLES.

DISCUSSION

Qualitative analysis is concerned with identification of the constituents contained in a sample of unknown composition (Sections 17.6 and 17.7). Inorganic qualitative analysis deals with detection and identification of the elements that are present in a sample of material. Frequently this is accomplished by making an aqueous solution of the sample and then determining which cations and anions are present on the basis of chemical and physical properties. In this experiment you will become familiar with some of the chemistry of ten cations (Ag^+, Fe^{3+}, Cr^{3+}, Al^{3+}, Ca^{2+}, Mg^{2+}, Ni^{2+}, Zn^{2+}, Na^+, NH_4^+) and six anions (SO_4^{2-}, NO_3^-, Cl^-, Br^-, I^-, CO_3^{2-}). You also will learn how to test for their presence or absence. Because there are many other elements and ions than those you will consider, this experiment is called an "abbreviated" qualitative analysis scheme.

PART I: CATIONS

If a substance contains only a single cation (or anion), its identification is a fairly simple and straightforward process, as you may have witnessed in Experiments 7 and 10. However, even in this instance, additional confirmatory tests are sometimes required to distinguish between two cations (or anions) that have similar chemical properties. Detection of a particular ion in a sample that contains several ions is more difficult because the presence of the other ions may interfere with the test. For example, if you are testing for Ba^{2+} with K_2CrO_4 and obtain a yellow precipitate, you may draw an erroneous conclusion because if Pb^{2+} is present, it also will form a yellow precipitate. Thus, the presence of lead ions interferes with this test for barium ions. This problem can be circumvented by first precipitating the lead as PbS with H_2S, thereby removing the lead ions from solution prior to testing for Ba^{2+} because BaS is soluble.

The successful analysis of a mixture containing 10 or more cations centers about the systematic separation of the ions into groups containing only a few ions. It is a much simpler task to work with 2 or 3 ions than with 10 or more. Ultimately, the separation of cations depends upon the differences in their tendencies to form precipitates, to form complex ions, or to exhibit amphoterism.

The chart in Figure 31.1 illustrates how the 10 cations you will study are separated into groups. Three of the 10 cations are colored: Fe^{3+} (rust to yellow), Cr^{3+} (blue-green), and Ni^{2+} (green). Therefore, a preliminary examination of an unknown that can contain any of the 10 cations under consideration yields valuable information. If the solution is colorless, you know immediately that iron, chromium, and nickel are absent. You are to take advantage of all clues that will aid you in identifying ions. However, in your role as a detective in identifying ions, be aware that clues can sometimes be misleading. For example, if Fe^{3+} and Cr^{3+} are present together, what color would you expect this mixture to display? Would the color depend upon the proportions of Fe^{3+} and Cr^{3+} present? Could you assign a *definite* color to such a mixture?

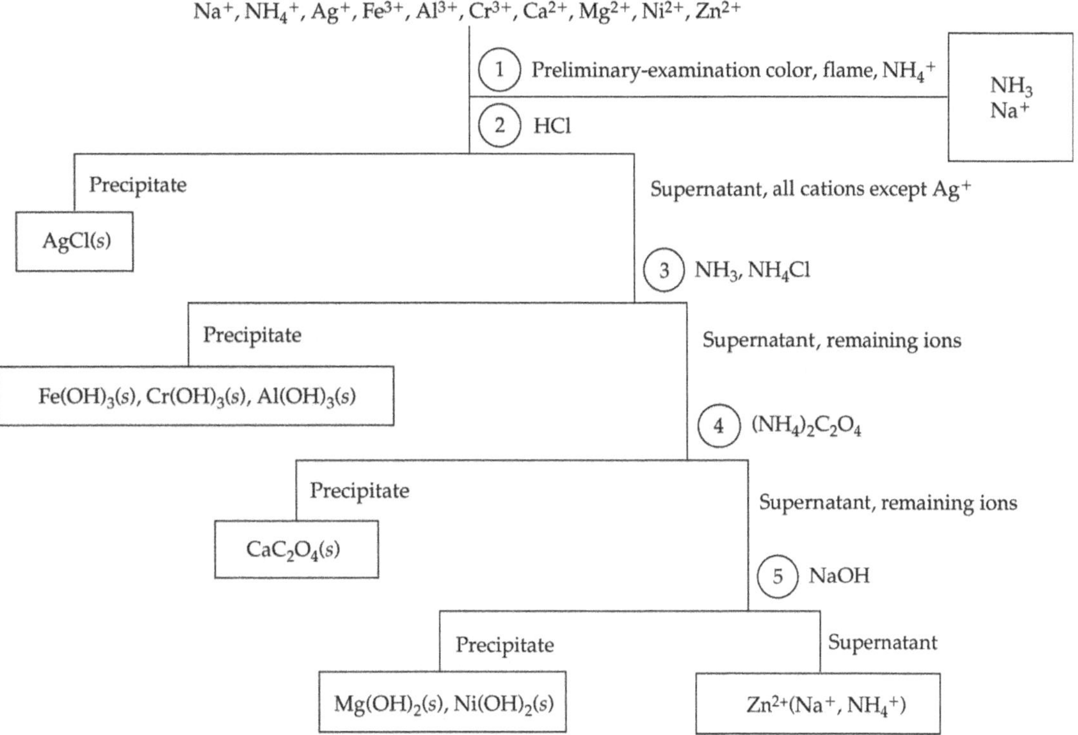

▲**FIGURE 31.1** Flow chart for group separation.

The group separation chart (Figure 31.1) shows that silver can be separated from all of the other cations (what are they?) as an insoluble chloride by the addition of hydrochloric acid. Iron, chromium, and aluminum are separated from the supernatant as their corresponding hydroxides by precipitating them from a buffered ammonium hydroxide solution. Next, calcium can be isolated by precipitating it as insoluble calcium oxalate. Finally, magnesium and nickel can be separated from the remaining ions (Zn^{2+}, Na^+, NH_4^+) as their insoluble hydroxides. Examination of the chart shows that in achieving these separations, reagents containing sodium and ammonium ions are used; therefore, tests for these ions must be made prior to their introduction into the solution.

GIVE IT SOME THOUGHT

If you look at the K_{sp} tables in Appendix E, how do the K_{sp} values verify these observations?

To derive the maximum benefit from this exercise, you should be thoroughly familiar with the group separation chart in Figure 31.1. You should know not only which 10 cations (by formula and charge) you are studying but also how they are separated into groups. More details regarding the identification of these cations are provided in the flowchart for cations (Figure 31.2) and in the following discussion about the chemistry of the analytical scheme. Refer to the flowcharts frequently while learning about the chemistry of the qualitative analysis scheme.

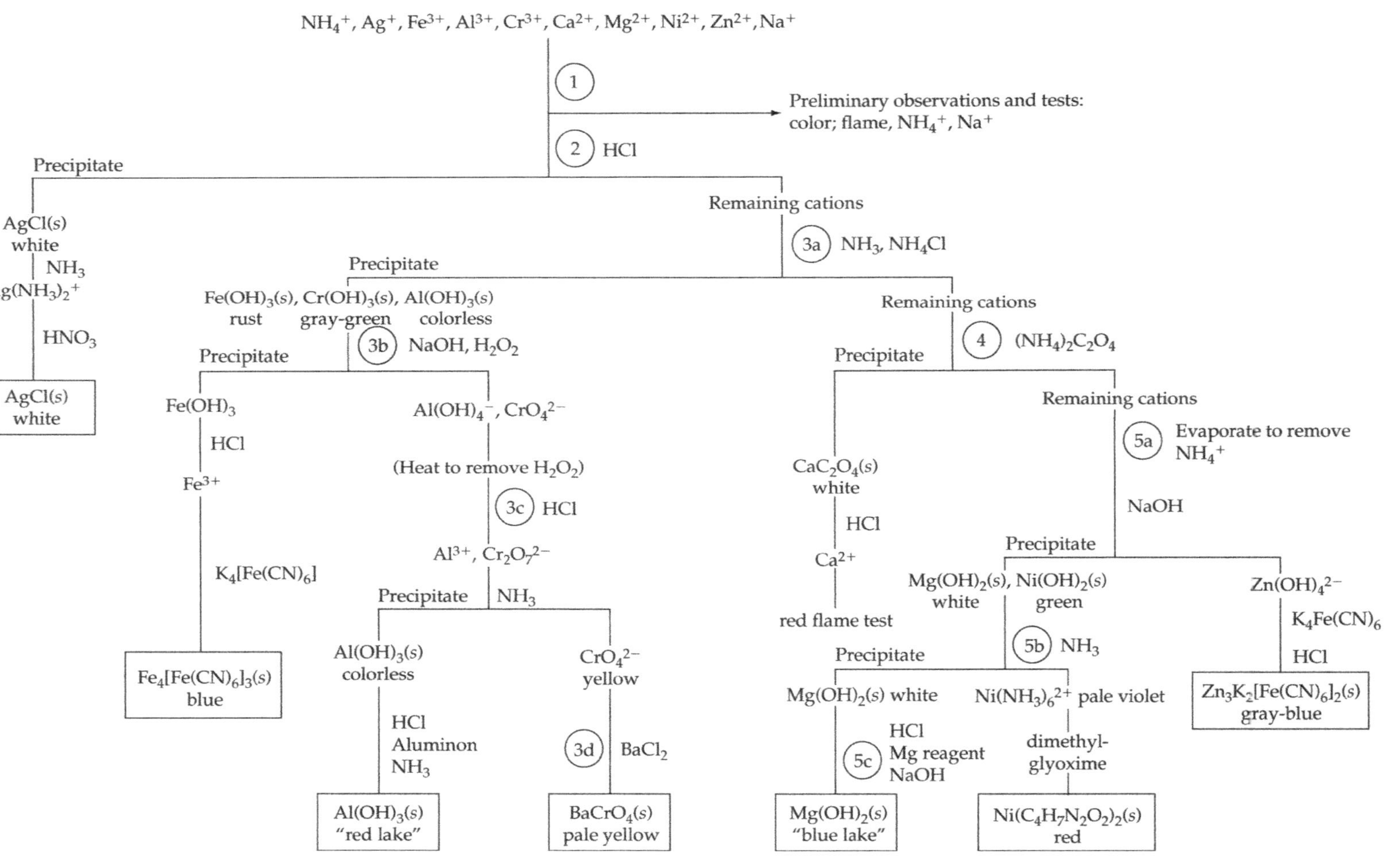

▲**FIGURE 31.2** Cation flowchart.

CHEMISTRY OF THE QUALITATIVE ANALYSIS SCHEME

1 Detection of Sodium and Ammonium

Sodium salts and ammonium salts are added as reagents in the analysis of your general unknown. Hence, tests for Na^+ and NH_4^+ must be made on the original sample before tests are performed for the other cations. Remember, your unknown may contain up to 10 cations, and you do not want to inadvertently introduce any of them into your unknown.

Sodium Most sodium salts are water-soluble. The simplest test for the sodium ion is a flame test. Sodium salts impart a characteristic yellow color to a flame. The test is *very* sensitive, and because of the prevalence of sodium ions, much care must be taken to keep equipment clean and free from contamination by these ions.

Ammonium The ammonium ion, NH_4^+, is the conjugate acid of the base ammonia, NH_3. The test for NH_4^+ takes advantage of the following equilibrium:

$$NH_4^+(aq) + OH^-(aq) \rightleftharpoons NH_3(g) + H_2O(l)$$

Thus, when a strong base such as sodium hydroxide is added to a solution of an ammonium salt and this solution is heated, NH_3 gas is evolved. The NH_3 gas can easily be detected by its effect upon moist red litmus paper.

GIVE IT SOME THOUGHT

a. If sodium is present, will it strongly influence the pH of the solution containing the NH_3?

b. Why or why not?

2 Separation and Detection of Silver

All chloride salts are soluble in water except those of Pb^{2+}, Hg_2^{2+}, and Ag^+. Silver can be precipitated and separated from the other nine cations, you are considering, by the addition of HCl to the original unknown.

$$Ag^+(aq) + Cl^-(aq) \longrightarrow \underset{\text{white}}{\mathbf{AgCl}(s)}$$

GIVE IT SOME THOUGHT

How do the solubility rules compare with the K_{sp} tables in terms of the solubility of the chloride salts?

A slight excess of HCl is used to ensure the complete precipitation of silver ions and to reduce their solubility by the common-ion effect; an excess of chloride ions drives the above equilibrium to the right. However, a large excess of chloride ions must be avoided because AgCl tends to dissolve by forming a *soluble-complex ion.*

$$\mathbf{AgCl}(s) + Cl^-(aq) \longrightarrow AgCl_2^-(aq)$$

To be absolutely certain that the white precipitate is AgCl ($PbCl_2$ and Hg_2Cl_2 are also insoluble, and they are white), NH_3 is added to the precipitate. If the precipitate is indeed AgCl, it will dissolve and then reprecipitate when the ammoniacal solution is made acidic.

$$\mathbf{AgCl}(s) + 2NH_3(aq) \longrightarrow [Ag(NH_3)_2]^+(aq) + Cl^-(aq)$$

$$[Ag(NH_3)_2]^+(aq) + 2H^+(aq) + Cl^-(aq) \longrightarrow \mathbf{AgCl}(s) + 2NH_4^+(aq)$$

The other two insoluble chlorides do not behave this way. Thus, you can be assured that the white chloride precipitate is silver chloride.

3a Separation and Detection of Iron, Aluminum, and Chromium

Iron, aluminum, and chromium can be separated from the other ions (Ca^{2+}, Mg^{2+}, Ni^{2+}, and Zn^{2+}) by making the solution alkaline with ammonium hydroxide (ammonia) buffer and precipitating these cations as their corresponding hydroxides.

$$Fe^{3+}(aq) + 3NH_3(aq) + 3H_2O(l) \longrightarrow \underset{\text{rust}}{\mathbf{Fe(OH)_3}}(s) + 3NH_4^+(aq)$$

$$Cr^{3+}(aq) + 3NH_3(aq) + 3H_2O(l) \longrightarrow \underset{\text{gray-green}}{\mathbf{Cr(OH)_3}}(s) + 3NH_4^+(aq)$$

$$Al^{3+}(aq) + 3NH_3(aq) + 3H_2O(l) \longrightarrow \underset{\text{colorless}}{\mathbf{Al(OH)_3}}(s) + 3NH_4^+(aq)$$

The hydroxide ion concentration required to precipitate these three ions must be carefully controlled because if it is too high, $Mg(OH)_2$ will also precipitate. An alkaline buffer (Section 17.2) of NH_3 and NH_4Cl provides a hydroxide ion concentration that is high enough to precipitate Fe^{3+}, Cr^{3+}, and Al^{3+}, yet low enough to prevent precipitation of $Mg(OH)_2$. Aqueous ammonia is a weak base.

$$NH_3(aq) + H_2O(l) \rightleftharpoons NH_4^+(aq) + OH^-(aq)$$

By itself, it would provide too high of a hydroxide ion concentration, and $Mg(OH)_2$ would precipitate along with the other cations. However, the NH_4^+ ions derived from the NH_4Cl causes this equilibrium to shift to the left; this reduces the hydroxide ion concentration sufficiently to prevent Mg^{2+} from precipitating.

3b Separation and Detection of Iron

Iron hydroxide can be separated from the other hydroxides by treating the precipitate with the strong base NaOH and hydrogen peroxide, H_2O_2. These reagents do not react with the insoluble $Fe(OH)_3$; however, $Al(OH)_3$ is amphoteric and dissolves, forming the soluble complex ion $Al(OH)_4^-$ (Section 23.2). The $Cr(OH)_3$ also dissolves, being oxidized by H_2O_2 to form soluble CrO_4^{2-}.

$$\mathbf{Al(OH)_3}(s) + OH^-(aq) \longrightarrow [Al(OH)_4]^-(aq)$$

$$\mathbf{2Cr(OH)_3}(s) + 3H_2O_2(aq) + 4OH^-(aq) \longrightarrow \underset{\text{yellow}}{2CrO_4^{2-}(aq)} + 8H_2O(l)$$

The rust-colored $Fe(OH)_3$ remains undissolved. That the rust-colored precipitate is in fact iron hydroxide can be confirmed by dissolving it in acid, adding potassium hexacyanoferrate(II), $K_4[Fe(CN)_6]$, and noting the formation of a dark blue precipitate (Prussian blue).

$$\mathbf{Fe(OH)_3}(s) + 3H^+(aq) \longrightarrow Fe^{3+}(aq) + 3H_2O(l)$$

$$4Fe^{3+}(aq) + 3Fe(CN)_6^-(aq) \longrightarrow \underset{\text{blue}}{\mathbf{Fe_4[Fe(CN)_6]_3}(s)}$$

3c and 3d Separation and Detection of Chromium and Aluminum

While iron was being precipitated as the hydroxide, chromium was oxidized to the yellow chromate ion, CrO_4^{2-}, and aluminum was converted to the soluble complex aluminate ion, $Al(OH)_4^-$, which is colorless. These two ions are in the supernatant liquid, which upon acidification, converts the chromate ion to the orange dichromate ion and the aluminate to the colorless solvated aluminum ion.

$$2CrO_4^{2-}(aq) + 2H^+(aq) \rightleftharpoons H_2O(l) + Cr_2O_7^{2-}(aq)$$

$$4H^+(aq) + [Al(OH)_4]^-(aq) \longrightarrow 4H_2O(l) + Al^{3+}(aq)$$

When this solution is treated with aqueous ammonia, aluminum precipitates as $Al(OH)_3$, which can be separated from the chromate ion in the supernatant liquid. (Note: The CrO_4^{2-} ion is stable in neutral or alkaline solution but is reversibly converted to the dichromate ion, $Cr_2O_7^{2-}$, upon acidification.) The formation of a yellow precipitate, $BaCrO_4$, upon the addition of barium chloride confirms the presence of chromium.

$$Ba^{2+}(aq) + CrO_4^{2-}(aq) \longrightarrow \underset{\text{yellow}}{\mathbf{BaCrO_4}(s)}$$

Aluminum hydroxide is a clear, colorless substance that is difficult to see in this analysis. A confirmatory test for aluminum involves dissolving the aluminum hydroxide in acid and then reprecipitating it with ammonia in the presence of Aluminon reagent. As the aluminum hydroxide precipitates, it absorbs the Aluminon reagent and assumes a red coloration known as a "lake."

$$Al^{3+}(aq) + 3NH_3(aq) + 3H_2O(l) + \underset{\text{reagent}}{\text{Aluminon}} \longrightarrow \underset{\text{"red lake"}}{\mathbf{Al(OH)_3}(s)} + 3NH_4^+(aq)$$

4 Separation and Detection of Calcium

Calcium Calcium can be separated from the remaining cations (Mg^{2+}, Ni^{2+}, and Zn^{2+}) by precipitating it as an insoluble oxalate salt.

$$Ca^{2+}(aq) + C_2O_4^{2-}(aq) \longrightarrow \mathbf{CaC_2O_4}(s)$$

If magnesium is present, it may precipitate and be mistaken for calcium. To confirm that the precipitate is that of calcium and not magnesium, it is dissolved in acid and a flame test is performed on the solution.

$$\mathbf{CaC_2O_4}(s) + 2H^+(aq) \longrightarrow Ca^{2+}(aq) + H_2C_2O_4(aq)$$

A transient brick-red flame verifies the presence of calcium ions. Magnesium ions do not impart any color to a flame.

5a Separation and Detection of Zinc

Separation of zinc from magnesium and nickel can be accomplished by precipitating $Mg(OH)_2$ and $Ni(OH)_2$ from a strongly alkaline solution. Excess ammonium ions must be removed before precipitation of these hydroxides because ammonium ions interfere with the confirmatory test for magnesium. Ammonium ions are easily removed by evaporating the solution, which has been acidified with nitric acid to dryness.

$$NH_4Cl(s) \xrightarrow{\Delta} NH_3(g) + HCl(g)$$

$$NH_4NO_3(s) \xrightarrow{\Delta} N_2O(g) + 2H_2O(g)$$

Although magnesium and nickel form insoluble hydroxides in the presence of a strong base, zinc is amphoteric and forms the soluble complex ion $Zn(OH)_4{}^{2-}$.

$$Ni^{2+}(aq) + 2OH^-(aq) \longrightarrow \mathbf{Ni(OH)_2}(s)$$

$$Mg^{2+}(aq) + 2OH^-(aq) \longrightarrow \mathbf{Mg(OH)_2}(s)$$

$$Zn^{2+}(aq) + 4OH^-(aq) \longrightarrow [Zn(OH)_4]^{2-}(aq)$$

The presence of zinc is confirmed by precipitating it from an acidic solution as a gray-blue salt, $Zn_3K_2[Fe(CN)_6]_2$.

$$3Zn(OH)_4{}^{2-}(aq) + 12H^+(aq) + 2K^+(aq) + 2Fe(CN)_6{}^{4-}(aq) \longrightarrow \underset{\text{gray-blue}}{\mathbf{Zn_3K_2[Fe(CN)_6]_2}}(s) + 12H_2O(l)$$

5b Separation and Detection of Nickel

Because nickel forms the soluble complex ion hexaamminenickel(II), $[Ni(NH_3)_6]^{2+}$, in the presence of aqueous ammonia, it can be separated from $Mg(OH)_2$.

$$\mathbf{Ni(OH)_2}(s) + 6NH_3(aq) \longrightarrow [Ni(NH_3)_6]^{2+}(aq) + 2OH^-(aq)$$

It is confirmed by forming a strawberry red precipitate, $Ni(C_4H_7N_2O_2)_2$, with an organic reagent, dimethylglyoxime.

$$[Ni(NH_3)_6]^{2+}(aq) + 2HC_4H_7N_2O_2(aq) \longrightarrow \underset{\text{strawberry red}}{\mathbf{Ni(C_4H_7N_2O_2)_2}}(s) + 4NH_3(aq) + 2NH_4{}^+(aq)$$

5c Detection of Magnesium

Magnesium is confirmed by dissolving the hydroxide with acid and then re-precipitating it in the presence of an organic compound called Magnesium reagent. The presence of Mg^{2+} is indicated by formation of a "blue lake."

$$\mathbf{Mg(OH)_2}(s) + 2H^+(aq) \longrightarrow Mg^{2+}(aq) + 2H_2O(l)$$

$$Mg^{2+}(aq) + 2OH^-(aq) + \text{Mg reagent} \longrightarrow \underset{\text{"blue lake"}}{\mathbf{Mg(OH)_2}(s)}$$

PROCEDURE

First, you will analyze a known that contains all 10 cations. Record on your report sheet the reagents used in each step, your observations, and the equations for each precipitation reaction. After completing this practice analysis, obtain an unknown. Follow the same procedures as with the known, again recording reagents and observations. Also record your conclusions regarding the presence or absence of all cations. *Before beginning this experiment, review the techniques used in qualitative analysis found in Appendix J: heating solutions, precipitation, centrifugation, washing precipitates, and testing acidity.*

Waste Disposal Instructions All waste from this experiment should be placed in appropriate containers in the laboratory.

1 Initial Observations and Tests for Sodium and Ammonium

Note the color of your sample and on your report sheet, record any conclusions about what cations are present or absent.

GIVE IT SOME THOUGHT

What information can you obtain from the color of your sample?

The flame test for sodium is very sensitive, and traces of sodium ion will impart a characteristic yellow color to the flame. Just about every solution has a trace of sodium and thus will give a positive result. On the basis of the intensity and duration of the yellow color, you can decide whether Na^+ is merely a contaminant or is present in substantial quantity. To perform the flame test, obtain a piece of platinum or Nichrome wire that has been sealed in a piece of glass tubing. Clean the wire by dipping it in 12 *M* HCl that is contained in a small test tube and heat the wire in the hottest part of your Bunsen burner flame. Repeat this operation until you see no color when you place the wire in the flame. Several cleanings will be required before this is achieved. Then place 10 drops of the solution to be analyzed in a clean test tube and perform a flame test on it. If the sample being tested is your unknown, run a flame test on distilled water and then another flame test on a 0.2 *M* NaCl solution. Compare the tests; this should help you make a decision as to the presence of sodium in your unknown.

GIVE IT SOME THOUGHT

What part of the flame is the hottest?

Place 10 drops of the original sample to be analyzed in an evaporating dish or a crucible. Moisten a strip of red or neutral litmus paper with distilled water and place the paper on the bottom of a small watch glass. Add 10 drops of 3 *M* NaOH to the unknown, swirl the evaporating dish or crucible, and immediately place the watch glass on it with the litmus paper down. Let stand for a few minutes. The presence of NH_4^+ ions is confirmed if the paper turns blue.

2 Separation and Detection of Silver

Place 10 drops of the original solution to be analyzed in a small test tube and add five drops of distilled water and two drops of 6 *M* HCl. Stir well, centrifuge (consult Appendix J for techniques), and reserve the supernatant for Procedure 3. Wash the precipitate with 10 drops of distilled water, centrifuge (see Appendix J), and add the washings to the supernatant. Dissolve the precipitate in four drops of 6 *M* NH_3*; then add 6 *M* HNO_3 to the ammoniacal solution until it is acidic to litmus. **(CAUTION:** ***Nitric acid can cause severe burns. Avoid contact with it. If you come in contact with it, immediately wash the area with copious amounts of water.*****)** A curdy white precipitate of AgCl confirms the presence of Ag^+ ions.

3a Separation and Detection of Iron, Aluminum, and Chromium

To the supernatant from Procedure 2, add two drops of NH_4Cl solution and 6 *M* NH_3 until the solution is basic to litmus. Centrifuge and reserve the supernatant for Procedure 4. Wash the precipitate with 10 drops of distilled water, centrifuge, and add the washings to the supernatant.

GIVE IT SOME THOUGHT

a. When NH_3 is added here, what solid products form?

b. Which ions will remain in solution?

3b Separation and Detection of Iron

To the precipitate from Procedure 3a, add five drops of distilled water, 10 drops of 3 *M* NaOH, and 5 drops of 3% H_2O_2. Stir well, centrifuge, and reserve the supernatant for Procedure 3c. Wash the precipitate with 10 drops of distilled water and add the washings to the supernatant. Dissolve the reddish-brown precipitate in 2 drops of 6 *M* HCl and add 10 drops of $K_4[Fe(CN)_6]$ solution. A blue precipitate of $Fe_4[Fe(CN)_6]_3$ confirms the presence of Fe^{3+} ions.

*Bottles may be labeled "6 *M* NH_4OH."

3c Separation and Detection of Aluminum

If the supernatant from Procedure 3b is yellow, place it in an evaporating dish or a crucible and evaporate almost to dryness to remove excess H_2O_2 (see Note 1 below). Add 10 drops of distilled water and 6 *M* HCl until acid to litmus. Then add 6 *M* NH_3 until the solution is basic to litmus. Centrifuge and reserve the supernatant for Procedure 3d. Wash the precipitate with 10 drops of distilled water, centrifuge, and add the washings to the supernatant. Dissolve the precipitate in two drops of 6 *M* HCl and add two drops of Aluminon reagent and 6 *M* NH_3 until basic to litmus. Centrifuge the solution. A red "lake" confirms the presence of Al^{3+} ions.

GIVE IT SOME THOUGHT

What does the yellow color tell you about which ion(s) are present in your sample?

Note 1 Hydrogen peroxide is a reducing agent in acid solutions, so CrO_4^{2-} ions could be reduced to Cr^{3+} when the solution is acidified. When NH_3 is added, $Cr(OH)_3$ will be precipitated and could be incorrectly reported as $Al(OH)_3$. Also, Cr^{3+} would not be confirmed in the proper procedure.

3d Separation and Detection of Chromium

Add two drops of $BaCl_2$ solution to the supernatant from Procedure 3c. A yellow precipitate of $BaCrO_4$ confirms the presence of Cr^{3+} ions.

4 Separation and Detection of Calcium

Add three drops of $(NH_4)_2C_2O_4$ solution to the supernatant from Procedure 3a and centrifuge. Reserve the decantate for Procedure 5a. Wash the precipitate with 10 drops of distilled water, centrifuge, and add the washings to the decantate. Dissolve the precipitate of CaC_2O_4 in two drops of 6 *M* HCl and carry out a flame test with this solution. A brick-red flame confirms Ca^{2+} ions (see Note 2 below).

Note 2 If magnesium ions are present, they may be precipitated as MgC_2O_4. To determine if MgC_2O_4 is precipitated, test the acid solution for Mg^{2+} as described in Procedure 5c.

5a Separation and Detection of Zinc

Place the supernatant from Procedure 4 in an evaporating dish and carefully evaporate to dryness using a burner flame. **(CAUTION: *Concentrated* HNO_3 *can cause severe burns. Avoid contact with it. If you come in contact with it, immediately wash the area with copious amounts of water.*)** Add five or six drops of 16 *M* HNO_3 and heat again until no more fumes are observed. (You are removing excess NH_4^+ ions that would interfere with the tests for Mg^{2+} ions.) Dissolve the residue in five drops of 6 *M* HCl. Add 10 drops of distilled water to the HCl solution and transfer it to a small test tube. Wash out the evaporating dish with 10 drops of distilled water and add the wash to the acid solution. Add 3 *M* NaOH until the solution is basic to litmus; centrifuge and reserve the precipitate for Procedure 5b. Wash the precipitate with 10 drops of distilled water and add the washings to the supernatant. Add three drops of $K_4[Fe(CN)_6]$ to the supernatant and acidify with 6 *M* HCl; a gray-blue coloration or precipitate of $Zn_3K_2[Fe(CN)_6]_2$ confirms the presence of Zn^{2+} ions.

5b Separation and Detection of Nickel

(CAUTION: *Concentrated* NH_3 *has a strong irritating odor and causes severe burns. Avoid inhaling it. If you come in contact with it, immediately wash the area with copious amounts of water.*)

Add five drops of distilled water and five drops of 15 *M* NH_3 to the precipitate from Procedure 5a; centrifuge and reserve the precipitate for Procedure 5c. Wash the precipitate with 10 drops of distilled water and add the wash to the supernatant. Add one drop of dimethylglyoxime solution to the supernatant; a strawberry red precipitate of $Ni(C_4H_7N_2O_2)_2$ confirms the presence of Ni^{2+} ions.

5c Detection of Magnesium

If Ni^{2+} ions were confirmed in Procedure 5b, add 15 *M* NH_3 to the precipitate you saved from Procedure 5b until a negative result for Ni^{2+} is obtained. Dissolve the hydroxide precipitate in two drops of 6 *M* HCl; add two drops of Magnesium reagent and 3 *M* NaOH until the solution is basic. A "blue lake" confirms the presence of Mg^{2+} ions.

Name ______________________________ Desk ______________

Date ______________ Laboratory Instructor ______________________

Introduction to Qualitative Analysis | 31 Pre-lab Questions

Before beginning Part I of this experiment in the laboratory, you should be able to answer the following questions.

1. What are the names and formulas of the 10 cations you will identify?

2. Why are confirmatory tests necessary in identifying ions?

3. Which of the 10 cations are colored, and what are their colors?

4. Which salt is soluble: $ZnCl_2$, $AgCl$, or $PbCl_2$?

5. How could you separate Fe^{3+} from Ag^{+}?

6. How could you separate Cr^{3+} from Mg^{2+}?

7. How could you separate Al^{3+} from Ag^{+}?

8. Complete and balance the following:

$$NH_4^{+}(aq) + OH^{-}(aq) \longrightarrow$$

$$AgCl(s) + NH_3(aq) \longrightarrow$$

PART II: ANIONS

A systematic scheme based on the kinds of principles involved in cation analysis can be designed for the analysis of anions. Because you will limit your consideration to only six anions (SO_4^{2-}, NO_3^-, Cl^-, Br^-, I^-, and CO_3^{2-}) and will not consider mixtures of the ions, your method of analysis is simple and straightforward. It is based on specific tests for the individual ions and does not require special precautions to eliminate interferences that may arise in mixtures.

Initially, you will make a general test on a solid salt with concentrated sulfuric acid (H_2SO_4). The results of this test should strongly suggest what the anion is. You will then confirm your suspicions by performing a specific test for the ion you believe to be present.

Table 31.1 summarizes the behavior of anions (as dry salts) with concentrated sulfuric acid.

PROCEDURE

Perform the following general sulfuric acid test on the individual anions. Then perform the specific tests on each of the ions. Record your observations and equations for the reactions that occur. After completing these tests on the six anions, obtain a solid salt unknown and identify its anion. Record your observations and conclusion. Only one anion is present in the salt.

Waste Disposal Instructions All waste should be disposed of according to your instructor's directions.

TABLE 31.1 Behavior of Anions with Concentrated Sulfuric Acid, H_2SO_4

A. Cold H_2SO_4

SO_4^{2-}	No reaction.
NO_3^-	No reaction.
CO_3^{2-}	A colorless, odorless gas forms. $CO_3^{2-}(s) + 2H^+(aq) \longrightarrow H_2O(l) + CO_2(g)$
Cl^-	A colorless gas forms. It has a sharp pungent odor, gives an acidic test result with litmus, and fumes in moist air. $Cl^-(s) + H^+(aq) \longrightarrow HCl(g)$
Br^-	A brownish-red gas forms. It has a sharp odor, gives an acidic test result with litmus, and fumes in moist air. The odor of SO_2 may be detected. $2Br^-(s) + 4H^+(aq) + SO_4^{2-}(aq) \longrightarrow Br_2(g) + SO_2(g) + 2H_2O(l)$ (HBr is also liberated.)
I^-	Solid turns dark brown immediately with the slight formation of violet fumes. The gas has the odor of rotten eggs, gives an acidic test result with litmus, and fumes in moist air. $2I^-(s) + 4H^+(aq) + SO_4^{2-}(aq) \longrightarrow I_2(g) + SO_2(g) + 2H_2O(l)$ (HI and H_2S are also liberated.)

B. Hot Concentrated H_2SO_4

There are no additional reactions with any of the anions except NO_3^-, which forms brown fumes of NO_2 gas.

$$4NO_3^-(s) + 4H^+(aq) \longrightarrow 4NO_2(g) + O_2(g) + 2H_2O(l)$$

Sulfuric Acid Test

In a hood, place a small amount of the solid (about the size of a pea) in a small test tube. Add one or two drops of 18 *M* H_2SO_4 and observe everything that occurs, especially the color and odor of gas formed. **(CAUTION: *Concentrated* H_2SO_4 *causes severe burns. Do not get it on your skin or clothing. If you come in contact with it, immediately wash the area with copious amounts of water.*** *Do not* place your nose directly over the mouth of the test tube, but carefully fan gases toward your nose. Then *carefully* heat the test tube, but not so strongly as to boil the H_2SO_4. **(CAUTION: *If you heat the acid too strongly, it could come shooting out!*)** Note whether or not brown fumes of NO_2 are produced. **(CAUTION: *Do not look down into the test tube. Do not point the test tube at yourself or at any one else.* SAFETY GLASSES MUST BE WORN.)**

Specific Tests for Anions

When an anion is indicated by the preliminary test result with concentrated H_2SO_4, it is confirmed using the appropriate specific test. Make an aqueous solution of the solid unknown and perform the following tests on portions of this solution.

Sulfate Place 10 drops of a solution of the anion salt in a test tube, acidify with 6 *M* HCl, and add a drop of $BaCl_2$ solution. The formation of a white precipitate of $BaSO_4$ confirms SO_4^{2-} ions.

$$Ba^{2+}(aq) + SO_4^{2-}(aq) \longrightarrow \mathbf{BaSO_4}(s)$$

Nitrate In a hood, place 10 drops of a solution of the anion salt in a small test tube and add five drops of $FeSO_4$ solution; mix the solution. Carefully, without agitating the solution, pour concentrated H_2SO_4 down the inside of the test tube to form two layers. The formation of a brown ring between the two layers confirms NO_3^- ions.

$$(1)\ 3Fe^{2+}(aq) + NO_3^-(aq) + 4H^+(aq) \longrightarrow 3Fe^{3+}(aq) + NO(g) + 2H_2O(l)$$

$$(2)\ NO(g) + Fe^{2+}(aq)(\text{excess}) \longrightarrow Fe(NO)^{2+}(aq)(\text{brown})$$

Chloride Place 10 drops of a solution of the anion salt in a test tube and add a drop of $AgNO_3$ solution. A white, curdy precipitate confirms Cl^- ions.

$$Ag^+(aq) + Cl^-(aq) \longrightarrow \mathbf{AgCl}(s)$$

Bromide Place 10 drops of a solution of the anion salt in a test tube, add three drops of 6 *M* HCl, and add five drops of Cl_2 water and five drops of mineral oil. Shake well. Br^- ions are confirmed if the mineral-oil (top) layer is colored orange to brown. Wait 30 s for the layers to separate.

$$2Br^-(aq) + Cl_2(aq) \rightleftharpoons Br_2(aq) + 2Cl^-(aq)$$

Iodide Repeat the test as described for Br^- ions. If the layer of mineral oil is colored violet, I^- ions are confirmed.

$$2I^-(aq) + Cl_2(aq) \rightleftharpoons I_2(aq) + 2Cl^-(aq)$$

Carbonate Place a small amount of the solid anion salt in a small test tube and add a few drops of 6 *M* H_2SO_4. If a colorless, odorless gas evolves, hold a drop of $Ba(OH)_2$ solution by suspending the drop on the end of an eyedropper over the mouth of the test tube; CO_3^{2-} ions are confirmed if the drop turns milky.

$$(1)\ 2H^+(aq) + CO_3^{2-}(s) \longrightarrow CO_2(g) + H_2O(l)$$

$$(2)\ CO_2(g) + Ba(OH)_2(aq) \longrightarrow \mathbf{BaCO_3}(s) + H_2O(l)$$

NOTES AND CALCULATIONS

Name ______________________________ Desk ______________________

Date ____________________ Laboratory Instructor ______________________

Introduction to Qualitative Analysis | 31 Pre-lab Questions

Before beginning Part II of this experiment in the laboratory, you should be able to answer the following questions.

1. Give the names and formulas of the anions to be identified.

2. Describe the behavior of each solid containing the anions toward concentrated H_2SO_4.

3. If you had a mixture of NaCl and Na_2SO_4, would the action of dilute barium nitrate allow you to decide that both Cl^- and SO_4^- were present?

NOTES AND CALCULATIONS

Name ____________________ Desk ____________________

Date ____________________ Laboratory Instructor ____________________

Unknown no. ____________________

REPORT SHEET | EXPERIMENT

Introduction to Qualitative Analysis | 31

Part I: Cations

A. Known

Record the reagent used in each step, your observations, and the equations for each reaction.

	Reagent	*Observations*	*Equations*
1			
2			
3a			
3b			
3c			

	Reagent	*Observations*	*Equations*
3d			
4			
5a			
5b			
5c			

B. Unknown (Cation unknown no. _________)

Record the reagent used, your observations, and the equations for each precipitate formed.

Reagent *Observations* *Equations*

Cations present in unknown:

PART II: ANIONS

A. Known

Concentrated H_2SO_4 test

Ion	*Observations and equations*
SO_4^{2-}	
NO_3^-	
CO_3^{2-}	
Cl^-	
Br^-	
I^-	
NO_3^-	

B. Known

Specific tests

Ion	*Observations and equations*
SO_4^{2-}	
NO_3^-	
CO_3^{2-}	
Cl^-	
Br^-	
I^-	

C. Unknown (Anion unknown no. ________________)

Observations and equations

1 H_2SO_4 test:

2 Specific test(s):

Anions in unknown: ____

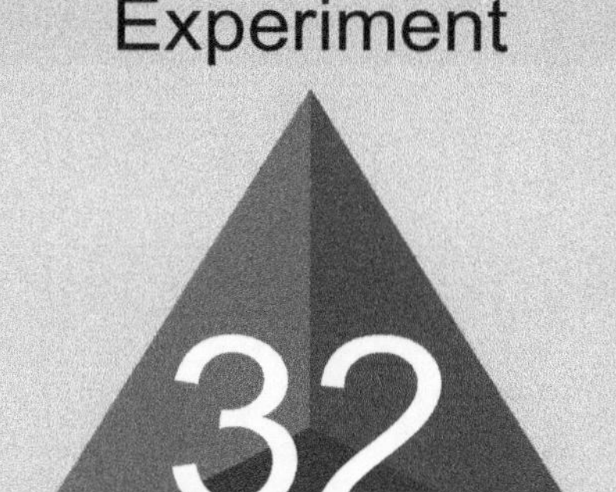

Abbreviated Qualitative Analysis Scheme

CATIONS: Ag^+, Pb^{2+}, Hg_2^{2+}, Cu^{2+}, Bi^{3+}, Sn^{4+}, Fe^{3+}, Mn^{2+}, Ni^{2+}, Al^{3+}, Ba^{2+}, NH_4^+, Na^+, Ca^{2+}

ANIONS: SO_4^{2-}, NO_3^-, CO_3^{2-}, Cl^-, Br^-, I^-, CrO_4^{2-}, PO_4^{3-}, S^{2-}, SO_3^{2-}

OBJECTIVE

To become acquainted with the chemistry of several elements and the principles of qualitative analysis.

APPARATUS AND CHEMICALS

Apparatus

small test tubes (12)
centrifuge
evaporating dish
litmus
droppers (6)
10 cm Nichrome wire with loop at end
Bunsen burner and hose
aluminum wire, 26-gauge

Chemicals

6 *M* CH_3COOH
2 *M* HCl
6 *M* HCl
12 *M* HCl
0.2 *M* $BaCl_2$
mineral oil
Cl_2 water
diethyl ether
1% dimethylglyoxime*
95% ethyl alcohol
0.2 *M* $Fe(NO_3)_2$
0.2 *M* $FeSO_4$
3% H_2O_2
6 *M* NH_3†
15 *M* NH_3†
Aluminon reagent‡
1 *M* NH_4CH_3COO
2 *M* NH_4Cl
$(NH_4)_2MoO_4$ §
$Ba(OH)_2$, saturated solution
6 *M* NaOH
0.2 M $SnCl_2$ (freshly prepared)
$(NH_4)_2S$ ‖
1 *M* K_2CrO_4
1 *M* $K_2C_2O_4$

---

*In 95% ethyl alcohol solution.
†Reagent bottles may be labeled "15 *M*" or "6 *M* NH_4OH."
‡One gram of ammonium aurintricarboxylic acid in 1 L H_2O.
§Dissolve 20 g of MoO_3 in a mixture of 60 mL of distilled water and 30 mL of 15 *M* NH_3. Add this solution slowly, stirring constantly, to a mixture of 230 mL of water and 100 mL of 16 *M* HNO_3.
‖Add 1 volume of reagent-grade ammonium sulfide liquid to two volumes of water or saturate 5 *M* NH_3 with H_2S.

$3\ M\ HNO_3$
$6\ M\ HNO_3$
$16\ M\ HNO_3$
$6\ M\ H_2SO_4$
$18\ M\ H_2SO_4$
Na_2O_2, solid
starch solution
$0.2\ M\ Pb(CH_3COO)_2$
$0.1\ M\ MnCl_2$
$0.1\ M\ HgCl_2$
$0.1\ M\ Hg(CH_3COO)_2$
1 *M* thioacetamide
$0.2\ M\ KNO_2$
$0.2\ M$ KSCN
$0.1\ M\ AgNO_3$
$NaBiO_3$, solid
Na_2CO_3, saturated solution
$0.2\ M$ NaCl
Zn, granulated
groups 1, 2, 3, and 4 known solutions
solid sodium salt anion knowns
unknown cation solution
unknown anion salt, solid mixture

ALL SOLUTIONS SHOULD BE PROVIDED IN DROPPER BOTTLES.

DISCUSSION

Qualitative analysis is concerned with identification of the constituents contained in a sample of unknown composition. Inorganic qualitative analysis deals with the detection and identification of the elements that are present in a sample of material (Sections 17.6 and 17.7). Frequently, this is accomplished by making an aqueous solution of the sample and then determining which cations and anions are present on the basis of chemical and physical properties. In this experiment, you will become familiar with some of the chemistry of 14 cations (Ag^+, Pb^{2+}, Hg_2^{2+}, Cu^{2+}, Bi^{3+}, Sn^{4+}, Fe^{3+}, Mn^{2+}, Ni^{2+}, Al^{3+}, Ba^{2+}, Ca^{2+}, NH_4^+, Na^+) and 10 anions (CrO_4^{2-}, PO_4^{3-}, S^{2-}, SO_3^{2-}, SO_4^{2-}, NO_3^-, Cl^-, Br^-, I^-, CO_3^{2-}) and you will learn how to test for their presence or absence. Because there are many elements and ions other than those you will consider, this experiment is called an "abbreviated" qualitative analysis scheme. These procedures are limited to these ions other ions present might present interferences.

GIVE IT SOME THOUGHT

Does qualitative analysis tell you anything about the amount of an ion present in a sample?

CATIONS

If a sample contains only a single cation (or anion), its identification is a fairly simple and straightforward process. However, even in this instance, additional confirmatory tests are sometimes required to distinguish between two cations (or anions) that have similar chemical properties. The detection of a particular ion in a sample that contains several ions is more difficult because the presence of the other ions may interfere with the test. For example, if you are testing for Ba^{2+} with K_2CrO_4 and obtain a yellow precipitate, you may draw an erroneous conclusion because if Pb^{2+} is present, it also will form a yellow precipitate. Thus, lead ions interfere with this test for barium ions. This problem can be circumvented by first precipitating the lead as PbS with H_2S, thereby removing the lead ions from solution prior to testing for Ba^{2+} because BaS is soluble.

The successful analysis of a mixture containing 14 or more cations centers upon the systematic separation of the ions into groups containing only a few ions.

It is much simpler to work with 2 or 3 ions than with 14 or more. Ultimately, the separation of cations depends upon the difference in their tendencies to form precipitates, to form complex ions, or to exhibit amphoterism.

The chart in Figure 32.1 illustrates how the 14 cations you will study are separated into groups. The format for this chart is that precipitates are shown on the left and supernatant on the right. Four of the 14 cations are colored: Fe^{3+} (rust to yellow), Cu^{2+} (blue), Mn^{2+} (very faint pink), and Ni^{2+} (green). Therefore, a preliminary examination of an unknown that can contain any of the 14 cations under consideration yields valuable information. If the solution is colorless, you know immediately that iron, copper, and nickel are absent. You are to take advantage of all clues that will aid you in identifying ions. However, in your role as a detective in identifying ions, be aware that clues can sometimes be misleading. For example, if Fe^{3+} and Ni^{2+} are present together, what color would you expect the mixture to display? Would the color depend upon the proportions of Fe^{3+} and Ni^{2+} present? Could you assign a *definite* color to such a mixture? You may not discern the extremely faint pink color of Mn^{2+} unless you carefully examine the solution, and the presence of any of the other colored ions will mask its color.

The group separation chart (Figure 32.1) shows that silver, mercury(I), and lead(II) can be separated from all of the other cations as insoluble chlorides by the addition of hydrochloric acid. Addition of hydrogen sulfide and hydrochloric acid to the supernatant results in the precipitation of the sulfides of lead, copper(II), bismuth(III), and tin(IV). The reason lead appears in both groups 1 and 2 is explained by the chemistry of these group cations.

GIVE IT SOME THOUGHT

How do these results compare with the K_{sp} tables in Appendix E?

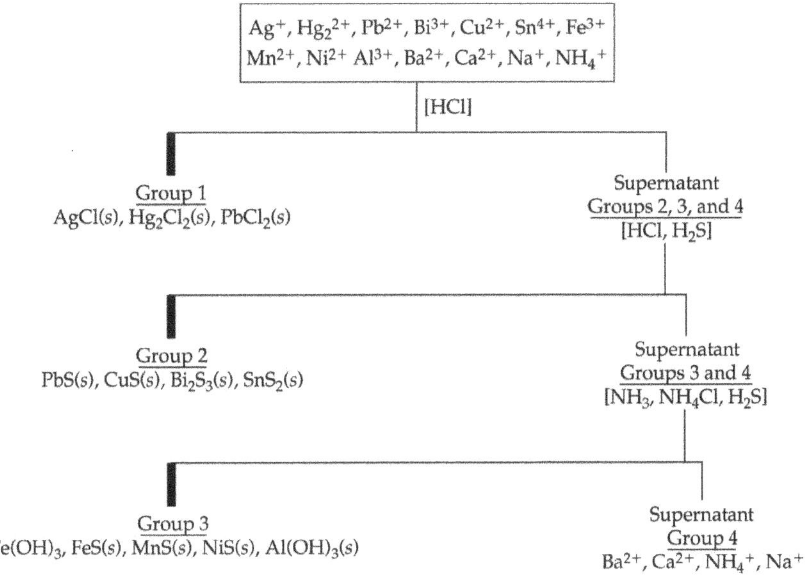

▲**FIGURE 32.1** Flowchart for group separation.

To derive any benefit from this exercise, you must become thoroughly familiar with the group separation chart in Figure 32.1. When you work with the four groups, you will see additional separation charts or flow schemes for the individual groups. Become familiar with these as well. Focus on the cations you are working with in each group. Know their formulas and charges and become familiar with how they are separated and identified.

You will analyze four known solutions, one for each group. Each of these solutions will contain all of the cations in the group you are studying. Then you will analyze a group unknown for each group. Your instructor may tell you to analyze the known and unknown side by side. If you do, be careful not to mix up your test tubes! As you proceed with your analysis, record on your report sheets the reagents used in each step, your observations, and the equations for each precipitation reaction. After completing the four groups, you will analyze a general cation unknown that may contain any number of the 14 ions studied. You may also analyze a simple salt and identify it.

PART I: CHEMISTRY OF GROUP 1 CATIONS

Pb^{2+}, Ag^{+}, Hg_2^{2+}

Because the chlorides of Pb^{2+}, Ag^{+}, and Hg_2^{2+} are insoluble, they may be precipitated and separated from the cations of groups 2, 3, and 4 by the addition of HCl. The following equations represent the reactions that occur:

$$Pb^{2+}(aq) + 2Cl^{-}(aq) \longrightarrow \mathbf{PbCl_2}(s) \qquad \text{white} \qquad [1]$$

$$Ag^{+}(aq) + Cl^{-}(aq) \longrightarrow \mathbf{AgCl}(s) \qquad \text{white} \qquad [2]$$

$$Hg_2^{2+}(aq) + 2Cl^{-}(aq) \longrightarrow \mathbf{Hg_2Cl_2}(s) \qquad \text{white} \qquad [3]$$

A slight excess of HCl is used to ensure complete precipitation of the cations and to reduce the solubility of the chlorides by the common-ion effect. However, a large excess of chloride must be avoided because both AgCl and $PbCl_2$ tend to dissolve by forming soluble complex anions.

$$\mathbf{PbCl_2}(s) + 2Cl^{-}(aq) \longrightarrow PbCl_4^{2-}(aq) \qquad [4]$$

$$\mathbf{AgCl}(s) + Cl^{-}(aq) \longrightarrow AgCl_2^{-}(aq) \qquad [5]$$

$PbCl_2$ is appreciably more soluble than either AgCl or Hg_2Cl_2. Thus, even when $PbCl_2$ precipitates, a significant amount of Pb^{2+} remains in solution and is subsequently precipitated with the group 2 cations as the sulfide PbS. Because of its solubility, Pb^{2+} sometimes does not precipitate as the chloride because its concentration is too small or the solution is too warm.

Lead (II) Lead chloride is more soluble in hot water than in cold. It is separated from the other two insoluble chlorides by being dissolved in hot water. The presence of Pb^{2+} is confirmed by the formation of a yellow precipitate, $PbCrO_4$, upon the addition of K_2CrO_4.

$$Pb^{2+}(aq) + CrO_4^{2-}(aq) \longrightarrow \mathbf{PbCrO_4}(s) \qquad \text{yellow} \qquad [6]$$

Mercury(I) Silver chloride is separated from Hg_2Cl_2 by the addition of aqueous NH_3. Silver chloride dissolves because Ag^+ forms a soluble complex cation with NH_3.

$$\textbf{AgCl}(s) + 2NH_3(aq) \longrightarrow Ag(NH_3)_2^+(aq) + Cl^-(aq) \qquad [7]$$

Mercury(I) chloride reacts with aqueous ammonia in a disproportionation reaction (oxidation and reduction of the same species) to form a dark gray precipitate.

$$\textbf{Hg}_2\textbf{Cl}_2(s) + 2NH_3(aq) \longrightarrow \textbf{HgNH}_2\textbf{Cl}(s) + Hg(l) + NH_4^+(aq) + Cl^-(aq) \quad [8]$$

Although $HgNH_2Cl$ is white, the precipitate appears dark gray because of a colloidal dispersion of $Hg(l)$.

Silver To verify the presence of Ag^+, the supernatant liquid is acidified and AgCl reprecipitates if Ag^+ is present. The acid decomposes $Ag(NH_3)_2^+$ by neutralizing NH_3 to form NH_4^+. The solution must be acidic; otherwise, the AgCl will not precipitate and Ag^+ can be missed.

$$Ag(NH_3)_2^+(aq) + 2H^+(aq) + Cl^-(aq) \longrightarrow \textbf{AgCl}(s) + 2NH_4^+(aq) \qquad [9]$$

The flowchart in Figure 32.2 shows how the group 1 cations are separated and identified. *You should become familiar with it and consult it often as you perform your analysis.*

PROCEDURE

First, you will analyze a known that contains all three cations of group 1. Record on your report sheet the reagents used in each step, your observations, and the equation for each precipitation reaction. After completing the analysis of a known, obtain an unknown. Follow the same procedures as you did with the known, again recording reagents and observations. Also record conclusions

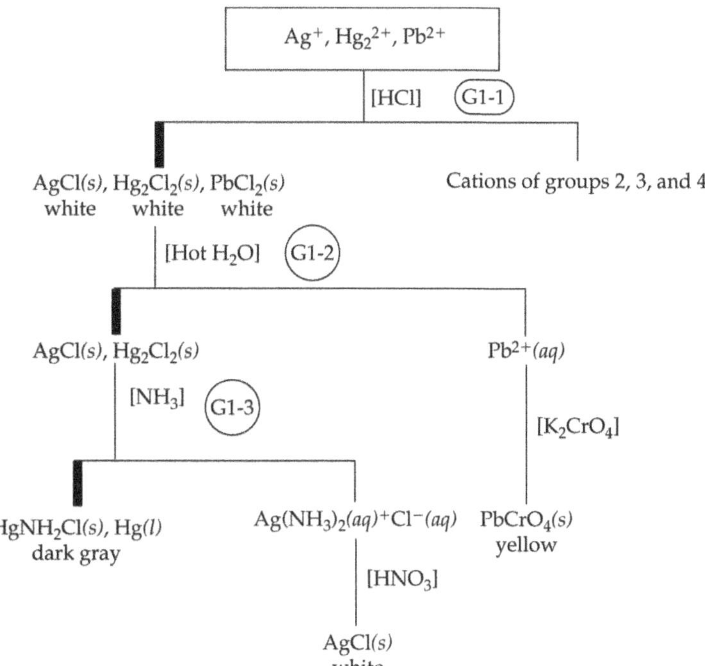

▲**FIGURE 32.2** Group 1 flowchart. Procedure numbers are circled.

regarding the presence or absence of all cations. *Before beginning this experiment, review the techniques used in qualitative analysis found in Appendix J: Centrifugation, Heating Solutions, Washing Precipitates, and Testing Acidity.*

Waste Disposal Instructions All waste from this experiment should be placed in appropriate waste containers in the laboratory. Follow the procedures specified by your instructor.

G1-1 Precipitation of Group 1 Cations

Measure out 10 drops (0.5 mL) of the test solution or the unknown into a small (10 mm × 75 mm) test tube. Add four drops of 6 *M* HCl, stir thoroughly, and then centrifuge. Test for completeness of precipitation by adding one drop of 6 *M* HCl to the clear supernatant. If the supernatant turns cloudy, not all of the group 1 cations have precipitated; add another two drops of 6 *M* HCl, stir, and centrifuge. Repeat this process until no more precipitate forms. All of the group 1 cations must be precipitated; otherwise, they will slip through and interfere with subsequent group analyses. If a general unknown is being analyzed, decant the supernatant into a clean test tube and save it for analysis of group 2 cations; otherwise, discard it. Dispose of any wastes in the designated receptacles. Wash the precipitate by adding five drops of cold distilled water and stir. Centrifuge (see Appendix J) and add the liquid to the supernatant. Why is the precipitate washed? See Appendix J.

GIVE IT SOME THOUGHT

Which cations are present in this precipitate?

G1-2 Separation and Identification of Pb^{2+}

Add 15 drops of distilled water to the precipitate and place the test tube in a hot water bath. Stir using a stirring rod and heat for 1 min or longer. Quickly centrifuge and decant the hot supernatant into a clean test tube. Repeat this procedure two more times, combining the supernatants, which should contain Pb^{2+} if it is present. Save the precipitate for Procedure G1-3. Add three drops of 1 *M* K_2CrO_4 to the supernate. The formation of a yellow precipitate, $PbCrO_4$, confirms the presence of Pb^{2+}. Dispose of all wastes as instructed by your instructor.

G1-3 Separation and Identification of Ag^+ and $Hg_2{}^{2+}$

Add 10 drops of 6 *M* NH_3 to the precipitate from Procedure G1-2. The formation of a dark gray precipitate indicates the presence of mercury. Centrifuge and decant the clear supernatant into a clean test tube. Add 20 drops of 6 *M* HNO_3 to the supernatant. Stir the solution and test its acidity. Continue to add HNO_3 dropwise until the solution is acidic. A white cloudiness confirms the presence of Ag^+.

Dispose of all wastes as directed by your instructor.

Name ______________________________ Desk ______________

Date ______________ Laboratory Instructor ______________________

Abbreviated Qualitative Analysis Scheme | 32 Pre-lab Questions

Before beginning Part I of this experiment in the laboratory, you should be able to answer the following questions.

1. What are the symbols and charges of the group 1 cations?

2. Which chloride salt is insoluble in cold water but soluble in hot water?

3. Which chloride salt dissolves in aqueous NH_3?

4. How could you distinguish the following?

   (a) NaCl from AgCl

   (b) H_2SO_4 from HCl

5. Complete and balance the following equations:

(a) $AgCl(s) + NH_3(aq) \longrightarrow$

(b) $Pb^{2+}(aq) + CrO_4^{2-}(aq) \longrightarrow$

(c) $Hg_2Cl_2(s) + NH_3(aq) \longrightarrow$

(d) $Ag(NH_3)_2^+(aq) + H^+(aq) + Cl^-(aq) \longrightarrow$

6. What can you conclude if no precipitate forms when HCl is added to the unknown solution?

7. Why are precipitates washed?

8. How do you decant supernatant liquids from small test tubes?

PART II: CHEMISTRY OF GROUP 2 CATIONS

Pb^{2+}, Cu^{2+}, Bi^{3+}, Sn^{4+}

Hydrogen sulfide is the precipitating agent for the group 2 cations, Pb^{2+}, Cu^{2+}, Bi^{3+}, and Sn^{4+}. You will generate H_2S from the hydrolysis of thioacetamide, CH_3CSNH_2.

$$CH_3CSNH_2(aq) + 2H_2O(l) \longrightarrow H_2S(aq) + CH_3CO_2^-(aq) + NH_4^+(aq) \quad [1]$$

By controlling the hydrogen ion concentration of the solution, you can control the sulfide ion concentration. You can see from Equation [2], which represents the overall ionization of H_2S, that the equilibrium will shift to the left if the hydrogen ion concentration is increased by the addition of some strong acid to the solution.

$$H_2S(aq) \rightleftharpoons 2H^+(aq) + S^{2-}(aq) \quad [2]$$

Under these conditions, the sulfide ion concentration is very small. Thus, through adjustment of the pH of the solution to about 0.5, only the more insoluble sulfides will precipitate. These sulfides are those of group 2—PbS, CuS, Bi_2S_3, and SnS_2.

$$Pb^{2+}(aq) + S^{2-}(aq) \longrightarrow \mathbf{PbS}(s) \quad \text{black} \quad [3]$$

$$Cu^{2+}(aq) + S^{2-}(aq) \longrightarrow \mathbf{CuS}(s) \quad \text{black} \quad [4]$$

$$2Bi^{3+}(aq) + 3S^{2-}(aq) \longrightarrow \mathbf{Bi_2S_3}(s) \quad \text{brown} \quad [5]$$

$$Sn^{4+}(aq) + 2S^{2-}(aq) \longrightarrow \mathbf{SnS_2}(s) \quad \text{yellow} \quad [6]$$

The sulfides of group 3 are soluble in such acidic solutions and therefore do not precipitate because the sulfide ion concentration is too small. Figure 32.3 summarizes how the cations of group 2 are separated and identified. Note the colors of the precipitates. *Consult this flowchart often to follow the remaining discussion.*

The cations are first treated with hydrogen peroxide (H_2O_2) and HCl to ensure that tin is in the $+4$ oxidation state.

$$Sn^{2+}(aq) + 2H^+(aq) + H_2O_2(aq) \longrightarrow Sn^{4+}(aq) + 2H_2O(l) \quad [7]$$

Tin must be in the +4 oxidation state to form the soluble SnS_3^{2-} ion, thereby allow SnS_3^{2-} it to be separated from the insoluble sulfides PbS, CuS, and Bi_2S_3.

$$\mathbf{SnS_2}(s) + S^{2-}(aq) \longrightarrow SnS_3^{2-} \quad [8]$$

The remaining insoluble sulfides are brought into solution through the treatment of hot nitric acid. Elemental sulfur is formed in the oxidation reactions.

$$\mathbf{3PbS}(s) + 8H^+(aq) + 2NO_3^-(aq) \longrightarrow 3Pb^{2+}(aq) + 3S(s) + 2NO(g) + 4H_2O(l) \quad [9]$$

$$\mathbf{3CuS}(s) + 8H^+(aq) + 2NO_3^-(aq) \longrightarrow 3Cu^{2+}(aq) + 3S(s) + 2NO(g) + 4H_2O(l) \quad [10]$$

$$\mathbf{Bi_2S_3} + 8H^+(aq) + 2NO_3^-(aq) \longrightarrow 2Bi^{3+}(aq) + 3S(s) + 2NO(g) + 4H_2O(l) \quad [11]$$

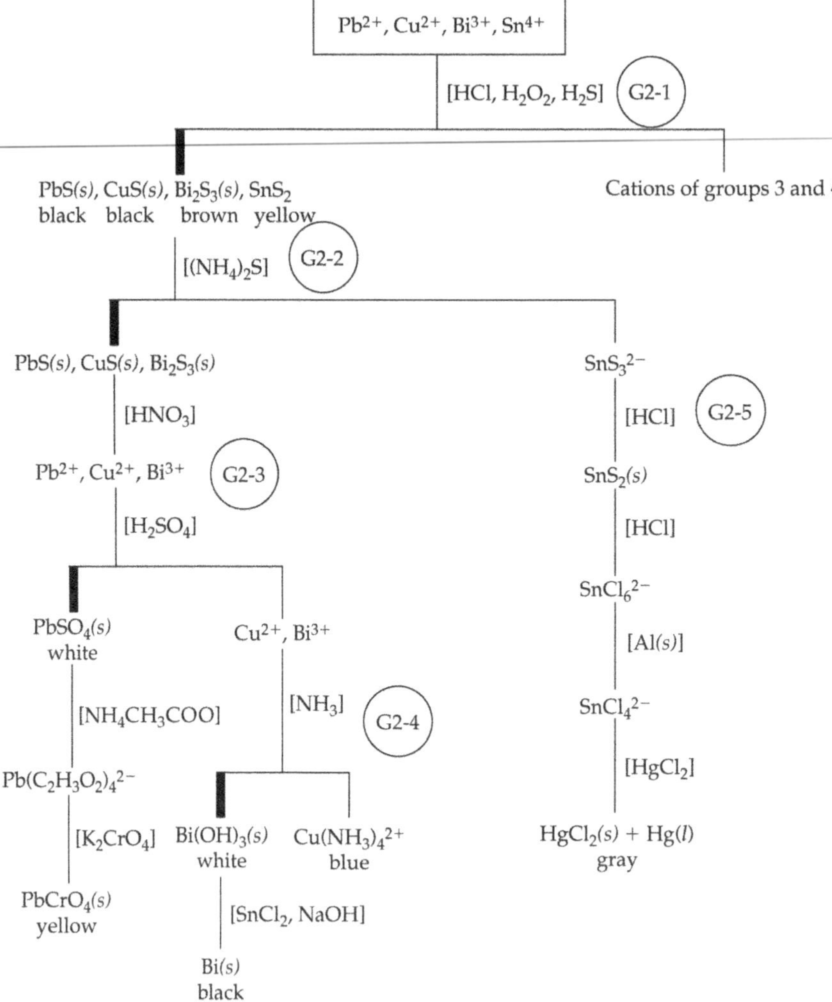

▲**FIGURE 32.3** Group 2 flowchart.

Lead(II) After the sulfides of lead, copper, and bismuth are brought into solution, lead is precipitated as the white sulfate by the addition of H_2SO_4.

$$Pb^{2+}(aq) + SO_4^{2-}(aq) \longrightarrow \mathbf{PbSO_4}(s) \quad \text{white} \qquad [12]$$

The solution is heated strongly to drive off HNO_3 because $PbSO_4$ is soluble in the presence of nitric acid. To confirm the presence of Pb^{2+}, the $PbSO_4$ is dissolved by complexation with NH_4CH_3COO and precipitated as yellow $PbCrO_4$.

$$\mathbf{PbSO_4}(s) + 4CH_3COO^-(aq) \longrightarrow Pb(CH_3COO)_4^{2-}(aq) + SO_4^{2-}(aq) \qquad [13]$$

$$Pb(CH_3COO)_4^{2-} + CrO_4^{2-}(aq) \longrightarrow 4CH_3COO^-(aq) + \mathbf{PbCrO_4}(s) \quad \text{yellow} \qquad [14]$$

GIVE IT SOME THOUGHT

The Pb^{2+} cation is part of group 1. Why is it also part of the group 2 analysis?

Copper(II) The flowchart of Figure 32.3 shows that the addition of aqueous NH_3 to the solution containing Cu^{2+} and Bi^{3+} precipitates Bi^{3+} as white $Bi(OH)_3$ and forms a soluble deep blue amine complex of copper, $Cu(NH_3)_4{}^{2+}$. The deep blue color confirms the presence of copper and can be seen even when the concentration of copper is very low.

$$Bi^{3+}(aq) + 3NH_3(aq) + 3H_2O(l) \longrightarrow 3NH_4{}^{+}(aq) + \mathbf{Bi(OH)_3}(s) \quad \text{white} \quad [15]$$

$$Cu^{2+}(aq) + 4NH_3(aq) \longrightarrow Cu(NH_3)_4{}^{2+}(aq) \quad \text{deep blue} \quad [16]$$

Bismuth(III) The $Bi(OH)_3$ precipitate is often difficult to observe when the solution is blue. Bismuth is confirmed by separating the $Bi(OH)_3$ from the solution and reducing it with $SnCl_2$ in an alkaline solution. A black powder of finely divided bismuth is formed.

$$2Bi(OH)_3(s) + 3Sn(OH)_4{}^{2-}(aq) \longrightarrow 3Sn(OH)_6{}^{2-} + \mathbf{2Bi}(s) \quad \text{black} \quad [17]$$

Tin Addition of concentrated HCl to a solution of $SnS_3{}^{2-}$ first yields the precipitate SnS_2, which eventually dissolves in the acid solution.

$$SnS_3{}^{2-}(aq) + 2H^{+}(aq) \longrightarrow \mathbf{SnS_2}(s) + H_2S(g) \quad [18]$$

$$\mathbf{SnS_2}(s) + 4H^{+}(aq) + 6Cl^{-}(aq) \longrightarrow SnCl_6{}^{2-}(aq) + 2H_2S(g) \quad [19]$$

Aluminum metal reduces tin from the +4 to the +2 oxidation state. The Sn^{2+} (in the form of the complex $SnCl_4{}^{2-}$) in turn reduces $HgCl_2$ to insoluble Hg_2Cl_2, which is white, and Hg(*l*) metal, which is black. Thus, the precipitate appears white to gray in color.

$$3SnCl_6{}^{2-} + 2Al \longrightarrow 3SnCl_4{}^{2-} + 2AlCl_3 \quad [20]$$

$$2SnCl_4{}^{2-}(aq) + 3HgCl_2(aq) \longrightarrow 2SnCl_6{}^{2-}(aq) + \mathbf{Hg_2Cl_2}(s) + \mathbf{Hg}(l) \quad [21]$$

PROCEDURE

G2-1 Oxidation of Sn^{2+} and Precipitation of Group 2

Place the supernatant from Procedure G1-1 or seven drops of known or unknown solution in an evaporating dish. Add four drops of 3% H_2O_2 and four drops of 2 *M* HCl. Carefully boil the solution by passing it back and forth over the flame of your burner. Formation of brown areas on the bottom of the evaporating dish indicates overheating. If brown areas appear, swish the solution around until the brown areas disappear. When the volume of the solution is reduced to about five drops, stop heating and allow the heat from the evaporating dish to complete the evaporation. About three drops of solution should be present after cooling. Add 10 drops of 6 *M* HCl. *In the hood*, evaporate the contents of the evaporating dish to a pasty mass, again being careful to avoid overheating. Let the evaporating dish cool; then add five drops of 2 *M* HCl and five drops of distilled water. Swish the contents to dissolve or suspend the residue and transfer it to a small test tube. The solution is evaporated to dryness or to a pasty mass to remove the unknown quantity of acid that is present. A known amount of HCl is then added, which is required for the group precipitation as sulfides.

GIVE IT SOME THOUGHT

Where does each part of the procedure fit in the flowchart in Figure 32.3?

Add 10 drops of 1 *M* thioacetamide to the solution in the test tube. Stir the mixture and heat the test tube in a boiling water bath for 10 min (see Appendix J for the use of a hot water bath). If excessive frothing occurs, temporarily remove the test tube from the bath. Occasionally stir the solution while it is being heated. After heating for 10 min, add 10 drops of *hot* water and 10 drops of 1 *M* thioacetamide and two drops of 1 *M* NH_4CH_3COO (ammonium acetate). Mix and heat in the boiling water bath for 10 more min, stirring occasionally. Cool, centrifuge, and decant using a pipet (see Appendix J for the technique) into a test tube. The precipitate contains the group 2 insoluble sulfides, while the supernatant liquid may contain cations from groups 3 and 4.

Test the supernatant for completeness of precipitation by adding three drops of H_2O, one drop of 1 *M* NH_4CH_3COO, and two drops of 1 *M* thioacetamide. Mix and heat in the boiling water bath for 1 min. If a colored precipitate forms, continue heating for 3 more min. A faint cloudiness may develop because of the formation of colloidal sulfur. Repeat the precipitation procedure until precipitation is complete. If the supernatant is to be analyzed for groups 3 and 4, transfer it to an evaporating dish and boil it to reduce the volume to about 0.5 mL. Otherwise, discard it in a designated receptacle. Transfer the supernatant to a labeled test tube and save it for Procedure G3-1. Rinse the evaporating dish with six drops of water and add the washing to the test tube and stopper it.

GIVE IT SOME THOUGHT

What does a colored precipitate indicate?

All precipitates should be combined in the test tube containing the original precipitate, using a few drops of water to aid in the transfer. Wash the precipitate three times, once with 10 drops of *hot* water and twice with 20-drop portions of a hot solution prepared using equal volumes of water and 1 *M* NH_4CH_3COO. Make sure you stir the washing liquid and precipitate with a stirring rod before each centrifugation. Should the precipitate form a colloidal suspension, add 10 drops of 1 *M* NH_4CH_3COO and heat the suspension in the boiling water bath. Discard the washings in a designated waste container. Note that NH_4CH_3COO is often used in washing precipitates. Its purpose is to help prevent the precipitate from becoming colloidal. Finely divided particles usually do not coagulate and precipitate from the solution.

G2-2 Separation of Sn^{4+} from PbS, CuS, and Bi_2S_3

To the precipitate from Procedure G2-1, add 10 drops of $(NH_4)_2S$ (ammonium sulfide) solution and stir well. Then heat for 3 to 4 min in the boiling water bath. Remove the test tube from the water bath as necessary to avoid excessive frothing. Centrifuge, decant, and save the supernatant. Repeat the treatment using seven drops of $(NH_4)_2S$. Centrifuge and combine the supernatant with the first. Stopper the combined supernatant and save for the analysis of tin (Procedure G2-5).

Wash the precipitate twice with 20-drop portions of a hot solution prepared by mixing equal volumes of water and 1 *M* NH_4CH_3COO. The precipitate may contain PbS, CuS, and Bi_2S_3 and should be analyzed according to Procedure G2-3.

GIVE IT SOME THOUGHT

What color do you expect this precipitate to be?

G2-3 Separation and Identification of Pb^{2+}

Add 1 mL (20 drops) of 3 *M* HNO_3 to the test tube containing the precipitate from Procedure G2-2. Mix thoroughly and transfer the contents to an evaporating dish. Boil the mixture gently for 1 min. If necessary, add more HNO_3 to keep the amount of liquid constant. Cool, centrifuge, and discard any free sulfur that forms. Transfer the supernatant to an evaporating dish and add six drops of 18 *M* H_2SO_4. **(CAUTION: *Concentrated* H_2SO_4 *causes severe burns. If you get any on yourself, immediately wash the area with copious amounts of water.*)** *In a hood*, evaporate the contents until the volume is about one drop and dense white fumes of SO_3 form. The fumes should be so dense that you cannot see the bottom of the evaporating dish. The appearance of dense white fumes of SO_3 ensures that all HNO_3 has been removed. Cool, add 20 drops of water, and stir. Quickly transfer the contents to a test tube before the suspended material settles. Cool the test tube. A finely divided white precipitate, $PbSO_4$, indicates the presence of lead. Centrifuge and save the supernatant for Procedure G2-4. Wash the precipitate twice with 10-drop portions of cold water and discard the washings as directed. Add six drops of 1 *M* NH_4CH_3COO to the precipitate and stir for about 15 sec. Then add one drop of 1 *M* potassium chromate, K_2CrO_4. The formation of a yellow precipitate ($PbCrO_4$) confirms the presence of lead(II). Dispose of waste as directed.

G2-4 Separation of Bi^{3+} and Identification of Bi^{3+} and Cu^{2+}

To the supernatant from Procedure G2-3, carefully add 15 *M* aqueous NH_3 dropwise, stirring constantly until the solution is basic to litmus. **(CAUTION: *Avoid inhalation of or skin contact with* NH_3. *If you come in contact with it, immediately wash the area with copious amounts of water.*)** The appearance of a deep blue color of $Cu(NH_3)_4{}^{2+}$ confirms the presence of Cu^{2+}. Centrifuge and discard the supernatant liquid as directed, but save the white precipitate of $Bi(OH)_3$. Carefully observe the contents of the test tube, the white gelatinous $Bi(OH)_3$ may be somewhat difficult to see when the solution is colored.

Wash the precipitate once with 10 drops of hot water and discard the washings as directed. Add six drops of 6 *M* NaOH and four drops of freshly prepared* 0.2 *M* $SnCl_2$ to the precipitate and stir. The formation of a jet-black precipitate confirms the presence of Bi^{3+}. Dispose of waste as directed.

G2-5 Identification of Sn^{4+}

Transfer the supernatant from Procedure G2-2 to an evaporating dish and boil for 1 min to expel H_2S; then add four drops of cold water. Add a 1 in. piece of 26-gauge aluminum wire and heat gently until the wire has dissolved. Continue to heat the solution gently for about two more min, replenishing the solution with 6 *M* HCl if necessary. There should be no dark residue at this stage; if there is, continue heating until it dissolves. (You cannot stop at this point.) Transfer the solution to a test tube and cool under running water. Immediately add three drops of 0.1 *M* mercuric chloride, $HgCl_2$, and mix. Allow the mixture to stand for 1 min. The formation of a white or gray precipitate confirms the presence of tin. Dispose of waste as directed.

*Solid tin should be present in the bottle to ensure that tin is in the +2 oxidation state.

Name ______________________________ Desk ______________

Date ______________ Laboratory Instructor ______________________________

Abbreviated Qualitative Analysis Scheme | 32 Pre-lab Questions

Before beginning Part II of this experiment in the laboratory, you should be able to answer the following questions.

1. What are the symbols and charges of the group 2 cations?

2. How could CuS be separated from SnS_2?

3. How are Cu^{2+} and Bi^{3+} separated?

4. How can Ag^{+} be separated from Cu^{2+}?

5. Complete and balance the following equations:

   (a) $Bi^{3+}(aq) + S^{2-}(aq) \longrightarrow$

   (b) $SnS_2(s) + S^{2-}(aq) \longrightarrow$

   (c) $PbS(s) + H^+(aq) + NO_3^-(aq) \longrightarrow$

   (d) $Bi^{3+}(aq) + NH_3(aq) + H_2O(l) \longrightarrow$

6. Why is H_2O_2 added in the initial step of the separation of group 2 cations?

7. What is the color of CuS? of SnS_2? of $PbSO_4$?

PART III: CHEMISTRY OF GROUP 3 CATIONS $Fe^{3+}, Ni^{2+}, Mn^{2+}, Al^{3+}$

The group 3 cations you will consider are Fe^{3+}, Ni^{2+}, Mn^{2+}, and Al^{3+}. These cations do not precipitate as insoluble chlorides (as do those of group 1) or as sulfides in acidic solution (like those of group 2). These ions can be separated from those of group 4 by precipitation as insoluble hydroxides or sulfides under slightly alkaline conditions. The separation is shown in the flowchart of Figure 32.4. As in group 2, the pH of the solution controls the sulfide ion concentration. The slightly alkaline conditions employed here favor a higher sulfide ion concentration than that used in the group 2 separation. FeS, NiS, and MnS are more soluble than the sulfides of group 2 and therefore require a higher sulfide ion concentration for their precipitation. Making the solution slightly alkaline reduces the hydrogen ion concentration. You can see that decreasing the hydrogen ion concentration causes the following equilibrium to shift to the right:

$$H_2S(aq) \rightleftharpoons 2H^+(aq) + S^{2-}(aq)$$

This results in an increase in the sulfide ion concentration.

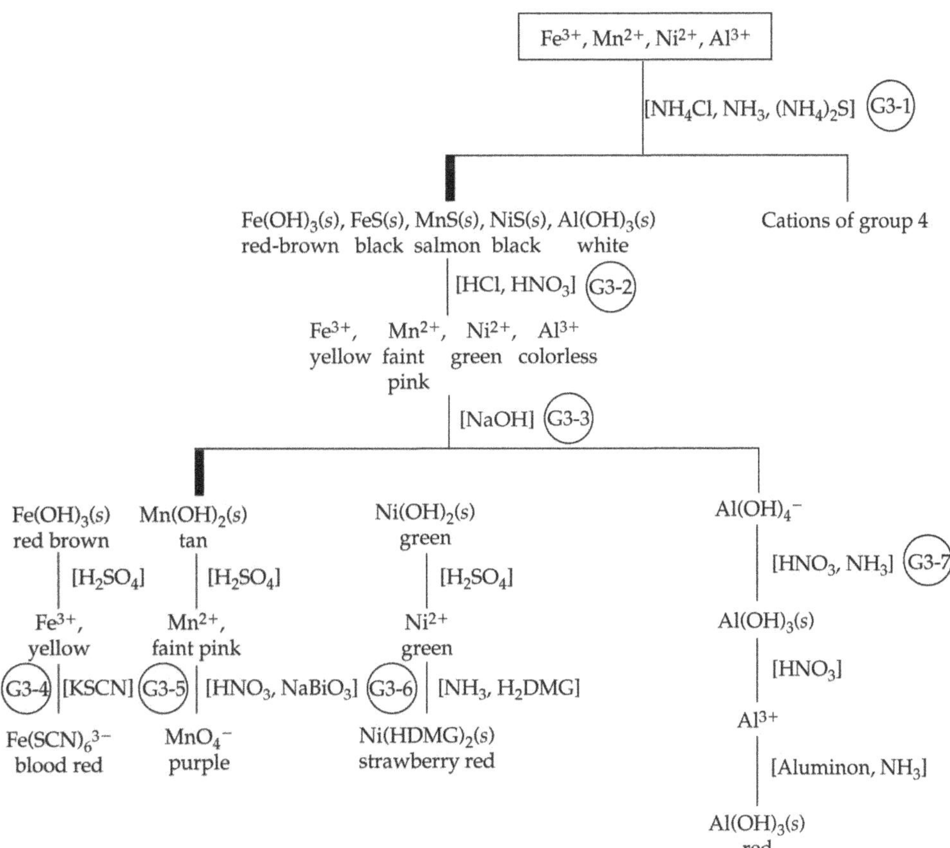

▲**FIGURE 32.4** Group 3 flowchart.

GIVE IT SOME THOUGHT

a. What are the colors of each of these ions in solution?
b. What would you expect the color to be if all four cations were present?

Aqueous ammonia is a weak base.

$$NH_3(aq) + H_2O(l) \rightleftharpoons NH_4^+(aq) + OH^-(aq)$$

A mixture of aqueous NH_3 and NH_4Cl makes a buffer solution (Section 17.2) whose pH allows the precipitation of FeS, MnS, and NiS. Moreover, in this slightly alkaline solution, the insoluble hydroxides $Fe(OH)_3$ and $Al(OH)_3$ also precipitate. Although $Mg(OH)_2$ is insoluble, it does not precipitate with the group 3 hydroxides because the OH^- ion concentration is too small. The common ion NH_4^+, from NH_4Cl, controls the OH^- concentration and keeps it sufficiently low, preventing the $Mg(OH)_2$ from precipitating.

GIVE IT SOME THOUGHT

If a buffer were not used, how would the results of the known solution be influenced?

As you analyze the group 3 ions, pay particular attention to the colors of the solutions and the precipitates. Aqueous solutions of Al^{3+} ions are colorless, while those of Fe^{3+} appear yellow to reddish-brown. Ni^{2+} solutions are green; Mn^{2+}, *very* faint pink.

The reactions involved in the precipitation of the group 3 cations are as follows:

$$Fe^{3+}(aq) + 3OH^-(aq) \longrightarrow \mathbf{Fe(OH)_3}(s) \quad \text{red-brown} \qquad [1]$$

$$Al^{3+}(aq) + 3OH^-(aq) \longrightarrow \mathbf{Al(OH)_3}(s) \quad \text{colorless} \qquad [2]$$

$$Ni(NH_3)_6^{2+}(aq) + S^{2-}(aq) \longrightarrow 6NH_3(aq) + \mathbf{NiS}(s) \quad \text{black} \qquad [3]$$

$$Mn^{2+}(aq) + S^{2-}(aq) \longrightarrow \mathbf{MnS}(s) \quad \text{salmon} \qquad [4]$$

In aqueous NH_3 solution, Ni^{2+} ions exist as the ammine complex $Ni(NH_3)_6^{2+}$ and addition of $(NH_4)_2S$ results in the precipitation of NiS. Some $Fe(OH)_3$ is reduced by sulfide.

$$\mathbf{2Fe(OH)_3}(s) + 3S^{2-}(aq) \longrightarrow 6OH^-(aq) + 2\mathbf{FeS}(s) + S(s) \quad \text{black} \qquad [5]$$

The group 3 precipitates are dissolved using HCl. Nitric acid is also added to help dissolve NiS by oxidizing S^{2-} ions to elemental sulfur. At the same time, nitric acid reoxidizes any Fe^{2+} to Fe^{3+}.

$$\mathbf{Fe(OH)_3}(s) + 3H^+(aq) \longrightarrow Fe^{3+}(aq) + 3H_2O(l) \qquad [6]$$

$$\mathbf{FeS}(s) + 2H^+(aq) \longrightarrow Fe^{2+}(aq) + H_2S(g) \qquad [7]$$

$$3Fe^{2+}(aq) + NO_3^-(aq) + 4H^+(aq) \longrightarrow 3Fe^{3+}(aq) + NO(g) + 2H_2O(l) \quad [8]$$

$$\mathbf{Al(OH)_3}(s) + 3H^+(aq) \longrightarrow Al^{3+}(aq) + 3H_2O(l) \quad [9]$$

$$\mathbf{3NiS}(s) + 2NO_3^-(aq) + 8H^+(aq) \longrightarrow 3Ni^{2+}(aq) + 3S(s) + 2NO(g) + 4H_2O(l) \quad [10]$$

$$\mathbf{MnS}(s) + 2H^+(aq) \longrightarrow Mn^{2+}(aq) + H_2S(g) \quad [11]$$

After all ions are in solution, addition of excess strong base allows the separation of Fe^{3+}, Ni^{2+}, and Mn^{2+} ions from Al^{3+} ions. $Fe(OH)_3$, $Ni(OH)_2$, and $Mn(OH)_2$ precipitate, while $Al(OH)_3$, being amphoteric, redissolves in excess base forming soluble aluminate ions:

$$Fe^{3+}(aq) + 3OH^-(aq) \longrightarrow \mathbf{Fe(OH)_3}(s) \quad \text{red-brown} \quad [12]$$

$$Ni^{2+}(aq) + 2OH^-(aq) \longrightarrow \mathbf{Ni(OH)_2}(s) \quad \text{green} \quad [13]$$

$$Mn^{2+}(aq) + 2OH^-(aq) \longrightarrow \mathbf{Mn(OH)_2}(s) \quad \text{tan} \quad [14]$$

$$Al^{3+}(aq) + 4OH^-(aq) \longrightarrow Al(OH)_4^-(aq) \quad \text{colorless} \quad [15]$$

The hydroxide precipitates are dissolved by the addition of H_2SO_4 to give a solution of Fe^{3+}, Ni^{2+}, and Mn^{2+} ions. After the solution is divided into three equal parts, tests for the individual ions are made as described below.

Iron(III) A very sensitive test for Fe^{3+} uses the thiocyanate ion, SCN^-. If Fe^{3+} is present, a blood-red solution results when SCN^- is added.

$$Fe^{3+}(aq) + SCN^-(aq) \longrightarrow Fe(NCS)^{2+}(aq) \quad \text{red} \quad [16]$$

Manganese Because of its intense purple color, the permanganate ion, MnO_4^-, affords a suitable confirmatory test for Mn^{2+}. When a solution of Mn^{2+} is acidified with HNO_3 and then treated with sodium bismuthate, $NaBiO_3$, Mn^{2+} is oxidized to MnO_4^-.

$$2Mn^{2+}(aq) + \mathbf{5NaBiO_3}(s) + 14H^+(aq) \longrightarrow 5Bi^{3+}(aq) + 5Na^+(aq) + 7H_2O(l) + 2MnO_4^-(aq) \quad \text{purple} \quad [17]$$

Nickel The presence of Ni^{2+} is confirmed by the formation of a bright red precipitate when an organic compound called dimethylglyoxime (abbreviated H_2DMG) is added to an ammoniacal solution.

$$Ni(NH_3)_6^{2+}(aq) + 2H_2DMG(aq) \longrightarrow 4NH_3(aq) + 2NH_4^+(aq) + \mathbf{Ni(HDMG)_2}(s) \quad \text{red} \quad [18]$$

Aluminum When a solution that contains $Al(OH)_4^-$ is acidified and then made slightly alkaline with the weak base NH_3, $Al(OH)_3$ precipitates.

$$Al(OH)_4^- + 4H^+(aq) \longrightarrow Al^{3+}(aq) + 4H_2O(l) \quad [19]$$

$$Al^{3+}(aq) + 3NH_3(aq) + 3NH_2O(l) \longrightarrow 3NH_4^+(aq) + \mathbf{Al(OH)_3}(s) \quad [20]$$

Aluminum hydroxide is not easily seen, however, for it is a gelatinous, translucent substance. So that the hydroxide can be seen, it is precipitated in the presence of a red dye called Aluminon reagent. The dye is adsorbed on the $Al(OH)_3$, giving it a cherry red color called a "red lake."

PROCEDURE

G3-1 Precipitation of Group 3 Cations

Place the supernatant from Procedure G2-1 or seven drops of known or unknown solution in a small (10 mm × 75 mm) test tube. If the supernatant from Procedure G2-1 has a precipitate, centrifuge, decant, and discard the precipitate as directed. Add five drops of 2 *M* NH_4Cl and stir. **(CAUTION: *Concentrated*** NH_3 ***has a strong irritating odor and causes severe burns. Avoid inhalation and contact. If you come in contact with it, immediately wash the area with copious amounts of water.*)** Add 15 *M* NH_3 dropwise while stirring until the solution is just basic to litmus paper. Usually, this requires only a few drops of the NH_3. Then add two additional drops of the 15 *M* NH_3 and 1 mL (20 drops) of water. Stir thoroughly. Next, add 10 drops of $(NH_4)_2S$ and mix thoroughly. Heat the test tube in a boiling water bath for about 5 min. If excessive frothing occurs, temporarily remove the test tube from the hot water bath. Centrifuge and test for completeness of precipitation using one drop of $(NH_4)_2S$. Note the color of the precipitate. Decant and save the supernatant for group 4 analysis or discard in a designated receptacle as directed by your instructor.

GIVE IT SOME THOUGHT

Where does each part of the procedure fit in the flowchart in Figure 32.4?

Wash the precipitate two times with 20 drops of a solution made by mixing equal portions of water and 1 *M* NH_4CH_3COO. For each washing, stir the precipitate with the wash solution and heat the mixture in the water bath before centrifuging. Discard the supernatant wash liquid.

G3-2 Dissolution of Group 3 Precipitate

Treat the precipitate from Procedure G3-1 with 12 drops of 12 *M* HCl; then cautiously add five drops of 16 *M* HNO_3 and carefully mix the solution. **(CAUTION:** HNO_3 ***can cause severe burns. If you come in contact with the acid, wash the area with copious amounts of water.*)** Heat the test tube in the hot water bath until the precipitate dissolves and a clear, but not necessarily colorless, solution is obtained. Add 10 drops of water, centrifuge to remove any sulfur that has precipitated, and decant into an evaporating dish. Note the color of the supernatant.

G3-3 Separation of Iron, Nickel, and Manganese from Aluminum

Make the solution in the evaporating dish from Procedure G3-2 strongly basic, using 6 *M* NaOH, mixing thoroughly. If the precipitate is pasty and nonfluid, add 12 drops of water. Note the color of the precipitate. Transfer to a test tube and centrifuge. Decant, saving the supernatant, which may contain aluminum, for Procedure G3-7. To the precipitate, add 20 drops of water and 10 drops of 6 *M* H_2SO_4. Stir and heat in a water bath for 3 min or until the precipitate dissolves. Add 12 drops of water and divide the solution into three approximately equal volumes.

G3-4 Test for Fe^{3+}

To one of the three samples from Procedure G3-3, add two drops of 0.2 *M* KSCN (potassium thiocyanate). A blood-red *solution* confirms the presence of iron as $Fe(NCS)^{2+}$. Traces of iron that have been introduced as impurities will give a weak test result. If you are in doubt about your results, perform this test on 10 drops of your original sample.

G3-5 Test for Mn^{2+}

To the second portion of a solution from Procedure G3-3, add an equal volume of water and four drops of 3 *M* HNO_3. Mix and then add a few grains of solid sodium bismuthate, $NaBiO_3$. Mix thoroughly with a stirring rod and centrifuge. A pink or purple color is due to MnO_4^- and confirms the presence of manganese.

G3-6 Test for Ni^{2+}

To a third portion of the sample from Procedure G3-3, add 6 *M* NH_3 until the solution is basic. If a precipitate forms, remove it by centrifuging and decanting, keeping the supernatant. Add about four drops of dimethylglyoxime reagent, mix, and allow to stand. The formation of a strawberry red *precipitate* indicates the presence of nickel.

G3-7 Test for Al^{3+}

Treat only half the supernatant from Procedure G3-3 with 16 *M* HNO_3 until the solution is slightly acidic. **(CAUTION: HNO_3 *can cause severe burns. Wash with water immediately if you come in contact with the acid.*)** Then while stirring, add 15 *M* NH_3 until the solution is distinctly alkaline. Allow at least 1 min for the formation of $Al(OH)_3$. Centrifuge and carefully remove the supernatant liquid with a capillary pipet without disturbing the gelatinous $Al(OH)_3$. Discard the supernatant as directed. Wash the precipitate two times with 20 drops of hot water, discarding the supernatant as directed. Dissolve the precipitate in seven drops of 3 *M* HNO_3. Add three drops of Aluminon reagent, which colors the solution; stir; and add 6 *M* NH_3 dropwise until the solution is just alkaline to litmus paper (avoid an excess). Stir and centrifuge. The formation of a cherry red *precipitate*, not solution, confirms the presence of aluminum.

NOTES AND CALCULATIONS

Name ______________________________ Desk ______________

Date ______________ Laboratory Instructor ______________________________

Abbreviated Qualitative Analysis Scheme | 32 Pre-lab Questions

Before beginning Part III of this experiment in the laboratory, you should be able to answer the following questions.

1. What are the symbols and charges of the group 3 cations?

2. What are the colors of the following aqueous ions: Mn^{2+}, Ni^{2+}, MnO_4^-, and Fe^{3+}?

3. What are the colors of the following solids: $Fe(OH)_3$, MnS, $Al(OH)_3$, and $Ni(OH)_2$?

4. How can Fe^{3+} be separated from Al^{3+}?

5. How can $Ni(OH)_2$ be separated from $Al(OH)_3$?

6. Complete and balance the following equations:

(a) $Ni^{2+}(aq) + OH^{-}(aq) \longrightarrow$

(b) $Fe(OH)_3(s) + H^{+}(aq) \longrightarrow$

(c) $FeS(s) + H^{+}(aq) \longrightarrow$

(d) $NiS(s) + NO_3^{-}(aq) + H^{+}(aq) \longrightarrow$

7. Give the formula for a reagent that precipitates the following:

(a) Pb^{2+} but not Ni^{2+}

(b) Fe^{3+} but not Al^{3+}

8. What cation forms the following?

(a) a blood-red solution with thiocyanate ion

(b) a bright red precipitate with dimethylglyoxime

9. If solid NH_4Cl is added to 3 M NH_3, does the pH increase, decrease, or remain the same? Explain?

PART IV: CHEMISTRY OF GROUP 4 CATIONS
$Ba^{2+}, Ca^{2+}, NH_4^+, Na^+$

In addition to the ammonium ion, the cations of group 4 consist of ions of the alkali and alkaline earth metals. The cations you will consider in this group are Ba^{2+}, Ca^{2+}, NH_4^+, and Na^+. Because their chlorides and sulfides are soluble, these ions do not precipitate with groups 1, 2, or 3.

Sodium ions are a common impurity and were even introduced (as was ammonium ion) in some of the reagents used in the analysis of groups 1, 2, and 3. Hence, in the analysis of a general unknown mixture, tests for these ions must be made on the original sample before the group analysis is performed. The flowchart for group 4 is shown in Figure 32.5.

Barium Because barium chromate, $BaCrO_4$ ($K_{sp} = 1.2\times10^{-10}$), is less soluble than calcium chromate, $CaCrO_4$ ($K_{sp} = 7.1\times10^{-4}$), Ba^{2+} can be separated from Ca^{2+} by precipitation as the insoluble yellow chromate salt.

$$Ba^{2+}(aq) + CrO_4^{2-}(aq) \longrightarrow \mathbf{BaCrO_4}(s) \quad \text{yellow} \qquad [1]$$

$BaCrO_4$ is insoluble in the weak acid CH_3COOH, but it is soluble in the presence of the strong acid HCl because it is the salt of a weak acid, H_2CrO_4. After $BaCrO_4$ is dissolved in HCl, a flame test is performed on the resulting solution. A green-yellow flame is indicative of Ba^{2+}. Further confirmation of Ba^{2+} is precipitation of $BaSO_4$, which is white.

$$Ba^{2+}(aq) + SO_4^{2-}(aq) \longrightarrow \mathbf{BaSO_4}(s) \quad \text{white} \qquad [2]$$

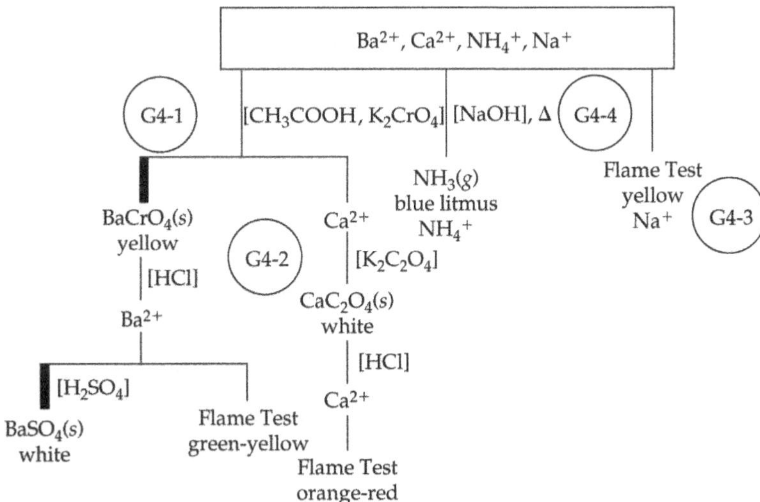

▲**FIGURE 32.5** Group 4 flowchart.

Calcium Calcium oxalate, CaC_2O_4, is very insoluble ($K_{sp} = 4.0 \times 10^{-9}$). The formation of a white precipitate when oxalate ion is added to a slightly alkaline solution confirms the presence of Ca^{2+}.

$$Ca^{2+}(aq) + C_2O_4^{2-}(aq) \longrightarrow \mathbf{CaC_2O_4(s)} \quad \text{white} \qquad [3]$$

Additional evidence for the calcium ion is obtained from a flame test. Dissolution of CaC_2O_4 with HCl, followed by a flame test, produces a transient orange-red flame that is characteristic of calcium ions.

Sodium Most sodium salts are soluble. The simplest test for sodium ion is a flame test. Sodium salts impart a characteristic yellow color to a flame; the test is very sensitive, and because of the prevalence of sodium ions, much care must be exercised to keep equipment clean and free from contamination by these ions.

Ammonium The ammonium ion, NH_4^+, is the conjugate acid of the base ammonia, NH_3. The test for NH_4^+ takes advantage of the following equilibrium:

$$NH_4^+(aq) + OH^-(aq) \rightleftharpoons NH_3(aq) + H_2O(l) \qquad [4]$$

Thus, when a strong base is added to a solution of an ammonium salt and this solution is heated, NH_3 gas is evolved. The NH_3 can easily be detected by its effect on red litmus.

PROCEDURE

G4-1 Separation and Identification of Ba^{2+}

If the solution, known or unknown, contains only cations of group 4, place seven drops of the solution in a small test tube; otherwise, use the supernatant from group 3 analysis, Procedure G3-1. Add eight drops of 6 *M* acetic acid, CH_3COOH, and one drop of 1 *M* K_2CrO_4 and mix. The formation of a yellow precipitate indicates the presence of Ba^{2+}. Centrifuge and save the supernatant for Procedure G4-2 to test for calcium. Dissolve the precipitate with 6 *M* HCl and perform a flame test as described below.

GIVE IT SOME THOUGHT

Where does each part of the procedure fit in the flowchart in Figure 32.5?

To perform the flame test, obtain a piece of platinum or Nichrome wire that has been sealed in a piece of glass tubing. Clean the wire by dipping it in 12 *M* HCl that is contained in a small test tube and heat the wire in the hottest part of your Bunsen burner flame. Repeat this operation until you see no color when you place the wire in the flame. Several cleanings will be required before this is achieved. Then dip the wire in the solution to be tested and place the wire in the flame. A pale green flame confirms the presence of Ba^{2+}. If the concentration of Ba^{2+} is very low, you may not detect the green color.

As further confirmation of barium, add 10 drops of 6 *M* H_2SO_4 to the solution on which the flame test was performed. A white precipitate confirms the presence of Ba^{2+}. Dispose of waste as directed.

G4-2 Test for Ca^{2+}

Make the supernatant from Procedure G4-1 alkaline to litmus with 15 *M* NH_3. If a precipitate forms, centrifuge and discard the precipitate. **(CAUTION: *Avoid inhalation of or skin contact with* NH_3. *If you come in contact with it, immediately wash the contacted area with copious amounts of water.*)** Add seven drops of 1 *M* $K_2C_2O_4$ (potassium oxalate) and stir. The formation of a white precipitate indicates the presence of calcium ion. Should no precipitate form immediately, warm the test tube briefly in the hot water bath and then cool.

Additional evidence for Ca^{2+} is obtained from a flame test. Dissolve the precipitate in 6 *M* HCl and perform a flame test on a freshly cleaned piece of platinum of Nichrome wire. A transitory red-orange color that appears when you first place the wire in the flame and appears somewhat redder as the wire is heated is characteristic of the calcium ion. If the concentration of Ca^{2+} is very low, you may not observe the red color.

G4-3 Test for Na^+

The flame test for sodium is very sensitive, and traces of sodium ion will impart a characteristic yellow color to the flame. Just about every solution has a trace of sodium and thus will give a positive test. On the basis of the intensity and duration of the yellow color, you can decide whether Na^+ is merely a contaminant or is present in substantial quantity. Using a clean wire, perform a flame test on your original (untreated) unknown. To help you make a decision as to the presence of sodium, run a flame test on distilled water and then on a 0.2 *M* NaCl solution. Compare the results.

G4-4 Test for NH_4^+

Place 2 mL of the original (untreated) unknown or known in a 100 mL beaker (or evaporating dish) and add 2 mL of 6 *M* NaOH. Moisten a piece of red litmus paper with water and stick it to the convex side of a small watch glass. Cover the beaker or evaporating dish with the watch glass, convex side down. (The litmus paper must not come in contact with any NaOH.) Gently warm the beaker (or evaporating dish) with a small burner flame; do not boil. Allow the covered beaker (or evaporating dish) to stand for 3 min. A change in the color of the litmus paper from red to blue confirms the presence of ammonium ion.

NOTES AND CALCULATIONS

Name ______________________________ Desk ______________

Date ______________ Laboratory Instructor ______________________

Abbreviated Qualitative Analysis Scheme | 32 Pre-lab Questions

Before beginning Part IV of this experiment in the laboratory, you should be able to answer the following questions.

1. What are the symbols and charges of the group 4 cations?

2. Which is less soluble: $BaCrO_4$ or $CaCrO_4$?

3. What is the color of each of the following?
   (a) $BaSO_4$

   (b) $BaCrO_4$

   (c) CaC_2O_4

4. What color do the following ions impart to a flame?
   (a) Ba^{2+}

   (b) Ca^{2+}

   (c) Na^{+}

5. What reagent will precipitate the following?
   (a) Cu^{2+} but not Ba^{2+}

   (b) Ag^{+} but not Ca^{2+}

   (c) Ba^{2+} but not NH_4^{+}

6. Complete and balance the following equations:
   (a) $Ba^{2+}(aq) + SO_4^{2-}(aq) \longrightarrow$

   (b) $Ca^{2+}(aq) + C_2O_4^{2-}(aq) \longrightarrow$

PART V: CHEMISTRY OF ANIONS SO_4^{2-}, NO_3^-, CO_3^{2-}, Cl^-, Br^-, I^-, CrO_4^{2-}, PO_4^{3-}, S^{2-}, SO_3^{2-}

DISCUSSION

A systematic scheme based on the kinds of principles involved in cation analysis can be designed for the analysis of anions. This would involve separating and then identifying the anions. However, it is generally easier to take another approach to the identification of anions. An effort is made to either eliminate or verify the presence of certain anions on the basis of the color and solubility of the samples; then the material being analyzed is subjected to a series of preliminary tests. From the results of the preliminary tests and observations, certain anions are shown to be present or absent; then specific tests are performed for those anions not eliminated in the preliminary tests and observations. The preliminary tests include treating the solid with concentrated sulfuric acid and using silver nitrate and $BaCl_2$ as precipitating agents. The behavior of the anions in these tests is conveniently summarized in Tables 32.1 and 32.2.

In this experiment, characteristics of the following 10 anions are considered: sulfate (SO_4^{2-}), nitrate (NO_3^-), carbonate (CO_3^{2-}), chloride (Cl^-), bromide (Br^-), iodide (I^-), chromate (CrO_4^{2-}), phosphate (PO_4^{3-}), sulfide (S^{2-}), and sulfite (SO_3^{2-}).

As in the case of cation identification, the physical and chemical properties of the compounds formed by the anions provide the basis for their identification. You will proceed by using the following general procedures.

PROCEDURE

Examination of the Solid

The color of a substance offers a clue to its constituents. For example, many transition metal salts are colored. Those of nickel, Ni^{2+}, are generally green; iron(III), Fe^{3+}, reddish-brown to yellow; iron(II), Fe^{2+}, grayish-green; chromium(III), Cr^{3+}, green to bluish-gray to black; copper(II), Cu^{2+}, blue to green to black; cobalt(II), Co^{2+}, wine-red to blue; and manganese(II), Mn^{2+}, pink to tan. By contrast, only a few anions are colored, and these contain transition metals as well. The colored anions are chromate, CrO_4^{2-}, yellow; dichromate, $Cr_2O_7^{2-}$, orange-red; and permanganate, MnO_4^{2-}, violet-purple. Hence, the color of the solid may be used as a reliable indicator of the presence of these ions. Observe the color of your unknown and note it on the report sheet.

GIVE IT SOME THOUGHT

a. Could the color of your initial sample be compared to the results in a reference such as the *Handbook of Chemistry and Physics*?

b. How would this help you narrow down your unknown?

Solubility of Salt

The solubility of a substance in water and in acidic or basic solutions, together with a knowledge of the cations present, often allows you to narrow down the choice of anions present. For example, a white substance containing Ag^+ that is insoluble in acid solution (HNO_3) but soluble in aqueous ammonia (NH_3) is most likely AgCl. There is some ambiguity, however, because AgBr behaves similarly. Thus, tests for both chloride and bromide would be necessary to confirm which anion is present. Test the solubility of your solid unknown anion sample in water, 6 *M* HNO_3, 6 *M* HCl, and 6 *M* NH_3 and note your results on your report sheet. Compare your results with the solubility properties of ions and solids given in Appendix D.

Reactions with $AgNO_3$ and $BaCl_2$

If your unknown substance is soluble in water, you may obtain additional clues as to the presence of certain anions by treating a solution of your unknown separately with solutions of silver nitrate ($AgNO_3$) and barium chloride ($BaCl_2$). The silver salts of all of your possible anions except nitrate and sulfate are insoluble in water and have the following colors (see Table 32.1):

TABLE 32.1 Summary of Preliminary Tests

		$AgNO_3$		$BaCl_2$	
Anion	**Concentrated H_2SO_4 solid salt**	**Neutral**	**Acidic**	**Neutral**	**Acidic**
NO_3^-	Cold: no observable reaction Hot: brown gas	No reaction		No reaction	
Cl^-	$HCl(g)$, colorless; pungent	$AgCl(s)$ white	Insol.	No reaction	
Br^-	Br_2, red-brown; pungent	$AgBr(s)$ cream	Insol.	No reaction	
I^-	Solids turn dark brown, violet vapors; H_2S odor	$AgI(s)$ yellow	Insol.	No reaction	
S^{2-}	$H_2S(g)$, odor of rotten eggs free S deposited	$Ag_2S(s)$ black	Insol.	No reaction	
SO_4^{2-}	No evidence of reaction	No reaction		$BaSO_4(s)$ white	Insol.
SO_3^{2-}	$SO_2(g)$, colorless; choking odor	$Ag_2SO_3(s)$ white	Sol.	$BaSO_3(s)$ white	Sol.
CO_3^{2-}	$CO_2(g)$, colorless; odorless	$Ag_2CO_3(s)$ white $\rightarrow$ black	Sol.	$BaCO_3(s)$ white	Sol.
PO_4^{3-}	No evidence of reaction	$Ag_3PO_4(s)$ yellow	Sol.	$Ba_3(PO_4)_2(s)$ white	Sol.
CrO_4^{2-}	Solid changes from yellow to orange-red	$Ag_2CrO_4(s)$ red-brown	Sol.	$BaCrO_4(s)$ yellow	Sol.

AgCl: white	Ag_2CrO_4: red-brown
AgBr: cream	Ag_2S: black
AgI: yellow	Ag_2CO_3: white
Ag_3PO_4: yellow	Ag_2SO_3: white

All of these solids except AgCl, AgBr, AgI, and Ag_2S are soluble in dilute nitric acid. Dissolve some of your solid unknown, about the size of a small pea, in water and add four drops of 0.1 *M* $AgNO_3$ solution. Note the color of any precipitate that forms and enter your observation on your report sheet. Dissolve another small pea-size portion of your sample in water, add a few drops of 6 *M* HNO_3 to make the solution acidic, add four drops of 0.1 *M* $AgNO_3$, and stir the mixture. Record your observation on your report sheet.

The barium salts $BaCl_2$, $BaBr_2$, BaI_2, BaS, and $Ba(NO_3)_2$ are soluble in water and in slightly basic solution, but $BaSO_4$, $BaSO_3$, $BaCO_3$, $BaCrO_4$, and $Ba_3(PO_4)_2$ are insoluble (see Table 32.1). Only $BaSO_4$ is insoluble in acidic solution, whereas the other barium salts that are insoluble in water dissolve in acidic solution. All of the barium salts are white except $BaCrO_4$, which is yellow. Dissolve a small pea-size portion of your unknown in water; then add a few drops of 0.2 *M* barium chloride and stir. Record your observations on your report sheet. Acidify the mixture with a few drops of 6 *M* HNO_3 and stir thoroughly. Record your observations on your report sheet.

Reactions of the Solid Unknown with Concentrated H_2SO_4

Several of the anions form volatile weak acids with or are oxidized by concentrated sulfuric acid. Careful observation of the reaction of concentrated sulfuric acid with your solid unknown can give you further clues as to the presence of specific anions. A summary of the pertinent reactions of concentrated sulfuric acid with various anions is given in Table 32.2.

Place a small pea-size amount of the solid in a dry, small test tube. Add one or two drops of 18 *M* H_2SO_4 and observe everything that occurs, especially the color and odor of gas formed. **(CAUTION: *Concentrated* H_2SO_4 *causes severe burns. Do not get it on your skin or clothing. If you come in contact with it, immediately wash the area with copious amounts of water. You must wear eye protection, as you should at all times when in the laboratory!*)** *Do not* place your nose directly over the mouth of the test tube, but carefully fan gases toward your nose. Then carefully heat the test tube, but not so strongly as to boil the H_2SO_4. **(CAUTION: *If you heat the acid too strongly, it could come shooting out!*)** Note whether brown fumes of NO_2 are produced. **(CAUTION: *Do not look down into the test tube. Do not point the test tube at yourself or at anyone else.*)** SAFETY GLASSES MUST BE WORN.

Specific Tests for Anions

When an anion is indicated by the preliminary tests with $AgNO_3$, $BaCl_2$, or concentrated H_2SO_4, it is confirmed using the appropriate specific test. Make an aqueous solution of the solid unknown and perform the following tests on portions of this solution:

TABLE 32.2 Behavior of Anions with Concentrated Sulfuric Acid, H_2SO_4

A. Cold H_2SO_4	
SO_4^{2-}	No reaction
NO_3^-	No reaction
PO_4^{3-}	No reaction
CO_3^{2-}	A colorless, odorless gas forms. $CO_3^{2-} + 2H^+ \longrightarrow H_2O + CO_2$
Cl^-	A colorless gas forms. It has a sharp, pungent odor; gives an acid test result with litmus; and fumes in moist air. $Cl^- + H^+ \longrightarrow HCl$
Br^-	A brownish-red gas forms. It has a sharp odor, gives an acid test result with litmus, and fumes in moist air. The odor of SO_2 may be detected. $Br^- + H_2SO_4 \longrightarrow HSO_4^- + HBr$ $H_2SO_4 + 2HBr \longrightarrow 2H_2O + SO_2 + Br_2$
I^-	Solid turns dark brown immediately, with a slight formation of violet fumes. The gas has the odor of rotten eggs, gives an acidic test result with litmus, and fumes in moist air. $I^- + H_2SO_4 \longrightarrow HSO_4^- + HI$ $H_2SO_4 + 8HI \longrightarrow H_2S + 4H_2O + 4I_2$ $H_2SO_4 + 2HI \longrightarrow 2H_2O + SO_2 + I_2$
S^{2-}	A colorless gas with the odor of rotten eggs forms, and some free sulfur is deposited. $S^{2-} + H_2SO_4 \longrightarrow SO_4^{2-} + H_2S$ $H_2SO_4 + H_2S \longrightarrow 2H_2O + SO_2 + S$
SO_3^{2-}	A colorless gas with a sharp, choking odor forms. $SO_3^{2-} + H_2SO_4 \longrightarrow SO_4^{2-} + H_2O + SO_2$
CrO_4^{2-}	Color changes from yellow to orange-red. $2K_2CrO_4 + H_2SO_4 \longrightarrow K_2Cr_2O_7 + H_2O + K_2SO_4$
B. Hot Concentrated H_2SO_4	
There are no additional reactions with any of the anions except NO_3^-, which forms brown fumes of NO_2 gas. $4NO_3^- + 4H^+ \longrightarrow 4NO_2 + O_2 + 2H_2O$	

Sulfate Place 10 drops of a solution of the anion unknown in a test tube, acidify with 6 *M* HCl, and add a drop of $BaCl_2$ solution. A white precipitate of $BaSO_4$ confirms SO_4^{2-} ions (see Note 1).

$$Ba^{2+}(aq) + SO_4^{2-}(aq) \longrightarrow \mathbf{BaSO_4}(s)$$

Note 1 Sulfites are slowly oxidized to sulfates by atmospheric oxygen. Consequently, sulfites commonly show a positive test for sulfates.

$$2SO_3^{2-}(aq) + O_2(g) \longrightarrow 2SO_4^{2-}(aq)$$
$$SO_4^{2-}(aq) + Ba^{2+}(aq) \longrightarrow \mathbf{BaSO_4}(s)$$

Sulfite Place 10 drops of a solution of the anion unknown in a test tube, acidify with 6 *M* HCl, add two or three drops of 0.2 *M* $BaCl_2$, and mix thoroughly. If a precipitate ($BaSO_4$) forms, remove it by centrifuging and decanting. To the clear supernatant, add a drop of 3% H_2O_2 (see Note 2). The formation of a white precipitate of $BaSO_4$ confirms SO_3^{2-} ions.

Note 2 H_2O_2 oxidizes sulfite to sulfate.

$$SO_3^{2-}(aq) + H_2O_2(aq) \longrightarrow SO_4^{2-}(aq) + H_2O(l)$$
$$Ba^{2+}(aq) + SO_4^{2-}(aq) \longrightarrow \mathbf{BaSO_4}(s)$$

Any $SO_4^{\ 2-}$ originally present was previously removed as $BaSO_4$ by centrifugation.

Chromate Place two drops of a solution of the anion unknown in a test tube, add 10 drops of water, and make the solution just acidic with 3 *M* HNO_3. In the hood, add five or six drops of ether. (KEEP THE ETHER AWAY FROM FLAMES. It is very flammable.) Then and 1 drop of 3% H_2O_2, stir well, and allow the precipitate to settle. A blue coloration of the top ether layer (see Note 3) confirms CrO_4^{2-} ions (see Note 4).

Note 3 The blue coloration in the ether layer is due to the presence of chromium oxide peroxide, CrO_5.

$$2CrO_4^{2-}(aq) + 2H^+(aq) \longrightarrow Cr_2O_7^{2-}(aq) + H_2O(l)$$
$$Cr_2O_7^{2-}(aq) + 4H_2O_2(aq) + 2H^+(aq) \longrightarrow \mathbf{2CrO_5}(aq) + 5H_2O(l)$$

Note 4 If your anion unknown is not colored or did not indicate CrO_4^{2-} in the H_2SO_4 reaction or if both conditions apply, you may omit this test.

Iodide Place five drops of a solution of the anion unknown in a test tube, add five drops of 6 *M* CH_3COOH, and add two drops of 0.2 *M* KNO_2. A reddish-brown coloration due to the presence of I_2 confirms I^-. If the brown color is very faint, add a few drops of mineral oil and shake well. A violet color in the top (mineral oil) layer confirms I^-.

$$NO_2^-(aq) + H^+(aq) \longrightarrow HNO_2(aq)$$
$$2HNO_2(aq) + 2I^-(aq) + 2H^+(aq) \longrightarrow 2NO(g) + \mathbf{I_2}(aq) + 2H_2O(l)$$

Bromide Iodides will interfere with this test (see Note 5). Place five drops of a solution of the anion unknown in a test tube and add five drops of chlorine water. A brown coloration due to the liberation of Br_2 confirms Br^-. If you shake the solution with a few drops of mineral oil, the brown color will concentrate in the top layer, which is mineral oil. Allow about 20 s for the layers to separate.

$$Cl_2(aq) + 2Br^-(aq) \longrightarrow 2Cl^-(aq) + Br_2(aq)$$

Note 5 Iodide ions, if present, must be removed before you test for bromide. To remove iodide, acidify the solution with 3 *M* HNO_3 and add 2 *M* KNO_2 dropwise, stirring constantly, until there is no further increase in the depth of the brown color. Extract once by shaking with five drops of mineral oil. Discard the layer of mineral oil in a designated receptacle. Boil the water layer carefully until the iodine has been largely driven off. Test the colorless, or nearly colorless, solution for bromide as directed previously.

Nitrate Iodides, bromides, and chromates interfere with this test and must be removed (see Note 6) if they are present. Place 10 drops of a solution of the anion salt in a small test tube, add five drops of $FeSO_4$ solution, and mix the solution. Carefully, without agitating the solution, pour concentrated H_2SO_4 down the inside of the test tube to form two layers. Allow to stand for 1 or 2 min. The formation of a brown ring between the two layers confirms NO_3^- ions. **(CAUTION: *Do not get the H_2SO_4 on yourself or on your clothing. If you do, wash immediately with copious amounts of water.*)**

$$3Fe^{2+}(aq) + NO_3^-(aq) + 4H^+(aq) \longrightarrow 3Fe^{3+}(aq) + NO(aq) + 2H_2O(l)$$

$$NO(aq) + Fe^{2+}(aq)\ (\text{excess}) \longrightarrow \mathbf{Fe(NO)^{2+}}(aq) \qquad (\text{brown})$$

Note 6 Iodides and bromides react with concentrated H_2SO_4 to liberate I_2 and Br_2.

$$SO_4^{2-}(aq) + 8I^-(aq) + 10H^+(aq) \longrightarrow H_2S(g) + 4I_2(aq) + 4H_2O(l)$$

$$SO_4^{2-}(aq) + 2I^-(aq) + 4H^+(aq) \longrightarrow SO_2(g) + I_2(aq) + 2H_2O(l)$$

$$SO_4^{2-}(aq) + 2Br^-(aq) + 4H^+(aq) \longrightarrow SO_2(g) + Br_2(aq) + 2H_2O(l)$$

Chromate ions, if present, will be reduced by Fe^{2+} to green Cr^{3+}.

$$2CrO_4^{2-}(aq) + 2H^+(aq) \longrightarrow Cr_2O_7^{2-}(aq) + H_2O(l)$$

$$Cr_2O_7^{2-}(aq) + 6Fe^{2+}(aq) + 14H^+(aq) \longrightarrow 2Cr^{3+}(aq) + 6Fe^{3+}(aq) + 7H_2O(l)$$

The colors of I_2, Br_2, and Cr^{3+} will interfere with your ability to detect the brown color of $Fe(NO)^{2+}$. Consequently, you must remove I^-, Br^-, and CrO_4^{2-} as follows: Place four drops of the unknown anion solution in a test tube and add 0.2 *M* $Pb(CH_3COO)_2$ until precipitation is complete. Centrifuge and decant, discarding the precipitate ($PbCrO_4$). Treat the supernatant with 0.1 *M* $Hg(CH_3COO)_2$ until precipitation is complete. Centrifuge the solution to separate the precipitate (HgI_2, $HgBr_2$) and treat the supernatant as above to test for nitrate.

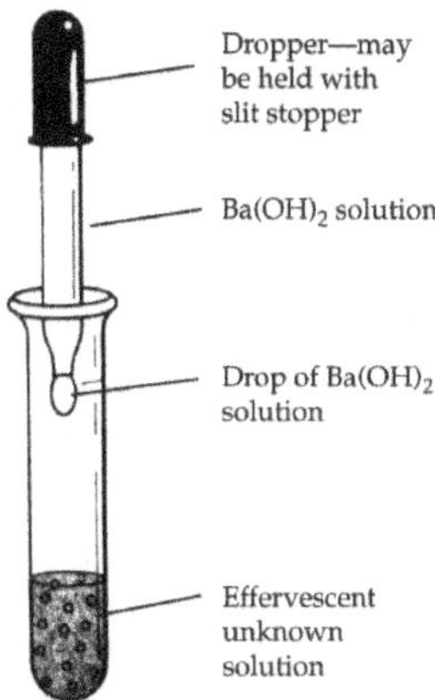

▲**FIGURE 32.6** Test for carbonate ion.

Carbonate Sulfites will interfere with this test for carbonates.

A When sulfites are absent. Place a small amount of the solid anion unknown in a small test tube and add a few drops of 6 *M* H_2SO_4. If a colorless, odorless gas evolves, hold a drop of $Ba(OH)_2$ solution over the mouth of the test tube, using an eyedropper or a Nichrome wire loop (see Figure 32.6). CO_3^{2-} ions are confirmed if the drop turns milky.

$$2H^+(aq) + CO_3^{2-}(aq) \longrightarrow CO_2(g) + H_2O(l)$$

$$CO_2(g) + Ba(OH)_2(aq) \longrightarrow \mathbf{BaCO_3}(s) + H_2O(l)$$

B When sulfites are present. Place a small amount of the solid anion unknown in a test tube and add an equal amount of solid Na_2O_2. This oxidizes SO_3^{2-} to SO_4^{2-}. **(CAUTION: *Sodium peroxide is a very strong oxidizing agent, and all contact with skin should be avoided. If you do get some on yourself, immediately wash it off with large volumes of water.*)** Add three or four drops of water, mix thoroughly, and proceed as in **A** above to test for carbonate.

Sulfide Place a small quantity of the solid unknown anion in a test tube and (*in a hood*) add 10 drops of 6 *M* HCl. Hold a piece of filter paper that has been moistened with 0.2 *M* $Pb(CH_3COO)_2$ over the mouth of the test tube so that any gas that escapes comes into contact with the paper. A brownish or silvery black stain (PbS) on the paper confirms the presence of S^{2-}. If no blackening of the lead acetate occurs after 1 min, heat the tube gently. If still no reaction occurs, add a small amount of granulated zinc to the contents of the tube. If the lead acetate is not darkened, S^{2-} is absent.

$$S^{2-}(aq) + 2H^+(aq) \longrightarrow H_2S(g)$$

$$Pb^{2+}(aq) + H_2S(aq) \longrightarrow \mathbf{PbS}(s) + 2H^+(aq)$$

$$\mathbf{Zn}(s) + HgS(aq) + 2H^+(aq) \longrightarrow Zn^{2+}(aq) + \mathbf{Hg}(l) + H_2S(aq)$$

Chloride Sulfides, bromides, and iodides will interfere with this test (see Note 7). Place 10 drops of a solution of the anion unknown in a test tube and add a drop of $AgNO_3$ solution. A white, curdy precipitate confirms Cl^- ions.

$$Ag^+(aq) + Cl^-(aq) \longrightarrow \mathbf{AgCl}(s)$$

Note 7 Because Ag_2S, AgBr, and AgI are also insoluble in acid solution, this test is not conclusive unless S^{2-}, Br^-, and I^- are definitely shown to be absent. Chromate ions, if present in high concentrations, may also interfere.

Interference from chromate can be eliminated by dilution with 3 *M* HNO_3. Sulfide ions can be removed by boiling the solution after adding two drops of 6 *M* H_2SO_4 until the escaping vapors give no test for H_2S with lead acetate paper. If you suspect the presence of chloride, bromide, iodide, or any combination of these, you may confirm all of them as follows:

Place 10 drops of a solution of the anion unknown in a test tube and add five drops of $AgNO_3$ solution. Centrifuge and discard the supernatant. To the precipitate, add 10 drops of concentrated NH_3 solution and 3 drops of yellow ammonium sulfide solution. Stir the mixture with a glass rod and warm gently until the black Ag_2S coagulates. Centrifuge and discard the precipitate as directed. Transfer the solution to a 50 mL beaker and boil to expel NH_3 and to decompose the ammonium sulfide. When the solution becomes cloudy, add five or six drops of 6 *M* HNO_3 and continue heating until H_2S is completely removed. Then carry out the following tests:

A Place one drop of the solution on a piece of filter paper; add one drop of iodide-free starch solution and one drop of 0.2 *M* KNO_2. A blue color confirms iodide.

$$\mathbf{AgI}(s) + 2NH_3(aq) \longrightarrow Ag(NH_3)_2^+(aq) + I^-(aq)$$

$$2Ag(NH_3)_2^+(aq) + (NH_4)_2S(aq) \longrightarrow \mathbf{Ag_2S}(s) + 2NH_4^+(aq) + 4NH_3(g)$$

$$I^-(aq) + HNO_2(aq) + 2H^+(aq) \longrightarrow 2NO(g) + \mathbf{I_2}(aq) + 2H_2O(aq)$$

$$I_2 + \text{starch} \longrightarrow \text{starch} \cdot I_2 \text{ complex (blue)}$$

B If iodide is present, add three or four drops of 0.2 *M* KNO_2 to the solution in the beaker and boil until no more brown fumes are evolved. If iodide is absent, proceed directly with the test for bromide. Cool the beaker to room temperature by running cold water over its outer surface. To the room temperature beaker, add four or five drops of 3 *M* HNO_3, then a small pea-size piece of solid Na_2O_2. A brown coloration due to the presence of Br_2 confirms Br^-.

$$4H^+(aq) + O_2^{2-}(aq) + 2Br^-(aq) \longrightarrow 2H_2O(l) + \mathbf{Br_2}(aq)$$

C If bromide is present, boil the uncovered contents of the beaker for 30 s to expel the remainder of the bromine. Allow to stand for 30 s and decant the solution into a test tube. Centrifuge if necessary. Add five drops of 0.1 *M* $AgNO_3$. A white precipitate of AgCl confirms chloride.

$$Ag^+(aq) + Cl^-(aq) \longrightarrow \mathbf{AgCl}(s)$$

Phosphate

A When iodides are absent. Place four or five drops of a solution of the anion unknown in a test tube; then add two drops of 6 *M* HNO_3 and three or four drops of $(NH_4)_2MoO_4$ (ammonium molybdate). Mix thoroughly and heat almost to boiling in a water bath for 2 min. Formation, sometimes very slowly, of a finely divided yellow precipitate confirms the presence of phosphate.

$$PO_4^{3-}(aq) + 12MoO_4^{2-}(aq) + 24H^+(aq) + 3NH_4^+(aq) \longrightarrow \mathbf{(NH_4)_3PO_4 \cdot 12MoO_3}(s) + 12H_2O(l)$$

B When iodides are present. The phosphate test gives a green solution when iodide is present. If iodide is known to be present, acidify the solution with 6 *M* HNO_3, add four drops of 0.1 *M* $AgNO_3$, centrifuge to remove AgI, and test the supernatant for phosphate as in **A** above.

Waste Disposal Instructions All waste from this experiment should be placed in appropriate waste containers in the laboratory as directed by your instructor.

NOTES AND CALCULATIONS

Name ______________________________ Desk ______________________

Date ____________________ Laboratory Instructor ______________________

Abbreviated Qualitative Analysis Scheme | 32 Pre-lab Questions

Before beginning Part V of this experiment in the laboratory, you should be able to answer the following questions.

1. What are the names and formulas of the anions to be identified?

2. Describe the behavior of each solid containing the anions toward concentrated H_2SO_4.

3. If you had a mixture of NaCl and Na_2CO_3, would the action of dilute HCl allow you to decide whether both Cl^- and CO_3^{2-} were present?

4. Identify each of the following anions from the information given (consult Appendix D).
   (a) Its barium salt is insoluble in water, but its copper(II) salt is soluble.

   (b) Its copper(II) salt is insoluble in water, but its sodium salt is soluble.

   (c) Its mercury(II) salt is soluble in water, but its mercury(I) salt is insoluble.

5. A mixture of barium and silver salts is insoluble in water. What anion may not be present in this mixture?

6. A white solid unknown is readily soluble in water, and upon treatment of this solution with HCl, a colorless, odorless gas is evolved. This gas reacts with $Ba(OH)_2$ to give a white precipitate. What anion is indicated?

---

7. How would treatment with concentrated H_2SO_4 allow you to distinguish between the following?
   (a) $BaSO_4$ and $Hg(NO_3)_2$

   (b) $CaBr_2$ and Na_3PO_4

   (c) ZnS and $BaSO_4$

   (d) HgI_2 and K_2CrO_4

8. Five different salts each imparted a yellow color to a flame and reacted with cold H_2SO_4 as follows. Identify each salt.
   (a) Effervescence was observed, and the evolved gas was colorless and odorless and did not fume in moist air.

   (b) Effervescence was observed, and the evolved gas was pale reddish-brown, had a sharp odor, and fumed strongly in moist air.

   (c) No effervescence was observed, and the solid changed color from yellow to orange.

   (d) Effervescence was observed, and the evolved gas was colorless, had a sharp odor, and fumed in moist air.

   (e) Effervescence was observed, and the evolved colorless gas had a very sharp odor but did not fume in moist air and did not discolor a piece of filter paper that had been moistened with a solution of lead acetate.

9. How would you test for NO_3^- in the presence of I^-?

10. How would you test for CO_3^{2-} in the presence of SO_3^{2-}?

PART VI: ANALYSIS OF A SIMPLE SALT

(In the procedure that follows note all previous cautions and waste disposal instructions.)

PROCEDURE

Cation Identification

Note the color of your solid and eliminate any cations on the basis of the color. Place a small pea-size portion of your unknown in a test tube and add just enough distilled water to dissolve it. Follow the procedures for the identification of group 1, 3, and 4 cations and identify the cation. Once the cation has been identified, proceed to the next section, "Anion Identification."

Anion Identification

Test the solubility of your solid in distilled water and note its reactions with 18 *M* H_2SO_4, $AgNO_3$, and $BaCl_2$ solutions. Compare your results with the information in Table 32.2. Once the anion has been identified, perform the specific test for that anion to confirm its presence.

NOTES AND CALCULATIONS

Name ______________________________ Desk ______________

Date ______________ Laboratory Instructor ______________________

Unknown no. ______________

REPORT SHEET | EXPERIMENT

Abbreviated Qualitative Analysis Scheme | 32

PART I: GROUP 1 CATIONS

Record the reagent used in each step, your observations, and the equations for each precipitation reaction.

Procedure	*Reagent*	*Observations*	*Equations*	*Mark (+) if observed in unknown*
G1-1				
G1-2				
G1-3				

Cations in group 1 unknown ______________________

NOTES AND CALCULATIONS

Name ______________________________ Desk ______________

Date ______________ Laboratory Instructor ______________________

Unknown no. ______________

REPORT SHEET | EXPERIMENT

Abbreviated Qualitative Analysis Scheme | 32

PART II: GROUP 2 CATIONS

Record the reagent used in each step, your observations, and the equations for each precipitation reaction.

Procedure	*Reagent*	*Observations*	*Equations*	*Mark (+) if observed in unknown*
G2-1				
G2-2				
G2-3				
G2-4				
G2-5				

Cations in group 2 unknown ______________________________

NOTES AND CALCULATIONS

Name ____________________ Desk ____________________

Date ____________________ Laboratory Instructor ____________________

Unknown no. ____________________

REPORT SHEET | EXPERIMENT

Abbreviated Qualitative Analysis Scheme

32

PART III: GROUP 3 CATIONS

Record the reagent used in each step, your observations, and the equations for each precipitation reaction.

Procedure	*Reagent*	*Observations*	*Equations*	*Mark (+) if observed in unknown*
G3-1				
G3-2				
G3-3				
G3-4				
G3-5				

Procedure	*Reagent*	*Observations*	*Equations*	*Mark (+) if observed in unknown*
G3-6				
G3-7				

Cations in group 3 unknown ______________________________

Name ______________________ Desk ______________

Date ______________ Laboratory Instructor ______________________

Unknown no. ______________

REPORT SHEET | EXPERIMENT

Abbreviated Qualitative Analysis Scheme | 32

PART IV: GROUP 4 CATIONS

Record the reagent used in each step, your observations, and the equations for each precipitation reaction.

Procedure	*Reagent*	*Observations*	*Equations*	*Mark (+) if observed in unknown*
G4-1				
G4-2				
G4-3				
G4-4				

Cations in group 3 unknown ______________________

NOTES AND CALCULATIONS

Name ______________________ Desk ______________

Date ______________ Laboratory Instructor ______________________

Unknown no. ______________

REPORT SHEET | EXPERIMENT

Abbreviated Qualitative Analysis Scheme

32

PARTS I–IV: GENERAL CATION UNKNOWN

Record the reagent used in each step, your observations, and the equations for each precipitation reaction.

Procedure	*Reagent*	*Observations*	*Equations*

Cations in general unknown ______________________

NOTES AND CALCULATIONS

Name ______________________ Desk ______________

Date ______________ Laboratory Instructor ______________________

Unknown no. ______________

REPORT SHEET | EXPERIMENT

Abbreviated Qualitative Analysis Scheme | 32

PART V: ANIONS

1. **EXAMINATION OF THE SOLID**

   Color? ______________ Homogeneous? ______________

2. **SOLUBILITY**

	Water	**6 *M* HNO_3**	**6 *M* HCl**	**6 *M* NH_3**
	______	______	______	______

3. **REACTIONS WITH $AgNO_3$ AND $BaCl_2$ SOLUTIONS**

$AgNO_3$ *Observations*

$AgNO_3$ + 6 *M* HNO_3 ______________________

Anions indicated ______________________

Anions eliminated ______________________

$BaCl_2$ *Observations*

$BaCl_2$ + 6 *M* HNO_3 ______________________

Anions indicated ______________________

Anions eliminated ______________________

4. **REACTION OF SOLID WITH CONCENTRATED H_2SO_4**

Observations

Anions indicated ______________________

Anions eliminated ______________________

5. SPECIFIC TEST RESULTS

Test made	*Observation and equation*	*Anion confirmed*

6. ANIONS CONFIRMED IN UNKNOWN SOLID ____________________

Name ______________________________ Desk ______________________

Date ____________________ Laboratory Instructor ______________________

REPORT SHEET | EXPERIMENT

Abbreviated Qualitative Analysis Scheme | 32

PART VI: SIMPLE SALTS

Cation in Unknown Solid

1. Color of the solid ______________________

   Cations indicated ______________________

   Cations eliminated ______________________

2. For each step, record the procedure number, the reagent(s) used, your observations, and the corresponding equation(s) for each reaction.

Procedure	*Reagent*	*Observation(s)*	*Equation(s)*

Anion in Unknown Solid

1. Solubility of solid (see Appendix C):

   Water ________ 6 *M* HNO_3 ________ 6 *M* HCl ________ 6 *M* NH_3 ________

2. Reactions of solid with concentrated (18 M) H_2SO_4 (observations):

   With cold acid ________________

   With hot acid (only if necessary) ________________

   Anions indicated ________________

   Anions eliminated ________________

3. Reactions of unknown solution with $AgNO_3$ and $BaCl_2$ solutions (observations):

   With 0.1 *M* $AgNO_3$ (obs) ________________

   With 0.1 *M* $AgNO_3 + HNO_3$ (obs) ________________

   Anions indicated ________________

   Anions eliminated ________________

   With 0.2 *M* $BaCl_2$ ________________

   With 0.2 *M* $BaCl_2 + HNO_3$ ________________

   Anions indicated ________________

   Anions eliminated ________________

4. Specific test results

   Test made for (anion) ________________

   Observation(s) and equation(s) ________________

   ____

   ____

   ____

   The anion in unknown solid ________________

   The chemical formula of unknown solid ________________

QUESTIONS

1. Write balanced molecular and net ionic equations for the reactions occurring in each of the following solutions.

   (a) HCl is added to solid $Mg_3(PO_4)_2$.

(b) HBr is added to solid FeS.

(c) H_2O_2 is added to a solution containing SO_3^{2-}.

(d) $(NH_4)_2S$ solution is added to solid AgBr.

(e) HCl is added to a solution of Na_2CrO_4.

2. What conclusions can be drawn from the following observations?
   (a) An acidic solution of unknown anions forms no precipitate when $BaCl_2$ solution is added.

   (b) A solution of unknown anions forms a pale yellow precipitate when $AgNO_3$ solution is added.

   (c) Addition of 6 *M* NH_3 to a pale blue solution of an unknown gave a deep blue solution and a white precipitate.

3. Why is it more difficult to identify all of the components of a mixture than to identify the cation and anion in a simple salt?

4. The observation that a solid is colorless allows you to suggest that a broad category of elements is probably not present in this solid. What is the general name for this category of elements?

5. A white compound dissolves in water to give an acidic solution. Addition of either NaOH or NH_3 to this colorless solution produces a flocculent white precipitate that is soluble in acid. If an excess of NaOH is added to a water solution of this compound, a colorless precipitate forms that later dissolves as more NaOH is added. Addition of $AgNO_3$ to a water solution of the compound gives no reaction. The solid shows no reaction with concentrated H_2SO_4, and addition of a barium chloride solution to a solution of the compound gives a white precipitate that is insoluble in both acid and base. Identify the compound.

6. A white salt dissolves in water to give a neutral solution. No precipitate is formed when the aqueous solution is treated with the buffer NH_4Cl, NH_3. No precipitate forms when $(NH_4)_2S$ is added to the buffer solution. However, a yellow precipitate forms when K_2CrO_4 is added to a solution of the salt that is acidified with CH_3COOH, and the precipitate dissolves in concentrated HCl. When $AgNO_3$ is added to a solution of the salt that is acidified with HNO_3, a white precipitate forms. What is the salt?

Experiment

Molarity, Dilutions, and Preparing Solutions

OBJECTIVE

To become familiar with molarity and the methods of preparing solutions.

APPARATUS AND CHEMICALS

Apparatus

spectrophotometer and plastic cuvettes
100-mL volumetric flask, with cap
50-mL volumetric flask, with caps (4)
25-mL volumetric flask, with caps (2)
10-mL volumetric flask
volumetric pipets (5 mL, 10 mL, 15 mL, 20 mL)
pipet bulb
analytical balance, capable of massing a minimum of 100 g
top-loading balance
500-mL beaker
weigh boat
spatula
plastic funnel
label tape

Chemicals

copper(II) sulfate anhydrous

DISCUSSION

A *solution* is a homogeneous mixture of two or more substances. A solution is composed of a *solute* and a *solvent*. The solute is a dissolved solid, liquid, or gas. The solvent, which dissolves the solute, may also be a solid, liquid, or gas. For much of the field of chemistry, the solvent is often a liquid. Water is a common liquid solvent and will be used in this experiment to form *aqueous solutions*.

The solvent is present in a greater amount than the solute. The term *concentration* is used to designate the quantity of solute dissolved in a certain amount of solvent or solution. A solution with a greater amount of solute in a specific amount of solvent has a larger concentration than a solution with a smaller amount of solute in a specific amount of solvent. There are several ways to quantitatively express concentrations of solutions. The term *molarity* (symbol *M*) expresses the concentration as the number of moles of solute per liter of solution, as shown in equation [1].

$$Molarity = \frac{\text{moles of solute(mol)}}{\text{volume of solution(L)}} \qquad [1]$$

A 0.50 molar solution has 0.50 mole of solute per liter of solution.

GIVE IT SOME THOUGHT

If the temperature of the solution is raised and the volume of the solution increases, what happens to the molarity of the solution?

EXAMPLE 33.1

Calculate the molarity of a solution prepared by dissolving 12.2 g of barium chloride ($BaCl_2$) in enough water to form a 150.0-mL solution.

SOLUTION: The solute is barium chloride and the mass is given. The moles of $BaCl_2$ can be calculated. The volume of the solution is given in milliliters, which needs to be converted to liters.

The molar mass of $BaCl_2$is $137.33+(2\times35.453)=208.236$ g/mol.

The moles of $BaCl_2$ is obtained by dividing the mass by the molar mass:

$$12.2 \text{ g BaCl}_2 \times \frac{1 \text{ mol BaCl}_2}{208.232 \text{ g BaCl}_2} = 0.0586 \text{ mol BaCl}_2$$

The volume of the solution is converted to liters:

$$150 \text{ mL} \times \frac{1 \text{ L}}{1000 \text{ mL}} = 0.150 \text{ L}$$

The molarity is calculated as follows:

$$\frac{0.0586 \text{ mol BaCl}_2}{0.150 \text{ L solution}} = 0.391M$$

The concentration of a solution may be changed by changing the volume of the solution. If an aqueous solution of sodium chloride were left open to the air and some water evaporated, then it would become more concentrated. If water were added to the solution, then it would become less concentrated. The process of adding additional solvent to a solution is called *dilution*. Adding or removing solvent does not change the amount of the solute in the solution. The moles (mol) of solute before dilution equals the moles of solute after dilution. Given M = moles/volume (V) of solution, moles $= M \times V$. Considering a concentrated and a diluted solution, $\text{mol}_{\text{conc}} = \text{mol}_{\text{dil}}$. Then Equation [2], which is known as a *dilution equation*, may be derived by substitution.

$$M_{\text{conc}}V_{\text{conc}} = M_{\text{dil}}V_{\text{dil}} \qquad [2]$$

EXAMPLE 33.2

How many milliliters of 8.0 M HCl are needed to make 250 mL of 0.25 M HCl?

SOLUTION: The more concentrated solution is diluted to form a less concentrated solution. The concentration of the more concentrated solution is known, and the concentration and volume of the more dilute solution are given. Three of the four variables in the equation $M_{\text{conc}}V_{\text{conc}} = M_{\text{dil}}V_{\text{dil}}$ are therefore known. Calculate the volume of the more concentrated solution needed to prepare the more dilute solution.

Rearrange the equation to solve for $V_{\text{conc}} = \dfrac{M_{\text{dil}}V_{\text{dil}}}{M_{\text{conc}}}$.

$$M_{\text{dil}} = 0.25M, V_{\text{dil}} = 250 \text{ mL}, M_{\text{conc}} = 8.0M$$

$$V_{\text{conc}} = \frac{M_{\text{dil}}V_{\text{dil}}}{M_{\text{conc}}} = \frac{(0.25M)(250 \text{ mL})}{(8.0M)} = 7.8 \text{ mL}$$

Use the dilution equation cautiously, since it works only for dilutions and should not be used for calculations that involve mixing two or more solutions or for any type of titration or chemical reaction.

The preparation of low-concentration and/or low-volume solutions may be challenging because of the difficulty of accurately obtaining a small amount of a solute. Consider the following example and preparation of a low-concentration solution.

EXAMPLE 33.3

Cisplatin is a drug for treatment of cancer with the chemical formula $Pt(NH_3)_2Cl_2$. The maximum concentration is reported as 0.5 mg/mL. Convert this concentration to molarity. Calculate the amount of $Pt(NH_3)_2Cl_2$ in grams that is needed to make a 10.0-mL solution at that maximum concentration.

SOLUTION: The molar mass of $Pt(NH_3)_2Cl_2$ is $195.08+(2\times14.007)+(6\times1.0079)+(2\times35.453)=300.047$ g/mol.

The desired concentration is 0.5 mg/mL, or 0.5 mg per 1 mL.

Convert the concentration to units of molarity. First, convert from milligrams to moles of $Pt(NH_3)_2Cl_2$:

$$0.5 \text{ mg } Pt(NH_3)_2Cl_2 \times \frac{1\times10^{-3}\text{ g}}{1\text{ mg}} \times \frac{1\text{ mol}}{300.047\text{ g}} = 1.67\times10^{-6}\text{ mol } Pt(NH_3)_2Cl_2$$

The volume is given in milliliters, so convert to liters.

$$1.00\text{ mL} \times \frac{1\times10^{-3}\text{ L}}{1\text{ mL}} = 0.00100\text{ L}$$

The molarity is

$$M = \frac{1.67\times10^{-6}\text{ mol}}{0.00100\text{ L}} = 0.00167\ M$$

Next, find the amount of $Pt(NH_3)_2Cl_2$ needed to prepare a 10.0-mL solution with a concentration of 0.00167 *M*. The volume of the solution is $(10.0\text{ mL})\times(1\times10^{-3}\text{ L/1 mL}) = 0.0100\text{ L}$.

Rearranging the molarity equation yields $M = \text{mol}/V$, so

$$\text{mol} = M\times V = (0.00167\ M)\times(0.0100\text{ L}) = 1.67\times10^{-5}\text{ mol } Pt(NH_3)_2Cl_2$$

Determine the mass from the moles of $Pt(NH_3)_2Cl_2$ using the molar mass.

$$1.67\times10^{-5}\text{ mol } Pt(NH_3)_2Cl_2 \times \frac{300.047\text{ g } Pt(NH_3)_2Cl_2}{1\text{ mol } Pt(NH_3)_2Cl_2} = 0.0050\text{ g } Pt(NH_3)_2Cl_2$$

Accurately obtaining small amounts of reagents can be difficult. For this reason, it is easier to prepare a solution of higher concentration and then perform multiple dilutions. This is known as *serial dilution*.

EXAMPLE 33.4

A stock solution "S" is prepared by obtaining 0.125 g $Pt(NH_3)_2Cl_2$ and dissolving to form a 10.00-mL solution. Another solution ("A") is prepared by taking 2.00 mL of the stock solution and diluting to form a 10.00-mL solution. Another solution ("B") is prepared by taking 2.00 mL of solution "A" and dissolving to form a 10.00-mL solution. A schematic of the serial dilution is shown in Figure 33.1. Determine the concentration of $Pt(NH_3)_2Cl_2$ in the "B" solution.

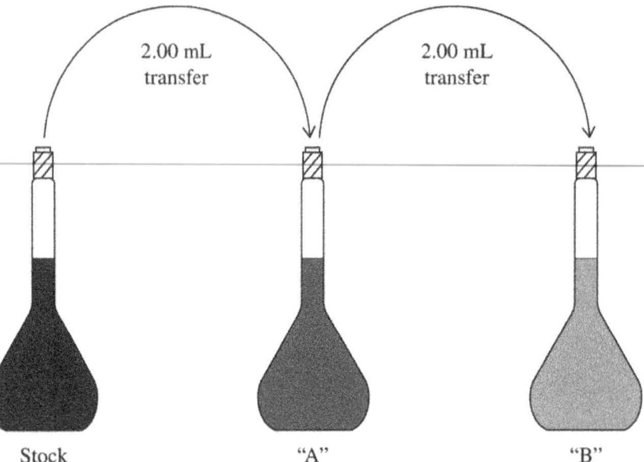

▲**FIGURE 33.1** Serial dilution. A quantitative portion of the more concentrated solution is transferred and diluted with the solvent to make a solution of lower concentration.

SOLUTION: The molar mass of $Pt(NH_3)_2Cl_2$ is

$$195.08 + (2\times 14.007) + (6\times 1.0079) + (2\times 35.453) = 300.047 \text{ g/mol}$$

Determine the moles of $Pt(NH_3)_2Cl_2$ from the mass:

$$0.125 \text{ g } Pt(NH_3)_2Cl_2 \times \frac{1 \text{ mol}}{300.047 \text{ g}} = 0.000417 \text{ mol}$$

Calculate the molarity of the stock solution. The volume is

$$10.00 \text{ mL} \times \frac{1\times 10^{-3} \text{ L}}{1 \text{ mL}} = 0.01000 \text{ L}$$

The molarity of the stock solution is

$$M = \frac{\text{mol}/V = 0.000417 \text{ mol}}{0.01000 \text{ L}} = 0.0417\,M$$

Calculate the molarity of solution "A." It is being prepared using the stock solution and diluting. $M_{conc}V_{conc} = M_{dil}V_{dil}$, so solve for M_{dil}.

$$M_{dil} = \frac{M_{conc}V_{conc}}{V_{dil}} = 0.0417\,M \times \frac{2.00 \text{ mL}}{10.00 \text{ mL}} = 0.00833\,M$$

Calculate the molarity of solution "B." It is being prepared by using solution "A" and diluting.

For this dilution, M_{conc} is the concentration of the "A" solution.

$$M_{dil} = \frac{M_{conc}V_{conc}}{V_{dil}} = 0.00833\,M \times \frac{2.00 \text{ mL}}{10.00 \text{ mL}} = 0.00167\,M$$

The concentration solution "B" is 0.00167 M, the same as that found in Example 33.3. It is much easier to accurately obtain 0.125 g of $Pt(NH_3)_2Cl_2$ and perform serial dilutions than to try to accurately obtain 0.0050 g and prepare a single low-concentration solution.

There are several methods to measure the concentration of a solution. One method relies on the fact that solutions of some compounds absorb light. A solution with a larger concentration absorbs more light than a solution with a smaller concentration. The relationship between the absorbance of a solution and its concentration is given by Beer's Law, equation [3], where A is the *absorbance* of the solution (unitless), ε is the molar absorptivity constant $(\text{L mol}^{-1}\text{cm}^{-1})$, b is the path length the light travels through the sample (cm^{-1}), and c is the molar concentration of the solution (mol L^{-1}).

$$A = \varepsilon bc \qquad [3]$$

The absorbance of a solution is directly proportional to its concentration. The molar absorptivity constant is a property inherent in a given chemical compound.

The absorbance of a solution may be experimentally measured using a spectrophotometer, which measures the quantity of light that passes through a solution. The *percent transmission* (%T) is the amount of light transmitted; at 100% T no light is absorbed, and at 0% all of the light is absorbed. The relationship between absorbance and percent transmission is given in Equation [4].

$$A = 2 - \log(\%\text{T}) \qquad [4]$$

EXAMPLE 33.5

The percent transmission of a solution is determined to be 23.5%. Calculate the absorbance of the solution.

SOLUTION: The relationship between absorbance and percent transmission is $A = 2 - \log(\%\text{T})$.

$$A = 2 - \log(\%\text{T}) = A = 2 - \log(23.5) = 0.629$$

Using a spectrophotometer involves placing the solution in a transparent plastic or glass cuvette. The light source is a tungsten lamp that emits white light. A specific wavelength can be selected using optics of the spectrophotometer. The wavelength is generally chosen at the wavelength of maximum absorption for the solution.

A standard curve can be made by making a series of solutions of known concentration and preparing a graph of absorbance versus molarity. The slope of this line may be used to determine the concentration of an unknown solution from its absorbance.

EXAMPLE 33.6

A series of $NiCl_2$ solutions were prepared at different concentrations, and their percent transmissions were measured, as shown in the following table. A 1-cm-path-length cuvette was used to contain the solution. The percent transmission of a solution of $NiCl_2$ with an unknown concentration was 22.7%. Determine the concentration of the unknown $NiCl_2$ solution.

	"A"	"B"	"C"	"D"	**Unknown**
Molarity, M	0.0330	0.0206	0.0070	0.0023	–
Percent transmission, %T	16.3	33.3	68.5	88.1	22.7

SOLUTION: Calculate the absorbance of each, using the equation $A = 2 - \log(\%\text{T})$.

$$\text{Calculation for solution "A": } A = 2 - \log(16.3) = 0.788$$

The absorbance values are shown in the following table.

	"A"	"B"	"C"	"D"	**Unknown**
Absorbance	0.788	0.477	0.164	0.0549	0.644

Plot absorbance versus concentration for the solutions of known concentration, and include the trend line equation.

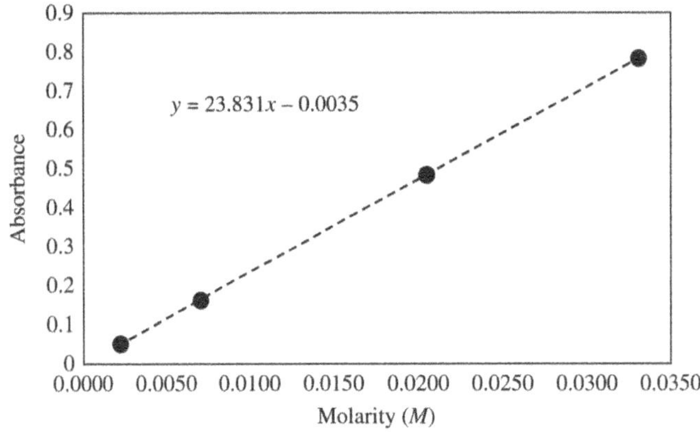

The slope of the line divided by the path length (1 cm^{-1}) is the molar absorptivity, $\varepsilon = 23.83\ M^{-1}\ cm^{-1}$.

Solve $A = \varepsilon bc$ for concentration.

$$c = \frac{A}{\varepsilon b} = \frac{0.644}{23.83\ M^{-1}\ cm^{-1} \times 1\ cm} = 0.0270\,M$$

To check this result, note that this concentration is between the concentrations of solutions A and B.

PROCEDURE

A. Preparing a Stock Solution of Known Concentration

Prior to preparing the stock solution, turn on the spectrophotometer so it may warm up. Obtain approximately 4 g of $CuSO_4$ using a top-loading balance. Obtain a clean and dry 100-mL volumetric flask and determine the mass of the empty volumetric flask. Add the $CuSO_4$ to the flask with the aid of a funnel. Determine the mass of the volumetric flask and $CuSO_4$. Add approximately 75 mL deionized (DI) H_2O to the flask, and swirl to dissolve the $CuSO_4$. Depending on the size of the crystals of $CuSO_4$, it may take several minutes to fully dissolve. When all of the crystals have dissolved, finish filling the volumetric flask to the calibration mark. The use of a plastic transfer pipet or dropper makes dropwise control possible near the mark. Add a cap to the flask, and invert and swirl several times to ensure the solution is entirely mixed. Label the stock solution as "S." The concentration of the stock solution may be calculated from the moles of the copper sulfate and the volume of the solution. Calculate the molarity of the stock solution.

B. Preparing the Spectrophotometer for use

Adjust the wavelength setting to 625 nm on the spectrophotometer, which is the analytical wavelength and the wavelength of maximum absorption for $CuSO_4$. Obtain a plastic cuvette and, using a marker, make a small mark on the top of the cuvette. This will aid in inserting the cuvette into the spectrophotometer in the same orientation for each measurement. Fill the cuvette with DI H_2O until it is about 80% full, and then insert it into the sample holder of the spectrophotometer. Adjust the spectrophotometer to 100.0% transmission; for some models this may be done with the push of a button.

C. Measuring the Absorbance of the Stock Solution

Empty the DI H_2O from the cuvette. The cuvette probably has residual water on the walls or bottom of the cuvette, so the cuvette should be rinsed several times with several small portions of the stock solution. Fill the cuvette with the stock $CuSO_4$ until it is about 80% full, and then insert it into the sample holder of the spectrophotometer, using the same orientations as before. Record the percent transmission. Calculate the absorbance of the stock solution of $CuSO_4$ using Equation [4]. Remove the cuvette and empty the contents into a waste beaker.

D. Preparing Volumetric Pipets for use

For the preparation of solutions with an accurate concentration, the highest-quality results are obtained when clean and accurate glassware is used. Glassware such as a beaker or Erlenmeyer flask is not typically used, because those are less accurate than a volumetric flask and volumetric pipet. The volumetric glassware has a mark around the circumference to indicate the fill point. For aqueous solutions, the bottom of the meniscus should be level with the mark. Before use, the pipet should be washed with soapy water and rinsed with DI H_2O. Use the pipet bulb to blow out any of the rinse water, and wipe the pipet tip dry. There might be some residual water inside the pipet, which would dilute the solution slightly. Into a clean beaker, pour a small amount of the solution that is to be transferred by pipet. Immerse the tip of the pipet in the solution, and fill the pipet about 10% full. Remove the pipet bulb and quickly place your finger over the top of the pipet. Dry the pipet tip, move the pipet to a horizontal position, and then rotate the pipet to allow the inside of the pipet to be coated by the solution. Return the pipet to a vertical orientation, and allow the solution to drain into a waste beaker. The inside of the pipet is now wetted with the solution to be transferred and is ready for use.

E. Construction of a Calibration Graph

A plot of absorbance (A) versus molarity (M) will be prepared to construct a calibration graph. A total of five solutions will be used, the stock solution and four dilutions, as data points. Obtain and clean four 50-mL volumetric flasks, and individually label each (A, B, C, D). Using the appropriate pre-rinsed pipets and stock solution, add 5.00 mL to volumetric flask A, 10.00 mL to volumetric flask B, 15.00 mL to volumetric flask C, and 20.00 mL to volumetric flask D. Dilute each flask to the calibration mark using DI H_2O. Cover with a cap, and then mix each thoroughly by inverting and swirling for a minute or so. If you are sharing a spectrophotometer, be sure to re-zero the instrument with DI H_2O in your cuvette before each use. Rinse the cuvette with solution D, and then measure and record the percent transmission. Repeat this for solutions C, D, and A. Calculate the molarity of each solution. Calculate the absorbance of each solution using Equation [4]. After the laboratory session, prepare a plot of absorbance (A) versus molarity (M) using the five data points. Include the trend line equation for a linear fit.

GIVE IT SOME THOUGHT

Does the "lightest" or the "darkest" solution have the highest percent transmission? The highest absorbance?

F. Serial Dilution Preparation of a Solution of Lower Concentration

A serial dilution will be done by using the stock solution and performing two dilutions to obtain a solution of lower concentration. Label two cleaned 25.00-mL volumetric flasks (X and Y). To a 25.00-mL volumetric flask "X" add 10.00 mL of the stock solution, using the appropriate pipet. Dilute to the calibration mark with DI H_2O and then cover. Swirl and invert to ensure thorough mixing. Pipet 10.00 mL of the solution in "X" to volumetric flask "Y." Dilute to the calibration mark with DI H_2O and then cover. Swirl and invert to ensure thorough mixing. Measure the percent transmission of the solution in volumetric flask "Y." Calculate the concentration of the serial-diluted sample in volumetric flask "Y." Determine the molarity of the solution using the calibration graph and the measured absorbance. Assuming the calibration graph to be the accepted value, calculate a percent error for the molarity of the solution in "Y."

G. Directly Preparing a Solution of Lower Concentration

Using an analytical balance, obtain 0.0650 g $CuSO_4$ in a small plastic weigh boat. Add it to a clean 10.00-mL volumetric flask labeled "Z." Fill with about 7 mL DI H_2O and swirl to dissolve the $CuSO_4$. When all of the $CuSO_4$ crystals have dissolved, finish filling the volumetric flask to the calibration mark. Swirl and invert to ensure thorough mixing. Measure the percent transmission of the solution in volumetric flask "Z." Calculate the concentration of the solution, using the mass of the $CuSO_4$ and the volume of the solution. Determine the molarity of the solution, using the equation obtained from the calibration graph and the absorbance calculated from the measured percent transmission. Assuming the calibration graph to be the accepted or true value, calculate a percent error for the molarity of the solution in "Z."

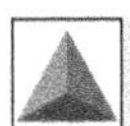

GIVE IT SOME THOUGHT

Why is it difficult to precisely obtain small quantities of a sample?

H. Determining the Concentration of an Unknown Solution

Obtain an unknown solution and label it with the ID given to you by your instructor (e.g., "U1"). Measure the percent transmission of the unknown solution. Determine the molarity of the solution, using the equation obtained from the calibration graph and the absorbance calculated from the measured percent transmission.

Clean and rinse all glassware thoroughly.

Waste Disposal Instructions: Dispose of all waste as indicated by your instructor.

Name ______________________________ Desk ______________

Date ______________ Laboratory Instructor ______________________

Molarity, Dilutions, and Preparing Solutions | 33 Pre-lab Questions

Before beginning this experiment in the laboratory, you should be able to answer the following questions.

1. Sodium chloride is dissolved in water to form a solution. Identify the solute and the solvent.

2. An aqueous solution of sucrose ($C_{12}H_{22}O_{11}$) is prepared by dissolving 6.5532 g in sufficient deionized water to form a 50.00-mL solution. Calculate the molarity of the solution.

3. An aqueous solution of iron(II) sulfate ($FeSO_4$) is prepared by dissolving 2.85 g in sufficient deionized water to form a 25.00-mL solution. Calculate the molarity of the solution.

4. Determine the absorbance if the percent transmission is 68.6%.

5. A pipet is used to transfer 5.00 mL of a 1.25 *M* stock solution in flask "S" to a 25.00-mL volumetric flask "B," which is then diluted with DI H_2O to the calibration mark. The solution is thoroughly mixed. Next, 2.00 mL of the solution in volumetric flask "A" is transferred by pipet to a 50.00-mL volumetric flask "B" and then diluted with DI H_2O to the calibration mark. Calculate the molarity of the solution in volumetric flask "B."

Name ______________________ Desk ______________

Date ______________ Laboratory Instructor ______________________

REPORT SHEET | EXPERIMENT

Molarity, Dilutions, and Preparing Solutions | 33

A. Preparing a Stock Solution of Known Concentration

1. Mass of 100-mL volumetric flask, g ______________
2. Mass of 100-mL volumetric flask and $CuSO_4$, g ______________
3. Mass of $CuSO_4$, g ______________
4. Volume of stock solution, mL ______________
5. Molarity of stock solution "S," *M* ______________

   (Show calculations.)

B. Preparing the Spectrophotometer for use

1. Percent transmission of DI H_2O, %T ______________
2. Absorbance of DI H_2O, *A* ______________

C. Measuring the Absorbance of the Stock Solution

1. Percent transmission of stock solution, %T ______________
2. Absorbance of stock solution, *A* ______________

E. Construction of a Calibration Graph

	"A"	"B"	"C"	"D"
Volume of stock solution, mL	5.00	10.00	15.00	20.00
Volume of prepared solution, mL	50.00	50.00	50.00	50.00
Molarity, *M*				
Percent transmission, %T				
Absorbance, *A*				

F. Serial Dilution Preparation of a Solution of Lower Concentration

1. Percent transmission of solution "Y," %T ______________________
2. Absorbance of solution "Y," *A* ______________________
3. Molarity of solution "Y" calculated from dilutions, *M* ______________________
4. Molarity of solution "Y" determined from calibration graph, *M* ______________________
5. Percent error of molarity of solution "Y," % ______________________

G. Directly Preparing a Solution of Lower Concentration

1. Mass of $CuSO_4$, g ______________________
2. Molarity of solution "Z," *M* ______________________
3. Percent transmission of solution "Z," %T ______________________
4. Absorbance of solution "Z," *A* ______________________
5. Molarity of solution "Z" determined from calibration graph, *M* ______________________
6. Percent error of molarity of solution "Z," % ______________________

H. Unknown Solution

1. Identity label of unknown ______________________
2. Percent transmission of unknown solution, %T ______________________
3. Absorbance of unknown solution, A ______________________
4. Molarity of solution determined from calibration graph, M ______________________

QUESTIONS

1. A stock solution of potassium permanganate ($KMnO_4$) was prepared by dissolving 13.0 g $KMnO_4$ with DI H_2O in a 100.00-mL volumetric flask and diluting to the calibration mark. Determine the molarity of the solution.

2. By pipet, 15.00 mL of the stock solution of potassium permanganate ($KMnO_4$) from Question 1 was transferred to a 50.00-mL volumetric flask and diluted to the calibration mark. Determine the molarity of the resulting solution.

3. Describe how to prepare 100.00 mL of a 0.250 M sodium chloride (NaCl) solution using sodium chloride powder.

4. Describe how to prepare 50.00 mL of a 3.00 M hydrochloric acid solution using 12.0 M HCl.

Experiment

Solubility and Thermodynamics

OBJECTIVE

To become familiar with the relationship between free energy and the equilibrium constant by determining the solubility product constant of calcium hydroxide in aqueous solution and to use the K_{sp} to calculate its thermodynamic quantities.

APPARATUS AND CHEMICALS

Apparatus

Bunsen burner and hose
Ring stand, iron ring, and wire gauze
100 mL beaker
125 mL Erlenmeyer flask
10 mL pipet
25 mL buret

Chemicals

saturated calcium hydroxide solution
standardized 0.01*XX* M hydrochloric acid solution
bromothymol blue indicator solution

DISCUSSION

Thermodynamics is the study of energy and how it is converted from one form to another (Section 19.1). Thermodynamic principles are included in physics, chemistry, biology, engineering, and material science courses, as well as in many other scientific disciplines. In any situation where heat or energy are considered, thermodynamics plays a major role. Thermodynamic principles are considered in a variety of applications—from heating your apartment or home to designing materials to line your jacket to keep you warm on a cold winter day to finding the proper mixture of fuel to power your automobile.

In this experiment, an acid–base titration will be performed to determine the K_{sp} of $Ca(OH)_2$. This experimentally determined value will be used to investigate the thermodynamic properties of the equilibrium established between the calcium and hydroxide ions in equilibrium with solid calcium hydroxide.

Under any set of conditions, the free-energy change (Section 19.7) is given by the expression

$$\Delta G = \Delta G^\circ + RT \ln Q \qquad [1]$$

If a saturated solution is prepared, an equilibrium exists between the solid and the ions in solution. Calcium hydroxide is a slightly soluble salt, and in a saturated solution, this equilibrium may be represented as follows:

$$Ca(OH)_2(s) \rightleftharpoons Ca^{2+}(aq) + 2\ OH^-(aq) \qquad [2]$$

The solubility-product constant expression (Section 17.4) for Equation [2] is

$$K_{sp} = [Ca^{2+}][OH^-]^2 \qquad [3]$$

At equilibrium, several relationships exist that enable Equation [1] to be simplified. The ΔG is equal to 0, and the reaction quotient, Q, is equal to the K_{sp}. Plugging these values into Equation [1] gives:

$$0 = \Delta G° + RT \ln K_{sp} \qquad [4]$$

or

$$\Delta G° = -RT \ln K_{sp} \qquad [5]$$

These relationships illustrate how the Gibbs free energy is determined using the solubility-product constant (Section 19.7).

GIVE IT SOME THOUGHT

a. When is the ln x positive?
b. When is the ln x negative?
c. When does the ln x = 0?
d. How does the magnitude of the K_{sp} influence the spontaneity of a reaction?

EXAMPLE 34.1

A titration was performed, and it was determined that the concentration of hydroxide ions in a saturated barium hydroxide solution at 25 °C was 0.2154 *M*. What is $\Delta G°$ at this temperature?

SOLUTION: First, write out the equilibrium expression for dissolution of barium hydroxide in aqueous solution.

$$Ba(OH)_2(s) \rightleftharpoons Ba^{2+}(aq) + 2\ OH^-(aq)$$

If the concentration of $[OH^-]$ is known, the $[Ba^{2+}]$ can be calculated using the balanced equilibrium expression.

$$0.2154 \text{ mol } OH^-/L \times 1 \text{ mol } Ba^{2+}/2 \text{ mol } OH^- = 0.1077 \text{ mol } Ba^{2+}/L$$

$$[Ba^{2+}] = 0.1077\ M$$

From this, the solubility-product, or K_{sp}, can be written and its value determined as follows:

$$K_{sp} = [Ba^{2+}][OH^-]^2$$
$$K_{sp} = [0.1077][0.2154]^2$$
$$K_{sp} = 4.997 \times 10^{-3}$$

Recall from Equation [5] that if the K_{sp} is known, the $\Delta G°$ can be calculated.

$$\Delta G^\circ = -RT \ln K_{sp}$$
$$\Delta G^\circ = -RT \ln 4.997\times10^{-3}$$
$$\Delta G^\circ = -8.314 \frac{\text{J}}{\text{mol K}}\times(298\text{ K})\ln 4.997\times10^{-3}$$
$$\Delta G^\circ = +13.13 \frac{\text{kJ}}{\text{mol}}$$

Under standard conditions, ΔG° can also be expressed in terms of enthalpy and entropy (Section 19.5).

$$\Delta G^\circ = \Delta H^\circ - T\Delta S^\circ \qquad [6]$$

Assuming that ΔH° and ΔS° are constant in the temperature range of this experiment, the ΔG° must be determined at two different temperatures to simultaneously solve for ΔH° and ΔS°, giving the following set of equations:

$$\Delta G^\circ{}_1 = \Delta H^\circ - T_1\Delta S^\circ \qquad [7]$$

$$\Delta G^\circ{}_2 = \Delta H^\circ - T_2\Delta S^\circ \qquad [8]$$

When K_{sp} is determined at two different temperatures, two values for ΔG° are obtained, and using Equations [7] and [8], both ΔH° and ΔS° can be calculated.

GIVE IT SOME THOUGHT

How does the temperature influence the K_{sp} of a slightly soluble salt?

EXAMPLE 34.2

A saturated barium hydroxide solution was prepared at 60 °C, and it was determined that the concentration of hydroxide ions was 0.1956 *M*, giving a ΔG° of 15.47 kJ/mol. Assuming that the change in enthalpy and the change in entropy are negligible between 25 °C and 60 °C, calculate ΔH° and ΔS°.

SOLUTION: Utilizing Equations [7] and [8] gives the following:

$$\Delta G^\circ{}_{298\text{ K}} = \Delta H^\circ - T_{298\text{ K}}\Delta S^\circ$$
$$\Delta G^\circ{}_{333\text{ K}} = \Delta H^\circ - T_{333\text{ K}}\Delta S^\circ$$

When the given experimental values are plugged into these equations, ΔH° and ΔS° can be solved utilizing simultaneous equations (Example 34.1).

$$13.13 \frac{\text{kJ}}{\text{mol}} = \Delta H^\circ - 298\text{ K}\times\Delta S^\circ$$
$$15.47 \frac{\text{kJ}}{\text{mol}} = \Delta H^\circ - 333\text{ K}\times\Delta S^\circ$$

The ΔH° for the first equation can be solved for $\Delta H^\circ = 13.13 \frac{\text{kJ}}{\text{mol}} + 298\text{ K}\times\Delta S^\circ$, and this expression can be inserted into the second equation to solve for ΔS°.

$$15.47\frac{\text{kJ}}{\text{mol}} = 13.13\frac{\text{kJ}}{\text{mol}} + 298\text{ K}\times\Delta S^\circ - 333\text{ K}\times\Delta S^\circ$$
$$2.34\frac{\text{kJ}}{\text{mol}} = -35\text{ K}\times\Delta S^\circ$$
$$\Delta S^\circ = -67\frac{\text{J}}{\text{mol}}\times\text{K}$$

This allows you to calculate ΔH° using either original equation. For example,

$$13.13\,\frac{\text{kJ}}{\text{mol}} = \Delta H^\circ - 298\ \text{K} \times \left(-67\,\frac{\text{kJ}}{\text{mol}} \times \text{K}\right)\left(\frac{1\ \text{kJ}}{1000\ \text{J}}\right)$$

$$\Delta H^\circ = 13.13\,\frac{\text{kJ}}{\text{mol}} + 298\ \text{K} \times \left(-67\,\frac{\text{kJ}}{\text{mol}} \times \text{K}\right)\left(\frac{1\ \text{kJ}}{1000\ \text{J}}\right)$$

$$\Delta H^\circ = -6.84\,\frac{\text{kJ}}{\text{mol}}$$

GIVE IT SOME THOUGHT

Do both ΔH° and ΔS° favor dissolution of $Ba(OH)_2$?

PROCEDURE

Using a beaker, acquire approximately 40 mL of the saturated calcium hydroxide solution prepared by your laboratory instructor. To ensure that all of the $Ca(OH)_2(s)$ is removed from the solution, remove by filtration the solid using a clean, dry long stem funnel and medium-porosity filter paper. Measure the temperature of this solution and record it in your lab notebook. Pipet 10.00 mL of the saturated solution into a clean 125 mL Erlenmeyer flask. Add roughly 25 mL of distilled water and a few drops of bromothymol blue indicator to the flask. Using a buret, titrate with a standard HCl solution (making sure you record the exact concentration of the HCl solution in your lab notebook) until the endpoint is reached. When bromothymol blue is used as an indicator, your solution should initially turn blue. As the endpoint is approached, it will turn green, and a yellow solution indicates that the endpoint has been passed. Record the volume of HCl required to reach the endpoint and repeat this titration two more times to give you a total of three titrations. Record all data in your lab notebook.

Now that you have performed titrations at that temperature, the K_{sp} and ΔG° can be calculated at one temperature. To determine ΔH° and ΔS°, you must perform the same set of experiments at another temperature. Place roughly 100 mL of distilled water in a 250 mL beaker and bring it to a boil. While the water is boiling (and it should be boiling for at least 2 minutes before you add any solid), add approximately 2 g of solid calcium hydroxide and keep your solution as close to boiling temperature as possible, stirring occasionally for about 5 minutes. Turn off your Bunsen burner and measure the temperature of the saturated calcium hydroxide solution. Record the temperature in your notebook and in the space provided below. As quickly as you can, filter about 40–50 mL of the hot solution using a clean, dry long stem funnel and medium-porosity filter paper. (Note: The reason you are asked to do this filtration quickly is because as the solution begins to cool, the concentration of the hydroxide ions will change if solid $Ca(OH)_2$ is still present in the solution.) After all of the solid $Ca(OH)_2$ has been removed, allow the saturated solution to cool and pipet 10.00 mL of this solution into a clean, dry 125 mL Erlenmeyer flask. Add 25 mL of water and a few drops of the bromothymol blue indicator. Titrate with the standardized HCl solution and record the volume of HCl required to reach the endpoint. Repeat this titration two more times to give you a total of three titrations. Record all data in your lab notebook.

The neutralized titration solutions can be disposed of as indicated by your instructor.

Name ______________________________ Desk ______________________________

Date ______________________________ Laboratory Instructor ______________________________

Solubility and Thermodynamics | 34 Pre-lab Questions

Before beginning this experiment in the laboratory, you should be able to answer the following questions.

1. Write out the Gibbs free energy equation and define the terms in it.

2. Write out the equilibrium expression for the dissolution of lead(II) iodide. What factor(s) can affect the magnitude of this equilibrium constant? How does it change?

3. How do the concentrations of HCl, OH^-, and Ca^{2+} relate to each other at the endpoint of an acid–base titration? How is this information used to determine K_{sp} and $\Delta G°$?

4. A saturated solution of manganese(II) hydroxide was prepared, and an acid–base titration was performed to determine its K_{sp} at 25 °C. The endpoint was reached when 50.00 mL of the manganese(II) hydroxide solution was titrated with 3.42 mL of 0.0010 M HCl solution. What is the K_{sp} of manganese(II) hydroxide?

5. Make a sketch of the graph for the function $y = \ln x$. For what values of x is y positive? What values of x give a negative result for y? Now consider $\Delta G° = -RT \ln K$. For what values of K is $\Delta G°$ positive? What values of K give a negative result for $\Delta G°$? What does this tell you about the spontaneity of a reaction?

6. Based on the $\Delta G° = -RT \ln K$ expression, would you predict that AgCl(s) dissolving into $Ag^{+}(aq) + Cl^{-}(aq)$ [$K_{sp} = 1.8\times10^{-10}$] is a spontaneous process? Would you expect $Co^{3+}(aq) + 6NH_3(aq)$ forming the $[Co(NH_3)_6]^{3+}$ complex ion [$K_f = 4.5\times10^{33}$] to be a spontaneous process? How does the magnitude of the equilibrium constant relate to the $\Delta G°$?

7. If a saturated solution of calcium hydroxide is prepared and its K_{sp} is determined, how can the $\Delta G = \Delta G° + RT \ln Q$ equation be modified to determine $\Delta G°$?

Name ______________________ Desk ______________

Date ______________ Laboratory Instructor ______________________

REPORT SHEET | EXPERIMENT

Solubility and Thermodynamics | 34

DETERMINING THE K_{sp} OF $Ca(OH)_2$ AT ROOM TEMPERATURE

Temperature ______________

Concentration of HCl solution ______________

	TRIAL 1	TRIAL 2	TRIAL 3
Volume of HCl used			
Moles of OH^-			
Volume of sample			
$[OH^-]$			
$[Ca^{2+}]$			
K_{sp}			

Experimentally determined value of K_{sp} for trial 1 ______________ ± ______________

DETERMINING THE K_{sp} OF $Ca(OH)_2$ AT AN ELEVATED TEMPERATURE

Temperature ______________

Concentration of HCl solution ______________

	TRIAL 1	TRIAL 2	TRIAL 3
Volume of HCl used			
Moles of OH^-			
Volume of sample			
$[OH^-]$			
$[Ca^{2+}]$			
K_{sp}			

Experimentally determined value of K_{sp} ______________ ± ______________

ΔG° ______________

ΔH° ______________ ΔS° ______________

QUESTIONS

Consider the following data for questions 1–6: Two solutions of an unknown slightly soluble salt, $A(OH)_2$, were allowed to equilibrate—one at 25 °C and the other at 80 °C. A 15.00 mL-aliquot of each solution is titrated with 0.200 *M* HCl. 8.00 mL of the acid is required to reach the endpoint of the titration at 25 °C, while 69.15 mL are required for the 80 °C solution.

1. Calculate the K_{sp} at 25 °C.

2. Calculate the K_{sp} at 80 °C.

3. Calculate the Gibbs free energy (in kJ/mol) at 25 °C.

4. Calculate the Gibbs free energy (in kJ/mol) at 80 °C.

5. Assuming that the change in enthalpy is negligible over this temperature range, calculate $\Delta H°$.

6. Assuming that the change in entropy is negligible over this temperature range, calculate $\Delta S°$.

7. An acid–base titration was performed to determine the K_{sp} of a slightly soluble salt. The K_{sp} was experimentally determined at 70 °C, and the $\Delta G°$ was determined to be 29.05 kJ/mol. If $\Delta S°$ is −115 J/mol·K, what is $\Delta H°$ (in kJ/mol) for this slightly soluble salt?

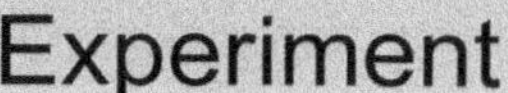

Experiment 35

Analysis of Water for Dissolved Oxygen

OBJECTIVE

To gain a basic understanding of quantitative techniques of volumetric analysis by determining the dissolved oxygen content of a water sample.

APPARATUS AND CHEMICALS

Apparatus

balance	barometer
50 mL buret	1 L beaker
500 mL Erlenmeyer flask	2 mL pipets (2)
250 mL Erlenmeyer flasks (3)	25 mL pipet
100 mL graduated cylinder	thermometer
250 mL narrow-mouth, glass-stoppered bottle or 500 mL bottle	1 L volumetric flask
buret clamp and ring stand	Bunsen burner and hose

Chemicals

alkaline iodide–azide reagent*	NaOH
chloroform	1 *M* H_2SO_4
2.15 *M* $MnSO_4$ (freshly prepared)	conc. H_2SO_4
1% boiled starch solution	NaN_3 solution*
water sample (unknown)	$Na_2S_2O_3 \cdot 5H_2O$
KIO_3	

DISCUSSION

The oxygen normally dissolved in water is indispensable to fish and other water-dwelling organisms. Certain pollutants deplete the dissolved oxygen during the course of their decomposition. This is particularly true of many organic compounds that are present in sewage or dead algae. These are decomposed by the aerobic metabolism of microorganisms, which use these organic compounds for food. The metabolic process is an oxidation of the organic compounds—the dissolved oxygen is the oxidizing agent. Thus, while these microorganisms are removing the pollutants, they are also removing the dissolved oxygen that otherwise would be present to support aquatic life. Because the solubility of most gases in solution decreases as the temperature of the solution increases, thermal pollution also decreases the dissolved oxygen content.

As a logical consequence of this, one empirical standard for determining water quality is the dissolved oxygen content (DO). The survival of aquatic life depends upon the water's ability to maintain certain minimum concentrations of

---

*The alkaline iodide–azide solution is prepared by dissolving 500 g of NaOH and 135 g of NaI in 1 L of distilled water and adding a solution of 10 g NaN_3 in 40 mL of distilled water.

the vital dissolved oxygen. Fish require the highest levels; invertebrates, lower levels; and bacteria, the lowest level. For a diversified warm water biota, including game fish, the DO concentration should be at least 5 mg/L (5 ppm). Another water quality standard is the biological oxygen demand (BOD). The BOD is the amount of oxygen needed by the microorganisms to remove the pollutant. To determine BOD, a sample containing organic pollutants is incubated with its microorganisms for a definite time, usually 5 days, and the amount of oxygen removed is measured. The BOD is taken as the difference in DO before and after incubation. A BOD of 1 ppm is characteristic of nearly pure water. Water is considered fairly pure with a BOD of 3 ppm and of doubtful purity when the BOD level reaches 5 ppm. Therefore, monitoring of water quality logically includes analysis for dissolved oxygen.

This experiment outlines the analysis of water samples for their total dissolved oxygen (DO) content using the azide modification of the iodometric (Winkler) method. This procedure is most commonly used for analysis of sewage, effluents, and streams. It is based upon the use of Mn(II) compounds that are oxidized to Mn(IV) compounds by the oxygen in the water sample. In turn, the Mn(IV) compound reacts with NaI to produce iodine, I_2. The released I_2 is then titrated with standardized sodium thiosulfate solution, $Na_2S_2O_3$, using starch as an indicator. The chemical reactions involved are as follows:

$$MnSO_4(aq) + 2KOH(aq) \longrightarrow Mn(OH)_2(s) + K_2SO_4(aq)$$

$$2Mn(OH)_2(s) + O_2(aq) \longrightarrow 2MnO(OH)_2(s)$$

$$MnO(OH)_2(s) + 2H_2SO_4(aq) \longrightarrow Mn(SO_4)_2(aq) + 3H_2O(l)$$

$$Mn(SO_4)_2(aq) + 2NaI(aq) \longrightarrow MnSO_4(aq) + Na_2SO_4(aq) + I_2(aq)$$

$$2Na_2S_2O_3(aq) + I_2(aq) \longrightarrow Na_2S_4O_6(aq) + 2NaI(aq)$$

The net overall chemical equation for this sequence of reactions is

$$\tfrac{1}{2}O_2(g) + 2S_2O_3^{2-}(aq) + 2H^+(aq) \longrightarrow S_4O_6^{2-}(aq) + H_2O(l)$$

PROCEDURE

A. Standard Sodium Thiosulfate Solution

First, prepare and standardize a 0.025 *M* sodium thiosulfate solution as follows: Begin boiling about 800 mL of distilled water. Determine the mass of 6.205 g of $Na_2S_2O_3 \cdot 5H_2O$ and dissolve it in 500 mL of water that has been boiled to remove CO_2. Cool to room temperature and add 5 mL of chloroform and 0.4 g of NaOH to retard bacterial decomposition and dilute to 1 L in a 1 L volumetric flask. **(CAUTION: *Avoid contact with* NaOH. *It causes severe burns. If you come in contact with it, immediately wash the contacted area with copious amounts of water.***

Accurately determine the mass of out about 0.223 g of dry potassium iodate and transfer to a 250 mL volumetric flask. Add previously boiled water and mix to dissolve then dilute to 250 mL. Mix the diluted solution by inversion of the flask. To each of three 25 mL aliquots of this solution delivered via a pipet in labeled 250 mL Erlenmeyer flasks, add 0.5 g of potassium iodide and about 2 mL of 1 *M* sulfuric acid. Titrate each of these solutions with your thiosulfate solution, stirring constantly. When the color of the solution becomes a pale yellow, dilute to approximately 200 mL with previously boiled distilled water, add about 2 mL of 1% starch solution, and continue the titration until the color changes from blue to colorless for the first time. Ignore any return of color. Rinse the sides of the flask, with a wash bottle containing previously boiled distilled water, just prior to the end point. Record the final buret reading and subtract that value from the initial reading to give the amount of thiosulfate used. The reactions involved in this standardization are

$$IO_3^-(aq) + 5I^-(aq) + 6H^+(aq) \rightleftharpoons 3I_2(aq) + 3H_2O(l)$$

$$2Na_2S_2O_3(aq) + I_2(aq) \rightleftharpoons Na_2S_4O_6(aq) + 2NaI(aq)$$

Note the stoichiometry in these two reactions.

GIVE IT SOME THOUGHT

What species is oxidized and which species is reduced in each reaction?

You are actually titrating with your thiosulfate the iodine formed by the first reaction. Calculate the molarity of your thiosulfate from the following equations:

$$\text{moles } Na_2S_2O_3 = 6 \times (\text{moles } KIO_3)$$

$$\text{moles } Na_2S_2O_3 = V_{Na_2S_2O_3} \times M_{Na_2S_2O_3}$$

$$\text{moles } KIO_3 = \frac{(\text{g } KIO_3)}{(214.02 \text{ g/mol})} \times \frac{25.00 \text{ mL}}{250.0 \text{ mL}}$$

Thus,

$$M_{Na_2S_2O_3} = \frac{(6)(\text{g } KIO_3)(25.00 \text{ mL})}{(214.02 \text{ g/mol})(\text{volume } Na_2S_2O_3 \text{ in liters})(250.0 \text{ mL})}$$

GIVE IT SOME THOUGHT

a. What is the key relationship at the end point of this titration?
b. How does this differ from a typical acid/base titration?

EXAMPLE 35.1

A 0.2264 g sample of KIO_3 was dissolved in 250.0 mL of water. To each 25.00 mL aliquot of this solution, 0.500 g of KI and 2.00 mL of 1.00 *M* H_2SO_4 were added. Titration of the liberated iodine required 25.30 mL of $Na_2S_2O_3$ solution. Calculate the molarity of the $Na_2S_2O_3$ solution.

SOLUTION: From the analytical reactions

$$IO_3^-(aq) + 5I^-(aq) + 6H^+(aq) \longrightarrow 3I_2(aq) + 3H_2O(aq)$$

$$2Na_2S_2O_3(aq) + I_2(aq) \longrightarrow Na_2S_4O_6(aq) + 2NaI(aq)$$

exactly 3 mol of iodine are liberated for each mole of iodate and 1 mol of iodine reacts with 2 mol $S_2O_3^{2-}$ where, in a 25.00 mL aliquot,

$$\text{moles KIO}_3 = \left(\frac{0.2264\ \text{g}}{214.02\ \text{g/mol}}\right)\left(\frac{25.00\ \text{mL}}{250.0\ \text{mL}}\right)$$

$$= 1.058 \times 10^{-4}\ \text{mol}$$

And because

$$\text{moles Na}_2\text{S}_2\text{O}_3 = 6\frac{(\text{moles Na}_2\text{S}_2\text{O}_3)}{(\text{mole KIO}_3)} \times (\text{moles KIO}_3) = 6(1.058 \times 10^{-4}\ \text{mol})$$

$$= 6.347 \times 10^{-4}\ \text{mol}$$

then

$$M_{\text{Na}_2\text{S}_2\text{O}_3} = \frac{\text{moles Na}_2\text{S}_2\text{O}_3}{\text{liters of solution}}$$

$$= \frac{6.347 \times 10^{-4}\ \text{mol}}{0.02530\ \text{L}}$$

$$= 0.02509\ M$$

B. Water Sample Analysis

Collection of Sample Collect the sample in a narrow-mouthed, glass-stoppered bottle (250 to 300 mL capacity). Avoid entrapment or dissolution of atmospheric oxygen. Allow the bottle to overflow its volume and replace the stopper so that no air bubbles are entrained; avoid excessive agitation, which will dissolve atmospheric oxygen. Record the temperature of the water sample in °C. You should analyze the sample as soon as possible. If the method outlined below is used, you may preserve the sample for four to eight hours by adding 0.7 mL of concentrated H_2SO_4 and 1 mL of sodium azide solution (2 g NaN_3/100 mL H_2O) to the DO bottle. **(CAUTION: *Concentrated H_2SO_4 causes severe burns. If you come in contact with it, immediately wash the area with copious amounts of water.*)** This will arrest biologic activity and maintain the DO if the bottle is stored at the temperature of collection or at 10° to 20 °C while tightly sealed in a refrigerator.

Release of Iodine Open the sample bottle with great care to avoid aeration and add 2.00 mL of 2.15 *M* manganous sulfate solution from a volumetric pipet. Similarly, add 2 mL of alkaline iodide–azide reagent using another pipet. The neck of the bottle will again have excess liquid, so replace the stopper carefully to avoid splashing. Thoroughly mix the contents of the bottle by inverting the bottle several times. A milky precipitate will form and will gradually change to a yellowish-brown color. Allow the precipitate to settle so that the clear solution occupies the top third of the bottle. Carefully remove the stopper and immediately add 2 mL of concentrated sulfuric acid. You should make this addition by bringing the pipet tip against the neck of the bottle just slightly below the surface of the liquid. Stopper the bottle and then mix the contents by gentle inversion until the precipitate dissolves. At this point, the yellowish-brown color due to liberated iodine should appear. You do not need to titrate the sample immediately, but if titration is delayed, you should store the sample in darkness. Keep in mind that the titration should be done within several hours.

Titration Accurately measure 200 mL of the sample into a 500 mL Erlenmeyer flask. Titrate the sample with the standardized thiosulfate solution, stirring constantly. When the color of the solution becomes a pale yellow, and remains as such for about 2 minutes add about 2 mL of 1% starch solution and continue titrating until the color changes from blue to colorless for the first time. Record the volume of titrant necessary. If time permits, analyze another sample.

GIVE IT SOME THOUGHT

What is the key molar relationship at the end point of this titration?

Calculation of Dissolved Oxygen Content According to the equations on the previous page, 1 mol of O_2 (32 g) requires 4 mol of $Na_2S_2O_3$ in reaching an end point. The number of moles of $Na_2S_2O_3$ is equal to the volume of $Na_2S_2O_3$ (in liters) times the concentration of the $Na_2S_2O_3$ (in molarity).

$$\text{moles } Na_2S_2O_3 = \text{volume}_{Na_2S_2O_3} \times \text{molarity}_{Na_2S_2O_3}$$

Using the preceding information, calculate the number of grams of O_2 in your 200 mL sample. From this, calculate the number of milligrams of O_2 per liter of solution. Because a liter of solution weighs approximately 1000 g (the bulk of the solution is water), 1 mg/L is equivalent to 1 mg in 10^6 mg, or 1 million mg, of solution. Therefore, the number of milligrams of O_2 per liter is equivalent to parts per million (ppm).

EXAMPLE 35.2

To a 200.0 mL water sample, 0.50 g of KI, 2.00 mL of 1.000 *M* H_2SO_4, and 2.00 mL of starch solution were added. The liberated iodine required 7.88 mL of 0.0251 *M* $Na_2S_2O_3$. Calculate the O_2 concentration in the sample in ppm.

SOLUTION: From the analytical reactions

$$MnSO_4 + 2KOH \longrightarrow Mn(OH)_2 + K_2SO_4$$

$$2Mn(OH)_2 + O_2 \longrightarrow 2MnO(OH)_2$$

$$MnO(OH)_2 + 2H_2SO_4 \longrightarrow Mn(SO_4)_2 + 3H_2O$$

$$Mn(SO_4)_2 + 2KI \longrightarrow MnSO_4 + K_2SO_4 + I_2$$

$$2Na_2S_2O_3 + I_2 \longrightarrow Na_2S_4O_6 + 2NaI$$

1 mol of O_2 requires 4 mol of $Na_2S_2O_3$.

$$\begin{aligned}\text{moles } Na_2S_2O_3 &= (\text{volume } Na_2S_2O_3)\times(\text{molarity } Na_2S_2O_3)\\ &= (0.00788\text{ L})(0.0251\text{ mol/L})\\ &= 1.98\times10^{-4}\text{ mol}\\ \text{moles } O_2 &= \tfrac{1}{4}\text{ mol } Na_2S_2O_3\\ &= 4.95\times10^{-5}\text{ mol}\\ \text{mass } O_2 &= (4.95\times10^{-5}\text{ mol})\times(32.0\text{ g/mol})\\ &= 1.58\times10^{-3}\text{ g}\\ 1.58\times10^{-3}\text{ g}\times1000\text{ mg/g} &= 1.58\text{ mg}\\ \text{concentration } O_2 &= \frac{1.58\text{ mg}}{0.200\text{ L}} = 7.90\text{ mg/L}\\ &= 7.90\text{ ppm}\end{aligned}$$

You may apply a correction factor to your answer to correct for solution loss during the addition of manganous sulfate and sulfuric acid. This amounts to multiplication by 204/200 if these reagents were added to a 200 mL aliquot.

Comparisons in Dissolved Oxygen Contents The amount of oxygen dissolved in water depends not only upon the amount of chemical pollution but also upon such factors as water temperature and the atmospheric pressure above the water (Appendix L). At temperatures between 0° and 39 °C, the amount of O_2 that will be present in oxygen-saturated distilled water is given by the equation

$$\text{ppm dissolved } O_2 = \frac{(P-p)\times5.09}{35+T} = \text{SLDO}$$

where P is the barometric pressure in kPa, T is the temperature of the water in °C, and p is the vapor pressure of water in kPa at the temperature of the water (Appendix L). Calculate the saturation level (SL) for your water sample. The percent saturation, %SL, is given by

$$\%\text{SL} = \frac{100\%(\text{DO in ppm})}{(\text{SLDO in ppm})}$$

EXAMPLE 35.3

A water sample at 12 °C (where the vapor pressure of water in 1.40 kPa) and 86.9 kPa was found to contain 7.90 ppm O_2. Calculate the percent saturation of this sample.

SOLUTION:

$$\begin{aligned}\text{SLDO} &= \frac{(86.9\text{ kPa}-1.4\text{ kPa})\times5.09\text{ ppm-K/kPa}}{(35+12)\text{K}}\\ &= 9.3\text{ ppm}\\ \%\text{SL} &= \frac{(100\%)(7.90\text{ ppm})}{9.3\text{ ppm}} = 85\%\end{aligned}$$

Waste Disposal Instructions All solutions should be disposed of in the appropriately labeled waste containers.

Name ______________________ Desk ______________________

Date ______________________ Laboratory Instructor ______________________

Analysis of Water for Dissolved Oxygen | 35 Pre-lab Questions

Before beginning this experiment in the laboratory, you should be able to answer the following questions.

1. A 0.120 g sample of KIO_3 required 9.80 mL of $Na_2S_2O_3$ solution for its standardisation. What is the molarity of $Na_2S_2O_3$ solution?

2. Suggest how chloroform helps preserve the water sample.

3. Which species is being oxidized and which is being reduced in the following reaction?

$$5Fe^{2+}(aq) + MnO_4^-(aq) + 8H^+(aq) \longrightarrow 5Fe^{3+}(aq) + Mn^{2+}(aq) + 4H_2O(l)$$

4. During the course of the standardization of $S_2O_3^{2-}$ with KIO_3, the text tells you to ignore the return of the blue color with time after the end point has been reached in the reaction. Suggest a reason for the return of the blue color. Write a chemical reaction that is consistent with your suggestion.

5. Calculate the concentration in mg/L (ppm) of a solution that is 4×10^{-3} M in O_2, assuming a density of 1 g/mL for the solution.

6. Some characteristic BOD levels are given below.

Source	**BOD range (ppm)**
Untreated municipal sewage	100–400
Runoff from barnyards and feed lots	100–10,000
Food processing wastes	100–10,000

Assuming the minimum values above, by what factor must these waters be diluted with pure water to reduce the BOD to a value sufficiently low to support aquatic life (BOD $\leq$ 5 ppm)?

7. In determining the BOD of a water sample, it was found that the DO level was 9.3 ppm before incubation and 6.7 ppm after five days of incubation. What is the BOD of this sample?

8. Would the water sample in question 7 support aquatic life?

Name ______________________ Desk ______________

Date ______________ Laboratory Instructor ______________________

Unknown no. ______________

REPORT SHEET | EXPERIMENT

Analysis of Water for Dissolved Oxygen | 35

A. Standard Sodium Thiosulfate Solution

1. Mass in grams of KIO_3 in 250 mL __________ g Aliquot volume __________ mL

	Sample 1	*Sample* 2	*Sample* 3
2. Final buret reading	________ mL	________ mL	________ mL
3. Initial buret reading	________ mL	________ mL	________ mL
4. Volume of $Na_2S_2O_3$ solution	________ mL	________ mL	________ mL
	________ L	________ L	________ L
5. Molarity of $Na_2S_2O_3$ (show calculations)	________ *M*	________ *M*	________ *M*

Avg. *M* __________ ± __________

6. Standard deviation (show calculations) __________

B. Water Sample Analysis

Water temperature ________ °C

	Sample 1	*Sample 2*
1. Volume of $Na_2S_2O_3$ required to titrate water sample		
Final buret reading	________ mL	________ mL
Initial buret reading	________ mL	________ mL
Volume of $Na_2S_2O_3$ (used standard $Na_2S_2O_3$)	________ mL	________ mL

	Sample 1	*Sample 2*
2. Moles of $Na_2S_2O_3$ required to titrate water sample (show calculations)	________ mol	________ mol

	Sample 1	*Sample 2*
3. Moles of O_2 present in 200 mL of water sample (show calculations)	________ mol	________ mol
4. Grams of O_2 present in 200 mL of water sample	________ g	________ g
5. Milligrams of O_2 present in 200 mL of water sample	________ mg	________ mg
6. Milligrams of O_2 present in 1000 mL of water sample	________ ppm	________ ppm
7. Saturation level for your water sample (show calculations)	________ ppm	

8. Percent saturation level of your water sample ____________
(show calculations)

9. Do you think this water has sufficient DO to sustain aquatic life?

10. Calculate the SLDO for water at 20°, 30°, 40°, 50°, and 100 °C and 93.3 kPa and plot the data on the graph paper provided. How would you expect the solubility of a gas in water to change with temperature? Are your calculations in agreement with your expectations? Consult Appendix L.

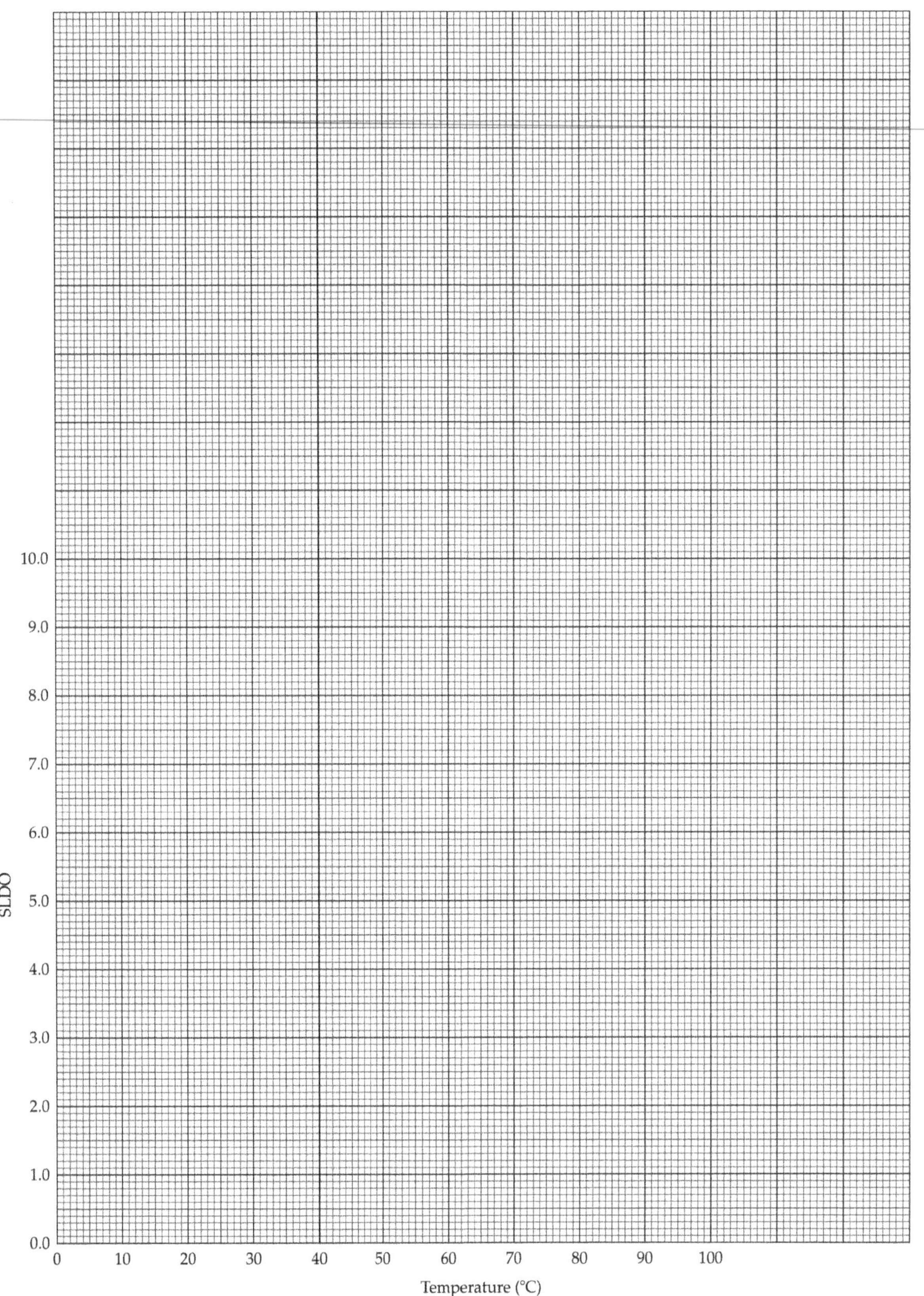
SLDO
10.0
9.0
8.0
7.0
6.0
5.0
4.0
3.0
2.0
1.0
0.0
0
10
20
30
40
50
60
70
80
90
100
Temperature (°C)

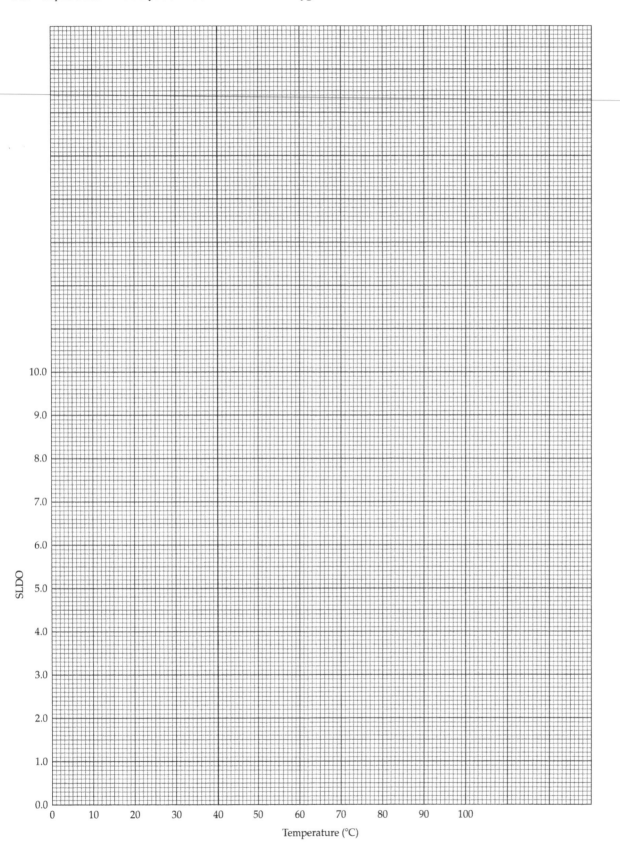

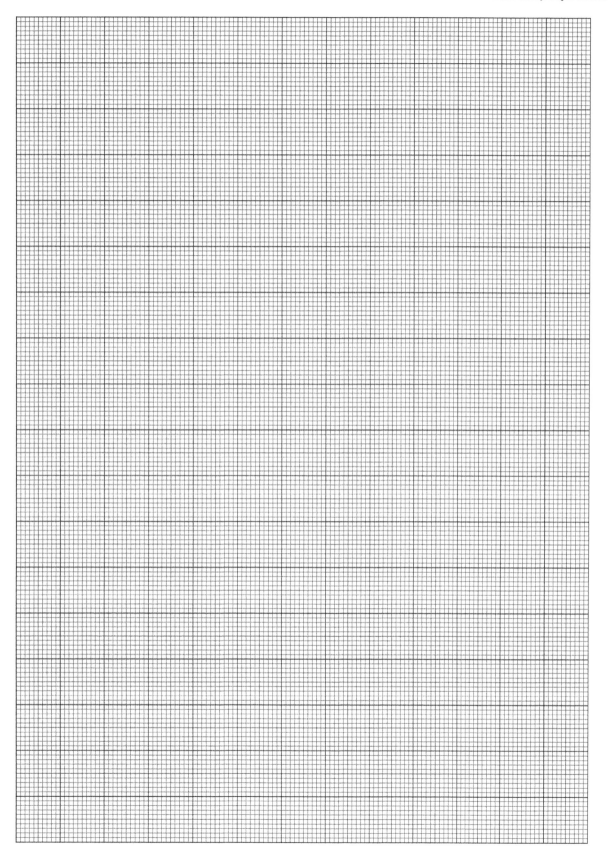

NOTES AND CALCULATIONS

Preparation and Reactions of Coordination Compounds: Oxalate Complexes

Experiment 36

OBJECTIVE

To gain familiarity with coordination compounds by preparing a representative compound and witnessing some typical reactions.

APPARATUS AND CHEMICALS

Apparatus

balance
Bunsen burner and hose or hotplate
250 mL wide-mouth bottle
No. 6 two-hole rubber stopper
glass stirring rod
9 cm filter paper
wire gauze
vial
9 cm Büchner funnel
250 mL suction flask
aspirator
100 and 250 mL beakers
thermometer
ring stand and iron ring
glass wool

Chemicals

cis- and *trans*-$K[Cr(C_2O_4)_2(H_2O)_2]$ (prep given)
6 *M* NH_3
$K_2C_2O_4 \cdot H_2O$ (potassium oxalate monohydrate)
acetone
ferrous ammonium sulfate
6 *M* H_2SO_4
ice
aluminum powder
6 *M* KOH
$H_2C_2O_4$ (oxalic acid)
$K_2Cr_2O_7$ (potassium dichromate)
50% ethanol
95% ethanol
absolute ethanol
$CuSO_4 \cdot 5(H_2O)$
6% H_2O_2
6 *M* HCl

DISCUSSION

When gaseous boron trifluoride, BF_3, is passed into liquid trimethylamine, $(CH_3)_3N$, a highly exothermic reaction occurs and a creamy white solid, $(CH_3)_3N{:}\,BF_3$, separates. This solid, which is an adduct of trimethylamine and boron trifluoride, is a coordination compound. It contains a coordinate covalent, or dative, bond that unites the Lewis acid BF_3 with the Lewis base trimethylamine. Numerous coordination compounds are known; in fact, nearly all compounds of the transition elements are coordination compounds wherein the metal is a Lewis acid and the atoms or molecules joined to the metal are Lewis bases. These Lewis bases are called *ligands*, and the coordination compounds are usually denoted by square brackets when their formulas are written (Section 23.2). The metal and the ligands bound to it constitute what is termed the coordination sphere. When writing chemical formulas for coordination compounds, you use square brackets to set off the coordination sphere from other parts of the compound. For example, the salt $NiCl_2 \cdot 6H_2O$ is, in reality, the coordination compound $[Ni(H_2O)_6]Cl_2$, with an octahedral geometry, as shown in Figure 36.1.

GIVE IT SOME THOUGHT

What is the coordination number of nickel in this complex?

The apexes of a *regular* octahedron are all equivalent positions (Section 23.2). Thus, each of the monodentate (one donor site or Lewis base electron pair) H_2O molecules in the $[Ni(H_2O)_6]^{2+}$ ion and the three bidentate (two donor sites) oxalate ions, $C_2O_4^{2-}$, in $[Co(C_2O_4)_3]^{2-}$ are in identical environments. The water molecules in the two isomeric compounds, *cis*- and *trans*-$[Cr(C_2O_4)_2(H_2O)_2]^-$, are in equivalent environments within each complex ion (coordination compound), but the two isomeric ions are not equivalent to each other. The two water molecules are adjacent in the *cis* isomer and opposite each other in the *trans* isomer. These two isomers are termed *geometric isomers* (Section 23.3), and although they have identical empirical and molecular formulas, their geometrical arrangements in space are different (Section 23.4).

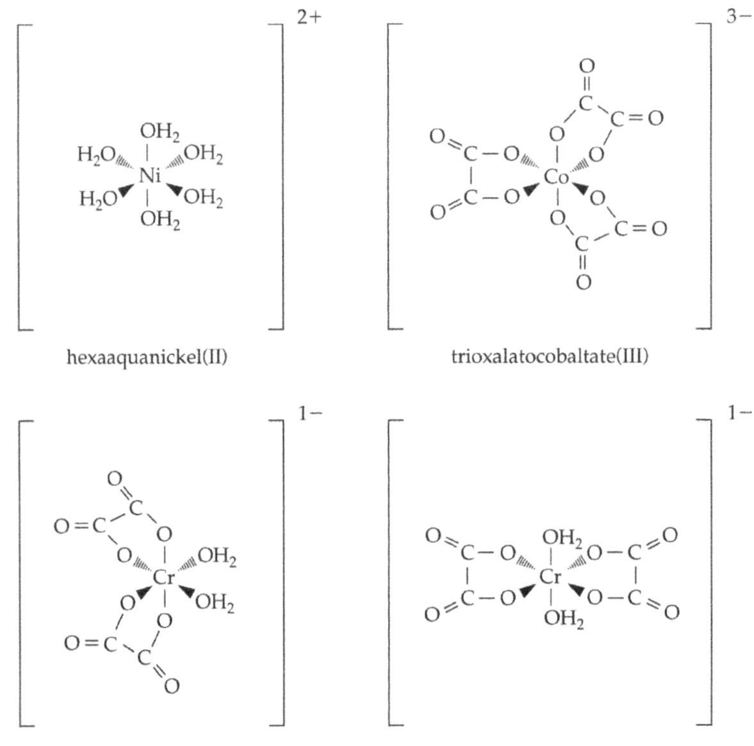

▲**FIGURE 36.1** Typical octahedral (6-coordinate) coordination compounds.

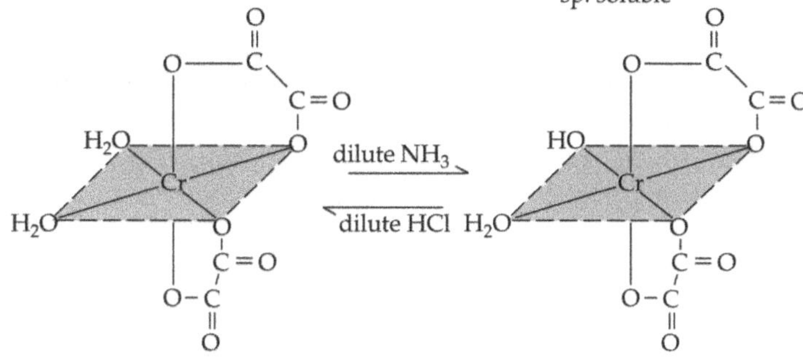

▲**FIGURE 36.2** Reaction of geometric isomers: *cis*- and *trans*-$[Cr(C_2O_4)_2(H_2O)_2]^-$ ions. The procedure is to place a small amount of an aqueous solution of each complex on separate pieces of filter paper on a watch glass. Let them dry; then add a drop of reagent. Observe the results.

Consequently, they have different chemical and physical properties, as your laboratory instructor will demonstrate through the reactions shown in Figure 36.2.

Your goal in this experiment is to prepare an oxalate-containing coordination compound. It may be analyzed for its oxalate concentration by the procedure described in Experiment 37. You will prepare *one* of the following compounds:

- $K_3[Cr(C_2O_4)_3]\cdot 3H_2O$
- $K_2[Cu(C_2O_4)_2]\cdot 2H_2O$
- $K_3[Fe(C_2O_4)_3]\cdot 3H_2O$
- $K_3[Al(C_2O_4)_3]\cdot 3H_2O$

Your laboratory instructor will tell you which one to prepare. Someone in your lab section will prepare each of these compounds so that you can compare their properties. **(CAUTION: *Oxalic acid is a toxic compound and is absorbed through the skin. Should any come in contact with your skin, wash it off immediately with copious amounts of water.*)**

PROCEDURE

Prepare one of the complexes whose synthesis is given below.

1. Preparation of $K_3[Cr(C_2O_4)_3]\cdot 3H_2O$

$$K_2Cr_2O_7 + 7H_2C_2O_4 \cdot 2H_2O + 2K_2C_2O_4 \cdot H_2O \longrightarrow 2K_3[Cr(C_2O_4)_3]\cdot 3H_2O + 6CO_2 + 17H_2O$$

GIVE IT SOME THOUGHT

What is the coordination number of chromium in the $K_3[Cr(C_2O_4)_3]\cdot 3H_2O$ complex?

In the hood, slowly add 3.6 g of potassium dichromate to a suspension of 10 g of oxalic acid in 20 mL of H_2O in a 250 mL beaker. The orange-colored mixture should spontaneously warm up almost to boiling as a vigorous evolution of gas commences. When the reaction has subsided (about 15 min), dissolve 4.2 g of potassium oxalate monohydrate in the hot green-black liquid and heat to boiling for 10 min. Allow the beaker and its contents to cool to room temperature. Add about 10 mL of 95% ethanol, with stirring, into the cooled solution in the beaker. Further cool the beaker and its contents in ice. The cooled liquid should thicken with crystals. After cooling the liquid in ice for 15 to 20 min, collect the crystals by filtration with suction using a Büchner funnel (See Figure 36.3) and filter flask. Wash the crystals on the funnel with three 10 mL portions of 50% aqueous ethanol followed by 25 mL of 95% ethanol and dry the product in air. Determine the mass of the air-dried material and store it in a vial. You should obtain about 9 g of product. Calculate the theoretical yield and determine your percentage yield. Reactions of chromium(III) are slow, and your yield will be low if you work too fast (Section 3.7).

$$\%\ \text{yield} = \frac{\text{actual yield in grams}}{\text{theoretical yield in grams}} \times 100\%$$

Your instructor may tell you to save your sample for analysis in Experiment 37 or dispose of it as directed.

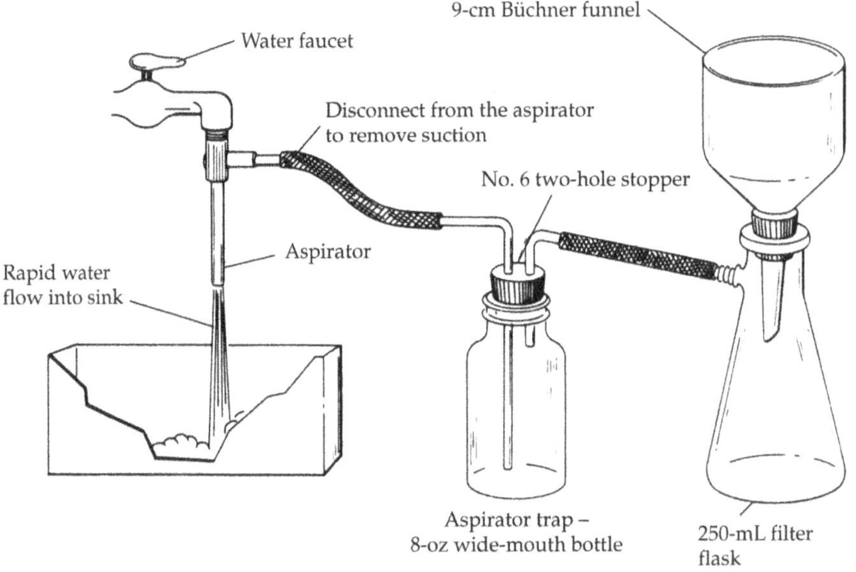

▲**FIGURE 36.3** Suction filtration assembly.

EXAMPLE 36.1

In the preparation of *cis*-$K[Cr(C_2O_4)_2(H_2O)_2]\cdot 2H_2O$, 12.0 g of oxalic acid was allowed to react with 4.00 g of potassium dichromate and 8.20 g of *cis*-$K[Cr(C_2O_4)_2(H_2O)_2]\cdot 2H_2O$ was isolated. What is the percent yield in this synthesis?

SOLUTION:

$$K_2Cr_2O_7 + 7H_2C_2O_4 \cdot 2H_2O \longrightarrow 2K[Cr(C_2O_4)_2(H_2O)_2]\cdot 2H_2O + 6CO_2 + 13H_2O$$

From the above reaction, you can see that 1 mol of $K_2Cr_2O_7$ reacts with 7 mol of $H_2C_2O_4$ to produce 2 mol of $K[Cr(C_2O_4)_2(H_2O)_2]\cdot 2H_2O$. In the synthesis, you used the following:

$$\text{moles } K_2Cr_2O_7 = \frac{4.00 \text{ g}}{294.19 \text{ g/mol}} = 0.0136 \text{ mol}$$

$$\text{moles } H_2C_2O_4 \cdot 2H_2O = \frac{12.0 \text{ g}}{126.07 \text{ g/mol}} = 0.0952 \text{ mol}$$

The reaction requires a 7:1 molar ratio of oxalic acid to $K_2Cr_2O_7$, but you actually used a 6.999 molar ratio (or within experimental error, the stoichiometric amount of each reagent and there is no limiting reagent). Hence, the number of moles of $K[Cr(C_2O_4)_2(H_2O)_2]\cdot 2H_2O$ formed should be twice the number of moles of $K_2Cr_2O_7$ or 2/7 of the moles of oxalic acid reacted.

$$\text{moles } K[Cr(C_2O_4)_2(H_2O)_2]\cdot 2H_2O(\text{expected}) = (2)(0.0136 \text{ mol}) = 0.0272 \text{ mol}$$

The theoretical yield of $K[Cr(C_2O_4)_2(H_2O)_2]\cdot 2H_2O$ is

$$(0.0272 \text{ mol})(339.2 \text{ g/mol}) = 9.23 \text{ g}$$

The percentage yield is then

$$\% \text{ yield} = \frac{(8.20 \text{ g})}{(9.23 \text{ g})}(100\%) = 88.8\%$$

2. Preparation of $K_2[Cu(C_2O_4)_2]\cdot 2H_2O$

$$CuSO_4 \cdot 5H_2O + 2K_2C_2O_4 \cdot H_2O \longrightarrow K_2[Cu(C_2O_4)_2]\cdot 2H_2O + K_2SO_4 + 5H_2O$$

Heat a solution of 6.2 g of copper sulfate pentahydrate in 12 mL of water to about 90 °C and add it rapidly, while stirring vigorously, to a hot (~90 °C) solution of 10.0 g of potassium oxalate monohydrate ($K_2C_2O_4 \cdot H_2O$) in 50 mL of water contained in a 100 mL beaker. Cool the mixture by setting the beaker in an ice bath for 15 to 30 min. Suction-filter the resultant crystals using a Büchner funnel (see Figure 36.3) and filter flask. Wash the crystals successively with about 12 mL of cold water, then 10 mL of absolute ethanol, and finally 10 mL of acetone and air-dry. Determine the mass of the air-dried material and store it in a vial. You should obtain about 7 g of product. Calculate the theoretical yield and determine your percentage yield. Your instructor may tell you to save your sample for analysis in Experiment 37 or dispose of it as directed. **(CAUTION: *Keep ethanol away from flames.*)**

3. Preparation of $K_3[Fe(C_2O_4)_3]\cdot 3H_2O$

$$Fe(NH_4)_2(SO_4)_2\cdot 6H_2O + H_2C_2O_4\cdot 2H_2O \longrightarrow FeC_2O_4 + H_2SO_4 + (NH_4)_2SO_4 + 8H_2O$$

$$H_2C_2O_4\cdot 2H_2O + 2FeC_2O_4 + 3K_2C_2O_4\cdot H_2O + H_2O_2 \longrightarrow 2K_3[Fe(C_2O_4)_3]\cdot 3H_2O + H_2O$$

This preparation contains two separate parts. Iron(II) oxalate is prepared first and then converted to $K_3[Fe(C_2O_4)_3]\cdot 3H_2O$ by oxidation with hydrogen peroxide, H_2O_2, in the presence of potassium oxalate.

Prepare a solution of 10 g of ferrous ammonium sulfate hexahydrate in 30 mL of water containing a few drops of 6*M* H_2SO_4 (to prevent premature oxidation of Fe^{2+} to Fe^{3+} by O_2 in the air). Then add, while stirring, a solution of 6 g of oxalic acid in 50 mL of H_2O. Yellow iron(II) oxalate forms. Carefully heat the mixture to boiling while stirring constantly to prevent bumping. Decant and discard the supernatant liquid as directed and wash the precipitate several times by adding about 30 mL of hot water, stirring, and decanting the liquid. Filtration is not necessary at this point.

To the wet iron(II) oxalate, add a solution of 6.6 g of $K_2C_2O_4\cdot H_2O$ in 18 mL of water and heat the mixture to about 40 °C. *Slowly and cautiously* add 17 mL of 6% H_2O_2 while stirring constantly and maintaining the temperature at 40 °C. After the addition of H_2O_2 is complete, heat the mixture to boiling and add a solution containing 1.7 g of oxalic acid in 15 mL of water. When adding the oxalic acid solution, add the first 8 mL all at once and the remaining 5 mL dropwise, keeping the temperature near boiling. Remove any solid by gravity filtration and add 20 mL of 95% ethanol to the filtrate. Cover the beaker with a watch glass and store it in your lab desk until the next laboratory period. Filter by suction using a Büchner funnel and filter flask (Figure 36.3) and wash the green crystals with a 50% aqueous ethanol solution, then with acetone, and air-dry. Determine the mass of the product and store it in a vial in the dark. This complex is photosensitive and reacts with light as follows:

$$[Fe(C_2O_4)_3]^{3-} \xrightarrow{h\nu} [Fe(C_2O_4)_2]^{2-} + 2CO_2 + e^-$$

GIVE IT SOME THOUGHT

What is the purpose of adding the hydrogen peroxide in this step?

To demonstrate this, place a small specimen on a watch glass near the window and observe any changes that occur during the lab period. You should obtain about 8 g of product. Calculate the theoretical yield and determine your percent yield. Your instructor may tell you to save your sample for analysis in Experiment 37 or dispose of it as directed. Keep it away from light by wrapping the vial with aluminum foil.

4. Preparation of $K_3[Al(C_2O_4)_3]\cdot 3H_2O$

$$Al + 3KOH + 3H_2C_2O_4 \cdot 2H_2O \longrightarrow K_3[Al(C_2O_4)_3]\cdot 3H_2O + 6H_2O + \frac{3}{2}H_2$$

Place 1 g of aluminum powder in a 200 mL beaker and cover with 10 mL of hot water. Add 20 mL of 6 *M* KOH solution in small portions to regulate the vigorous evolution of hydrogen. Finally, heat the liquid almost to boiling on a hot plate to dissolve any residual metal. Maintain the heating and in small portions, add a solution of 13 g of oxalic acid in 100 mL of water. During the neutralization, hydrated alumina will precipitate, but it will redissolve at the end of the addition after gentle boiling. Cool the solution in an ice bath and add 50 mL of 95% ethanol. If oily material separates, stir the solution and scratch the sides of the beaker with your glass rod to induce crystallization. Suction-filter the product using the Büchner funnel (see Figure 36.3) and suction flask and wash with a 20 mL portion of ice-cold 50% aqueous ethanol and then with small portions of absolute ethanol. Air-dry the product, determine its mass and store it in a stoppered bottle. You should obtain about 11 g of product. Calculate the theoretical yield and determine your percent yield. Your instructor may tell you to save your sample for analysis in Experiment 37 or dispose of as directed.

Preparation of Materials for the Demonstration

(These preparations should be done a week before the laboratory period.)

cis-$K[Cr(C_2O_4)_2(H_2O)_2]\cdot 2H_2O$

$$K_2Cr_2O_7 + 7H_2C_2O_4 \cdot 2H_2O \rightleftharpoons 2K[Cr(C_2O_4)_2(H_2O)_2]\cdot 2H_2O + 6CO_2 + 13H_2O$$

In a hood, separately powder in a *dry* mortar 12 g of oxalic acid dihydrate and 4 g of potassium dichromate. Mix the powders as intimately by grinding gently in the mortar. Moisten a large evaporating dish (10 cm) with water and pour off all of the water but do not wipe dry. Place the powdered mixture in the evaporating dish as a *compact heap*; it will become moistened by the water that remains in the evaporating dish. Cover the evaporating dish with a large watch glass and warm it gently on a hot plate. A vigorous spontaneous reaction will soon occur and will be accompanied by frothing as steam and CO_2 escape. The mixture should then liquefy to a deep-colored syrup. Pour about 20 mL of 95% ethanol on the hot liquid and continue to warm it gently on the hot plate. Triturate (grind or crush) the product with a spatula until it solidifies. If you cannot effect complete solidification with one portion of 95% alcohol, decant the liquid, add another 20 mL of 95% alcohol, warm gently, and resume the trituration until the product is entirely crystalline and granular. The yield is essentially quantitative at about 9 g. This compound is intensely dichroic (appears different colors when viewed from different directions or lighting conditions), appearing in the solid state as almost black in diffuse daylight and deep purple in artificial light.

***trans*-$K[Cr(C_2O_4)_2(H_2O)_2]\cdot 3H_2O$** Dissolve 12 g of oxalic acid dihydrate in a minimum of boiling water in a 300 mL (or larger) beaker. In a hood, add to this in small portions a solution of 4 g of potassium dichromate in a minimum of hot water and cover the beaker with a watch glass while the violent reaction proceeds. After the addition is complete, cool the contents of the beaker and allow spontaneous evaporation at room temperature to occur so that the solution reduces to about one-third its original volume (this takes 36 to 48 hours). Collect the deposited crystals by suction filtration, wash several times with cold water and 95% alcohol, and air-dry. The yield is about 6.5 g. The complex is rose-colored with a violet tinge and is not dichroic.

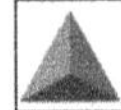

GIVE IT SOME THOUGHT

What are the differences in the physical properties of the cis and trans isomers?

Waste Disposal Instructions All oxalate- metal-containing solutions, and ethanol containing solutions should be disposed of in appropriate waste containers.

Name ______________________________ Desk ______________

Date ______________ Laboratory Instructor ______________________

Preparation and Reactions of Coordination Compounds: Oxalate Complexes | 36 Pre-lab Questions

Before beginning this experiment in the laboratory, you should be able to answer the following questions.

1. Give an example of a *Lewis acid* and a *Lewis base*.

2. Give three examples of molecules that can act as *ligands*.

3. Define and give an example of a coordination compound.

4. Define the term *geometric isomer*.

5. Draw structures for all possible isomers of the six-coordinate compounds $[FeCl_2(NH_3)_2(CO)_2]$ and $[NiCl_4(NO)_2]$.

6. Are the water molecules in equivalent environments in the following compounds $[Cu(NH_3)_4(H_2O)_2]^{2+}$ and $[Cu(NH_3)_2(H_2O)_4]^{2+}$?

7. What is the meaning of *dichroism*?

8. What is the meaning of *trituration*?

9. Look up the preparation of an oxalate complex of Ni, Mn, or Co. Cite your reference and state whether this preparation would be suitable to add to this experiment. Explain.

10. Find an analytical method to determine the amount of Fe, Cu, Cr, or Al in your oxalate complex. Cite the reference to the method. Could you determine the amount using the chemicals and equipment available in your laboratory? Why or why not?

11. Oxalic acid is used to remove rust and corrosion from automobile radiators. How do you think it works?

Name ______________________________ Desk ______________

Date ______________ Laboratory Instructor ______________________

REPORT SHEET | EXPERIMENT

Preparation and Reactions of Coordination Compounds: Oxalate Complexes

36

1. Complex prepared ______________
2. Chemical reaction for its preparation
3. Theoretical yield of oxalate complex (show calculations):
4. Experimental yield of oxalate complex ______________
5. Percent yield of oxalate complex (show calculations):
6. Color and general appearance of complex:
7. Describe the reactions of *cis*- and *trans*-$K[Cr(C_2O_4)_2(H_2O)_2]$ with NH_3 and the reverse reactions with HCl using chemical equations. List any observations, such as color changes and apparent solubilities.
8. Is your complex soluble in H_2O ? __________ Alcohol? __________ Acetone? __________

QUESTIONS

1. Sodium trioxalatocobaltate(III) trihydrate is prepared by the following reactions:

$$[Co(H_2O)_6]Cl_2 + K_2C_2O_4 \cdot H_2O \longrightarrow CoC_2O_4 + 2KCl + 7H_2O$$

$$2CoC_2O_4 + 4H_2O + H_2O_2 + 4Na_2C_2O_4 \longrightarrow 2Na_3[Co(C_2O_4)_3] \cdot 3H_2O + 2NaOH$$

What is the percent yield of $Na_3[Co(C_2O_4)_3] \cdot 3H_2O$ if 7.6 g is obtained from 12.5 g of $[Co(H_2O)_6]Cl_2$?

2. Why are $K_3[Cr(C_2O_4)_3] \cdot 3H_2O$, $K_2[Cu(C_2O_4)_2] \cdot 2H_2O$ and $K_3[Fe(C_2O_4)_3] \cdot 3H_2O$ colored, whereas $K_3[Al(C_2O_4)_3] \cdot 3H_2O$ is colorless?

3. What are the names of the following compounds?

   a. $K_3[Cr(C_2O_4)_3] \cdot 3H_2O$
   b. $K_2[Cu(C_2O_4)_2] \cdot 2H_2O$
   c. $K_3[Fe(C_2O_4)_3] \cdot 3H_2O$
   d. $K_3[Al(C_2O_4)_3] \cdot 3H_2O$

4. What is the percent oxalate in each of the following compounds?

   a. $K_3[Cr(C_2O_4)_3] \cdot 3H_2O$
   b. $K_2[Cu(C_2O_4)_2] \cdot 2H_2O$
   c. $K_3[Fe(C_2O_4)_3] \cdot 3H_2O$
   d. $K_3[Al(C_2O_4)_3] \cdot 3H_2O$

Oxidation–Reduction Titrations I: Determination of Oxalate

Experiment

37

OBJECTIVE

To gain familiarity with redox chemistry through analysis of an oxalate sample.

APPARATUS AND CHEMICALS

Apparatus

50 mL buret	balance
buret clamp	weighing bottle
400 mL beakers (3)	thermometer
glass stirring rods	Bunsen burner and hose
ring stand	Wash bottle

Chemicals

an oxalate sample (either an unknown or one of the complexes from Experiment 36)	1.0 *M* H_2SO_4
	$Na_2C_2O_4$ (primary standard)
	~0.02 *M* $KMnO_4$

DISCUSSION

Potassium permanganate reacts with oxalate ions to produce carbon dioxide and water in an acidic solution, and the permanganate ion is reduced to manganese(II) as follows:

$$5C_2O_4^{2-}(aq) + 2MnO_4^-(aq) + 16H^+(aq) \longrightarrow 10CO_2(g) + 8H_2O(l) + 2Mn^{2+}(aq) \quad [1]$$

Because this reaction proceeds slowly at room temperature, the solution must be heated gently to ensure that satisfactory reaction rates are obtained. No indicators are necessary in permanganate titrations because the end points are easily observed. The permanganate ion is intensely purple, whereas the manganese(II) ion is nearly colorless. The first slight excess of permanganate imparts a pink color to the solution, signaling that all of the oxalate has been consumed.

The balanced half-reactions associated with Equation [1] are (Section 20.2) as follows:

Oxidation: $$C_2O_4^{2-}(aq) \longrightarrow 2CO_2(g) + 2e^- \quad [2]$$

Reduction: $$MnO_4^-(aq) + 8H^+(aq) + 5e^- \longrightarrow Mn^{2+}(aq) + 4H_2O(l) \quad [3]$$

In this experiment, you will standardize a $KMnO_4$ solution—that is, you will determine its exact molarity—by titrating it against a very pure sample of sodium oxalate, $Na_2C_2O_4$. You will then use your standardized $KMnO_4$ to determine the percentage of oxalate ion, $C_2O_4^{2-}$, in either an unknown sample or the complex you prepared in Experiment 36. The basis of the determination is that the reagents react in a molar ratio of 5:2.

$$5\times(\text{mol } MnO_4^-) = 2\times(\text{mol } C_2O_4^{2-}) \qquad [4]$$

GIVE IT SOME THOUGHT

a. How do you go from the balanced half-reactions in Equations [2] and [3] to the overall balanced reaction in Equation [1]?

b. How does balancing the equation relate to the 5:2 ratio in Equation [4]?

Measuring the volume of $KMnO_4$ that reacts with a known mass of sodium oxalate, $Na_2C_2O_4$, allows you to calculate the molarity of the $KMnO_4$ solution. Recall that the product of the volume in liters and the molarity of a solution is moles, as shown in Equation [5] (Section 4.5).

$$M \times V_L = \text{moles} = \frac{mass}{molar\ mass} \qquad [5]$$

Because volumes are measured in milliliters, a more convenient form of Equation [5] is

$$M \times V_{mL} = \text{mmol} = \frac{mass(10^3)}{molar\ mass} \qquad [6]$$

where, V_{mL} is the volume in mL and mmol is millimoles (that is, moles times 10^{-3}). Equation [6] expresses mass in mg. Example 37.1 illustrates the calculation of molarity for a standardization; Example 37.2 illustrates analysis for oxalate.

EXAMPLE 37.1

What is the molarity of a $KMnO_4$ solution if 40.41 mL is required to titrate 0.2538 g of $Na_2C_2O_4$?

SOLUTION: The reaction proceeds according to Equation [1]. At the equivalence point

$$\text{mmol } KMnO_4 = 2/5 \text{ mmol } Na_2C_2O_4$$

The number of mmoles $Na_2C_2O_4$ is

$$\frac{0.2538\text{ g}}{134.0\text{ g/mol}} \times \frac{10^3\text{ mmol}}{1\text{ mol}} = 1.894\text{ mmol } Na_2C_2O_4$$

Now $M \times V_{mL}$ = the number of mmoles for $KMnO_4$. Thus,

$$M \times 40.41\text{ mL} = \frac{2\text{ mmol } Na_2C_2O_4}{5\text{ mmol } KMnO_4}(1.894\text{ mmol) } KMnO_4$$

and

$$M = \frac{0.7576\text{ mmol}}{40.41\text{ mL}} = 0.01875\ M\ KMnO_4$$

EXAMPLE 37.2

What is the percent oxalate, $C_2O_4^{2-}$ by mass, in a 1.429 g sample if 34.21 mL of 0.02000 *M* $KMnO_4$ solution is required for titration?

SOLUTION: You want to find the following:

$$\% \ C_2O_4^{2-} = \frac{\text{mass of } C_2O_4^{2-}}{\text{mass of sample}} \times 100\%$$

Thus, you need to know the mass of $C_2O_4^{2-}$ in the sample with a mass of 1.429 g. At the equivalence point

$$\begin{aligned}\text{mmol } C_2O_4^{2-} &= \frac{5 \text{ mmol } Na_2C_2O_4}{2 \text{ mmol } KMnO_4}(\text{mmol}) \ KMnO_4 \\ &= 5/2\ (0.02000 \text{ mmol/mL})(34.21 \text{ mL}) \\ &= 1.711 \text{ mmol } C_2O_4^{2-}\end{aligned}$$

The corresponding mass of $C_2O_4^{2-}$ is

$$m = 1.711 \text{ mmol } C_2O_4^{2-} \times \frac{1 \text{ mol } C_2O_4^{2-}}{10^3 \text{ mmol } C_2O_4^{2-}} \times \frac{88.02 \text{ g } C_2O_4^{2-}}{\text{mol } C_2O_4^{2-}}$$

$$= 0.1506 \text{ g } C_2O_4^{2-}$$

and

$$\% C_2O_4^{2-} = \frac{0.1506 \text{ g } C_2O_4^{2-}}{1.429 \text{ g}} \times 100\% = 10.54\%$$

PROCEDURE

A. Preparation of $KMnO_4$ Solution

Potassium permanganate solutions are not stable over long periods of time because they react slowly with the organic matter present even in most distilled water. The solution should be protected from heat and light as much as possible because both induce decomposition, producing manganese dioxide, which seems to act as a catalyst for further decomposition. Because commercial samples of the permanganate usually contain some manganese dioxide, it is advisable to boil and filter the permanganate solutions before standardizing. Almost any form of organic matter will reduce permanganate solutions, so care must be taken to keep the solutions out of contact with rubber, filter paper, dust particles, etc. For this reason, permanganate solutions are filtered through sintered glass.

GIVE IT SOME THOUGHT

a. What is the oxidation state of Mn in $KMnO_4$?
b. How does this oxidation state influence its stability?
c. Is your answer consistent with the details of this paragraph?

For this determination, you will be provided with a filtered potassium permanganate solution that is approximately 0.02 *M*, which will have to be standardized.

B. Standardization of Permanganate Solution—Sodium Oxalate Method

Several reducing agents can be used as primary standards for permanganate solutions, but sodium oxalate is commonly used.

The oxidation of the oxalate ion is carried out in an acid solution that is maintained at between 80° and 90 °C. This oxidation is catalyzed by the Mn^{2+} ion, which is a product of the oxidation–reduction reaction. The intense purple color of the permanganate ion may persist until several milliliters of the permanganate reagent have been added; however, when sufficient Mn^{2+} ions have been formed to catalyze the reaction, the pale purple color suddenly disappears and continues to do so with each drop of reagent until the equivalence point is reached, at which point one drop in excess will turn the solution pale purple.

The titration can be accomplished more quickly if you know the approximate volume of $KMnO_4$ required. Assume that a 0.20 g sample of $Na_2C_2O_4$ is used. This corresponds to the following:

$$\text{mmol } Na_2C_2O_4 = 0.20 \text{ g} \times \frac{1 \text{ mol}}{134.0 \text{ g}} \times \frac{10^3 \text{ mmol}}{1 \text{ mol}} = 1.5 \text{ mmol}$$

The reaction requires 2/5 mmol $KMnO_4$ for each mmol $Na_2C_2O_4$. If a 0.020 *M* $KMnO_4$ solution is used, the number of mL required in a titration is (see Equations [4] and [5])

$$\frac{2}{5}(1.5 \text{ mmol}) = V_{\text{mL}} \times 0.020 \text{ mmol/mL}$$

$$V_m = 30 \text{ mL}$$

A volume of $KMnO_4$ solution up to within about 5.0 mL of this amount (that is, up to 25 mL) can be rapidly added to the hot acidified oxalate solution without much danger of overrunning the end point. The permanganate solution is then added dropwise until one drop in excess turns the solution pale purple.

Procedure Using a weighing bottle determine the mass to the nearest 0.1 mg of triplicate portions of about 0.20 g each of pure sodium oxalate. Add the portions to three separate 400-mL beakers and label the beakers. Run each trial as follows: Add about 250 mL of 1.0 *M* sulfuric acid; then stir the solution (not with the thermometer) and warm the mixture until the oxalate has dissolved and the temperature has been brought to 80° to 90 °C. After you take the initial reading of the permanganate in the buret (see Note 1), titrate with the permanganate, stirring constantly, while keeping the solution above 70 °C at all times. Add the permanganate dropwise when near the end point, allowing each drop to decolorize before adding the next. The end point is reached when the faintest visible shade of pale purple remains even after the solution has been allowed to stand for 15 s (see Note 2). Again, read the buret and record the volume of permanganate solution used.

GIVE IT SOME THOUGHT

a. At the equivalence point of an acid–base titration, moles of acid = moles of base. Is this relationship present at the end point?

b. If not, what is the key relationship at the end point of this titration?

From the data obtained, calculate the molarity of the potassium permanganate solution. The three results should agree within 3 ppt (parts per thousand). If they do not, titrate another sample.

Note 1 A 0.02 *M* permanganate solution is so deeply colored that the bottom of the meniscus is difficult to read. Thus, you must read the top of the surface of the solution, taking the usual precautions to have your eyes level with the solution surface when making the reading. Remember, the buret reading can be estimated to ±0.02 mL.

Note 2 If there is any doubt as to whether an end point has been reached, a good practice is to take the buret reading and then add another drop of the permanganate. The development of an intense color indicates that the reading did correspond to the true end point and the previous reading is a better estimate.

Ordinarily, a permanganate end point is not permanent. Due to the action of dissolved reducing species in the water, the permanganate is slowly reduced and the color fades. A pink color that remains after you have been stirring the solution for 15 s should be taken as the true end point.

C. The Determination of Oxalate in Your Oxalate Complex or Unknown

The unknown for this determination may be one that your instructor provides or the oxalate complex you prepared in Experiment 36, in which case you should read Note 3 before starting the procedure.

Procedure Using a weighing bottle determine the mass to the nearest 0.1 mg of triplicate portions of about 0.2 to 0.3 g of pure sodium oxalate or about 0.5 g of unknown. Add the portions to three separate 400-mL beakers and label the beakers. Add about 250 mL of 1.0 *M* sulfuric acid and mix to dissolve the sample. Titrate with the standardized permanganate solution as described in Procedure B. Calculate the mass percent of $C_2O_4^{2-}$. A standard deviation of 0.3 or better represents acceptable results.

Note 3 Chromium oxalate complexes are difficult to analyze by titration with MnO_4^- because of their dark colors and slow rates of decomposition. To analyze them, decompose the complex by boiling about 0.6 to 1.0 g of the complex in 10 mL of water and 10 mL of 3 *M* KOH for 15 min. Filter off the green solid $Cr(OH)_3$ and wash with distilled water. Combine the filtrate and washings, dilute to 25 mL in a volumetric flask, pipet 5 mL of this solution into a 400 mL beaker and titrate with MnO_4^- as above.

Waste Disposal Instructions Dispose of all oxalate- and metal-containing solutions in appropriate containers in the laboratory.

NOTES AND CALCULATIONS

Name ____________________ Desk ____________________

Date ____________________ Laboratory Instructor ____________________

Oxidation–Reduction Titrations I: Determination of Oxalate

37 Pre-lab Questions

Before beginning this experiment in the laboratory, you should be able to answer the following questions.

1. If 0.5543g of sodium oxalate, $Na_2C_2O_4$, requires 35.65 mL of a $KMnO_4$ solution to reach the end point. What is the molarity of the $KMnO_4$ solution?

2. Titration of an oxalate sample gave the following percentages: 15.53%, 15.55%, and 15.56%. Calculate the average and the standard deviation.

3. What is the color change at the end point of permanganate titrations?

4. Why is the $KMnO_4$ solution filtered, and why should it not be stored in a rubber-stoppered bottle?

5. What volume of 0.100 M $KMnO_4$ would be required to titrate 0.55 g of $K_2[Cu(C_2O_4)_2] \cdot 2H_2O$?

---

6. Calculate the percent $C_2O_4^{2-}$ in each of the following: $K_2[Cu(C_2O_4)_2] \cdot 2H_2O$, CuC_2O_4, $Na_2[Cu(C_2O_4)_2] \cdot 4H_2O$ and $NaK[Cu(C_2O_4)_2] \cdot 2H_2O$.

7. If 28.60 mL of 0.0200 M $KMnO_4$ is required to titrate a 0.255 g sample of $K_2[Cu(C_2O_4)_2] \cdot 2H_2O$, what is the percent $C_2O_4^{2-}$ in the complex?

8. What is the percent purity of the complex in question 7?

Name ______________________ Desk ______________

Date ______________ Laboratory Instructor ______________________

Unknown no. or complex ______________

REPORT SHEET | EXPERIMENT

Oxidation–Reduction Titrations I: Determination of Oxalate

37

B. Standardization of $KMnO_4$

Mass of $Na_2C_2O_4$	*Trial 1*	*Trial 2*	*Trial 3*
Weighing bottle initial mass	______	______	______
Weighing bottle final mass	______	______	______
Mass of $Na_2C_2O_4$	______	______	______
Titration volume of $KMnO_4$			
Final reading	______	______	______
Initial reading	______	______	______
Volume of $KMnO_4$	______	______	______
Calculations			
mMoles of $Na_2C_2O_4$	______	______	______
Volume of $KMnO_4$	______	______	______
mMoles of $KMnO_4$	______	______	______
Molarity of $KMnO_4$	______	______	______

Average molarity ______ See example 37.1

Standard deviation (show calculations) ______

C. Analysis of Oxalate Complex or Unknown

Mass of sample	*Trial 1*	*Trial 2*	*Trial 3*
Weighing bottle initial mass	______	______	______
Weighing bottle final mass	______	______	______
Mass of sample	______	______	______
Titration			
Final reading	______	______	______
Initial reading	______	______	______
Volume of $KMnO_4$	______	______	______
Calculations			
Millimoles of oxalate, $C_2O_4^{2-}$	______	______	______
Mass of oxalate, $C_2O_4^{2-}$	______	______	______
Percent oxalate, $C_2O_4^{2-}$	______	______	______

Average percent oxalate ______ Standard deviation ______

If you analyzed your oxalate complex from Experiment 36, complete the following:

Theoretical percent oxalate in your complex (show calculations) ______

Determine the purity of your complex as $\frac{\text{experimental \% oxalate}}{\text{theoretical \% oxalate}} \times 100\% = \%$ purity.

QUESTIONS

1. Balance the following half reactions:

   (a) $MnO_4^- + e^- + H^+ \longrightarrow Mn^{2+} + H_2O$

   (b) $MnO_4^- + e^- + H^+ \longrightarrow MnO_2 + H_2O$

   (c) $MnO_4^- + e^- + H^+ \longrightarrow Mn^{3+} + H_2O$

2. How many grams of $KMnO_4$ are required to prepare 2.000 L of 0.0300 *M* $KMnO_4$ solution that is used in titrating $Na_2C_2O_4$?

3. MnO_4^- reacts with Fe^{2+} in acid solution to produce Fe^{3+} and Mn^{2+}. Write a balanced equation for this reaction.

Enthalpy of Vaporization and Clausius–Clapeyron Equation

Experiment

38

OBJECTIVE

To determine the enthalpy of vaporization of a compound by measuring the vapor pressure as a function of temperature.

APPARATUS

Apparatus

distilled water	clamp, three prong
10 mL graduated cylinder	clamp, thermometer
thermometer, alcohol	magnetic stirring hot plate
1000 mL tall form beaker	stir bar
25 mL beaker	plastic transfer pipets
ring stand	

WORK IN GROUPS OF TWO OR THREE, DEPENDING ON EQUIPMENT AVAILABILITY, BUT ANALYZE THE DATA INDIVIDUALLY

DISCUSSION

A liquid in an open container undergoes the process of evaporation which involves molecules escaping from the surface of a liquid into the gas phase. A liquid placed into an evacuated and closed container will begin to evaporate and generate a pressure above the liquid. After sufficient time, the pressure of the vapor attains a constant value called the vapor pressure. At equilibrium the rate of molecules from the liquid entering into the vapor phase by evaporation is equal to the rate of molecules in the gas phase entering the liquid phase in the process of condensation. The process is not static and molecules are continuously entering and leaving the liquid and vapor states. The liquid and vapor are in a *dynamic equilibrium* (Section 11.5).

The vapor pressure of a substance increases with temperature in a nonlinear manner. The vapor pressure for four liquids as a function of temperature is shown in Figure 38.1. At a higher temperature the molecules have a higher average kinetic energy and a larger fraction of the molecules at the surface have sufficient energy to overcome the attractive forces in the liquid and escape into the gas phase. A liquid with molecules that have weaker intermolecular forces has molecules that can more easily escape into the gas phase and therefore has a higher vapor pressure at a given temperature. The tendency of a liquid phase to enter the vapor phase can be qualitatively described as its volatility. A more volatile compound has a higher equilibrium vapor pressure at a given temperature.

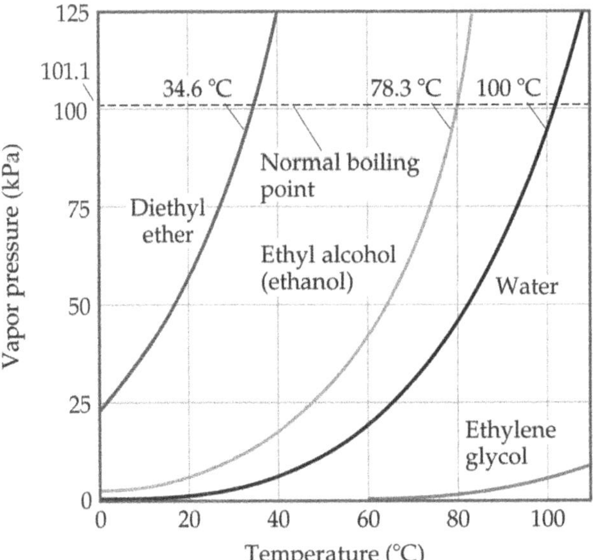

▲**FIGURE 38.1** Vapor pressure (1 torr = 0.1333 kPa) for four liquids as a function of temperature.

The scientists Rudolf Clausius and Benoît Paul Émile Clapeyron showed that for a pure liquid a linear relation exists between the reciprocal of the absolute temperature in Kelvin ($1/T$) and the natural logarithm of the vapor pressure. We denote this equation as the Clausius-Clapeyron equation.

$$\ln P = \frac{-\Delta H_{vap}}{RT} + C$$

where P is the vapor pressure, T is the absolute temperature, R is the ideal gas constant ($R = 8.314$ J/mol·K), ΔH_{vap} is the molar enthalpy of vaporization (in J/mol), and C is a constant. Rearranging the equation, one can determine that a plot of ln P (as the ordinate, vertical axis) versus $1/T$ (as the abscissa, horizontal axis) gives a line with slope equal to $-\Delta H_{vap}/R$

$$\ln P = \frac{-\Delta H_{vap}}{R}\left(\frac{1}{T}\right) + C$$

Note the slope of the line has units of kelvins.

The volume of a gas above a liquid can be measured using an eudiometer, which is available in two common sizes of 50 mL and 100 mL. The volume changes in this experiment are relatively small, so an inverted 10 mL graduated cylinder will be used in place of an eudiometer. The graduated cylinder is calibrated for the meniscus of an aqueous solution in a standard vertical orientation. The inversion of the graduated cylinder in this experiment requires a correction to the measured volume, which is obtained by subtracting 0.2 mL from the measured volume.

Example 38.1

In Figure 38.2 the recorded volume of the vapor is 3.4 mL, thus the corrected volume is 3.2 mL.

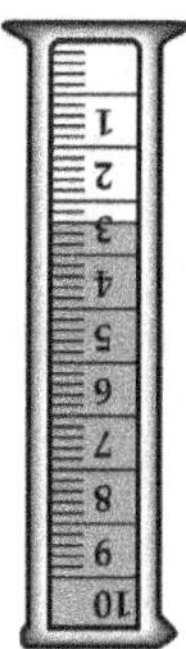

▲**FIGURE 38.2** Inverted graduated cylinder.

EXAMPLE 38.2

The vapor pressure of a pure liquid was measured at several temperatures and is given in the table below. Determine the enthalpy of vaporization for the pure liquid in kJ.mol.

P **(kPa)**	***T*** **(°C)**
0.1333	−15.0
1.33	14.7
5.32	36.4
13.3	52.8
53.3	82.0

SOLUTION: Create a table of ln *P*, temperature in K, and reciprocal temperature in K^{-1}. Graph ln *P* versus 1/*T*. Obtain the slope in units of K.

T **(°C)**	***T*** **(K)**	**1/*T*** **(K^{-1})**	**P_{vap} (kPa)**	**ln P_{vap}**
−15	258.2	0.00387	0.1333	-2
14.7	287.9	0.00347	1.33	0.29
36.4	309.6	0.00323	5.32	1.67
52.8	326.0	0.00307	13.3	2.58
82	355.2	0.00282	53.3	4.00

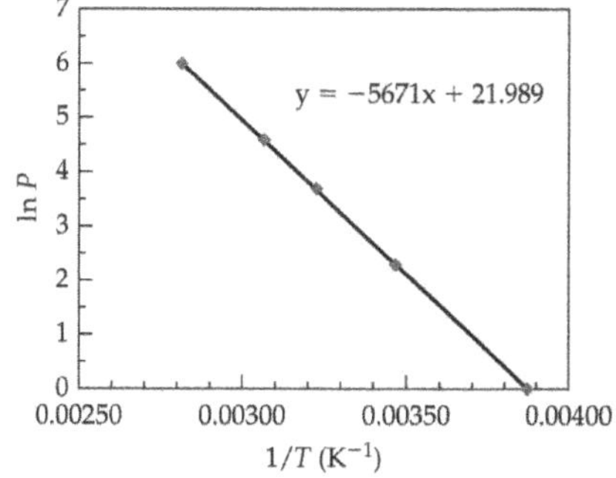

Solve for $\Delta H_{vap} = -m \times R$, where m is the slope of the line. $\Delta H_{vap} = -(-5671\text{ K}) \times (8.314\text{ J/mol}\cdot\text{K})(1\text{ kJ}/1000\text{ J}) = 47.1\text{ kJ/mol}$.

PROCEDURE

A. Determination of enthalpy of vaporization of water

Construct the experimental apparatus similar to the one shown in Figure 38.3. Add about 700 mL distilled water to partly fill about 3/4 of a 1000 mL tall form beaker. Fill a 10 mL graduated cylinder with distilled water and, while plugging the end, invert in the tall form beaker. The bubble inside the graduated cylinder needs to be about 4 mL. If the bubble is smaller than 4 mL, use a transfer pipet to force air bubbles under the graduated cylinder until the desired volume is obtained. Gently clamp a 25 mL beaker on top of the graduated cylinder to aid in leveling and to keep the graduated cylinder stable. Ensure the stirbar does not impact the inverted graduated cylinder.

GIVE IT SOME THOUGHT

What will happen to the volume of the bubble inside the cylinder if the temperature of the water reaches the boiling point?

Begin heating the water bath while stirring. Ensure that the rate of stirring doesn't create cavitation and introduce bubbles near the opening of the graduated cylinder. Continue heating until the air bubble inside the graduated cylinder just surpasses the 10 mL reading or the temperature reaches 80 °C. Turn off the heating and allow the apparatus to cool under constant stirring. Collect a data point on cooling approximately every 5 °C. Measure and record the volume to the nearest 0.02 mL and temperature to the nearest 0.1 °C. After collecting a data point, the cooling process may be advanced by removing a portion of the water in the beaker and adding cold water or ice in small portions. Ensure the ice is completely melted and the temperature of the water has approached equilibrium before conducting the next measurement. Continue collecting data until the temperature approaches 4 or 5 °C. The ice floating at the top of the beaker may not melt completely for the final data point.

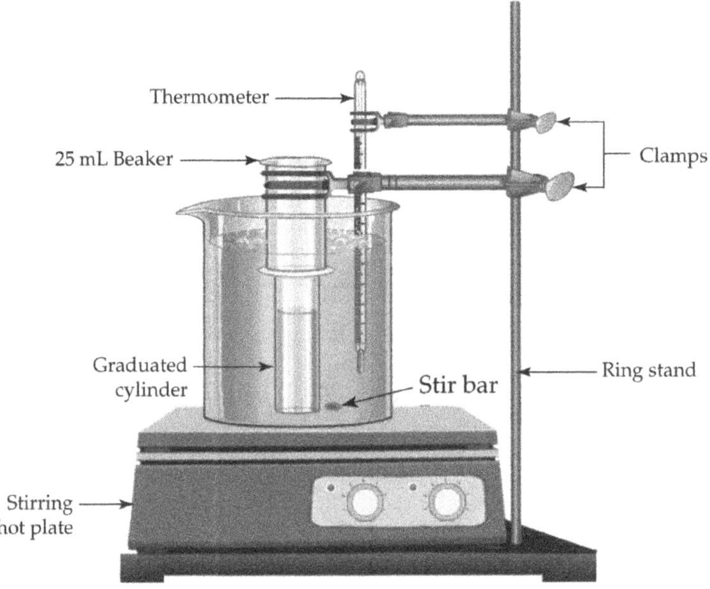

▲FIGURE 38.3 Experimental apparatus to measure the volume of the gas bubble as a function of temperature.

Obtain a value of the ambient room temperature in kelvins and barometric pressure from your lab instructor. Be sure to include units on the numbers.

CALCULATIONS

As the freezing point of a liquid is approached, the vapor pressure of a liquid approaches zero, as shown in the vapor pressure versus temperature graph in Figure 38.1. Near 4 °C, the vapor pressure of water is minimal (<1% of the total pressure) and the bubble is primarily composed of air. To convert pressure between units of atmospheres and kPa, use the relationship 1 atm = 101.3 kPa.

The ideal gas equation, $PV = nRT$, solved for moles of air is $n = PV/RT$, where P is the atmospheric pressure in kPa, V is the corrected volume in liters at the lowest collected temperature (~4–5 °C), R is the ideal gas constant (8.314 L-kPa/mol-K), and T is the absolute temperature in kelvins.

The moles of air are constant inside the cylinder, but the partial pressure of the air varies with temperature. Calculate the partial pressure of air at each temperature by solving the ideal gas equation for $P = nRT/V$, where n is the calculated moles of air, R is the ideal gas constant (8.314 L-kPa/mol-K), T is the absolute temperature in Kelvin, and V is the corrected volume in liters at the measured temperature.

Dalton's Law of partial pressure states that the total pressure is the sum of the individual pressures. Thus, $P_{\text{atmospheric}} = P_{\text{air}} + P_{H_2O}$, and $P_{H_2O} = P_{\text{atmospheric}} - P_{\text{air}}$ for each temperature.

From these data, prepare a table of ln P_{H_2O} and the reciprocal of the absolute temperature ($1/T$) with units of K^{-1}.

Prepare a graph of ln P_{H_2O} (as the ordinate, vertical axis) versus reciprocal temperature in K^{-1} (as the abscissa, horizontal axis). Add a trendline and determine the slope of the line, with proper units. Calculate ΔH_{vap} using the Clausius-Clapeyron equation. Ensure units are used in all calculations, where appropriate.

The accepted enthalpy of vaporization for water is 40.68 kJ/mol. Calculate the percent error.

$$\text{Percent Error} = \frac{(\text{experimental value} - \text{accepted value})}{\text{accepted value}} \times 100\%$$

NOTES AND CALCULATIONS

Name ______________________________ Desk ______________________

Date ____________________ Laboratory Instructor ______________________

Enthalpy of Vaporization and Clausius–Clapeyron Equation | 38 Pre-lab Questions

Before beginning this experiment in the laboratory, you should be able to answer the following questions.

1. Read the volume to the nearest 0.1 mL on the graduated cylinder illustrated below. Determine the corrected volume for the inverted graduated cylinder.

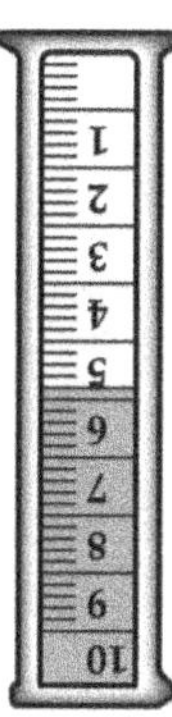

2. Why does a correction to the volume need to be made for the inverted graduated cylinder?

3. At 332 K the bubble volume was measured to be 2.6 mL in an inverted graduated cylinder. After correcting the volume, determine the moles of air in the bubble assuming a pressure of 101.3 kPa.

4. Given the following corrected volumes and temperatures, determine the enthalpy of vaporization ΔH_{vap}. Attach the labeled graph.

P (kPa)	T (°C)
0.1995	−44.0
1.6	−16.2
6.66	5.0
16.00	21.2
60.00	49.9

Name ____________________ Desk ____________________

Date ____________________ Laboratory Instructor ____________________

REPORT SHEET | EXPERIMENT

Enthalpy of Vaporization and Clausius–Clapeyron Equation | 38

A. Determination of enthalpy of vaporization of water

Atmospheric pressure ____________________

Ambient temperature ____________________

Volume at 4 °C ____________________ Corrected Volume at 4 °C ____________________

Moles of air ____________________

Volume (mL)	Corrected Volume (mL)	T (°C)	P_{air} (kPa)	P_{H_2O} (kPa)	ln P_{H_2O}	$1/T$ (K^{-1})

Slope ____________________

ΔH_{vap} ____________________

Percent error ____________________

Show your calculations on the next page.

Calculations:

QUESTIONS

1. Would you expect the vapor pressure of diethyl ether to be smaller or larger than that of water at a particular temperature below the boiling point?

2. Why does the partial pressure of air change with temperature?

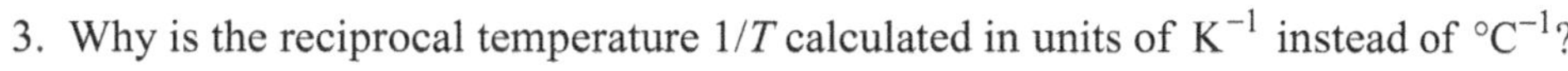

3. Why is the reciprocal temperature $1/T$ calculated in units of K^{-1} instead of $°C^{-1}$?

4. Using your experimental data and slope equation, determine the vapor pressure of water at 90 °C.

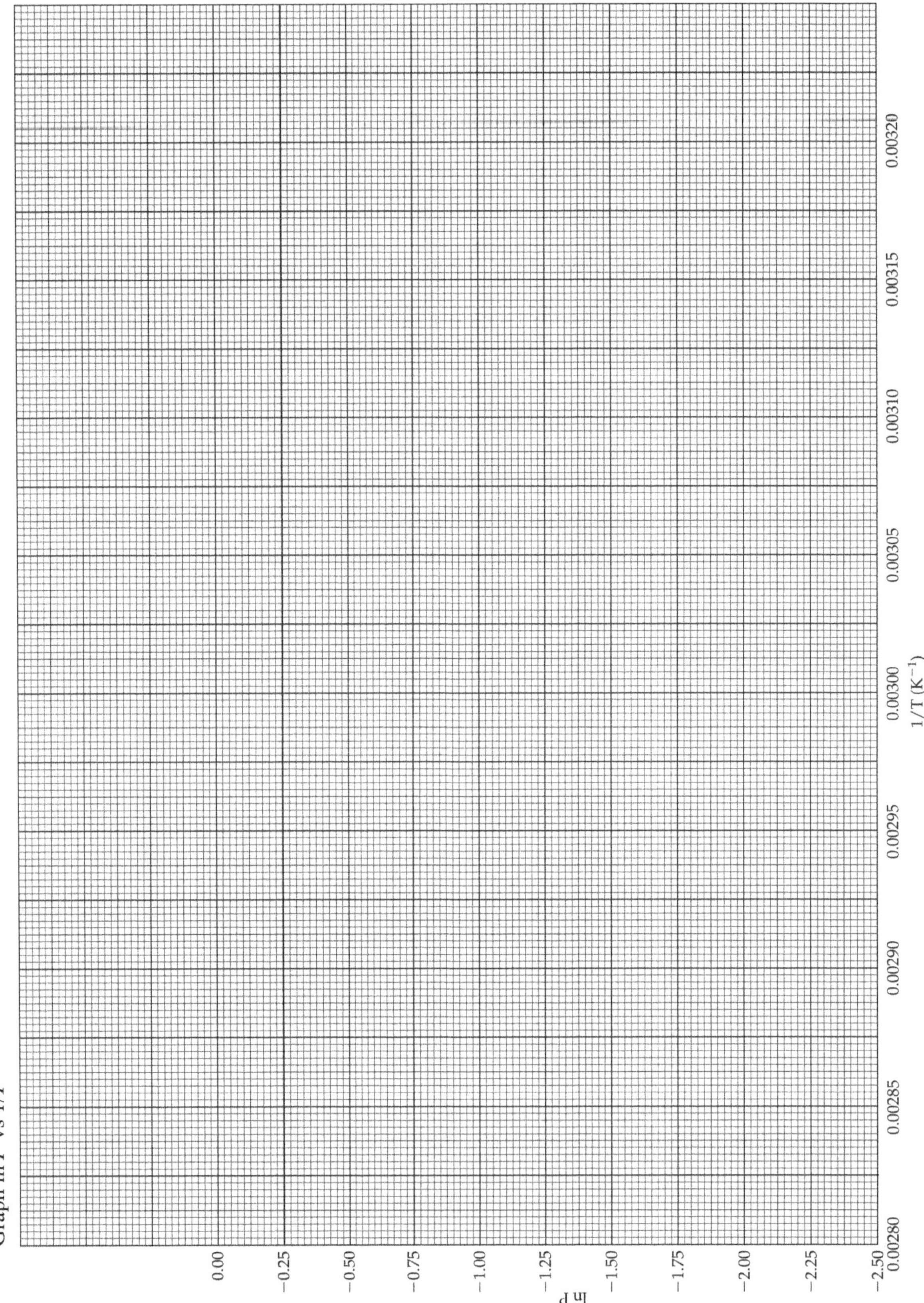
Graph ln P vs 1/T
ln P
0.00
−0.25
−0.50
−0.75
−1.00
−1.25
−1.50
−1.75
−2.00
−2.25
−2.50
1/T (K⁻¹)
0.00280
0.00285
0.00290
0.00295
0.00300
0.00305
0.00310
0.00315
0.00320

Experiment

Oxidation–Reduction Titrations II: Analysis of Bleach

OBJECTIVE

To show how redox reactions can be used to determine the amount of hypochlorite in household bleach.

APPARATUS AND CHEMICALS

Apparatus

50 mL burets (2)
250 mL Erlenmeyer flasks (3)
500 mL Erlenmeyer flask
100 mL beaker
buret clamp
balance
16 cm test tubes (3)
10 and 100 mL graduated cylinders
wash bottle
ring stand

Chemicals

3 *M* H_2SO_4
0.0100 *M* KIO_3
starch indicator (freshly prepared)
3 *M* KI
1 *M* $Na_2S_2O_3$
3% ammonium molybdate

DISCUSSION

Laundering white clothes to remove dirt and stains is an everyday chore that is often accomplished with the aid of bleach. The effectiveness of a bleach to whiten and remove stains is related to its oxidizing (bleaching) strength. The use of detergents helps remove grease by an emulsification action, while agitation helps loosen grime and dirt. Most liquid "chlorine" bleaches, such as Clorox® and Purex®, contain the hypochlorite ion OCl^-, as the oxidizing agent (note the last syllable in Clorox and the first syllable that is relevant to chlorine). Hypochlorite is generally present as the sodium salt, NaOCl, or as the calcium salt, $Ca(OCl)_2$. Nonchlorine bleaches that are used in washing colored as well as white fabrics utilize the oxidative properties of hydrogen peroxide.

GIVE IT SOME THOUGHT

a. What are the oxidation states of oxygen and chlorine in the hypochlorite ion?
b. Why do you think hypochlorite is a good oxidizing agent?

This experiment illustrates how redox reactions can be used to quantitatively determine the amount of oxidizing agent in liquid hypochlorite bleaches (Section 20.1).

Two redox reactions are involved in this experiment in analyzing the oxidizing capacity of a liquid bleach. Initially, you will add an excess of potassium iodide solution to the bleach. The iodide ions, I^-, are oxidized to iodine I_2 after the solution has been acidified at the same time the hypochlorite is reduced.

$$HOCl(aq) + 2I^-(aq) + H^+(aq) \longrightarrow I_2(aq) + Cl^-(aq) + H_2O(l) \qquad [1]$$

GIVE IT SOME THOUGHT

In the presence of a strong acid, the hypochlorite ion is converted to hypochlorous acid. Why?

The iodine that is formed is then titrated with a standardized sodium thiosulfate, $Na_2S_2O_3$, solution that quantitatively reduces the iodine to iodide as follows:

$$2S_2O_3^{2-}(aq) + I_2(aq) \longrightarrow 2I^-(aq) + S_4O_6^{2-}(aq) \qquad [2]$$

Starch is used as the indicator for this reaction; the starch solution is not added until the dark brownish color, due to the iodine, has changed to a pale yellow. When the starch is added, the yellow color changes to blue-black. The end point in the titration is reached when a drop of the thiosulfate causes the solution to become colorless. Starch and iodine (actually the triiodide ion, I_3^-) form a blue-black complex. If the starch is added too soon in the titration, the formation of the blue-black complex is not easily reversed, making the end point very slow and difficult to detect.

$$I_3^- + \text{starch} \rightleftharpoons \underset{\text{blue-black}}{\text{starch} \cdot I_3^-(\text{complex})} \qquad [3]$$

GIVE IT SOME THOUGHT

a. What is the key molar relationship at the end point of this redox titration?

b. How is this different from an acid/base titration?

Adding Equations [1] and [2] yields Equation [4]

$$HOCl(aq) + 2S_2O_3^{2-}(aq) + H^+(aq) \longrightarrow Cl^-(aq) + S_4O_6^{2-}(aq) + H_2O(l) \quad [4]$$

which shows that for every mole of hypochlorite, 2 moles of thiosulfate are required. Thus, from the volume of standardized thiosulfate that is required to react with the liberated iodine and from the mass of bleach, you can calculate the percentage oxidizing agent by mass. You may assume the oxidizing agent to be NaOCl. This is illustrated in Example 39.1.

EXAMPLE 39.1

A 0.501 g sample of bleach was treated with an excess of KI. The iodine liberated required 10.21 mL of 0.0692 *M* $Na_2S_2O_3$ for titration. What is the percentage of NaOCl in the bleach?

SOLUTION: The number of moles of $Na_2S_2O_3$ used in the titration is twice the number of moles of hypochlorite; or alternatively, the number of moles of hypochlorite that react is half the number of moles of $Na_2S_2O_3$.

$$\text{moles NaOCl} = \frac{1 \text{ mol NaOCl}}{2 \text{ mol } Na_2S_2O_3} \text{moles } Na_2S_2O_3$$

$$= \frac{1 \text{ mol NaOCl}}{2 \text{ mol } Na_2S_2O_3} (0.0692 \text{ moles } Na_2S_2O_3/\text{L})(0.01021 \text{ L}) = 3.53 \times 10^{-4} \text{ mol}$$

Changing this number of moles to grams

$$\text{grams NaOCl} = 3.53\times10^{-4}\ \text{mol NaOCl}\times\frac{74.5\ \text{g NaOCl}}{\text{mol NaOCl}} = 0.0263\ \text{g NaOCl}$$

And the percentage NaOCl is

$$\%\text{NaOCl} = \frac{0.0263\ \text{g}}{0.501\ \text{g}}\times100\% = 5.25\%$$

Your $Na_2S_2O_3$ solution used to titrate the I_2 that formed according to Equation [1] will be standardized using potassium iodate KIO_3 as the primary standard. As in the analysis of the bleach solution described previously, two redox reactions, Equations [5] and [6], are involved in the standardization. Notice in Equation [5] that IO_3^- plays an analogous oxidative role to HOCl in Equation [1].

$$IO_3^-(aq) + 5I^-(aq) + 6H^+(aq) \longrightarrow 3I_2(aq) + 3H_2O(l) \qquad [5]$$

$$3I_2(aq) + 6S_2O_3^{2-}(aq) \longrightarrow 6I^-(aq) + 3S_4O_6^{2-}(aq) \qquad [6]$$

The standardization procedure is similar to the procedure you used to analyze bleach. An excess of KI is allowed to react with a known amount of KIO_3 in acidic solution. The iodine formed according to Equation [5] will be titrated with your $Na_2S_2O_3$ solution with starch used as the indicator. Once again, the starch is not added until the iodine solution has turned a pale yellow. The end point in the titration is signaled by the disappearance of the blue-black color.

The stoichiometric relation between the primary standard KIO_3 and $Na_2S_2O_3$ can easily be seen by adding Equations [5] and [6] to give Equation [7]. (NOTE: Equation [6] is obtained by multiplying Equation [2] by the factor 3 to coincide with the stoichiometry of Equation [5].)

$$IO_3^-(aq) + 6S_2O_3^{2-}(aq) + 6H^+(aq) \longrightarrow I^-(aq) + 3S_4O_6^{2-}(aq) + 3H_2O(l) \qquad [7]$$

Equation [7] shows that for each mole of KIO_3 used in the titration, 6 moles of $Na_2S_2O_3$ are required. Example 39.2 illustrates how the concentration of a $Na_2S_2O_3$ solution can be determined using KIO_3 as the primary standard.

EXAMPLE 39.2

What is the concentration of an $Na_2S_2O_3$ solution if 21.21 mL of the solution was required to titrate the iodine formed from 20.95 mL of 0.0100 *M* KIO_3 and excess KI?

SOLUTION: First, determine the number of moles of KIO_3 that react.

$$\text{moles KIO}_3 = (0.02095\ \text{L})(0.0100\ \text{mol KIO}_3/\text{L}) = 2.10\times10^{-4}\ \text{mol}$$

According to Equation [7], the number of moles of $Na_2S_2O_3$ that react is six times the number of moles of KIO_3.

$$\text{moles Na}_2\text{S}_2\text{O}_3 = 6\ (\text{moles KIO}_3)$$
$$= (6\ \text{mol KIO}_3/1\ \text{mol Na}_2\text{S}_2\text{O}_3)\times2.10\times10^{-4}\ \text{mol KIO}_3 = 1.26\times10^{-3}\ \text{mol}$$

Hence, the molarity of the $Na_2S_2O_3$ is

$$\text{molarity } Na_2S_2O_3 = \frac{1.26 \times 10^{-3}\text{ mol}}{0.02121\text{ L}} = 0.0594\ M$$

In this experiment, you will standardize a sodium thiosulfate solution and use it to analyze a liquid bleach.

PROCEDURE

You can perform this experiment most expeditiously if you procure the following solutions at the beginning of the laboratory period:

3 M H_2SO_4	35 mL	one-quarter of a test tube of 3% ammonium molybdate
0.0100 M KIO_3	50 mL	
1 M $Na_2S_2O_3$	15 mL	one-half of a test tube starch solution
distilled water	400 mL	
3 M KI	15 mL	

The flask or test tube for each solution *must* be clean and labeled. Each clean vessel should be rinsed with a small amount of the required solution before it is filled, except for the flask for the $Na_2S_2O_3$. It should be rinsed with distilled water. Avoid using the 250 mL Erlenmeyer flask, for it will be needed as a titration flask. Dispose of all chemicals as instructed.

A. Standardization of 0.05 M $Na_2S_2O_3$ Solution

Prepare about 300 mL of 0.05 M $Na_2S_2O_3$ solution by diluting 15 mL of 1 M $Na_2S_2O_3$ solution with distilled water. Make sure you mix the solution thoroughly. The approximately 0.05 M $Na_2S_2O_3$ will be standardized with potassium iodate. Rinse and fill a buret with the standard KIO_3 solution. Rinse and fill a second buret with the 0.05 M sodium thiosulfate solution to be standardized.

To carry out the standardization, accurately dispense about 15 mL of the KIO_3 solution into a 250 mL Erlenmeyer flask. Record the initial and final readings of the buret to the nearest 0.02 mL. Add 25 mL of distilled water and 3 mL of 3 M KI to the flask and swirl the flask. Add 2 mL of 3 M H_2SO_4 and swirl the flask. A deep brown color should appear, indicating the presence of iodine. After recording the initial buret reading, immediately* begin titration with the sodium thiosulfate solution. As titrant is added, the color of the solution in the flask will fade to light brown and then light yellow. When the solution reaches a light yellow, rinse the sides of the flask with distilled water and add 0.5 mL of starch indicator and mix. The solution will turn a dark blue-black color. Slowly continue to add titrant drop by drop. The dark blue-black color will fade, and when the solution becomes a transparent bright blue, the end point is generally a drop or two away. Rinse the sides of the flask again and complete the titration.

---

*Iodide is oxidized by oxygen in air.

$$4I^-(aq) + O_2(g) + 4H^+(aq) \longrightarrow 2I_2(aq) + 2H_2O(l)$$

The reaction is slow in neutral solution but is faster in acid and is accelerated by sunlight. After the solution has been acidified, it must be titrated immediately. Moreover, after reaching the end point, the solution may darken after standing for an extended length of time.

The end point is the transition from a blue to a colorless solution that remains colorless for about 2 minutes. Record the final buret reading. Repeat the titration two more times. Calculate the molarity of the $Na_2S_2O_3$ for the three trials and calculate the average molarity and standard deviation.

GIVE IT SOME THOUGHT

What does the color change from blue to clear at the end point indicate?

(NOTE: Solid iodine crystals may form during each titration, as the iodine concentration exceeds the solubility product. All of the iodine must be redissolved and reacted before the titration is complete. As the titration proceeds and the solution becomes pale yellow, the flask should be swirled until all of the iodine is dissolved. Over time, the iodine will clump together and become more difficult to dissolve. For this reason, the titrations should be performed as rapidly as practical.)

GIVE IT SOME THOUGHT

If the iodine crystals were not dissolved, how would this effect the titration results?

B. Determination of the Oxidizing Capacity of an Unknown Liquid Bleach

The oxidizing capacity of an unknown liquid bleach will be determined as follows: Place a clean test tube in a 100 mL beaker, determine the mass of the test tube and beaker, and record the mass. Add about 0.5 mL of bleach to the test tube. **(CAUTION: *Liquid bleaches containing sodium hypochlorite are corrosive to the skin. When handling the liquid bleach, take care not to get any on yourself. If you do, immediately wash the area with a copious amount of water.*)** Determine the mass of the test tube, beakers, and bleach and record the mass. The total mass of bleach should be 0.4 to 0.6 g. Pour the bleach into a 250 mL titration flask and rinse the test tube several times with distilled water to ensure that all of the bleach is transferred. (The total volume of water added to the flask should be about 25 mL, including the water used to rinse the test tube.) Add 3 mL of 3 *M* KI to the flask and swirl the flask. Add 2 mL of 3 *M* H_2SO_4 and swirl the flask. Add 5 drops of 3% ammonium molybdate catalyst immediately after you've added the acid. The molybdate ion catalyzes the reaction between the iodide and oxidizing agent. Proceed with the titration as in the preceding standardization of the $Na_2S_2O_3$ solution.

GIVE IT SOME THOUGHT

What effect does a catalyst have on a reaction?

Perform two more titrations. From the concentration and volume of the added sodium thiosulfate solution used to titrate the bleach, calculate the mass of sodium hypochlorite, NaOCl, present. For each titration, calculate the strength of the bleach as the effective percentage mass of the bleach that is sodium hypochlorite. Report the average percentage mass of sodium hypochlorite in the bleach. This is the oxidizing capacity of the bleach. Calculate the standard deviation.

Name ______________________________ Desk ______________

Date ______________ Laboratory Instructor ______________

Oxidation–Reduction Titrations II: Analysis of Bleach | 39 Pre-lab Questions

Before beginning this experiment in the laboratory, you should be able to answer the following questions.

1. Write the formula for hypochlorite.

2. Complete and balance the following equations:

   (a) $HOCl + I^- + H^+ \longrightarrow$

   (b) $S_2O_3^{2-} + I_2 \longrightarrow$

3. Is thiosulfate ion, $S_2O_3^{2-}$, a reducing agent or an oxidizing agent?

4. In the standardization process, when the iodate–iodide solution is acidified with H_2SO_4, it should be titrated immediately. Why?

5. What is the color of the iodate–iodide solution being titrated with $Na_2S_2O_3$

   (a) before the starch solution is added?

   (b) after the starch solution is added?

   (c) at the end point?

6. How many moles of KIO_3 react with one mole of $Na_2S_2O_3$ used in the standardization?

7. If 15.43 mL of a $Na_2S_2O_3$ solution were required to titrate the iodine formed from 21.10 mL of a 0.0121 *M* KIO_3 solution and excess KI, what would the molarity of the $Na_2S_2O_3$ solution be?

8. What is meant by the term *standard solution*?

9. What are the oxidation states of the following elements?
   (a) oxygen in OCl^-

   (b) sodium in $Na_2S_2O_3$

Name ____________________ Desk ____________________

Date ____________________ Laboratory Instructor ____________________

Unknown no. ____________________

REPORT SHEET | EXPERIMENT

Oxidation–Reduction Titrations II: Analysis of Bleach

39

A. Standardization of 0.05 M $Na_2S_2O_3$ Solution

KIO_3 concentration ______

	Trial 1	*Trial 2*	*Trial 3*
Volume of KIO_3			
Final buret reading	______	______	______
Initial buret reading	______	______	______
mL KIO_3 used	______	______	______
Volume of $Na_2S_2O_3$			
Final buret reading	______	______	______
Initial buret reading	______	______	______
mL $Na_2S_2O_3$ used	______	______	______
Molarity of $Na_2S_2O_3$	______	______	______

Average molarity ______ Standard deviation ______

(show calculations)

B. Determination of the Oxidizing Capacity of an Unknown Liquid Bleach

	Trial 1	*Trial 2*	*Trial 3*
Mass of beaker, test tube, and bleach	________	________	________
Mass of beaker and test tube	________	________	________
Mass of unknown bleach	________	________	________
Volume of $Na_2S_2O_3$			
Final buret reading	________	________	________
Initial buret reading	________	________	________
mL $Na_2S_2O_3$	________	________	________
Mass NaOCl	________	________	________
Percent NaOCl	________	________	________

(show calculations)

QUESTIONS

1. In the standardization of the $Na_2S_2O_3$ solution, if you did not titrate the liberated iodine immediately, how would this likely have affected the value of the molarity you calculated? Explain.

2. Some cleansers contain bromate salts as the oxidizing agent. The bromate ion, BrO_3^-, oxidizes the iodide ion as follows:

$$BrO_3^- + 6I^- + 6H^+ \longrightarrow 3I_2 + Br^- + 3H_2O$$

On a molar basis, which is a more effective bleaching agent— $KBrO_3$ or NaOCl?

NOTES AND CALCULATIONS

Molecular Geometry: Experience with Models

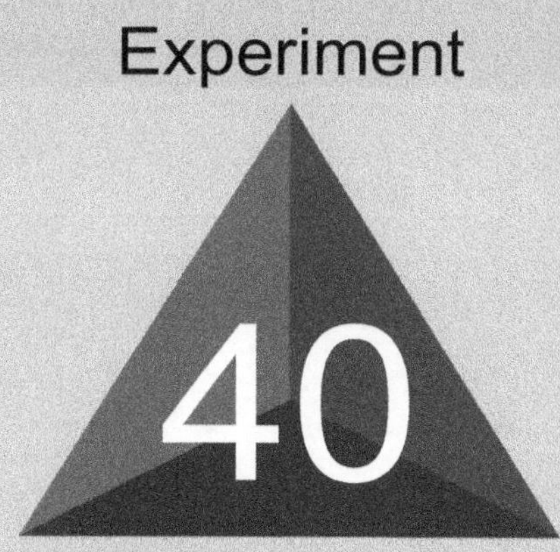

OBJECTIVE

To become familiar with the three-dimensional aspects of organic molecules.

MATERIALS

Prentice Hall Molecular Model Set for General and Organic Chemistry or comparable molecular model kit

DISCUSSION

Organic compounds are numerous—in fact, there are approximately 1×10^8 known organic compounds. The chemical and physical properties of these compounds depend upon what elements are present, how many atoms of each element are present, and how these atoms are arranged in the molecule (Section 24.1). Molecular formulas often allow someone to distinguish between two compounds. For example, even though there are eight atoms in both C_2H_6 and C_2H_5Cl, you know immediately that these are different substances on the basis of their molecular formulas. Similarly, inspection of the molecular formulas C_2H_6 and C_3H_8 reveals that these are different compounds. However, many substances have identical molecular formulas but are different compounds. Consider the molecular formula C_2H_6O. Two compounds correspond to this formula: ethyl alcohol and dimethyl ether. While the molecular formula gives no clue as to which compound some one may be referring to, examination of the *structural formula* immediately reveals a different arrangement of atoms for these substances.

```
    H   H               H       H
    |   |               |       |
H — C — C — O — H   H — C — O — C — H
    |   |               |       |
    H   H               H       H
```

Ethyl alcohol   Dimethyl ether

In addition, when molecular models (ball-and-stick type) are used, trial and error will show just two ways that two carbons, six hydrogens, and one oxygen can be combined to form molecules. Compounds that have the same molecular formula but different structural formulas are termed *isomers* (Section 24.2). This difference in molecular structure results in differences in chemical and physical properties of isomers. In the case of ethyl alcohol and dimethyl ether, whose molecular formulas are C_2H_6O, these differences are very pronounced (Table 40.1). In other cases, the differences may be more subtle.

The importance of the use of structural formulas in organic chemistry becomes evident when you consider that 35 known isomers correspond to the formula

TABLE 40.1 Properties of Ethyl Alcohol and Dimethyl Ether

Property	**Ethyl alcohol**	**Dimethyl ether**
Boiling point	78.5 °C	−24 °C
Melting point	−117 °C	−139 °C
Solubility in H_2O	Infinite	Slight
Behavior toward sodium	Reacts vigorously, liberating hydrogen	No reaction

C_9H_{20}! For the sake of convenience, *condensed structural formulas* are often used. The structural and condensed structural formulas for ethyl alcohol and dimethyl ether are as follows:

```
                         H   H                  H         H
                         |   |                  |         |
Structural formulas:  H— C — C — O — H     H — C — O — C — H
                         |   |                  |         |
                         H   H                  H         H
```

Condensed structural formulas:	CH_3CH_2OH Ethyl alcohol	CH_3OCH_3 Dimethyl ether

A quick glance at these formulas readily reveals their difference. The compounds differ in their **connectivity**. The atoms in ethyl alcohol are connected, or bonded, in a different sequence from those in dimethyl ether.

You must learn to mentally translate these condensed formulas to three-dimensional structures and to translate structures represented by molecular models to condensed structural formulas.

PROCEDURE

Construct models of the molecules as directed and determine the number of isomers by trial and error as directed below. Use a black ball with four holes for carbon, a white ball with one hole for hydrogen, and a green ball with one hole for chlorine. Answer all questions on the report sheet at the end of this experiment.

A. Methane

Construct a model of methane, CH_4. Place the model on your desk top and note the symmetry of the molecule. Note that the molecule looks the same regardless of which three hydrogens are resting on the desk. All four hydrogens are said to be *equivalent*. Now grasp the top hydrogen and tilt the molecule so that only two hydrogens rest on the desk and the other two are in a plane parallel to the desk top (Figure 40.1). Imagine pressing this methane model flat on the desk top. The resulting imaginary projection in the plane of the desk is the conventional representation of the structural formula of methane.

```
    H
    |
H — C — H
    |
    H
```

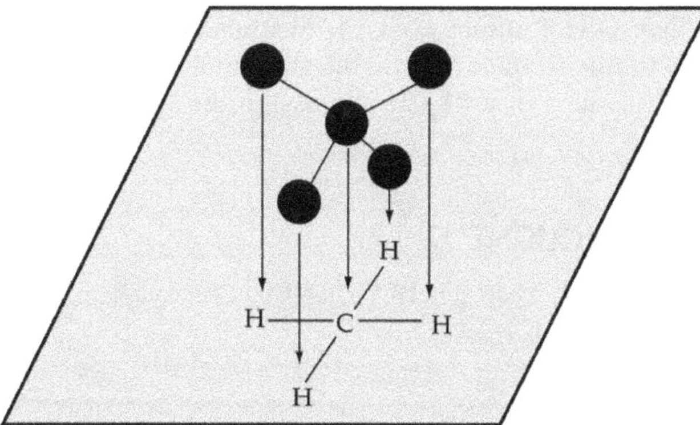

▲**FIGURE 40.1** Model of methane.

GIVE IT SOME THOUGHT

How does the three-dimensional shape of methane compare to its drawing in Figure 40.1?

Replace one of the hydrogen atoms with a chlorine atom to construct a model of chloromethane (or methyl chloride), CH_3Cl. Replace a second hydrogen atom with a chlorine atom to make dichloromethane, CH_2Cl_2. Convince yourself that the two formulas

```
      H                  Cl
      |                  |
 Cl — C — Cl   and  Cl — C — H
      |                  |
      H                  H
```

represent the same three-dimensional structure and are not isomers. Replace another hydrogen to make $CHCl_3$, chloroform (or trichloromethane). Finally, make CCl_4, carbon tetrachloride.

GIVE IT SOME THOUGHT

How would you convert the molecule on the left to the molecule on the right?

B. Ethane

Make a model of ethane, C_2H_6, from your model of CH_4 by replacing one of the hydrogens with a CH_3 unit; the —CH_3 unit is called the *methyl group*. Note that all of the hydrogens of ethane are equivalent. Replace one of the hydrogens in your ethane model with a chlorine. Now examine your model of C_2H_5Cl and note how many different hydrogen atoms are present. If another hydrogen of C_2H_5Cl is replaced with a chlorine atom to yield $C_2H_4Cl_2$, how many isomers result? Does the formula $C_2H_4Cl_2$ distinguish the possible molecules corresponding to this formula? Write the structural and condensed structural formulas for all isomers of $C_2H_4Cl_2$ and assign the IUPAC names to each.

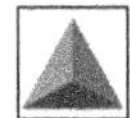

GIVE IT SOME THOUGHT

a. Does it matter which hydrogen of C_2H_5Cl you replace?

b. How many compounds result?

C. Propane

From your model of ethane, construct a molecular model of propane, C_3H_8, by replacing one of the hydrogen atoms with a methyl group, —CH_3. Examine your model of propane and note how many different hydrogen atoms are present in propane. If one of the hydrogens of propane is replaced with chlorine, how many isomers of C_3H_7Cl are possible? List them and give their names. How many isomers correspond to the formula $C_3H_6Cl_2$? Write their formulas and name them. Determine for yourself that there are five isomers with the formula $C_3H_5Cl_3$. Write both the structural and condensed structural formulas and IUPAC names for these isomers. By this stage in the experiment, you should realize that a systematic approach is most useful in determining the number of isomers for a given formula.

D. Butane

The formula of butane is C_4H_{10}. From your model of propane, C_3H_8, construct all of the possible isomers of butane by replacing a hydrogen atom with the methyl group, —CH_3. How many isomers of butane are there? List their structural formulas and IUPAC names. Four isomers correspond to the formula C_4H_9Cl. Write their structural formulas and name them. How many isomers of $C_4H_8Cl_2$ are there? Use your models to help you answer this question. Write and name all of the isomers of $C_4H_8Cl_2$.

E. Pentane

Use your models in a systematic manner to determine how many isomers there are for the formula C_5H_{12}. Write their structural formulas and name them. Write and name all isomers with the formula $C_5H_{11}Cl$.

F. Cycloalkanes

Cycloalkanes corresponding to the formula C_nH_{2n} exist. Without using too much force, try to construct models of cyclopropane, C_3H_6; cyclobutane, C_4H_8; cyclopentane, C_5H_{10}; and cyclohexane, C_6H_{12}. Although cyclopropane and cyclobutane exist, would you anticipate these to be highly stable molecules? How many isomers of 1,2-dichlorocyclopentane are there? The answer is not obvious. There are three. If you cannot determine for yourself that there are three, check with your laboratory instructor. This is another aspect of isomerization that is significant in biochemical systems. Although you may think this is trivial, such differences are of utmost importance in nature!

G. Alkenes

Using two of the longer, narrower, and more flexible bonds, construct a model of ethene (ethylene), C_2H_4. Note the rigidity of the molecule; note that there is no rotation about the carbon-carbon double bond as there is in the case of carbon-carbon single bonds such as ethane and propane. How many isomers correspond to the formulas C_2H_3Cl? $C_2H_2Cl_2$? Draw the formulas and name them.

NOTES AND CALCULATIONS

Name ______________________________ Desk ______________

Date ______________ Laboratory Instructor ______________________

Molecular Geometry: Experience with Models | 40 Pre-lab Questions

Before beginning this experiment in the laboratory, you should be able to answer the following questions.

1. Distinguish between molecular and structural formulas.

2. What is a condensed structural formula? Give an example.

3. What is the meaning of the term *isomer*?

4. Why should the properties of structural isomers differ?

5. Draw the structural formulae for butane and pentane.

6. Distinguish between molecular and empirical formulas.

7. The molecular formula of ethene is C_2H_4. What is the empirical formula of ethene?

8. Distinguish between geometric and structural isomers.

9. Carbon has a valence of 4; oxygen, 2; and hydrogen, 1. How many compounds of C, H, and O containing only one carbon and one oxygen can you make? Draw their structures.

10. Write the structures for CH_2Cl_2 and $CHCl_3$ and name them.

Name Grace Rademacher Desk ____________

Date 5/4/21 Laboratory Instructor Prof. Towle

A, B, C, D-1

REPORT SHEET | EXPERIMENT

Molecular Geometry: Experience with Models | 40

A. Methane

Write the structure for each chloromethane and name them.

Trichloromethane → chloroform

Dichloromethane

Chloromethane → methyl chloride

Carbon tetrachloride

B. Ethane

Write the structures for C_2H_5Cl and $C_2H_4Cl_2$ and name each compound.

C_2H_5Cl — Ethyl chloride

$C_2H_4Cl_2$

1) 1,2 dichloroethane

2) 1,2 dichloroethane

C. Propane

1. Write the structural and condensed formulas as well as the names for all isomers of C_3H_7Cl and $C_3H_6Cl_2$.

C_3H_7Cl — 1-monochloropropane

$C_3H_6Cl_2$

1. 1,2 dichloropropane

2. $CH_2-CH_2-CH_2$ (Cl on the first and third carbons) — 1,3 dichloropropane

3. 2,2 dichloropropane

4. 1,1 dichloropropane

2. Write the condensed and structural formulas as well as the names for all isomers of $C_3H_5Cl_3$.

$C_3H_5Cl_3$

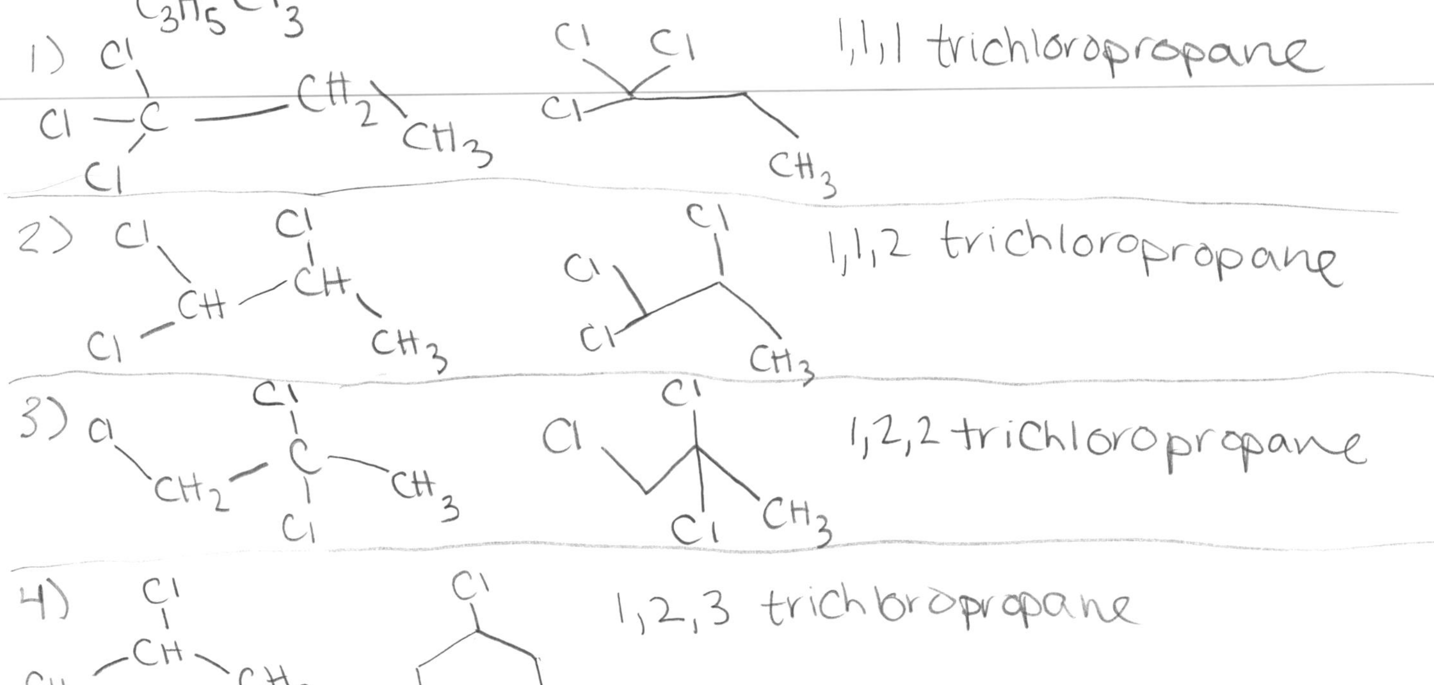

D. Butane

1. Write the structural formulas and names for all of the butanes. C_4H_{10}

Butane

iso-butane

2. Give the structural formulas and names for all isomers of the following:
   (a) C_4H_9Cl

$CH_3-CH_2-CH_2-Cl$ 1 chlorobutane

$CH_3-CH_2-CH(Cl)-CH_3$ 2 chlorobutane

$CH_3-C(Cl)(CH_3)-CH_3$ 2-2 chloromethyl propane

$Cl-CH_2-CH(CH_3)-CH_3$ 1-2 chloromethylpropane

~~(b) $C_4H_8Cl_2$~~

~~E. Pentane~~

1. Write the structural formulas and names for all isomers of C_5H_{12}.

2. Give the structural formula and names for all isomers of $C_5H_{11}Cl$.

~~F. Cycloalkanes~~

Using the following format for the structure of cyclopentane, give the structures of all isomers of 1,2-dichlorocyclopentane and 1,3-dichlorocyclopentane.

G. Alkenes

Give the structures and names of all isomers of $C_2H_2Cl_2$.

QUESTIONS

1. Write the structural and molecular formula for the following hydrocarbon compounds: propane, butane, but-1-ene, 2-methyl butane, and pentane.

2. Name the following four compounds:

C_6H_{14} - Hexane

C_7H_{16} - Heptane

C_4H_8 - But-1-ene

$C_5H_{12}O$ - Pentanol

3. Give the structural formulas for the following:

(a) 2-chloropentane

$C_5H_{11}Cl$

(b) 3-chloro-3-methylpentane

$C_6H_{13}Cl$

(c) 2,2,3-trimethylhexane

C_9H_{20}

(d) 4-ethyl-2,2-dimethylhexane

$C_{10}H_{22}$

NOTES AND CALCULATIONS

Experiment

Polymers

OBJECTIVE

To synthesize two common polymers and then test and observe their properties.

APPARATUS AND CHEMICALS

Apparatus

150-mL beaker
100-mL beaker
50-mL graduated cylinder
10-mL graduated cylinder
glass stir rod
gloves (chemical)
stirring hotplate
ring stand
thermometer clamp
thermometer, red alcohol
magnetic stir bar
stir bar retriever

Chemicals

polyvinyl alcohol, 5%
sodium borate decahydrate, $Na_2B_4O_7 \cdot 10H_2O$
sodium meta-silicate, $Na_2SiO_3 \cdot 9H_2O$
silica gel, 6-12 mesh
ethanol, 95%

DISCUSSION

Polymers are both found in nature and artificially prepared. Common naturally occurring polymers include silk, wool, starch, cellulose, natural rubber, proteins, and nucleic acids (e.g., DNA and RNA). There are a large number of synthetic polymers, including polyethylene, polyvinyl chloride, polystyrene, and polytetrafluoroethylene. Many synthetic polymers are solids with diverse and useful physical properties. Polymer composites, being lighter and more durable than aluminum, make up about half the weight of a Boeing 787 Dreamliner aircraft. An appealing property of being inert and resistant to corrosion leads to the use of polyethylene in bottles, polystyrene in disposable food containers, and polyvinyl chloride in pipes to carry water. Polytetrafluoroethylene is unreactive and slippery, finding use as a non-stick coating in cookware. Rubber in a rubber band is a polymer that is soft, flexible, and elastic, whereas rubber in car tires is a much tougher material. We need to examine the chemical bonding to explain how the properties of a polymers differ.

In 1920 Hermann Staudinger, a German chemist and winner of the 1953 Nobel Prize in chemistry, proposed that polymers were made of long chains of identical molecules linked together by strong chemical bonds. This idea was against the prevailing view at the time that polymers were loose aggregations of small molecules. Experimental data ultimately provided evidence that polymers are extremely large molecules ranging from hundreds to more than a million atoms, resulting in very large molar masses compared to molecular species. Polymers with these large molar masses are formed by the polymerization of small molecules called monomers. The monomers form covalent bonds to link together and form the polymer. Many synthetic polymers have strong carbon-carbon bonds, whereas others have silicon-oxygen bonds in the backbone of the polymer.

There are several general types of reactions that link monomers together to form a polymer. Monomers that contain multiple bonds may be linked together in an **addition polymerization**. The double bond is broken and forms new C – C single bonds with other styrene molecules. The equation for the polymerization of the styrene monomer can be written as follows:

$$n\ \text{C}_6\text{H}_5\text{CH}{=}\text{CH}_2 \longrightarrow \left[\ \text{CH}(\text{C}_6\text{H}_5)\text{–CH}_2\ \right]_n$$

where n represents a large number from hundreds to many thousands or more. The repeat unit shown inside the bracket is found along the entire polymer chain. The end of the chain is capped by an H or some other atom, and every C atom contains a total of four bonds. Many polymers are produced on an industrial sale; for example, over 10 million metric tons of polystyrene is produced every year.

Another type of polymerization occurs when two molecules join together and produce a smaller molecule (e.g., water); this is known as **condensation polymerization**. An example is the reaction between a molecule with carboxylic acid (–COOH) groups and a molecule with amine ($-NH_2$) groups. Unlike the addition polymerization, this involves two different monomers and forms a polymer called a **copolymer**. An example is the reaction between adipic acid (first reactant) and hexamethylenediamine (second reactant) shown below.

$$\text{HO}-\overset{\text{O}}{\overset{\|}{\text{C}}}-(\text{CH}_2)_4-\overset{\text{O}}{\overset{\|}{\text{C}}}-\text{OH} + \text{H}-\underset{\text{H}}{\underset{|}{\text{N}}}-(\text{CH}_2)_6-\underset{\text{H}}{\underset{|}{\text{N}}}-\text{H} \longrightarrow \text{HO}-\overset{\text{O}}{\overset{\|}{\text{C}}}-(\text{CH}_2)_4-\overset{\text{O}}{\overset{\|}{\text{C}}}-\underset{\text{H}}{\underset{|}{\text{N}}}-(\text{CH}_2)_6-\underset{\text{H}}{\underset{|}{\text{N}}}-\text{H} + \text{H}_2\text{O}$$

Water is formed, and a new C – N bond is formed between the two molecules. Further reactions with other monomers occur at either end, and the polymer increases in length. This polymer is named Nylon 6,6 and was one of the first commercial polymers produced on an industrial scale.

GIVE IT SOME THOUGHT

What is a common product of a condensation polymerization?

Polymer chains are long, but they are not linear because the atoms can rotate around the C – C single bonds. The polymer backbone is flexible, folding into different regions. This results in many polymers being flexible. The flexibility of a polymer may be modified in a process called cross-linking. Connecting the chains of polymers by forming covalent chemical bonds between them results in a stiffer material. Rubber erasers have a smaller degree of cross-linking than the rubber in the tire of an automobile. Sulfur is used to create carbon-sulfur bonds to connect chains in the polymerization of natural rubber. The cross-linking prevents the polymer chains from slipping past one another, so the polymer is harder and less flexible, but it still retains some elastic properties. There are a variety of cross-linking agents in addition to sulfur. In this experiment, sodium borate will be used to cross-link polyvinyl alcohol.

In addition to the many polymers based on carbon, there are inorganic polymers. The name *silicone* was given to polymers that have a $Si-O$ backbone and have two organic groups (R) bonded to the silicon atom. The general R group may be a methyl $(-CH_3)$, ethyl $(-C_2H_5)$ or some other organic group. The repeat unit is shown below.

$$\left(\begin{array}{c} R \\ | \\ Si \\ | \\ R \end{array} - O - \right)_n$$

A well-known silicone-based polymer is "Silly Putty," about 90 tons of which is sold per year. It is formed by reacting silicone oil with boric acid. Just as with organic polymers, long silicate chains may also be cross-linked to manipulate the physical properties.

In this experiment, a solution of "water glass" will be prepared according to the ratio of $Na_2O + 3.36\ SiO_2$ and is the source of the silicate chains. Ethanol (C_2H_5OH) is the source of the R group (i.e., $-OCH_2CH_3$).

PROCEDURE

A. Preparation of Water Glass

Obtain 6.0 g sodium meta-silicate $(Na_2SiO_3 \cdot 9H_2O)$ and 3.0 g silica gel (SiO_2) in separate weigh containers. Add the sodium meta-silicate to a 100-mL beaker. Add 30 mL deionized (DI) H_2O, add a stir bar, and then heat to near boiling. Continue to heat and, while stirring, periodically add portions of the silica gel over 30 minutes. Occasionally add additional water as needed to keep the volume near the 30-mL mark on the beaker. The preparation of water glass takes some time, nearly 1 hour, so part B of the experiment may be done concurrently with this water glass preparation in part A. If the silica gel is slow to dissolve, add additional small portions of DI H_2O to the beaker, keeping the total volume near or slightly above 30 mL.

B. Cross-Linked Polyvinyl Alcohol Polymer

Obtain 0.25 g sodium borate decahydrate $(Na_2B_4O_7 \cdot 10H_2O)$, and add it to a 30-mL beaker containing 5.0 mL DI H_2O. Gently heat and stir to dissolve and then cool to room temperature. Obtain 25 mL of 5% polyvinyl alcohol and pour it into a 100-mL beaker. Under constant stirring with a glass stir rod, add the sodium borate decahydrate solution to the polyvinyl alcohol solution. Continue to stir, and note any changes to the viscosity (Section 11.3). The kinetics of the cross-linking reaction are relatively slow, and it may take several seconds for the material to become gel-like. Stir for a couple of minutes, and then, while wearing gloves, roll the polymer into the shape of a ball. Estimate the diameter of the sphere with a ruler to the nearest 0.1 cm. Tare a balance with a weigh boat, and record the mass of the cross-linked polymer ball. Remove the gel and place it on a scrap sheet of paper. Leave it undisturbed while you clean the glassware, and then make an observation of the shape after several minutes.

Stretch it slowly and note what happens. Stretch the gel quickly and note what happens. Drop the gel onto a piece of paper on a hard surface and make an observation.

Discard the cross-linked polymer in the appropriate waste container as indicated by your instructor. Wash your hands after handling the cross-linked polyvinyl alcohol polymer.

C. Silicate Chain (Silicone)

When all or nearly all the silica gel has dissolved from part A to form the water glass, remove the stir bar with a stir bar retriever, and cool to room temperature. Decant into a different beaker to separate any undissolved components. Obtain 5 mL of 95% ethanol and pour it into the sodium silicate solution. Stir with a stirring rod until a solid polymer material forms. Gather the solid together, remove the polymer from the beaker, and place it in the palm of one of your gloved hands. Form the polymer into a spherical shape. Ethanol/water mixture may be pressed out of the polymer as you form it. If needed to prevent crumbling, add a small amount of water. Roll and pat the ball on a paper towel to dry. Measure the diameter of the sphere with a ruler to the nearest 0.1 cm. Tare a balance with a weigh boat, and record the mass of the cross-linked polymer ball. Drop the polymer onto a piece of paper on a hard surface and make an observation. Repeat this at a few different heights.

Clean all glassware thoroughly.

Waste Disposal Instructions: Dispose of all waste as indicated by your instructor.

Name ______________________________ Desk ______________

Date ______________ Laboratory Instructor ______________________

Polymers | 41 Pre-lab Questions

Before beginning this experiment in the laboratory, you should be able to answer the following questions.

1. Write a balanced reaction between sodium hydroxide (NaOH) and silica (SiO_2) to form a solution of sodium silicate (Na_2SiO_3). Include the states of matter.

2. $Na_2B_4O_7$ forms $NaB(OH)_4$ and $B(OH)_3$ when dissolved in water. Write a balanced chemical reaction, including the states of matter.

3. Write the reaction for $NaB(OH)_4$ forming a sodium ion and a tetrahydroxyborate anion.

4. Balance the following equilibrium reaction:

_______ $B(OH)_3(aq) +$ _______ $H_2O(l) \rightleftharpoons$ _______ $B(OH)_4^-(aq) +$ _______ $H_3O^+(aq)$

Name ______________________________ Desk ______________

Date ______________ Laboratory Instructor ______________

REPORT SHEET | EXPERIMENT

Polymers | 41

A. Preparation of Water Glass

Mass of $Na_2SiO_3 \cdot 9H_2O$, g ______________

Mass of silica gel, g ______________

Describe the water glass.

B. Polyvinyl Alcohol Cross-Linked Polymer

1. Describe the viscosity before the polyvinyl alcohol is crosslinked.

2. Describe the viscosity after the polyvinyl alcohol is crosslinked.

3. Density of the cross-linked polymer

Diameter of sphere, cm ______________

Mass of sphere, g ______________

Calculate the density of the cross-linked polymer to two decimal places in g/cm^3 (show calculations).

4. Describe what happens when the cross-linked polyvinyl alcohol is dropped.

5. Describe what happens when the cross-linked polyvinyl alcohol is left alone under the force of gravity.

C. Silicone Polymer

1. Describe the change when ethanol is added to the water glass solution to form the silicone polymer.

2. Density of the silicone polymer

   Diameter of sphere, cm ______________

   Mass of sphere, g ______________

Calculate the density of the silicone polymer to two decimal places in g/cm^3.

3. Describe what happens when the silicone polymer is dropped.

4. Describe what happens when the silicone polymer is left alone under the force of gravity.

QUESTIONS

1. What is the effect of absorbed moisture during the weighing process of the silica pellets on the calculated amount of SiO_2 in the water glass solution versus the actual amount?

2. What is a possible commercial use for the silicone polymer, given the properties you observed?

3. Draw the Lewis structure of $[B(OH)_4^-]$. Name the geometry of this polyatomic ion. What are the expected $O-B-O$ bond angles?

4. The cross-linking reaction of polyvinyl alcohol forms four water molecules for every $B(OH)_4^-$ that cross-links two polymer chains. Given the sections of polyvinyl alcohol chains below, sketch how $B(OH)_4^-$ cross-links the chains. Through what atom chain linkages are the $C-C$ polymer backbones connected?

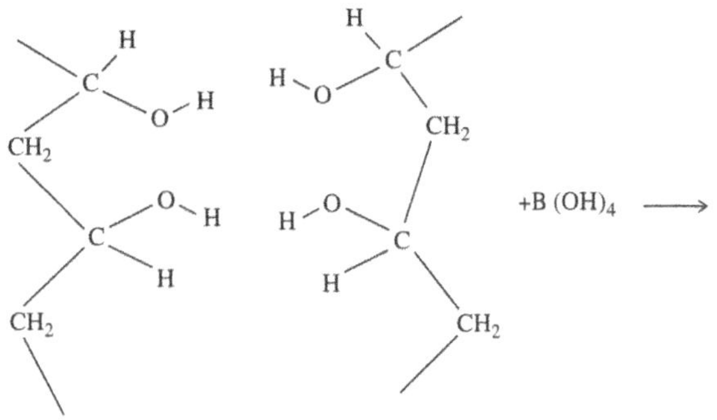

$+B(OH)_4 \longrightarrow$

5. Is it likely that the two sections of the polymer chain are in the same plane? Why or why not?

Experiment

42

Hydrates of Magnesium Sulfate and Its Reactions

OBJECTIVE

To identify the formula of a hydrate of magnesium and perform a precipitation reaction that forms a hydrate.

APPARATUS AND CHEMICALS

Apparatus

porcelain crucible, 30 mL, with lid
stirring hotplate
tongs
wire gauze square
weigh boats
analytical balance
beaker, 30 mL
glass stir rod (2)
watch glass, 4″
spatula
vacuum filtration apparatus
Buchner funnel
filter paper

Chemicals

hydrate of magnesium sulfate, $MgSO_4 \cdot 7H_2O$
sodium carbonate, Na_2CO_3
ethanol, 95%

DISCUSSION

Hydrates are crystalline solids that have water incorporated into the crystalline lattice. The hydrates may display different properties than the anhydrous form. For example, copper sulfate in the anhydrous form is white, whereas a common hydrate of copper sulfate is blue. The crystalline lattice of the hydrates incorporates the water in a weakly bound fashion. With moderate heating, water may be expelled from the lattice, and the remaining solid is known as the anhydrous form. With appropriate temperature control, these waters of hydration are removed to form the anhydrous phase without decomposing the anhydrate. The anhydrous samples may reversibly absorb moisture from the air to reform the hydrate.

One method to determine the composition of the hydrate experimentally is to use a gravimetric method. The mass of the hydrate is quantitatively determined. The sample is heated to form the anhydrous form and its mass quantitatively determined. The mass difference between the hydrate and the anhydrous form is the amount of water present in the sample. From the masses of the anhydrous compound and the water, the moles of each may be calculated using the appropriate molar masses. The ratio of the two mole amounts is used to determine the formula of the hydrate. The number of water molecules in a hydrate may be an integer, but this is not required, and non-integer hydrates are known.

Example 42.1

A hydrate of CuSO4 with a mass 1.0563 g is heated several times to a constant mass. The mass of anhydrous $CuSO_4$ is 0.6751 g. Determine the formula of the hydrate of $CuSO_4$.

SOLUTION: The hydrate of $CuSO_4$ is composed of $CuSO_4$ and H_2O. Assume the anhydrous $CuSO_4$ is all that remains after heating, and all of the mass loss is water. Convert the amounts of $CuSO_4$ and H_2O in grams to moles of each, and then find the ratio to determine the formula of the hydrate.

The mass of the water is

$$\text{mass of hydrate of } CuSO_4 - \text{mass of } CuSO_4 = 1.0563 \text{ g} - 0.6751 \text{ g} = 0.3812 \text{ g } H_2O$$

The molar mass of $CuSO_4$ is 159.610 g/mol, and the number of moles of $CuSO_4$ is

$$0.6751 \text{ g } CuSO_4 \times \frac{1 \text{ mol } CuSO_4}{159.610 \text{ g } CuSO_4} = 0.004229 \text{ mol } CuSO_4$$

The molar mass of H_2O is 18.015 g/mol, and the number of moles of H_2O is

$$0.3812 \text{ g } H_2O \times \frac{1 \text{ mol } H_2O}{18.015 \text{ g } H_2O} = 0.02116 \text{ mol } H_2O$$

The ratio of H_2O to $CuSO_4$ in $CuSO_4 \cdot xH_2O$ is

$$x = \frac{\text{mol } H_2O}{\text{mol } CuSO_4} = \frac{0.02116 \text{ mol}}{0.004229 \text{ mol}} = 5.004 \approx 5$$

The formula of the hydrate of $CuSO_4$ is as follows, where the waters on the right indicate the waters of hydration:

$$CuSO_4 \cdot 5\,H_2O$$

Magnesium sulfate is known to adopt an anhydrous form $MgSO_4$ and several hydrates, including $MgSO_4 \cdot 4H_2O$ and $MgSO_4 \cdot 7H_2O$. The two hydrates differ in the number of water molecules per formula unit of $MgSO_4$.

A hydrate of magnesium sulfate $MgSO_4 \cdot xH_2O$ will be analyzed to determine the value of x. It is important to minimize exposure of the anhydrous compound to humid air during the course of this experiment, to prevent uptake of water.

Hydrates may be synthesized in a precipitation reaction. Solubility rules (Section 4.2) indicate that the solubility of ionic compounds containing the sulfate ion are soluble, except for compounds of Sr^{2+}, Ba^{2+}, $Hg_2{}^{2+}$, and Pb^{2+}. The solubility of ionic compounds containing the carbonate ion are insoluble, except for compounds of $NH_4{}^+$ and the alkali metal compounds. A precipitation reaction of aqueous solutions of magnesium sulfate with sodium carbonate will be conducted. If the reaction occurs only between $MgSO_4(aq)$ and $Na_2CO_3(aq)$ via a metathesis reaction (Chapter 4), the reaction could be hypothesized to form magnesium carbonate and sodium sulfate. However, this reaction does not proceed as a simple metathesis reaction, and water is involved in the following overall reaction.

$$5MgSO_4(aq) + 5Na_2CO_3(aq) + 5H_2O(l) \rightarrow Mg(OH)_2 \cdot 3MgCO_3 \cdot 3H_2O(s) + Mg(HCO_3)_2(aq) + 5Na_2SO_4(aq)$$

The precipitate can be thought of as a combination of magnesium hydroxide, magnesium carbonate, and water.

EXAMPLE 42.2

A solution was made by dissolving 1.6289 g of anhydrous $MgSO_4$ in DI water. Consider the reaction of the magnesium sulfate solution with excess aqueous sodium carbonate to form a precipitate of $Mg(OH)_2 \cdot 3MgCO_3 \cdot 3H_2O(s)$. What is the theoretical yield in grams of $Mg(OH)_2 \cdot 3MgCO_3 \cdot 3H_2O(s)$?

SOLUTION: The mass of $MgSO_4$ is given, and the moles of $MgSO_4$ may be calculated. The balanced chemical reaction is used to find the moles of $Mg(OH)_2 \cdot 3MgCO_3 \cdot 3H_2O(s)$, which can be converted to mass using its molar mass.

$$\text{mol } MgSO_4 = 1.7122 \text{ g } MgSO_4 \times \frac{1 \text{ mol } MgSO_4}{120.3686 \text{ g } MgSO_4}$$
$$= 0.014225 \text{ mol } MgSO_4$$

There is 1 mole of $Mg(OH)_2 \cdot 3MgCO_3 \cdot 3H_2O$ formed from 5 moles $MgSO_4$. Some of the magnesium remains in solution in the form of aqueous $Mg(HCO_3)_2$.

$$\text{mol } Mg(OH)_2 \cdot 3MgCO_3 \cdot 3H_2O = 0.014225 \text{ mol } MgSO_4 \times \frac{1 \text{ mol } Mg(OH)_2 \cdot 3MgCO_3 \cdot 3H_2O}{5 \text{ moles } MgSO_4}$$
$$= 0.0028449 \text{ mol } Mg(OH)_2 \cdot 3MgCO_3 \cdot 3H_2O$$

The mass of $Mg(OH)_2 \cdot 3MgCO_3 \cdot 3H_2O$ may be calculated using its molar mass, 365.30722 g/mol.

$$(0.0028449 \text{ mol } Mg(OH)_2 \cdot 3MgCO_3 \cdot 3H_2O) \times (365.30722 \text{ g/mol})$$
$$= 1.0393 \text{ g } Mg(OH)_2 \cdot 3MgCO_3 \cdot 3H_2O$$

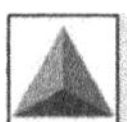

GIVE IT SOME THOUGHT

If the anhydrous $MgSO_4$ in the previous example absorbed water the air, what would be the effect on the percent yield?

PROCEDURE

A. Determination of Hydrate of Magnesium Sulfate

Obtain a clean and dry porcelain crucible and lid. Handle the crucible and lid with tongs throughout the procedure to minimize the influence of fingerprints on the mass measurements. Determine the mass of a crucible on an analytical balance. Heat the porcelain crucible on a hotplate to dryness for 5 minutes at high temperature (i.e., 9/10 on the stirring hotplate dial). Turn off the heat, remove the covered crucible with tongs, and place it on a wire gauze to cool. Determine the mass of the crucible on an analytical balance. Repeat the heating and cooling process, and determine the mass of the crucible. The mass should agree within 0.0010 g between the two data points, If it does, continue to the next step; otherwise, repeat the heating and cooling process for a third cycle.

Obtain approximately 1.0 g of a hydrate of magnesium sulfate in a plastic weigh boat. Place into a porcelain crucible, cover with the lid slightly jar, and quantitatively determine the mass on an analytical balance. Warm the crucible and hydrate of magnesium sulfate, increasing the temperature incrementally over about 5-7 minutes. Reach the maximum temperature setting, and heat at the high temperature. After about 10 minutes, remove the lid briefly and ensure no water is condensing on the bottom of the lid. Return the lid, and heat at the high temperature setting for about 15 minutes. Turn off the heat and transfer the crucible on a wire gauze next to a 50-mL beaker half filled with drierite. The drierite will reduced the amount of water vapor in the surrounding environment as the crucible cools. Cover the crucible and beaker with a single inverted 800-mL beaker, and cool to room temperature. Remove the cover, and promptly determine the mass of the anhydrate of magnesium sulfate. Repeat the heating process, heating rapidly to the high temperature, and cooling under the beaker with drierite. Determine the mass again, ideally until two mass measurements agree to 0.001 g. These data may be used to calculate the formula of the hydrate of magnesium sulfate.

B. Precipitation Reaction of Magnesium Sulfate with a Sodium Carbonate Solution

Obtain approximately 0.92 g Na_2CO_3 and quantitatively determine the mass. Transfer to a 30-mL beaker and dissolve in 5 mL H_2O. Add 5 mL DI H_2O to the crucible containing the previously heated magnesium sulfate. Gently heat each to dissolve. When both powders are fully dissolved, remove the crucible and beaker from the hotplate, and cool to room temperature. Add the magnesium sulfate solution in the crucible to the beaker containing the sodium carbonate solution. Stir with a glass stir rod for about a minute, then return the beaker to the benchtop and let the mixture settle. Determine the mass of a watch glass. Filter the precipitate using vacuum filtration and a Buchner funnel. Add 4-mL portions of 95% ethanol to the beaker, and scrape with a stir rod to transfer as much of the precipitate as possible to the funnel. Wash the filtrate several times with small portions of ethanol, ensuring coverage of the entire surface. Pull air through the filter paper to dry for about a minute, before transferring the filter paper with the precipitate to the watch glass. At your benchtop, gently scrape the precipitate onto the watch glass. Let the powder dry for a couple of minutes, and then determine the mass of the watch glass with precipitate. Transfer the watch glass to a beaker on top of a hotplate, and heat to speed the process of evaporation of remaining adsorbed ethanol or water. Cool and reweigh. Repeat this step as time allows until the mass agrees within 0.01 g between two data points. Calculate the actual yield, theoretical yield, and percent yield from these data.

Clean all glassware thoroughly.

Waste Disposal Instructions: Dispose of all waste indicated by your instructor.

Name ____________________ Desk ____________________

Date ____________________ Laboratory Instructor ____________________

Hydrates of Magnesium Sulfate and Its Reactions | 42 Pre-lab Questions

Before beginning this experiment in the laboratory, you should be able to answer the following questions.

1. A hydrate of $CaCl_2$ sample was analyzed using techniques from part A of this experiment. Assume that the hydrate was heated to its anhydrous form and that it had an integer number of waters of hydration. The following data were obtained:

   Mass of crucible and lid after second heating = 24.4272 g

   Mass of crucible, lid, and hydrate of $CaCl_2$ before heating = 26.4511 g

   Mass of crucible, lid, and $CaCl_2$ after third heating = 25.2280 g

Determine the formula of the hydrate of $CaCl_2$.

2. Write a balanced metathesis reaction, including states of matter, if the reaction of aqueous solutions of magnesium sulfate and sodium carbonate formed magnesium carbonate and sodium sulfate.

---

3. Determine how many of each type of atom is present in $Mg(OH)_2 \cdot 3MgCO_3 \cdot 3H_2O$.

   Mg:

   O:

   C:

   H:

4. Identify the limiting reactant if 0.9278 g of Na_2CO_3 and 1.0052 g $MgSO_4$ (anhydrous) are dissolved and reacted to form $Mg(OH)_2 \cdot 3MgCO_3 \cdot 3H_2O$. Determine the theoretical yield of $Mg(OH)_2 \cdot 3MgCO_3 \cdot 3H_2O$. If the actual yield of $Mg(OH)_2 \cdot 3MgCO_3 \cdot 3H_2O$ was 0.5422 g, determine the percent yield.

Name ______________________________ Desk ______________________

Date ____________________ Laboratory Instructor ______________________

REPORT SHEET | EXPERIMENT

Hydrates of Magnesium Sulfate and Its Reactions

42

A. Determination of Hydrate of Magnesium Sulfate

1. Mass of crucible and lid, (measurement 1) g ______________________
2. Mass of crucible and lid, (measurement 2) g ______________________
3. Mass of crucible and lid, (measurement 3) g ______________________
4. Mass of crucible, lid, and hydrate of magnesium sulfate, g ______________________
5. Mass of hydrate of magnesium sulfate, g ______________________
6. Mass of crucible, lid, and anhydrate of magnesium sulfate, (measurement 1) g ____________
7. Mass of crucible, lid, and anhydrate of magnesium sulfate, (measurement 2) g ____________
8. Mass of crucible, lid, and anhydrate of magnesium sulfate, (measurement 3) g ____________
9. Mass of anhydrous magnesium sulfate, g ______________________
10. Moles of magnesium sulfate, mol ______________________
11. Mass of water in hydrate magnesium sulfate, g ______________________
12. Moles of water, mol ______________________
13. Ratio of water to magnesium sulfate ______________________
14. Formula of hydrate of magnesium sulfate ______________________

(Show calculations.)

B. Precipitation Reaction of Magnesium Sulfate with a Sodium Carbonate Solution

1. Mass of sodium carbonate, g ______________________
2. Describe what happens when the solutions are mixed. ______________________
3. Mass of watch glass, g ______________________
4. Mass of watch glass with precipitate (measurement 1) g ______________________
5. Mass of watch glass with precipitate (measurement 2) g ______________________
6. Mass of watch glass with precipitate (measurement 3) g ______________________
7. Mass of precipitate, g ______________________
8. Moles of sodium carbonate, mol ______________________
9. Moles of magnesium sulfate (part A), mol ______________________
10. Limiting reactant ______________________
11. Theoretical yield of $[Mg(OH)_2(MgCO_3)_3 \cdot 3H_2O]$, mol ______________________
12. Theoretical yield of $[Mg(OH)_2(MgCO_3)_3 \cdot 3H_2O]$, g ______________________
13. Percent yield ______________________

(Show calculations.)

QUESTIONS

1. Why is the heated magnesium sulfate sample cooled under a beaker that contains a separate beaker with drierite desiccant?

2. What is a possible reason for the determination of a non-integer number of water molecules in the hydrate of magnesium sulfate?

3. Show the calculation for the molar mass of $Mg(OH)_2 \cdot 3MgCO_3 \cdot 3H_2O$.

4. What is the mass percent of water in $Mg(OH)_2 \cdot 3\,MgCO_3 \cdot 3\,H_2O$?

NOTES AND CALCULATIONS

Crystalline Solids

OBJECTIVE

To become familiar with crystal lattices, unit cells, and the three-dimensional structures of solids.

APPARATUS

ICE Solid-State Model Kit (can be ordered from http://ice.chem.wisc.edu/

SAFE-T® Templates Geometric Shapes (can be ordered from http://www.hand2mind.com/ or similar

DISCUSSION

Today's consumer products are becoming smaller, lighter, faster, and more intelligent. The demand that society puts on the rapid development of devices such as laptop computers, flat screen televisions, and smartphones has significantly advanced the research and development of solids and modern materials. All of these electronics function based on solid materials, and to improve their functionality, you must acquire a deep and comprehensive understanding of how atoms are arranged to form extended solids.

Unit Cells in Two Dimensions—Defining a Crystal Lattice

When describing the shape or geometry of a molecule, all of the atoms that make up the molecule must be considered (Section 9.1). Several key properties of solids require the formulation of a different set of rules to describe their shape/geometry. Solids consist of a very large number of atoms, and depending on the macroscopic size of the crystal, these numbers vary considerably. Atoms pack together in a three-dimensional repeat pattern to form a solid (similar to the way bricks are stacked to build a wall). This allows chemists to define the term **unit cell**, which is the three-dimensional repeat unit in which atoms pack in a crystal. The geometrical pattern of points on which the unit cells are arranged is called a **crystal lattice**. The resulting structure of an extended solid is defined by two main characteristics: (1) the size and shape of the unit cell and (2) the locations of the atoms within the unit cell (Section 12.2).

Although crystals form three-dimensional lattices, it is useful to consider two dimensional lattices first because they are simpler to describe and visualize. Figure 43.1 shows a two-dimensional array of **lattice points**. All lattice points have an identical environment, meaning that if you sat on a lattice point and looked out at your surroundings, the view would be identical regardless of which lattice point you were sitting on or how far you

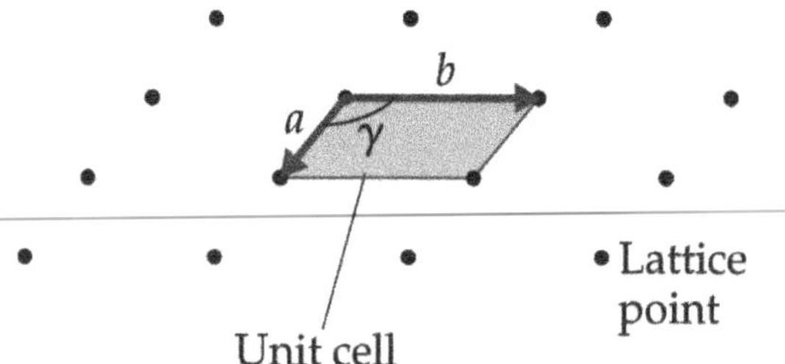

▲**FIGURE 43.1** An arbitrary two-dimensional lattice. The lattice vectors, ***a*** and ***b***, are denoted with bold arrows, and the unit cell is shaded in grey.

could see. In Figure 43.1, the lattice points are denoted by small circles and their positions are defined by the **lattice vectors, *a*** and ***b***. The lattice vectors allow you to start with any lattice point and move to any other lattice point through the translation of an integral number of the two lattice vectors. This property, called **translational symmetry**, is the defining characteristic of a crystal. The unit cell is defined by the parallelepiped formed by the lattice vectors, which is shaded gray in Figure 43.1. In two dimensions, the unit cells must tile all space, completely covering the entire area of the lattice, with no gaps. In three dimensions, the unit cells must stack together to fill all space.

Many different polygons exist; so it might make sense to think that many different two-dimensional lattices exist, but only a few unique lattices exist in two dimensions. The most general lattice is represented in Figure 43.1, where the lattice vectors, ***a*** and ***b***, are of different lengths and the angle between the two vectors, γ, is arbitrary. This lattice is referred to as an **oblique** lattice. In the following exercises, you will use the stencils provided by your instructor to determine the other lattices that exist in two dimensions.

1. An easy place to start identifying two-dimensional lattices is to know that they must tile all space (with no gaps). On your report sheet, tile space with each of the following shapes:

    a. Equilateral triangles
    b. Squares
    c. Rectangles
    d. Regular pentagons
    e. Regular hexagons
    f. Regular octagons

    Shade any areas that are not covered by the tiling.

GIVE IT SOME THOUGHT

a. Is a pentagon a good candidate for a lattice?
b. Why or why not?

2. Would any shapes not listed above tile all space? If so, identify those shapes and make a drawing to demonstrate your assertion.
3. Now you will transition from a lattice to a unit cell in two dimensions. A lattice is defined by a collection of points connected by the lattice vectors. In two dimensions, there are two lattice vectors, ***a*** and ***b***, which

form a unit cell that *must* be a parallelogram. To generate a two-dimensional lattice, consider all of the shapes that tiled all space from the previous exercise and place a point at the center of each polygon. Then draw the lattice vectors, ***a*** and ***b***, and shade the parallelogram that is the unit cell.

4. Based on your responses to questions 1–3, how many two-dimensional lattices are there? In addition to the oblique lattice described in Figure 43.1, there are three additional unique two-dimensional lattices. Use your sketches to complete the two-dimensional lattice table in the report sheet (Section 12.2).
5. Four different two-dimensional patterns of squares are illustrated on your report sheet (patterns A–D). These patterns will be referenced later in lab, but for now, consider these squares to be "atoms" in a crystal. For each pattern, (a) sketch the unit cell, (b) determine the lattice type, and (c) determine the number of squares per unit cell. (Hint: The unit cell should be as symmetric and as small as possible.)

If you complete the above questions properly, you will have determined a total of four unique two-dimensional **primitive lattices**, which are lattices containing lattice points only at the vertices of the unit cell.

Unit Cells in Three Dimensions—Filling a Crystal Lattice

When you consider the structures of metals in greater detail, the relationship between a lattice and a crystal structure becomes more apparent. In two dimensions, only two lattice vectors are required to define a unit cell. In three dimensions, you need a third vector to characterize each unit cell, leading to seven different possibilities, or **crystal systems**. There are seven different primitive lattices: cubic, tetragonal, orthorhombic, hexagonal, rhombohedral, monoclinic, and triclinic. (Section 12.2) Each of these seven primitive lattices has a different symmetry. The symmetry of each unit cell is defined by the length of the lattice vectors (unit cell edges): ***a***, ***b***, and ***c***, as well as the angles between the unit cell edges: α, β, and γ. Figure 43.2 represents the seven primitive lattices in three dimensions. The cubic unit cell is most symmetric. All of its unit cell edge lengths are the same ($a=b=c$), and all of the angles formed between these edges are 90° ($\alpha=\beta=\gamma=90°$). For any of the six other primitive lattices to be formed, the unit cell edges must be varied in length and/or the angles between these edges must deviate from 90°. In each case, the symmetry of the unit cell is lowered; as a result, the geometry needed to relate the physical properties of solids crystallizing in these lattices becomes fairly complex. However, the properties of solids can be properly illustrated through the analysis of structures that have cubic unit cells.

Three different types of cubic unit cells exist based upon the positioning of lattice points. Figure 43.3 shows these unit cells. If a lattice point is placed only on all eight corners of a unit cell, a **primitive cubic lattice** results. If in addition to lattice points on the corners of the unit cell there is a lattice point directly in the center of the unit cell, a **body-centered cubic lattice** is generated. The third cubic unit cell is a result of lattice points on the corners as well as a point in the center of each of the six faces. This arrangement of lattice points is known as a **face-centered cubic lattice**. To this point, the entire discussion has revolved around points and cells and vectors. The term *atom* has not even been mentioned yet! Long before the advent of X-ray diffraction,

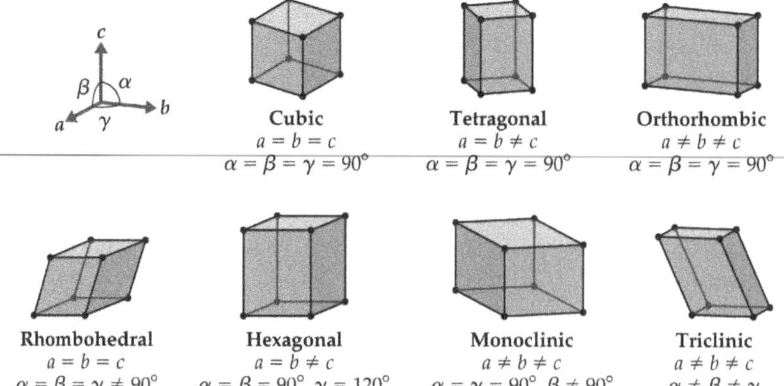

▲**FIGURE 43.2** The seven three-dimensional primitive lattices.

which allowed chemists to determine the position of atoms in a crystal, the mathematical theories (group theory) described in the preceding sections were well understood. Careful inspection of the X-ray diffraction data revealed that the repeat pattern formed when atoms arrange themselves in a crystal followed the same mathematical principles as lattice points and lattice vectors. When atoms, ions or molecules occupy the same positions described by the lattice points, the structures of elements and compounds are formed.

Arrangement of Atoms in Metallic Elements—Cubic and Hexagonal Close-Packing

The simplest representation of atoms in a crystal can be found in the metallic elements. In these structures, every atom is identical and a single atom is placed on each lattice point. The atoms in these structures pack together as closely as possible, with the atoms approximated as spheres that form close-packed layers. The first layer formed is shown in Figure 43.4(a), where each atom would have six neighbors depicted by the large gray spheres (Section 12.3). To form a three-dimensional structure, the next step would be to stack the layers together. The spheres can pack closest to each other if the next layer of spheres sits in the depressions marked with the white or black dots in Figure 43.4(a). Keep in mind that the spheres in a single layer are too large to sit over both sets of depressions. If the atoms in a crystal stopped packing after only two layers, you wouldn't have to worry about AB or AC stacking because these close-packed spheres would be indistinguishable. However, in reality, the atoms in a crystal form infinite layers in three dimensions, and it turns out that the stacking pattern does make a difference.

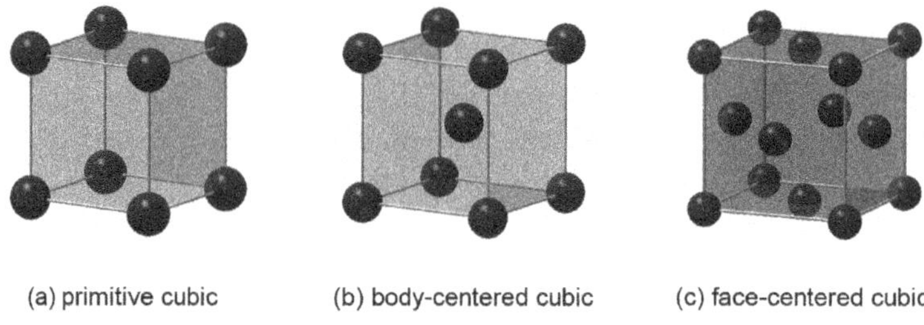

▲**FIGURE 43.3** The three different types of cubic lattices.

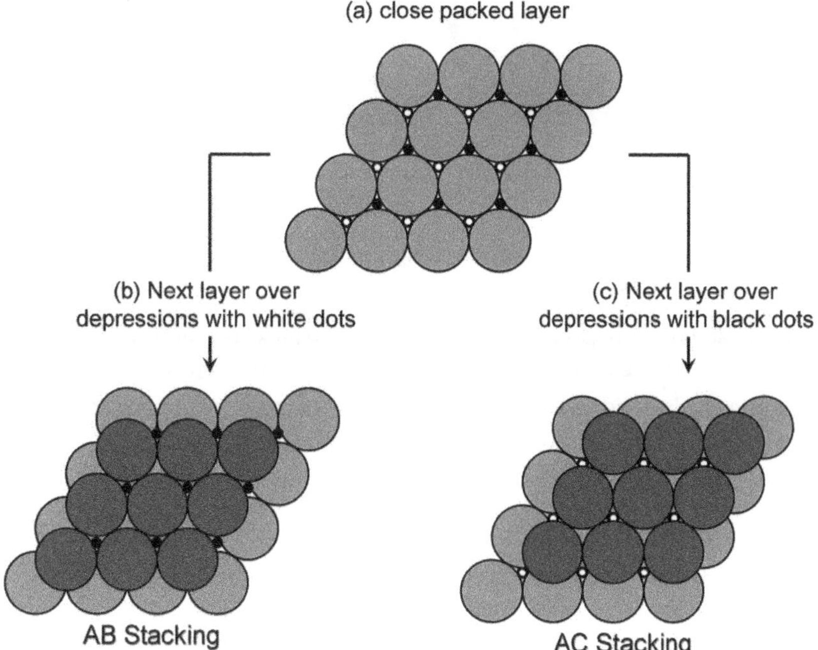

▲**FIGURE 43.4** (a) A close-packed layer of spheres. The white dots represent where an additional layer of spheres could sit, while the black dots represent an alternative set of depressions where the next layer could sit. The two different arrangements are described in (b) as AB stacking and in (c) as AC stacking.

The structures of most metals can be described by two simple stacking sequences depicted in Figures 43.5 and 43.6. If the third layer of spheres lies directly over the positions where the spheres in the first layer sit and the fourth layer lies directly over the layer where the spheres in the second layer sit, the stacking pattern repeats every other layer. This close-packing arrangement is given the ABABAB . . . stacking designation. This type of stacking gives a three-dimensional structure called **hexagonal close-packing** (hcp) and is shown in Figure 43.5.

When two layers of spheres form an AB layer, the third layer can form directly over the first layer. It can also offset in such a way that it lies over the black dots in Figure 43.3(b), which does not sit above layer 1 or layer 2. The three-layer sequence pattern repeats with the fourth layer directly on top

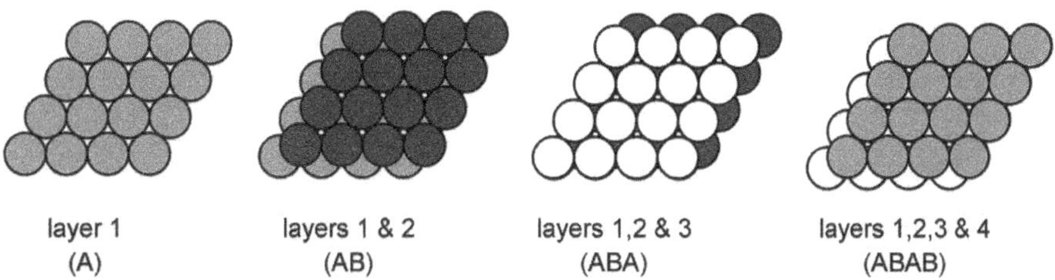

▲**FIGURE 43.5** The ABABAB stacking sequence giving rise to the hexagonal close-packing (hcp) of spheres.

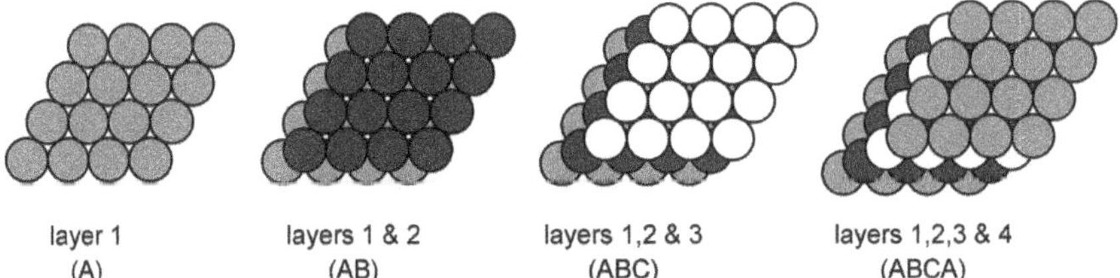

▲**FIGURE 43.6** The ABCABCABC stacking sequence giving rise to the cubic close-packing (ccp) of spheres.

of the first layer to give an ABCABCABC stacking pattern. This type of structure is called **cubic close-packing** (ccp), which is shown in Figure 43.6.

Now using the solid-state model kit, observe how these spheres, or atoms, pack.

6. Stack six layers of the clear spheres in the pattern seen for the hexagonal close-packed metal. Show your lab instructor. How many nearest neighbors does each atom have? (The term used for *nearest neighbors* is referred to as the coordination number.) Keep in mind that the lattice repeats in all directions. Repeat the same exercise for a cubic close-packed metal. What is the coordination number for each atom in the ccp structure?

Unit Cells of Metallic Elements—Defining the Simplest Repeat Pattern of Atoms

If you look back at Figure 43.3, you will see that some of the spheres representing each atom do not lie completely within a given unit cell. Only a fraction of the atom is within the unit cell depending on the lattice point it resides on. If you are interested in a physical property such as the density of a metal, you need to know the mass and the volume of each unit cell. To calculate the mass, you need to know how many atoms are inside the unit cell; to determine the volume, you need to know the unit cell edge length. Both the mass and the unit cell edge length are dictated by the arrangement of the atoms.

The general convention used to determine how many atoms are inside a unit cell is based upon how many unit cells share a particular atom. If an atom is located on a corner of a unit cell, it is shared by eight other unit cells and only one-eighth of that atom falls within a given unit cell. If an atom lies on a face of a unit cell, only half of that atom resides in the unit cell, while the other half is located within an adjacent unit cell. Table 43.1 lists all possible atom locations and their fraction residing within the unit cell (Section 12.3).

TABLE 43.1 The Fraction of an Atom Sitting on the Unit Cell Boundary That Resides within the Unit Cell

Atom location	# of unit cells sharing the atom	Fraction of atom within unit cell
Corner	8	1/8, or 12.5%
Edge	4	1/4, or 25%
Face	2	1/2, or 50%
Anywhere else within the unit cell	1	1, or 100%

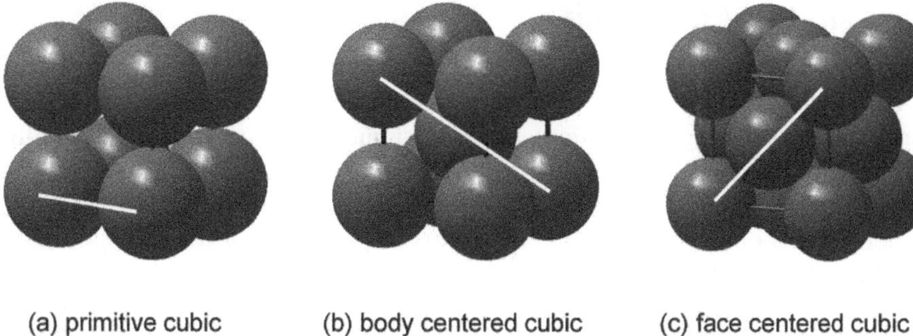

▲**FIGURE 43.7** The structures of (a) a primitive cubic metal, (b) a body-centered cubic metal, and (c) a face-centered cubic metal. The white lines denote the directions along which the atoms touch (a) along the edge, (b) along the body diagonal, running from the front top left to the back bottom right and (c) along the face diagonal.

The three cubic structures created from placing an atom on each lattice point are shown in Figure 43.7. Although it is fairly difficult to visualize, the structures shown in Figure 43.7 follow the same cubic close-packing scheme described earlier in this section. Figure 43.8 attempts to show this for a face-centered cubic metal. In this figure, the atoms have been shaded to indicate the different layers. In Figure 43.8(a), many atoms are shown to illustrate the packing order of the layers. Figure 43.8(b) depicts the face-centered cubic unit cell as it is seen in most textbooks. In this case, the layers stack perpendicular to the body diagonal of the cube.

This information can be used to determine the volume of each unit cell. Keep in mind that the spheres, or atoms, in each layer touch; so using geometry, the unit cell edge can be calculated using the radius of each sphere.

7. Using Figure 43.7 as a reference, work out expressions for the length of the unit cell vector, a, in terms of the radius of the metal atom for a primitive cubic metal, a body-centered cubic metal, and a face-centered cubic metal. Use these expressions to calculate the packing density (the percentage of space occupied by atoms) for each structure type.

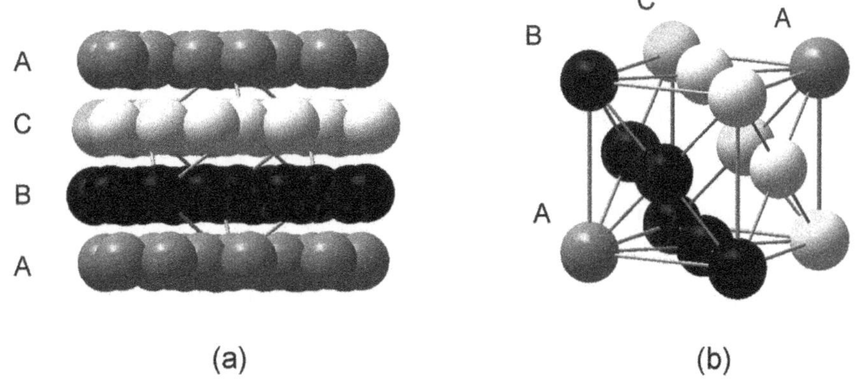

▲**FIGURE 43.8** The crystal structure of a cubic close-packed metal (a) emphasizing the ABC . . . layer stacking and (b) emphasizing the face-centered cubic unit cell. The close-packed layers are artificially shaded differently to emphasize the layers.

Structures of Ionic Solids

Ionic solids are similar to metallic solids in that they tend to prefer structures with a symmetric, close-packed arrangement of atoms. There is one key difference in the bonding interactions found in ionic solids compared to metallic solids, leading to a difference in the crystal structures that result. In a metal, the electron-sea model (Section 12.4) is used to describe bonding and structures where each atom having as many neighbors as possible is favored. Ionic solids, on the other hand, are held together by electrostatic interactions between the cation and the anion (Section 12.5). The preference for each atom now becomes to attract as many neighbors as possible provided they are the oppositely charged particles. Cations will seek to have as many anion neighbors as possible, but they want to distance themselves as far as they can from the other cations. Similarly, anions will seek to have as many cation neighbors as possible, but they want to distance themselves as far as they can from the other anions. The close-packing arrangement observed in ionic solids is the same as in metallic solids if you consider the larger anions to be close-packed, with the smaller cations occupying the cavities in the close-packing arrangement formed by the anions. To understand and visualize this statement, using your model kit, you will take a look at the cavities formed in the close-packed layers. Recall that anions have larger radii than the corresponding atom, while cations have smaller radii than the corresponding atom.

Figure 43.4 illustrated the stacking of layers and indicated that once a layer of atoms is formed, the layer above it fills in the depressions left in the first layer. In ionic solids, the layers alternate between cation and anion. In general, you can say that anions are typically larger than cations and that the cations fill in the "holes" in the close-packed anion structure. The two different types of holes in an anion lattice—octahedral holes and tetrahedral holes—are named based upon how many nearest neighbors each cation has in the resulting structure: six or four, respectively. Cations that fit into tetrahedral holes have four nearest neighbors, while cations residing in octahedral holes have six nearest neighbors.

8. Use your solid-state model kit to construct a close-packed layer of anions (clear spheres) as shown on the report sheet (layer 1).
   a. Fill in some or all of the depressions with cations (blue spheres); then add another anion layer (clear spheres). By trial and error, determine the positions of the octahedral holes. Mark their positions (layer 2a) on the report sheet.
   b. Repeat Part a and determine the positions of the tetrahedral holes. Mark their positions (layer 2b) on the report sheet.
   c. If you limit yourself to filling depressions in the first layer, you should reach the conclusion that the numbers of octahedral and tetrahedral holes are equal, but in actuality, there are twice as many tetrahedral holes as octahedral holes. Where are the remaining tetrahedral holes? Mark the positions (layer 2c) on the report sheet.

The different variations of layer stacking combined with the varying sizes of cations and anions contribute to a wide variety of structures for ionic compounds (Section 12.5). For example, the NaCl (or rock salt) structure is obtained by forming a cubic close-packing arrangement of chloride anions and filling all of the octahedral holes with sodium cations. The ZnS (or zinc blende) structure is obtained by forming a cubic close-packed arrangement of sulfide anions and filling half of the tetrahedral holes with zinc cations. Many other structures result from a hexagonal close-packing arrangement of anions, with cations filling all or a fraction of the tetrahedral or octahedral holes. The focus in this section will be on the most common ionic compounds with a unit cell that is cubic.

The crystal structures of ionic compounds are more complicated than those of close-packed metals because instead of an atom, you have a cation and an anion, and these ions are not the same size. As a result, at least two atoms must be present in each unit cell. The number of cations that will surround each anion (and vice versa) will vary depending upon the size of the cation and anion. If possible, ionic interactions favor structures with a high coordination number. If the cation and anion are similar in size, each ion can be surrounded with a relatively large number of ions of the opposite charge. For example, in cesium chloride, the cesium cation and the chloride anion have relatively similar radii. As a consequence, both the cation and the anion have a coordination number of 8. In comparison, the arrangement of the cations and anions in lithium chloride is different. Because lithium is much smaller than cesium, it cannot accommodate as many negative charges around it, and for it to be stable, the coordination number must be reduced. When the ratio of the cation radius to the anion radius becomes too small, it becomes favorable to form an alternative structure. Figure 43.9 represents the effect of making the r_{cation}/r_{anion} smaller. When the r_{cation}/r_{anion} decreases, the coordination number of the cation decreases from 8 to 6 to 4.

GIVE IT SOME THOUGHT

When the $r_{(cation)} >>> r_{(anion)}$, why must the coordination number decrease? (Hint: See Figure 43.9.)

Figure 43.10 shows the most common cubic crystal structures of ionic compounds. These three structures illustrate the effect of reducing the r_{cation}/r_{anion} ratio. Take note of the **coordination number** (the number of nearest neighbors) and **coordination geometry** (the shape formed by the nearest neighbors) for

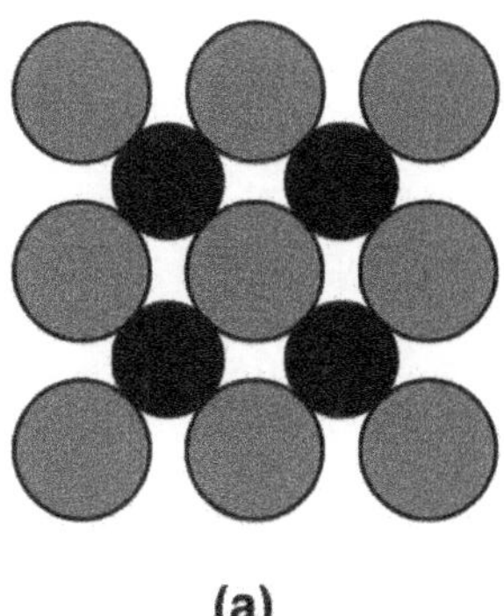

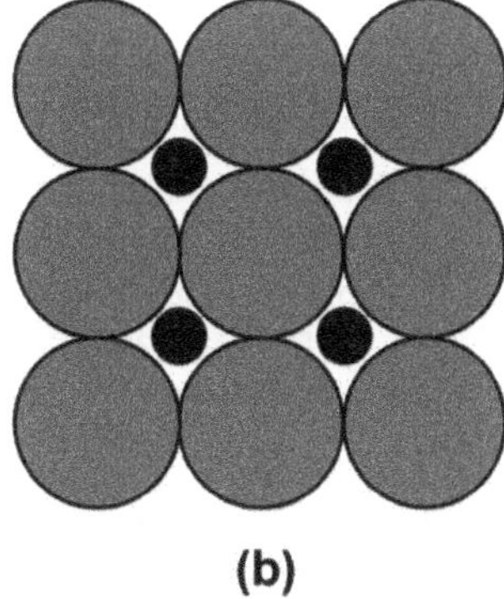

▲**FIGURE 43.9** The two-dimensional packing of cations (black circles) and anions (gray circles) when r_{cation}/r_{anion} is equal to (a) 0.71 and (b) 0.32. When r_{cation}/r_{anion} becomes too small, the cation–anion contacts cannot be maintained without reducing the coordination number.

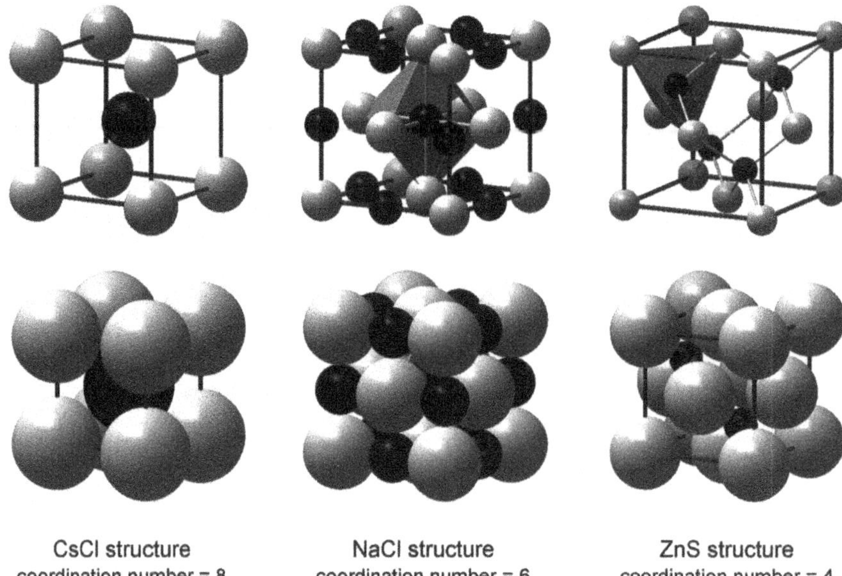

▲**FIGURE 43.10** The (a) CsCl, (b) rock salt (NaCl), and (c) zinc blende (ZnS) structures. The cations are represented with black spheres; the anions, with gray spheres. The sphere radii have been reduced in the upper set of figures to show all ions in the unit cell. In the lower set of figures, the sizes of ions are drawn to scale.

each structure. Also note that the number of anions surrounding each cation decreases from eight in the **cesium chloride (CsCl) structure** (cubic coordination geometry) to six in the **rock salt (NaCl) structure** (octahedral coordination geometry), to four in the **zinc blende (ZnS) structure** (tetrahedral coordination geometry). All other things being equal, you could expect the most stable structure to go from CsCl to NaCl to ZnS as the r_{cation}/r_{anion} decreases.

Each of these structures is difficult to visualize on paper. Read through the next paragraph before you begin the last exercise. Make sure you refer back to it while you are holding the models in your hand to fully visualize how the ions pack to form these structures.

GIVE IT SOME THOUGHT

a. Does it make sense that the cesium cation in CsCl has a coordination number of 8 while the sodium cation in NaCl has a coordination number of 6?

b. What accounts for the difference in coordination number?

The CsCl structure is derived from a primitive cubic arrangement of anions with 100% of the cubic holes filled by the cations (Figure 43.10a). The NaCl and ZnS structures both contain a cubic close-packed or face-centered cubic anion arrangement, but they differ in the way the cations fill the depression in this ccp layer. In the NaCl structure (Figure 43.10b), all of the octahedral holes are filled with cations. The octahedral holes are located on the edges and at the center of the unit cell. The octahedral coordination at the center of the unit cell is emphasized in this figure, but if you consider neighboring unit cells, you would see that all of the cations are surrounded by six anions. Remember that a unit cell is the simplest repeat unit in a crystal and that the repeat unit extends in three dimensions. This means that you must consider the adjacent unit cells in determining the coordination number of the cation and anion. In the NaCl structure, the anions are also surrounded by six cations.

GIVE IT SOME THOUGHT

Why isn't the unit cell for NaCl just?

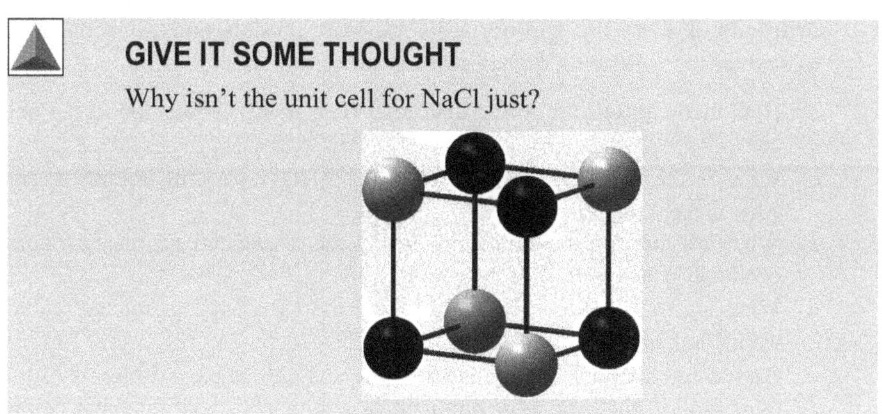

The easiest way to visualize the tetrahedral holes is to divide the unit cell into eight smaller cubes of equal size, one for each quadrant. The tetrahedral holes are located at the center of each of the smaller cubes. If half of the tetrahedral holes from the ccp anion layer are filled with cations, the zinc blende structure results. It is also possible to fill all of the tetrahedral holes with cations in the ccp anion layer; if this occurs, there are twice as many cations as there are anions in the unit cell. This forms the **antifluorite structure**, shown in Figure 43.11a. There are very few examples of this structure, but it is common to observe a structure where the positions of the cation and the anion are reversed. This arrangement, which has a close-packed array of cations with the anions filling all of the tetrahedral holes is known as the **fluorite structure**, which is shown in Figure 43.11b and is named after the mineral fluorite, or CaF_2.

To complete the next part of this lab, you will answer a series of questions by examining a preconstructed model of an ionic crystal. Each student will be assigned a different compound. Your lab instructor will give you the name of your compound (that is, scandium nitride) and the structure identifier that

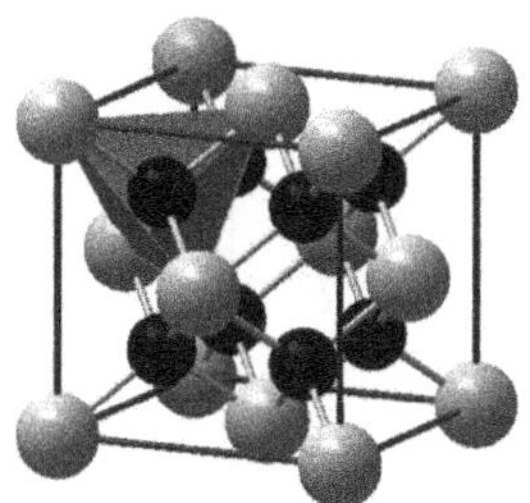

(a) antifluorite structure

(b) fluorite structure

▲**FIGURE 43.11** The crystal structures of (a) the antifluorite structure (that is, Na_2O) and (b) the fluorite structure (that is, CaF_2).

corresponds to one of the preconstructed models (A, B, C, or D). The possible structure types are zinc blende (ZnS), fluorite (CaF_2), rock salt (NaCl), or CsCl. **The cations are represented by the black spheres and the anions are represented by the grey spheres.**

9. For the next several questions, you will examine preconstructed models of ionic crystals. Get the name of your ionic compounds from your lab instructor. Locate the appropriate solid-state models and examine them to answer the following questions on your report sheet.
   a. How many anions are in the unit cell? How many cations are in the unit cell?
   b. What is the empirical formula (that is, Al_2O_3) of your unknown, and what is the structure type?
   c. What are the coordination number and the associated geometry (cubic, octahedral, tetrahedral) of the cation?
   d. What are the coordination number and the associated geometry (cubic, octahedral, tetrahedral) of the anion?
   e. Derive an equation for the length of the unit cell edge, *a*, in terms of the ionic radii of the cation and anion, $r_{(\text{cation})}$ and $r_{(\text{anion})}$. Remember that in ionic compounds, it is the anions and cations that touch. (Neither anions nor cations touch each other.)
   f. Estimate the length of the unit cell edge, *a*, for your solid. Use the ionic radii in Table 43.2 for assistance in this calculation.
   g. Using the estimated length of the unit cell edge, *a*, and the number of atoms in the unit cell, calculate the density of your solid. Consult the *CRC Handbook of Chemistry and Physics* to check the accuracy of your answer. (Note: Your value may not exactly match the handbook value, but it should be close.)

TABLE 43.2 Selected ionic radii (pm) for ions in a four or six coordination numbers (CN) environment.

	CN IV	CN VI		CN IV	CN VI		CN VI
Li^+	73	90	Ca^{2+}	-	114	F^-	119
Na^+	113	116	Sr^{2+}	-	132	Cl^-	167
K^+	151	152	Ba^{2+}	-	149	Br^-	182
Rb^+	-	166	Cu^+	74	91	I^-	206
Cs^+	-	181	Ag^+	114	129	O^{2-}	126
Tl^+	-	164	Fe^{2+}	77	92	S^{2-}	170
Cd^{2+}	92	109				Se^{2-}	184
Hg^{2+}	110	116				Te^{2-}	207

Note: Crystal radii from Shannon, R.D. *Acta Cryst.* (1976), **A32**, 751.

Name ____ Desk ____________________________

Date ____________________________ Laboratory Instructor ____________________________________

Crystalline Solids | 43 Pre-lab Questions

Before beginning this experiment in the laboratory, you should be able to answer the following questions.

1. Using Table 43.1 and Figure 43.7, calculate the number of atoms per unit cell for a metal crystallizing with a (a) primitive cubic, (b) body-centered cubic, and (c) face-centered cubic arrangement of atoms.

2. Palladium crystallizes with a face-centered cubic unit cell. The length of the unit cell edge for palladium is 388.8 pm. Determine the atomic radius (in pm) of palladium.

3. Based on your answer from question 2, determine the density (in g/cm^3) of palladium.

4. What is the coordination number of Na^+ in the NaCl (or rock salt) structure?

5. How does the coordination number differ for a close packed metal versus an ionic solid?

NOTES AND CALCULATIONS

Name ______________________________ Desk ______________________

Date ____________________ Laboratory Instructor ______________________________

REPORT SHEET | EXPERIMENT

Crystalline Solids | 43

Unit Cells in Two Dimensions—Determining a Crystal Lattice

1. Is it possible to tile all two-dimensional space with each of the following polygons (yes or no)?

   a. Equilateral triangles ______________________

   b. Squares ______________________

   c. Rectangles ______________________

   d. Regular pentagons ______________________

   e. Regular hexagons ______________________

   f. Regular octagons ______________________

2. Do you think that other shapes not investigated in question 1 would tile space? If so, identify those shapes and make a drawing to demonstrate your assertion. (Use another sheet of paper if necessary.) __________

3. A lattice is defined by a collection of points connected by the lattice vectors. In two dimensions, there are two lattice vectors, ***a*** and ***b***, which form a unit cell that must be a parallelogram. To generate a two-dimensional lattice for each of the shapes from step 1 that tiled all space, place a point at the center of each polygon. Then draw the lattice vectors ***a*** and ***b*** and shade the parallelogram that is the unit cell.

4. Based on your answers to questions 1–3, how many two-dimensional lattices are there? In addition to the oblique lattice described above, there are three more unique two-dimensional lattices.

Lattice	***a*, *b* Relationship**	γ

5. For each of the following patterns, (a) determine the lattice type, and (b) determine the number of squares per unit cell.

Pattern A

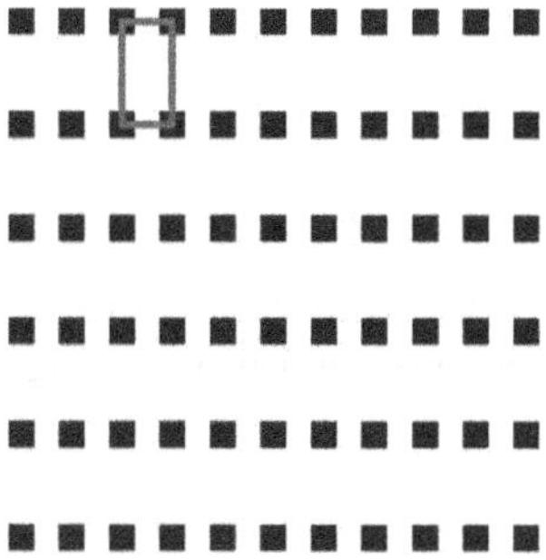

Lattice Type: ____________________
# of Squares: ________

Pattern B

Lattice Type: ____________________
# of Squares: ________

Pattern C

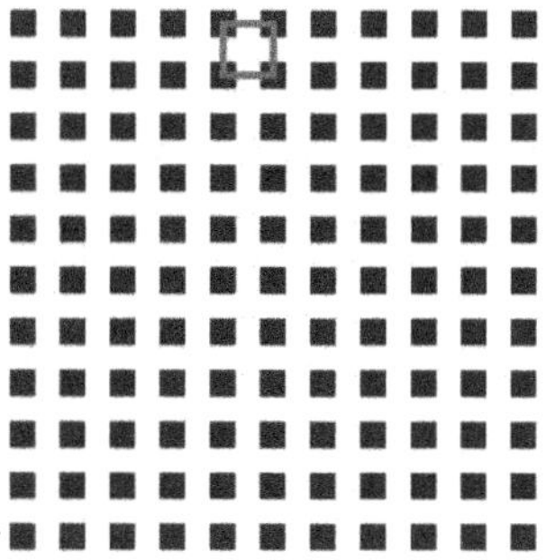

Lattice Type: ____________________
# of Squares: ________

Pattern D

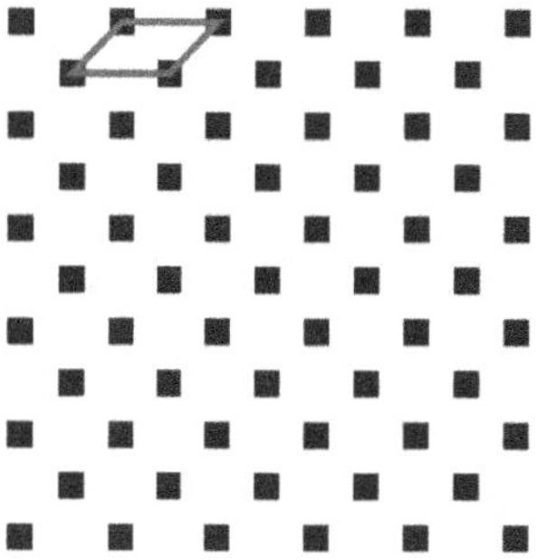

Lattice Type: ____________________
# of Squares: ________

Arrangement of Atoms in Metallic Elements—Cubic and Hexagonal Close-Packing

6. How many nearest neighbors (coordination number) does each atom have in the two types of close-packing?

   a. Hexagonal close-packing ______________

   b. Cubic close-packing ______________

Unit Cells in Three Dimensions—Filling a Crystal Lattice

7. Determine the length of the unit cell edge (that is, the lattice vector ***a***) in terms of the atomic radius, *r*, and the packing density (the percentage of space that is filled) for each of the following structures.

Structure	*a* =	**Packing density (%)**
Primitive cubic		
Body-centered cubic		
Face-centered cubic		

Structures of Ionic Substances

8. Close-packed layers:

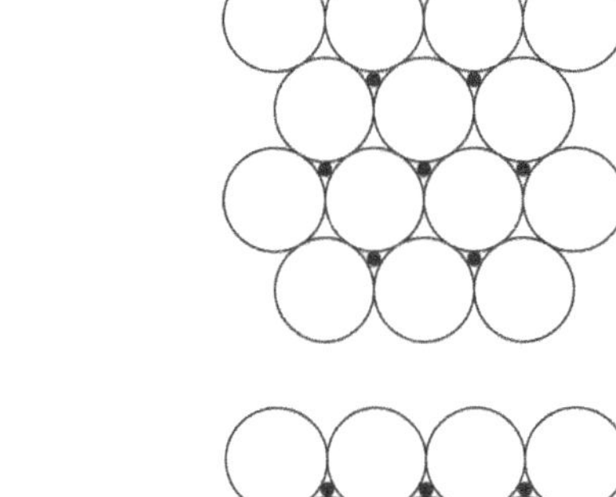

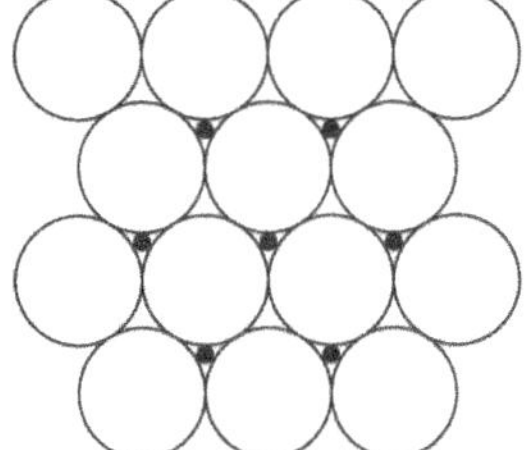

Layer 2a
Cations Octahedral Holes

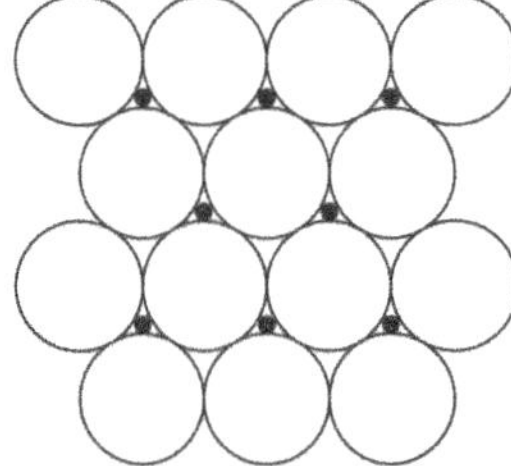

Layer 2b
Cations Tetrahedral Holes

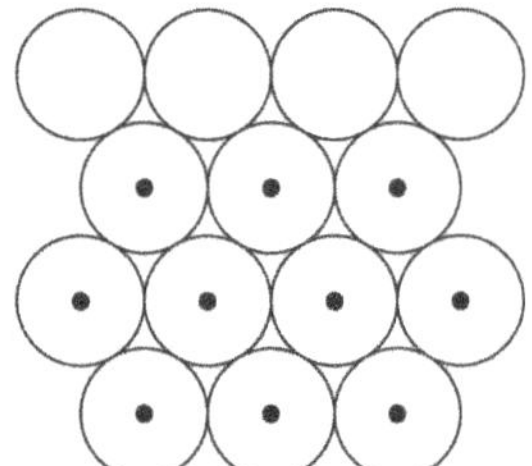

Layer 2c
Cations Tetrahedral Holes

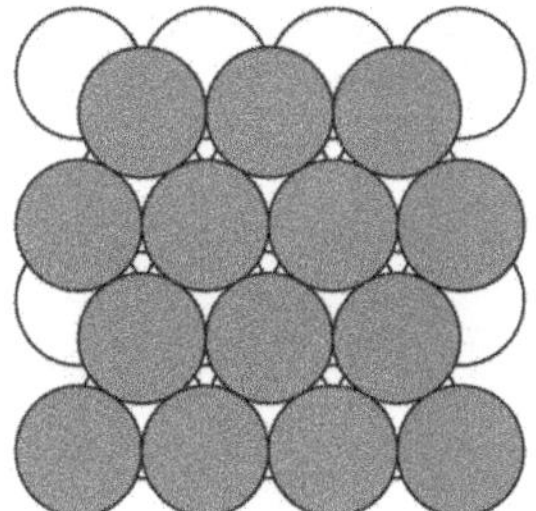

Layer 3
Anions

9. Based on the name of your compound and the model that represents its structure (get this information from your lab instructor), supply answers to the following. See following tables for answer key.

Name of first ionic compound: ______________________

Model (A, B, C, D): ______________________

Empirical formula: ______________________

Structure type: ______________________

	Anion	**Cation**
Number in unit cell		
Coordination number		
Coordination geometry		

e. What is the equation for the length of the unit cell edge, *a*, in terms of the ionic radii of the cation and anion, $r_{(\text{cation})}$ and $r_{(\text{anion})}$? Remember that in ionic compounds, it is the anions and cations that touch. (Neither anions nor cations touch each other.)

f. Use the equation you developed in Part e and the ionic radii given in the lab manual to estimate the length of the unit cell edge, *a*, in pm.

g. Estimate the density of your solid in g/cm^3.

Name of second ionic compound: ________________

Model (A, B, C, D): ________________

Empirical formula: ________________

Structure type: ________________

	Anion	**Cation**
Number in unit cell		
Coordination number		
Coordination geometry		

e. What is the equation for the length of the unit cell edge, a, in terms of the ionic radii of the cation and anion, $r_{(\text{cation})}$ and $r_{(\text{anion})}$? Remember that in ionic compounds, it is the anions and cations that touch. (Neither anions nor cations touch each other.)

f. Use the equation you developed in Part e and the ionic radii given in the lab manual to estimate the length of the unit cell edge, a, in pm.

g. Estimate the density of your solid in g/cm^3.

Name of third ionic compound: ______________________________

Model (A, B, C, D): ______________________________

Empirical formula: ______________________________

Structure type: ______________________________

	Anion	**Cation**
Number in unit cell		
Coordination number		
Coordination geometry		

e. What is the equation for the length of the unit cell edge, *a*, in terms of the ionic radii of the cation and anion, $r_{(\text{cation})}$ and $r_{(\text{anion})}$? Remember that in ionic compounds, it is the anions and cations that touch. (Neither anions nor cations touch each other.)

f. Use the equation you developed in Part e and the ionic radii given in the lab manual to estimate the length of the unit cell edge, *a*, in pm.

g. Estimate the density of your solid in g/cm^3.

Name of fourth ionic compound: ______________________

Model (A, B, C, D): ______________________

Empirical formula: ______________________

Structure type: ______________________

	Anion	**Cation**
Number in unit cell		
Coordination number		
Coordination geometry		

e. What is the equation for the length of the unit cell edge, a, in terms of the ionic radii of the cation and anion, $r_{(cation)}$ and $r_{(cation)}$? Remember that in ionic compounds, it is the anions and cations that touch. (Neither anions nor cations touch each other.)

f. Use the equation you developed in Part e and the ionic radii given in the lab manual to estimate the length of the unit cell edge, a, in pm.

g. Estimate the density of your solid in g/cm^3.

Structures of Solids Lab Compound List for Model Exercise

Instructor Name: ______________________________

Structure Type	Compound	Student's Name
A	lithium chloride	
B	copper(I) fluoride	
C	cesium chloride	
D	calcium fluoride	
A	potassium chloride	
B	copper(I) chloride	
C	cesium bromide	
D	strontium fluoride	
A	rubidium chloride	
B	copper(I) bromide	
C	cesium iodide	
D	strontium chloride	
A	silver chloride	
B	copper(I) iodide	
C	thallium(I) chloride	
D	barium fluoride	
A	iron(II) oxide	
B	silver iodide	
C	thallium(I) bromide	
D	barium chloride	
A	rubidium iodide	
B	cadmium selenide	
C	thallium(I) iodide	
D	cadmium fluoride	
D	mercury(II) fluoride	

Structure of Solids: Model Exercise					Unknown Key							
Group 1	**Compound**	**# cations**	**# anions**	**Emp. Formula**	**Struct. Type**	**Cation Coord #**	**Cation Geom**	**Anion Coord #**	**Anion Geom**	**Eq'n for a**	**unit cell edge, a (Å)**	**density (g/cm^3)**
A	lithium chloride											
B	copper(I) fluoride											
C	cesium chloride											
D	calcium fluoride, fluorite											
Group 2	**Compound**	**# cations**	**# anions**	**Emp. Formula**	**Struct. Type**	**Cation Coord #**	**Cation Geom**	**Anion Coord #**	**Anion Geom**	**Eq'n for a**	**unit cell edge, a (Å)**	**density (g/cm^3)**
A	potassium chloride											
B	copper(I) chloride											
C	cesium bromide											
D	strontium fluoride											
Group 3	**Compound**	**# cations**	**# anions**	**Emp. Formula**	**Struct. Type**	**Cation Coord #**	**Cation Geom**	**Anion Coord #**	**Anion Geom**	**Eq'n for a**	**unit cell edge, a (Å)**	**density (g/cm^3)**
A	rubidium chloride											
B	copper(I) bromide											
C	cesium iodide											
D	strontium chloride											

Group 4	Compound	# cations	# anions	Emp. Formula	Struct. Type	Cation Coord #	Cation Geom	Anion Coord #	Anion Geom	Eq'n for a	unit cell edge, a (Å)	density (g/cm^3)
A	silver chloride											
B	copper(I) iodide											
C	thallium(I) chloride											
D	barium fluoride											
Group 5	**Compound**	**# cations**	**# anions**	**Emp. Formula**	**Struct. Type**	**Cation Coord #**	**Cation Geom**	**Anion Coord #**	**Anion Geom**	**Eq'n for a**	**unit cell edge, a (Å)**	**density (g/cm^3)**
A	iron(II) oxide											
B	silver iodide											
C	thallium(I) bromide											
D	barium chloride											
Group 6	**Compound**	**# cations**	**# anions**	**Emp. Formula**	**Struct. Type**	**Cation Coord #**	**Cation Geom**	**Anion Coord #**	**Anion Geom**	**Eq'n for a**	**unit cell edge, a (Å)**	**density (g/cm^3)**
A	rubidium iodide											
B	cadmium selenide											
C	thallium(I) iodide											
D	cadmium fluoride											
D	mercury(II) fluoride											

Appendices

Appendix

A

The Laboratory Notebook

GENERAL FORMAT FOR A STUDENT LABORATORY NOTEBOOK

One fine day you or one of your classmates will make an important discovery in the laboratory and seek a patent for it. Your laboratory notebook provides the legal and scientific record of this discovery. Your laboratory notebook should state what you did and when and how you did it. It should be written in such a way that it is clearly understandable by anyone unfamiliar with the experiment. It should be a complete dated and *signed* account.

The notebook should have numbered pages; a table of contents; and sections for each experiment that state the *purpose*, describe the *methods* employed, lists the *results*, and state the *conclusions* reached. Arranging a notebook, prior to performing the experiment, to accept numerical data is an excellent way to prepare for an experiment. The report sheets at the end of each experiment in this laboratory manual serve as an example of how you might prepare your laboratory notebook for data entry shown in part E below.

A. The cover should contain the following:
   1. Your name
   2. Your course number
   3. Your laboratory desk drawer or locker number
B. All pages must be numbered consecutively, beginning with the first page, in the upper outside corner of the page.
C. Pages one (1) and two (2) should be used for the table of contents where experiments are listed with the following information:
   1. Experiment number
   2. Title
   3. Page numbers upon which the experiment is to be found
D. Each experiment should be written using a seven-part format. Each experiment must be headed with the following information:
   1. Experiment number and title
   2. The name(s) of your partner(s), if any
   3. The date the experiment is begun

   Each experiment ends with your *signature* and the date of completion.
E. ***Layout*** for each experiment.
   1. ***Purpose***—One or two sentences summarizing why, not how, you are doing the experiment.
   2. ***Chemicals and equipment***—A complete listing of all reagents used, including their concentrations. Also include a complete listing of the equipment and instruments used, including manufacturer and model number.

3. ***Method***—Two or three sentences summarizing the approach you use to complete the experiment. This is not a procedure section. You must include balanced chemical equations for all reactions taking place in the experiment. You should also list proper references to all sources consulted while preparing for the experiment and any deviations from the referenced procedures.
4. ***Completion of***—The pre-lab assignment.
5. ***Raw data***—In this section, you record all data—masses, volumes, etc.—as you obtain them in the laboratory, including observations for qualitative experiments. *Do not use* scratch paper. When you collect data for an experiment, you should try to obtain all numbers to at least *four* (4) significant figures, more if possible. *In any Case, all Apparatus must be Read to One-Tenth the Finest Division.*
6. ***Sample calculations, if applicable***—Here you show one full set of calculations for each type of calculation you do in the experiment to arrive at the results. (Note: Do not use this section to figure out your results—calculate your results elsewhere and show a *sample calculation* in this section.) These calculations must employ *dimensional analysis* and the proper use of significant figures. Sample calculations of statistics (standard deviation and relative standard deviation) need not be shown.
7. ***Results (for quantitative experiments)***—Here you list in tabular form your results from each trial. You report at least three results in this section because you do each experiment at least in triplicate to ensure statistically meaningful results. You must do a complete set of calculations for each trial. Never average the data and then do one calculation. You first calculate the results, then average them. Also report the average (or mean), the standard deviation, and the relative precision (in parts per thousand). Relative precision here is defined in terms of the standard deviation of the mean. In parts per thousand, the relative precision is:

   Relative precision = (standard deviation/mean)(1000)

8. ***Conclusions***—Here you report the final answer (that is, the average value of your results and its statistical analysis). You should also include observations on the accuracy and validity of your results. Any additional comments or suggestions about the experiment should be included here. This includes relisting the references cited in the method section. Give *the date you completed* the experiment (including calculations) and sign the page.

F. You should never tear pages out of your notebook. If you make a mistake, draw a single line through the mistake. If a whole page is in error, draw a large *X* across the page from corner to corner. In no case should anything be obliterated such that it cannot be read.

G. You are not to use pencil in your laboratory notebook for any reason; the notebook is to be a permanent record. You also should not use red ink.

H. When it becomes necessary to continue a portion of an experiment from one page to the next, you must write *continued on page . . .* at the bottom of the page you have just completed and *continued from page . . .* at the top of the next page.

I. No blank pages should appear in your notebook, except perhaps between the table of contents and the beginning of the first experiment. Write on both sides of each page. If you finish the lab work on one experiment and want to start another experiment before you do the calculations on the first one, do not leave room at the end of the first experiment for the calculations and any other information before beginning the write-up for the second experiment. Begin the experiment on the next page (two experiments should never appear on the same page), and when you get around to it, finish the write-up for the first experiment on the next available page. (See Part H.)

J. ***Do not record data on slips of paper.*** Enter data directly into your lab book as you obtain it. The notebook is intended to be primarily for your benefit and is not expected to be a flawless work of art. It must be written so that anyone can determine exactly what you did simply by referring to the notebook alone.

A sample notebook entry follows.

PM-1 Determination of Chloride by Titration with Hg^{2+}

Expt started: 1/27/83; no partners

1. Purpose: to determine the percentage Cl^- in an unknown sample

2. Chemicals & Equipment:
Triple beam balance
mercuric nitrate, $Hg(NO_3)_2 \cdot 2H_2O$
concentrated nitric acid, HNO_3
potassium chloride, KCl
10% nitroprusside, $Na_2[Fe(CN)_5NO]$
50-mL buret
Oven for drying reagents
weighing bottles
analytical balance

3. Method: Percentage Cl^- will be determined by titrating unknown against standardized $Hg(NO_3)_2$ soln using 10% nitroprusside as an indicator. The $Hg(NO_3)_2$ soln will be standardized by titrating it against primary-grade KCl:

$$Hg^{2+} + 2Cl^- \rightleftharpoons HgCl_2$$

Experimental procedures are described on the lab handout.

4. Raw Data
a. Standardization of $Hg(NO_3)_2$ soln (contains about 8 g $Hg(NO_3)_2$ / 500 mL)
Table 1 - KCl used to standardize $Hg(NO_3)_2$ soln

trial #	1S	2S	3S	4S	5S	6S
initial wt, g	32.5910	32.2982	31.9643	31.6394	31.3830	31.1002
final wt, g	32.2982	31.9643	31.6394	31.3830	31.1002	30.8037
wt of KCl, g	0.2928	0.3339	0.3249	0.2564	0.2828	0.2965

TABLE 2 – mL $Hg(NO_3)_2$ soln titrated against KCl

Trial #	1S	2S	3S	4S	5S	6S
final buret rdg	36.20	42.40	45.55	46.80	39.35	41.42
initial buret rdg	0.00	0.00	0.20	1.30	0.00	0.02
mL $Hg(NO_3)_2$	36.20	42.40	45.35	45.50	39.35	41.40

b. Analysis of Unknown

TABLE 3 – wt unknown

trial #	1A	2A	3A	4A	5A	6A
initial wt, g	18.2553	18.0280	17.6980	17.3846	17.1754	16.959
final wt, g	18.0280	17.6980	17.4285	17.1754	16.9633	16.7565
wt of unknown, g	0.2273	0.3300	0.2695	0.2092	0.2121	0.2027

TABLE 4 – mL $Hg(NO_3)_2$ titrated against unknown

trial #	4A	1A	3A	2A	5A	6A
final buret rdg	30.95	34.15	39.65	49.02	31.70	29.93
initial	0.00	0.00	0.10	0.22	0.35	0.00
mL $Hg(NO_3)_2$	30.95	34.15	39.55	48.80	31.35	29.93

5. Sample Calculations

Molarity of $Hg(NO_3)_2$ (for trial 6S)

$$M = \frac{\text{mol } Hg(NO_3)_2}{mL} = \left(\frac{0.2965\,g\ KCl}{41.40\,mL}\right)\left(\frac{1\ mol\ KCl}{74.56\,g\ KCl}\right)\left(\frac{1\ mol\ Hg(NO_3)_2}{2\ mol\ KCl}\right)\left(\frac{10^3\,mL}{L}\right)$$

$$= 0.04803$$

$$N = 2 \times M = 2(0.05) = 0.09605$$

wt of Cl^- in sample (for trial 2A)

$$Cl^- = (48.80\,mL)\left(\frac{0.0484\ mmol\ Hg}{mL}\right)\left(\frac{2\ mmol\ Cl}{1\ mmol\ Hg^{2+}}\right)\left(\frac{35.45\ mg\ Cl}{1\ mmol\ Cl}\right)\left(\frac{1\ g}{10^3\ mg}\right)$$

(using average Molarity for trials 3, 5, 6)

$$= 0.1662\ g\ Cl^-$$

% Cl^- in sample (for trial 2A)

$$\%\ Cl^- = \frac{wt\ Cl^-}{wt\ sample} \times 100$$

$$= \frac{0.1662\,g}{0.3300\,g} \times 100 = 50.36\ \%$$

6. Results

a. Standardization of $Hg(NO_3)_2$ Solution

Trial #	1S	2S	3S	4S	5S	6S
mL soln	36.20	42.40	45.35	45.50	39.35	41.40
wt KCl, g	0.2928	0.3339	0.3249	0.2564	0.2828	0.2965
M	0.05424	0.05281	0.04789	0.03779	0.04819	0.04803
N	0.1085	0.1056	0.09577	0.07558	0.09639	0.09605

Average of trials 3, 5 & 6S: 0.09607 ± 0.00031 N
(relative precision = 3.2 ppt)

b. % Cl^- in sample

trial #	4A	1A	3A	2A	5A	6A
mL $Hg(NO_3)_2$	30.95	34.15	39.58	48.80	31.35	29.93
wt sample	0.2092	0.2273	0.2695	0.3300	0.2121	0.2027
% Cl	50.39	51.17	49.98	50.36	50.34	50.29

Average of trials 2, 4, 5, 6 A: (50.34 ± 0.04) %
(relative precision = 0.8 ppt)

7. Conclusions

The sample contains 50.34 % Cl^-. My technique improved during the course of the experiment. Many early titrations (both for standardization & analysis) were not as good as latter titrations.

Joe Blow

finished: 2/10/83

report handed in 2/15/83

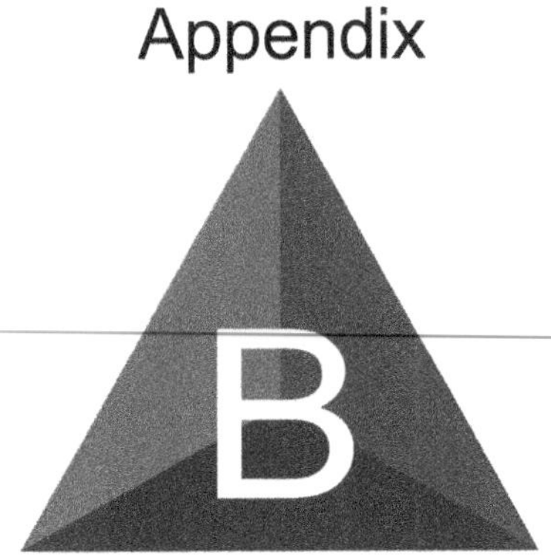

Appendix B Chemical Arithmetic

Elementary mathematics is frequently used in the study of general chemistry. Exponential arithmetic, significant figures, and logarithms are of particular importance and widespread application in these calculations. Each is discussed in this appendix.

EXPONENTIAL ARITHMETIC

Many quantities measured in chemistry are very large or very small. Because of this, it is convenient to express numbers in standard scientific notation. Scientific notation is a way of expressing all numbers as a product. The two members of the product are (1) a number between 1 and 10 (2) the power of 10 that places the decimal point. This second number is called the *exponential term* and is written as 10 with a right-hand superscript (the exponent). The exponent denotes the power of 10 (that is, how many times 10 is multiplied by 10). Some examples of the exponential method of expressing numbers are given below.

$$1000 = 1\times10^3 \qquad 0.1 = 1\times10^{-1}$$

$$100 = 1\times10^2 \qquad 0.01 = 1\times10^{-2}$$

$$10 = 1\times10^1 \qquad 0.001 = 1\times10^{-3}$$

$$1 = 1\times10^0 \qquad 2386 = 2.386\times1000 = 2.386\times10^3$$

$$0.123 = 1.23\times0.1 = 1.23\times10^{-1}$$

As should be evident from the above examples, the power (exponent) of 10 is equal to the number of places the decimal is shifted to give the digit number. The efficacy of using exponential numbers becomes readily apparent when you compare writing 1,230,000,000 with 1.23×10^9 or you compare writing 0.000,000,000,36 with 3.6×10^{-10}.

Once numbers have been expressed as exponentials, the question arises: How do you perform mathematical operations such as addition and multiplication with exponentials? The answers to this question are illustrated in the following examples.

Addition of Numbers in Scientific Notation

Convert all of the numbers to the same power of 10 and add the digit terms of the numbers.

EXAMPLE

$5.0\times10^{-2}+3\times10^{-3}$

$$\begin{array}{r} 50\times10^{-3} \\ \underline{+3\times10^{-3}} \\ 53\times10^{-3}=5.3\times10^{-2} \end{array} \quad \text{or} \quad \begin{array}{r} 5.0\times10^{-2} \\ \underline{+0.3\times10^{-2}} \\ 5.3\times10^{-2} \end{array}$$

Subtraction of Numbers in Scientific Notation

Convert all of the numbers to the same power of 10 and subtract the digit terms of the numbers.

EXAMPLE

$5.0\times10^{-6}-4\times10^{-7}$

$$\begin{array}{r} 5.0\times10^{-6} \\ \underline{-0.4\times10^{-6}} \\ 4.6\times10^{-6} \end{array} \quad \text{or} \quad \begin{array}{r} 50\times10^{-7} \\ \underline{-4\times10^{-7}} \\ 46\times10^{-7}=4.6\times10^{-6} \end{array}$$

Multiplication of Numbers in Scientific Notation

Multiply the digit terms in the usual way and add the exponents of the exponential terms (that is, $10^a\times10^b=10^{a+b}$). Be sure to play close attention to the signs associated with the exponents.

EXAMPLES

$$(4.2\times10^{-8})(2.0\times10^{3})=8.4\times10^{(-8+3)}=8.4\times10^{-5}$$
$$(4.2\times10^{-8})(2.0\times10^{-3})=8.4\times10^{[-8+(-3)]}=8.4\times10^{-11}$$
$$(4.2\times10^{8})(2.0\times10^{-3})=8.4\times10^{[8+(-3)]}=8.4\times10^{5}$$

Division of Numbers in Scientific Notation

Divide the digit terms of the numerator by the digit term of the denominator and algebraically subtract the exponents of the exponential terms (that is, $10^a/10^b=10^{a-b}$).

EXAMPLES

$$\frac{4.2\times10^{-8}}{2.0\times10^{3}}=2.1\times10^{[-8-3]}=2.1\times10^{-11}$$
$$\frac{4.2\times10^{-8}}{2.0\times10^{-3}}=2.1\times10^{[-8-(-3)]}=2.1\times10^{-5}$$
$$\frac{4.2\times10^{8}}{2.0\times10^{-3}}=2.1\times10^{[8-(-3)]}=2.1\times10^{11}$$

Squaring of Exponentials

Square the digit term in the usual way and multiply the exponent of the exponential term by 2 [that is, $(10^a)^2=10^{2a}$].

EXAMPLES

$$(4.0\times10^{-2})^2 = 16\times10^{-2\times2} = 16\times10^{-4} = 1.6\times10^{-3}$$
$$(5.0\times10^{4})^2 = 25\times10^{4\times2} = 25\times10^{8} = 2.5\times10^{9}$$
$$(1.20\times10^{3})^2 = 1.44\times10^{3\times2} = 1.44\times10^{6}$$

Raising Numbers in Scientific Notation to a General Power

Raise the digit term to the power in the usual way and multiply the exponent of the exponential term by the power [that is, $(10^a)^b = 10^{ab}$].

EXAMPLES

$$(2\times10^{-2})^3 = 8\times10^{-2\times3} = 8\times10^{-6}$$
$$(1\times10^{3})^5 = 1\times10^{3\times5} = 1\times10^{15}$$

Extraction of Square Roots of Numbers in Scientific Notation

Decrease or increase the exponential term so that the power of 10 is evenly divisible by 2. Extract the square root of the digit term by inspection, by logarithms, or by calculator, and divide the exponential term by 2 (that is, $\sqrt{10^a} = 10^{a/2}$).

EXAMPLE

Find the square root of 1.6×10^{-7}.

$$1.6\times10^{-7} = 16\times10^{-8}$$
$$\sqrt{16\times10^{-8}} = \sqrt{16}\times\sqrt{10^{-8}} = 4.0\times10^{-8/2} = 4.0\times10^{-4}$$

Combined Operations

Any combination of the preceding operations is performed by doing each operation individually and combining the results.

EXAMPLE

Solve the equation $a/(2.5\times10^{-3}) = (7.4\times10^{8})/(3.9\times10^{-6})$.

$$a = \frac{(2.5\times10^{-3})(7.4\times10^{8})}{(3.9\times10^{-6})} = 4.7\times10^{11}$$

Because the operations of multiplication and division are commutative, the order of operations is immaterial. But if addition and subtraction are involved, the order of operations is important.

$$a = \frac{(2.5\times10^{-3})(7.4\times10^{8})}{(3.9\times10^{-6})} = \frac{1.9\times10^{6}}{3.9\times10^{-6}} = 4.7\times10^{11}$$

SIGNIFICANT FIGURES

Many operations in the chemistry laboratory, such as determining the mass of a compound and measuring the volume of a liquid, involve measurements of some kind. It is important to record these data properly so that the value reported will correctly represent the accuracy of the measurement. The following is a brief guide for calculating and reporting numerical results.

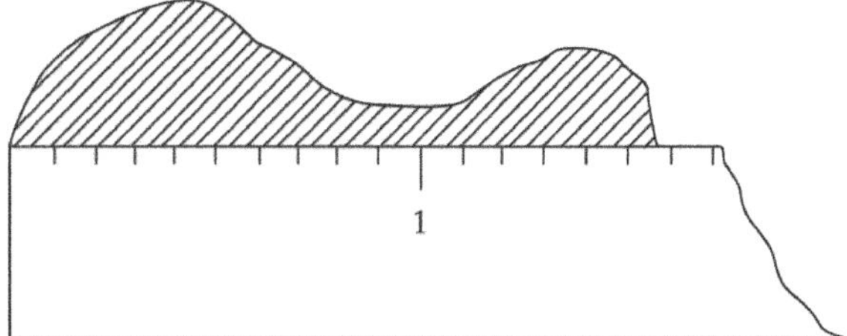

▲**FIGURE B.1** Measurement of the length of an irregular-shaped object.

Every observed measurement that is made is an approximation. For example, the length of the object in Figure B.1 is between 1.5 and 1.6 units. Its length is seen to be approximately 1.56 units. There is uncertainty in the last digit, 6; it is estimated. Consider the recorded mass of an object as 2.6 g. This means that the mass of the object was determined to the nearest tenth (0.1) of a gram and that its exact mass is between 2.55 g and 2.65 g. When a result is recorded, the last digit represents a degree of uncertainty; for example, in 2.6 g, the number 6 represents some degree of uncertainty. The number 2.6 g contains two significant figures—the numbers 2 and 6. If the recorded mass of the object were 2.63$\underline{3}$ g, there would be four significant figures (2, 6, 3, and 3), which would mean that the mass of the object was determined to the nearest thousandth (0.001 g) of a gram. Thus, it is the last (underscored) 3 that has been estimated. *Significant figures* refer to those digits you know with certainty in addition to the first doubtful or estimated digit.

Zeros

A zero may or may not represent a significant figure.* The number 5.00 represents three significant figures, the zeros being significant; this result implies that the measurement was made to the nearest one-hundredth part. Similarly, all of the zeros in 5.034, 7.206, 9.310, 10.20, and 2.001 are regarded as significant figures. Each of the five preceding numbers contains four significant figures.

A common error is to omit the zeros that indicate the reliability of a number. An object whose mass is 2 g when measured to the nearest 0.01 g on a triple beam balance should be recorded as 2.00 g, *not* 2 g.

When the zero is used to locate a decimal point, as in 0.023, the zero is *not* a significant figure. There are only two significant figures in each of the following numbers: 0.027, 0.00078, and 0.0010. These numbers could be written as 2.7×10^{-2}, 7.8×10^{-4}, and 1.0×10^{-3}, respectively. In numbers such as 10, 100, and 1000, the number of significant figures is uncertain. These numbers should be written as 1.0×10^{1}, 1.0×10^{2}, and 1.0×10^{3}, respectively if two significant figures are used. In this lab manual you should assume that numbers like 10, 20, 30 represent two significant figures.

Exact Numbers

Some numbers by their very nature are exact numbers. When you say that you are holding three test tubes in your hand, you mean exactly three test tubes, not two or four. Thus, if this figure 3 is used in a calculation, you may regard it as containing as many significant figures as desired: 3.0000....

---

* See *Journal of Chemical Education, 54*, 578 (1977), 57, 646 (1980) for further discussion.

Another example of exact numbers is a number used to relate quantities within the same system of units.

$$1 \text{ ft} = 12 \text{ in.}$$

$$1 \text{ cm} = 10 \text{ mm}$$

This quality of exactness follows from the definition of the equality. Note, however, that the relationship between units in two different systems is not exact. For example,

$$1 \text{ kg} = 2.205 \text{ lb}$$

One kilogram does not exactly equal 2.205 lb because the two systems have been defined independently of each other. Thus, 1 kg is approximately equal to 2.205 lb.

Rounding Off Numbers

A number may be rounded off to the desired number of significant figures by dropping one or more digits at the end of the number. The following rules should be observed when rounding off a number.

1. When the first digit dropped is less than 5, the last digit retained should remain unchanged (which is called rounding down). For example, 17.373 becomes 17.37 and 1.5324 becomes 1.53 when rounded off to three significant figures and 1.5 when rounded off to two significant figures.
2. When the first digit dropped is equal to or greater than 5, the last digit retained is increased by 1. Statistically, 5 s should be rounded up only half the time. Round up when the number preceding the 5 is even and round down when the number preceding the 5 is odd. For example, 19.765 becomes 19.77 and 8.1525 becomes 8.153 when rounded off to four significant figures and 8.2 when rounded off to two significant figures.

Rounding should be done at the end of the calculation.

Significant Figures and Calculations

In any calculation in which experimental results are used, the final result should contain only as many significant figures as are justified by the experiment. Thus, the least precise measurement dictates the number of significant figures that should be present in the final answer.

Addition and subtraction In addition and subtraction, retain only as many decimal places in the result as there are in that component with the least number of decimal places. For example:

$$\begin{array}{l} 21.1 \\ \ \ 2.035 \\ \ \ \underline{6.12} \\ 29.255 \text{ becomes } 29.3 \end{array}$$

Multiplication and division In multiplication and division, the answer should contain only as many significant figures as are contained in the factor with the least number of significant figures. For example:

$$21.71 \times 0.029 \times 89.2 = 56.159428$$

The number 0.029 contains only two significant figures; therefore, according to the rule above, the answer should be rounded off to contain two significant figures: 56.

Problems

1. How many significant figures are present in each of the following numbers?
   (a) 454 g **(b)** 22.01 cm **(c)** 18.00 mL **(d)** 0.020 g

2. Add each of the following.
   (a) 311 cm + 10.1 cm + 1.21 cm **(b)** 18.00 mL + 1.2 mL + 0.71 mL
   (c) 3.286 ft + 7.01 ft + 0.001 ft

3. Multiply each of the following.
   (a) 3.70×1.11 **(b)** 3.70×2.2 **(c)** 3.70×0.022

4. Divide each of the following.
   (a) $\frac{98.98}{4.90}$ **(b)** $\frac{75.24}{1.1}$ **(c)** $\frac{37.1}{2.5312}$

Answers: **1.** **(a)** 3 **(b)** 4 **(c)** 4 **(d)** 2
2. **(a)** 322 cm **(b)** 19.9 mL **(c)** 10.30 ft
3. **(a)** 4.11 **(b)** 8.1 **(c)** 0.081
4. **(a)** 20.2 **(b)** 68 **(c)** 14.7

THE USE OF LOGARITHMS AND EXPONENTIAL NUMBERS

The *common logarithm* of a number is the power to which the number 10 must be raised to equal that number. For example, the logarithm of 100 is 2 because the number 10 must be raised to the second power to be equal to 100, that is, $\log_{10} 100 = 2$, or $10^2 = 100$. Additional examples are given in Table B.1.

What is the logarithm of 60? Because 60 lies between 10 and 100, which have logarithms of 1 and 2, respectively, the logarithm of 60 must lie between 1 and 2. The logarithm of 60 is 1.78, that is, $60 = 10^{1.78}$, or $\log_{10} 60 = 1.78$.

Every logarithm is made up of two parts: the *characteristic* and the *mantissa.* The characteristic is that part of the logarithm that lies to the left of the decimal point; thus, the characteristic of the logarithm of 60 is 1. The mantissa is that part of the logarithm that lies to the right of the decimal point; thus, the mantissa of the logarithm of 60 is 0.78. The characteristic of the logarithm of a number greater than 1 is 1 less than the number of digits to the left of the decimal point of the number, as shown in the following table.

Number	Characteristic	Number	Characteristic
60	1	2.340	0
600	2	23.40	1
6,000	3	234.0	2
52,840	4	2340.0	3

The mantissa of the logarithm of a number is independent of the position of the decimal point. Thus, the logarithms of 2.340, 23.40, 234.0, and 2340.0 have the same mantissa, and the logarithms of these numbers are 0.3692, 1.3692, 2.3692, and 3.3692, respectively. You may verify this on your calculator.

The meaning of the mantissa and characteristic can be better understood by considering their relationship to exponential numbers. For example, 2340 may be written as 2.340×10^3. The logarithm of (2.340×10^3) = the logarithm of 2.340 plus the logarithm of 10^3. Recall that $(10^a) \times (10^b) = 10^{a+b}$. To find the logarithm of a product, $\log 10^{a+b} = a + b = \log 10^a + \log 10^b$.

TABLE B.1 Examples of Common Logarithms

Number	Number expressed exponentially	Logarithm
10,000	10^4	4
1000	10^3	3
10	10^1	1
1	10^0	0
0.1	10^{-1}	−1
0.01	10^{-2}	−2
0.001	10^{-3}	−3
0.0001	10^{-4}	−4

The logarithm of 2.340 is 0.3692, and the logarithm of 10^3 is 3. Thus, the logarithm of $2340 = 3 + 0.3692 = 3.3692$.

The logarithm of a number less than 1 has a negative value. A convenient method of obtaining the logarithm of such a number is given below.

Example

Obtain the logarithm of 0.00234. When expressed exponentially, $0.00234 = 2.34\times10^{-3}$. The logarithm of 2.34×10^{-3} equals the logarithm of 2.34 plus the logarithm of 10^{-3}. The logarithm of 2.34 is 0.369, and the logarithm of 10^{-3} is −3. Thus, the logarithm of $0.00234 = 0.369 + (-3) = -2.631$.

To multiply two numbers, add the logarithms of the numbers.

Example

Find 412×353.

$$\begin{array}{rl} \text{Logarithm of } 412 & = 2.615 \\ + \text{ Logarithm of } 353 & = 2.548 \\ \hline \text{Logarithm of product} & = 5.163 \end{array}$$

The number whose logarithm is 5.163 is called the *antilogarithm* of 5.163, and it is 1.45×10^5. Thus, $412\times353 = (4.12\times10^2)(3.53\times10^2) = 14.5\times10^4 = 1.45\times10^5$.

To divide two numbers. Subtract the logarithms of the numbers.

Example

Find 412/353.

$$\begin{array}{rl} \text{Logarithm of } 412 & = 2.615 \\ - \text{ Logarithm of } 353 & = 2.548 \\ \hline \text{Logarithm of quotient} & = 0.067 \end{array}$$

The antilogarithm of 0.0671 is 1.17. Thus,

$$\frac{412}{353} = 1.17 \quad \text{or} \quad \frac{4.12\times10^2}{3.53\times10^2} = 1.17$$

Combined operations are performed in the same manner.

Significant Figures and Common Logarithms

For the common logarithm of a measured quantity, the number of digits after the decimal point (the number of digits in the mantissa) equals the number of significant figures in the original number. For example, if 23.5 is a measured quantity (three significant figures), then $\log 23.5 = 1.371$ (three significant figures in the mantissa). The characteristic, 1, just places the decimal point and is not a significant figure.

EXAMPLE

Find $\dfrac{(353)(295)}{(412)}$.

$$\begin{array}{lr} \text{Logarithm of } 353 & = 2.548 \\ + \text{ Logarithm of } 295 & = 2.470 \\ \hline \text{Logarithm of } 353 \times 295 & = 5.018 \\ - \text{ Logarithm of } 412 & = 2.615 \\ \hline \text{Logarithm of quotient} & = 2.403 \\ \hline \end{array}$$

The antilogarithm of 2.403 is 253. Thus, $(353)(295)/(412) = 253$.

The extraction of roots of numbers by means of logarithms is a simple procedure.

EXAMPLE

Find $\sqrt[3]{7235} = (7235)^{1/3}$.

$$\text{Logarithm of } 7235 = 3.8594$$

$$\tfrac{1}{3}\text{logarithm of } 7235 = 1.2865$$

$$\text{Antilogarithm } 1.2865 = 1.934 \times 10^{1} = 19.34$$

Thus, $19.34 = (7235)^{1/3}$.

Powers are found in the same fashion.

EXAMPLE

$(353)^3 = ?$

$$\text{Logarithm } 353 = 2.548$$

$$3 \text{ logarithm } 353 = 7.644$$

$$\text{Antilogarithm } 7.644 = 4.40 \times 10^{7}$$

Thus, $(3.53 \times 10^2)^3 = 4.40 \times 10^7$.

Finding the antilogarithm of a negative logarithm is best illustrated by example.

EXAMPLE

Find the antilogarithm of −7.1594. First, convert the mantissa to a positive value because there are no tables of negative logarithms. This is done as follows:

$$-7.1594 = -8 + 0.8406$$

Then find the antilogarithm of this logarithm as follows:

$$\begin{aligned}\text{Antilog } -7.1594 &= \text{antilog}(-8) \times \text{antilog } (0.8406)\\ &= 10^{-8} \times 6.928\end{aligned}$$

Thus, antilog of $-7.1594 = 6.928 \times 10^{-8}$.

After a little practice, you should find that logarithms simplify the mathematical operations involving very large and very small numbers. Operations involving roots and powers are most easily performed using logarithms. Some basic rules to remember are the following:

$$10^a \times 10^b = 10^{(a+b)} \qquad \log(10^a \times 10^b) = \log 10^a + \log 10^b = a + b$$

$$\frac{10^a}{10^b} = 10^{(a-b)} \qquad \log(10^a / 10^b) = \log 10^a - \log 10^b = a - b$$

$$(10^a)^b = 10^{ab} \qquad \log(10^a)^b = b \log 10^a = ba$$

$$\sqrt{10^a} = (10^a)^{1/2} = 10^{a/2} \qquad \log(10^a)^{1/2} = \tfrac{1}{2}\log 10^a = a/2$$

Appendix

Graphical Interpretation of Data: Calibration Curves and Least-Squares Analysis

GRAPHICAL INTERPRETATION OF DATA

Relationships between two or more variables can easily be visualized when the data are plotted or graphed. In such a graph, the horizontal axis (x-axis) represents the experimentally varied variable, called the *independent variable.* The vertical axis (y-axis) represents the *dependent variable*, which responds to a change in the independent variable. For example, consider the data in Table C.1 for the mass of mercury as a function of volume. In this case, volume is the independent variable and mass is the dependent variable. A graph of these data is shown in Figure C.1. Clearly, the relationship between mass and volume of mercury is linear. The equation for a linear relationship is of the form $y = ax + b$, where a is the slope of the line, the change in y divided by the change in x ($\Delta y/\Delta x$), and b is the intercept with the y-axis. That is, b is the value of y when $x = 0$. Therefore, the equation for the data in Table C.1 is as follows:

$$\text{mass} = (a)\ \text{volume} + b$$
$$\text{where } a = 13.5\ \text{g/mL and } b = 0$$

or

$$\text{mass} = \left(\frac{13.5\ \text{g}}{\text{mL}}\right)\ \text{volume} + 0$$

In this case, the slope, 13.5 g/mL, has the units of mass per unit volume, which is density. In practice, experimental data have uncertainties; therefore, all points

TABLE C.1 Relationship between Mass and Volume of Mercury

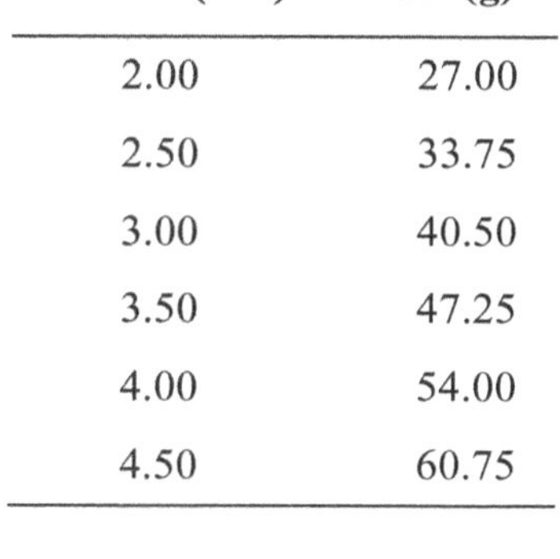

Volume (mL)	Mass (g)
2.00	27.00
2.50	33.75
3.00	40.50
3.50	47.25
4.00	54.00
4.50	60.75

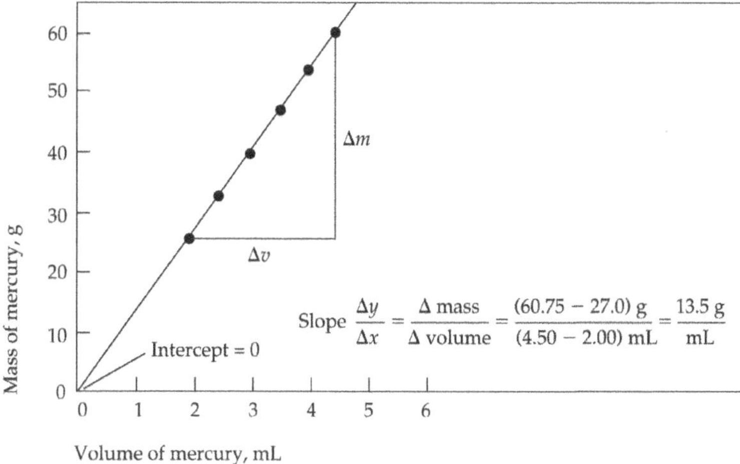

▲**FIGURE C.1** Graph of mass vs. volume of mercury.

usually do not lie precisely on the line. In these cases, draw the best straight line or calculate the best straight line from a least-squares analysis and determine the slope and intercept of this line as discussed.

CALIBRATION CURVES: LEAST-SQUARES ANALYSIS

Many analytical methods require a calibration step in which standards containing known amounts of analyte, x, are analyzed the same way as the unknown. The experimental result, y, is then plotted versus x to give a calibration curve such as that shown in Figure C.2. These plots often are straight lines or linear relations. Seldom, however, do the experimental data fall exactly on the line, usually because of experimental errors. The experimenter is then obligated to draw the "best" straight line through the experimental points. Statistical methods are available that allow for an objective determination of such a line and for an estimation of the uncertainties associated with this line. Statisticians call this technique *linear regression analysis.* You will use the simplest of these techniques, called the *method of least squares.*

Assume that a linear relation of the form

$$y = ax + b$$

does in fact exist.

Method

The line generated by a least-squares method is the line that minimizes the squares of the individual vertical displacements, or *residuals* (Figure C.2), from that line. In addition to providing the best fit of the experimental points to the line, the method calculates the slope, a, and intercept, b, of the line.

The following equations are defined:

$$S_{xx} = \Sigma(x_i - \overline{x})^2 = \Sigma x_i^2 - \frac{(\Sigma x_i)^2}{n} \qquad [1]$$

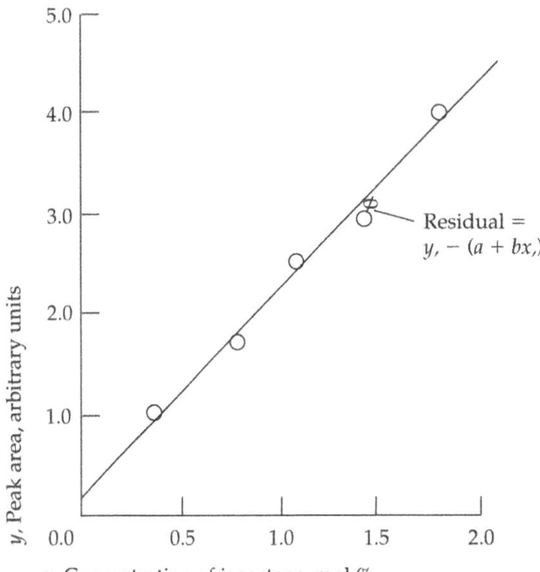

▲**FIGURE C.2** Graph of peak area vs. concentration of isooctane.

$$S_{yy} = \Sigma(y_i - \bar{y})^2 = \Sigma y_i^2 - \frac{(\Sigma y_i)^2}{n} \qquad [2]$$

$$S_{xy} = \Sigma(x_i - \bar{x})(y_i - \bar{y}) = \Sigma x_i y_i - \frac{\Sigma x_i \Sigma y_i}{n} \qquad [3]$$

where x_i and y_i are individual pairs of values for x and y defining each of the points to be plotted. The quantity n is the number of points, and $\bar{x}$ and $\bar{y}$ are the average values of x and y; that is,

$$\bar{x} = \frac{\Sigma x_i}{n} \qquad \bar{y} = \frac{\Sigma y_i}{n}$$

Note that S_{xx} and S_{yy} are simply the sums of the squares of the deviations from the means for individual values of x and y.

From S_{xx}, S_{yy}, and S_{xy}, you may calculate the following:

1. The slope of the line, a

$$a = \frac{S_{xy}}{S_{xx}}$$

2. The intercept of the line, b

$$b = \bar{y} - a\bar{x}$$

3. The standard deviation about the regression, S_r, which is based upon deviations of the individual points from the line

$$S_r = \sqrt{\frac{S_{yy} - a^2 S_{xx}}{n - 2}}$$

EXAMPLE C.1

Given the following data, carry out a least-squares analysis.

x_i	y_i	x_i^2	y_i^2	$x_i y_i$
0.352	1.09	0.12390	1.1881	0.38368
0.803	1.78	0.64481	3.1684	1.42934
1.08	2.60	1.16640	6.7600	2.80800
1.38	3.03	1.90140	9.1809	4.18140
1.75	4.01	3.06250	16.0801	7.01750
5.365	12.51	6.89901	36.3775	15.81992

SOLUTION:

$$S_{xx} = \Sigma x_i^2 - \frac{(\Sigma x_i)^2}{n} = 6.90201 - \frac{(5.365)^2}{5} = 1.145365$$

$$S_{yy} = \Sigma y_i^2 - \frac{(\Sigma y_i)^2}{n} = 36.3755 - \frac{(12.51)^2}{5} = 5.07748$$

$$S_{xy} = \Sigma x_i y_i - \frac{\Sigma x_i \Sigma y_i}{n} = 15.81992 - \frac{(5.365)(12.51)}{5} = 2.39669$$

$$a = \frac{2.39669}{1.145365} = 2.0925 = 2.09$$

$$b = \overline{y} - \overline{ax} = \frac{12.51}{5} - 2.09\frac{(5.365)}{5} = 0.259$$

Thus, the best linear equation is

$$y = 2.09x + 0.259$$

and

$$S_r = \sqrt{\frac{S_{yy} - b^2 S_{xx}}{n-2}} = \sqrt{\frac{5.07748 - (2.0925)^2(1.145365)}{5-2}}$$

$$= \pm 0.144$$

$$= \pm 0.14$$

All of this functionality may be found on programmable scientific calculators.

Appendix

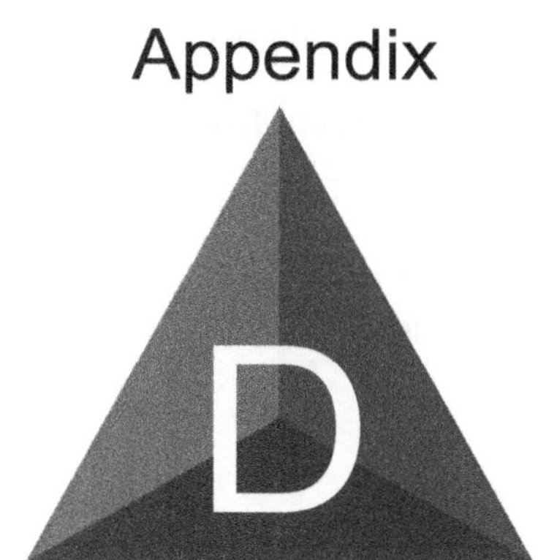

Summary of Solubility Properties of Ions and Solids

	Cl^-	SO_4^{2-}	CO_3^{2-} PO_4^{3-}	CrO_4^{2-}	OH^- O^{2-}	H_2S, pH = 0.5	S^{2-}, pH = 9
Li^+, Na^+, K^+, NH_4^+	S	S	S	S	S	S	S
Ba^{2+}	S	I	A	A	S^-	S	S
Ca^{2+}	S	S^-	A	S	S^-	S	S
Mg^{2+}	S	S	A	S	A	S	S
Fe^{3+}	S	S	A	A	A	S	A
Cr^{3+}	S	S	A	A	A	S	A
Al^{3+}	S	S	A, B	A, B	A, B	S	A, B
Ni^{2+}	S	S	A, N	A, N	A, N	S	A^+, O^+
Co^{2+}	S	S	A	A	A	S	A^+, O^+
Zn^{2+}	S	S	A, B, N	A, B, N	A, B, N	S	A
Mn^{2+}	S	S	A	A	A	S	A
Cu^{2+}	S	S	A, N	A, N	A, N	O	O
Cd^{2+}	S	S	A, N	A, N	A, N	A^+, O	A^+, O
Bi^{3+}	A	A	A	A	A	O	O
Hg^{2+}	S	S	A	A	A	O^+, C	O^+, C
Sn^{2+}, Sn^{4+}	A, B	A, B	A, B	A, B	A, B	A^+, C	A^+, C
Sb^{3+}	A, B	A, B	A, B	A, B	A, B	A^+, C	A^+, C
Ag^+	A^+, N	S^-, N	A, N	A, N	A, N	O	O
Pb^{2+}	HW, B, A^+	B	A, B	B	A, B	O	O
Hg_2^{2+}	O^+	S^-, A	A	A	A	O^+	O^+

Key:

S	soluble in water	I	insoluble in any common reagent
A	soluble in acid (6 *M* HCl or another nonprecipitating, nonoxidizing acid)	S^-	slightly soluble in water
		A^+	soluble in 12 *M* HCl
B	soluble in 6 *M* NaOH	O^+	soluble in aqua regia
O	soluble in hot 6 *M* HNO_3	C	soluble in 6 *M* NaOH containing excess S^{2-}
N	soluble in 6 *M* NH_3	HW	soluble in hot water

Example: For Cd^{2+} and OH^+, the entry is A, N. This means that $Cd(OH)_2(s)$, the product obtained when solutions containing Cd^{2+} and OH^- are mixed, will dissolve to the extent of at least 0.1 mol/L when treated with 6 *M* HCl or 6 *M* NH_3. Because 6 *M* HNO_3, 12 *M* HCl, and aqua regia are at least as strongly acidic as 6 *M* HCl is, $Cd(OH)_2(s)$ would also be soluble in those reagents.

Solubility Rules

Water-soluble salts	
Na^+, K^+, NH_4^+	All sodium, potassium, and ammonium salts are soluble.
NO_3^-, ClO_3^-, $C_2H_3O_2^-$	All nitrates, chlorates, and acetates are soluble.
Cl^-	All chlorides are soluble except AgCl, Hg_2Cl_2, and $PbCl_2$.*
Br^-	All bromides are soluble except AgBr, Hg_2Br_2, $PbBr_2$,* and $HgBr_2$.*
I^-	All iodides are soluble except AgI, Hg_2I_2, PbI_2, and HgI_2.
SO_4^{2-}	All sulfates are soluble except $CaSO_4$,* $SrSO_4$, $BaSO_4$, Hg_2SO_4, $PbSO_4$, and Ag_2SO_4.
Water-insoluble salts	
CO_3^{2-}, SO_3^{2-}, PO_4^{3-}, CrO_4^{2-}	All carbonates, sulfites, phosphates, and chromates are insoluble except those of alkali metals and NH_4^+.
OH^-	All hydroxides are insoluble except those of alkali metals and $Ca(OH)_2$,* $Sr(OH)_2$,* and $Ba(OH)_2$.
S^{2-}	All sulfides are insoluble except those of the alkali metals, the alkaline earth metals, and NH_4^+.

*Slightly soluble

Another way of stating the above rule is as follows:

Solubility Guidelines for Common Ionic Compounds in Water

Soluble ionic compounds		Important exceptions
Compounds containing	NO_3^-	None
	CH_3COO^-	None
	Cl^-	Compounds of Ag^+, Hg_2^{2+}, and Pb^{2+}
	Br^-	Compounds of Ag^+, Hg_2^{2+}, and Pb^{2+}
	I^-	Compounds of Ag^+, Hg_2^{2+}, and Pb^{2+}
	SO_4^{2-}	Compounds of Sr^{2+}, Ba^{2+}, Hg_2^{2+}, and Pb^{2+}
Insoluble ionic compounds		**Important exceptions**
Compounds containing	S^{2-}	Compounds of NH_4^+, the alkali metal cations, Ca^{2+}, Sr^{2+}, and Ba^{2+}
	CO_3^{2-}	Compounds of NH_4^+ and the alkali metal cations
	PO_4^{3-}	Compounds of NH_4^+ and the alkali metal cations
	OH^-	Compounds of NH_4^+, the alkali metal cations, Ca^{2+}, Sr^{2+}, and Ba^{2+}

Appendix

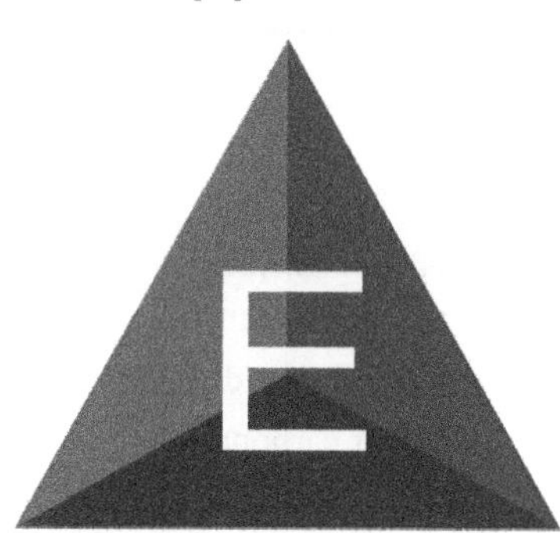

Solubility-Product Constants for Compounds at 25 °C

Name	Formula	K_{sp}	Name	Formula	K_{sp}
Barium carbonate	$BaCO_3$	5.1×10^{-9}	Copper(II) hydroxide	$Cu(OH)_2$	4.8×10^{-2}
Barium chromate	$BaCrO_4$	2.1×10^{-10}	Copper(II) phosphate	$Cu_3(PO_4)_2$	1.3×10^{-37}
Barium fluoride	BaF_2	1.7×10^{-6}	Copper(II) sulfide*	CuS	6×10^{-37}
Barium hydroxide	$Ba(OH)_2$	5×10^{-3}	Gold(I) chloride	$AuCl$	2.0×10^{-13}
Barium oxalate	BaC_2O_4	1.6×10^{-6}	Gold(III) chloride	$AuCl_3$	3.2×10^{-25}
Barium phosphate	$Ba_3(PO_4)_2$	3.4×10^{-23}	Iron(II) carbonate	$FeCO_3$	2.1×10^{-11}
Barium sulfate	$BaSO_4$	1.1×10^{-10}	Iron(II) hydroxide	$Fe(OH)_2$	7.9×10^{-11}
Cadmium carbonate	$CdCO_3$	1.8×10^{-14}	Iron(II) sulfide*	FeS	6×10^{-19}
Cadmium hydroxide	$Cd(OH)_2$	2.5×10^{-14}	Iron(III) hydroxide	$Fe(OH)_3$	4×10^{-38}
Cadmium sulfide*	CdS	8×10^{-28}	Lanthanum fluoride	LaF_3	2×10^{-19}
Calcium carbonate	$CaCO_3$	4.5×10^{-9}	Lanthanum iodate	$La(IO_3)_3$	7.4×10^{-11}
Calcium chromate	$CaCrO_4$	4.5×10^{-9}	Lead carbonate	$PbCO_3$	7.4×10^{-14}
Calcium fluoride	CaF_2	3.9×10^{-11}	Lead chloride	$PbCl_2$	1.7×10^{-5}
Calcium hydroxide	$Ca(OH)_2$	6.5×10^{-6}	Lead chromate	$PbCrO_4$	2.8×10^{-13}
Calcium oxalate	CaC_2O_4	1×10^{-9}	Lead fluoride	PbF_2	3.6×10^{-8}
Calcium phosphate	$Ca_3(PO_4)_2$	2.0×10^{-29}	Lead hydroxide	$Pb(OH)_2$	1.2×10^{-15}
Calcium sulfate	$CaSO_4$	2.4×10^{-5}	Lead sulfate	$PbSO_4$	6.3×10^{-7}
Cerium(III) fluoride	CeF_3	8×10^{-16}	Lead sulfide*	PbS	3×10^{-28}
Chromium(III) fluoride	CrF_3	6.6×10^{-11}	Magnesium hydroxide	$Mg(OH)_2$	1.8×10^{-11}
Chromium(III) hydroxide	$Cr(OH)_3$	6.3×10^{-31}	Magnesium oxalate	MgC_2O_4	8.6×10^{-5}
Cobalt(II) carbonate	$CoCO_3$	1.0×10^{-10}	Manganese carbonate	$MnCO_3$	3.5×10^{-8}
Cobalt(II) hydroxide	$Co(OH)_2$	1.3×10^{-15}	Manganese hydroxide	$Mn(OH)_2$	1.6×10^{-13}
Cobalt(III) hydroxide	$Co(OH)_3$	1.6×10^{-44}	Manganese(II) sulfide*	MnS	2×10^{-53}
Cobalt(II) sulfide*	CoS	5.0×10^{-22}	Mercury(I) chloride	Hg_2Cl_2	1.2×10^{-18}
Copper(I) bromide	$CuBr$	5.3×10^{-9}	Mercury(I) oxalate	$Hg_2C_2O_4$	2.0×10^{-13}
Copper(I) chloride	$CuCl$	1.2×10^{-6}	Mercury(I) sulfide	Hg_2S	1×10^{-47}
Copper(I) sulfide	Cu_2S	2.5×10^{-48}	Mercury(II) hydroxide	$Hg(OH)_2$	3.0×10^{-26}
Copper(II) carbonate	$CuCO_3$	1.4×10^{-10}	Mercury(II) sulfide*	HgS	2×10^{-53}
Copper(II) chromate	$CuCrO_4$	3.6×10^{-6}	Nickel carbonate	$NiCO_3$	1.3×10^{-7}

* For a solubility equilibrium of the type $MS(s) + H_2O(l) \rightleftharpoons M^{2+}(aq) + HS^-(aq) + OH^-(aq)$.

Name	Formula	K_{sp}	Name	Formula	K_{sp}
Nickel hydroxide	$Ni(OH)_2$	6.0×10^{-16}	Silver sulfate	Ag_2SO_4	1.5×10^{-5}
Nickel oxalate	NiC_2O_4	4×10^{-10}	Silver sulfide	Ag_2S	6×10^{-51}
Nickel sulfide*	NiS	3×10^{-20}	Strontium carbonate	$SrCO_3$	9.3×10^{-10}
Silver arsenate	Ag_3AsO_4	1.0×10^{-22}	Tin(II) hydroxide	$Sn(OH)_2$	1.4×10^{-28}
Silver bromide	AgBr	5.5×10^{-13}	Tin(II) sulfide	SnS	1×10^{-26}
Silver carbonate	Ag_2CO_3	8.1×10^{-12}	Zinc carbonate	$ZnCO_3$	1.0×10^{-10}
Silver chloride	AgCl	1.8×10^{-10}	Zinc hydroxide	$Zn(OH)_2$	3.0×10^{-16}
Silver chromate	Ag_2CrO_4	0.2×10^{-12}	Zinc oxalate	ZnC_2O_4	2.7×10^{-8}
Silver cyanide	AgCN	1.2×10^{-16}	Zinc sulfide*	ZnS	2×10^{-25}
Silver iodide	AgI	8.3×10^{-17}			

* For a solubility equilibrium of the type $MS(s) + H_2O(l) \rightleftharpoons M^{2+}(aq) + HS^-(aq) + OH^-(aq)$.

Appendix

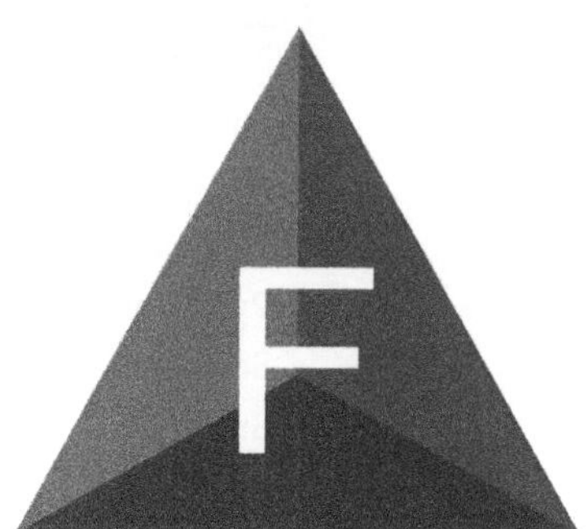

Dissociation Constants for Acids at 25 °C

Name	Formula	K_{a1}	K_{a2}	K_{a3}
Acetic	$HC_2H_3O_2$	1.8×10^{-5}		
Arsenic	H_3AsO_4	5.6×10^{-3}	1.0×10^{-7}	3.0×10^{-12}
Arsenous	H_3AsO_3	5.1×10^{-10}		
Ascorbic	$HC_6H_7O_6$	8.0×10^{-5}	1.6×10^{-12}	
Benzoic	$HC_7H_5O_2$	6.3×10^{-5}		
Boric	H_3BO_3	5.8×10^{-10}		
Butanoic	$HC_4H_7O_2$	1.5×10^{-5}		
Carbonic	H_2CO_3	4.3×10^{-7}	5.6×10^{-11}	
Chloroacetic	$HC_2H_2O_2Cl$	1.4×10^{-3}		
Chlorous	$HClO_2$	1.1×10^{-2}		
Citric	$H_3C_6H_5O_7$	7.4×10^{-4}	1.7×10^{-5}	4.0×10^{-7}
Cyanic	$HCNO$	3.5×10^{-4}		
Formic	$HCHO_2$	1.8×10^{-4}		
Hydrazoic	HN_3	1.9×10^{-5}		
Hydrocyanic	HCN	4.9×10^{-10}		
Hydrofluoric	HF	6.8×10^{-4}		
Hydrogen chromate ion	$HCrO_4^{-}$	3.0×10^{-7}		
Hydrogen peroxide	H_2O_2	2.4×10^{-12}		
Hydrogen selenate ion	$HSeO_4^{-}$	2.2×10^{-2}		
Hydrosulfuric acid	H_2S	9.5×10^{-8}	1×10^{-19}	
Hypobromous	$HBrO$	2.5×10^{-9}		
Hypochlorous	$HClO$	3.0×10^{-8}		
Hypoiodous	HIO	2.3×10^{-11}		
Iodic	HIO_3	1.7×10^{-1}		
Lactic	$HC_3H_5O_3$	1.4×10^{-4}		
Malonic	$H_2C_3H_2O_4$	1.5×10^{-3}	2.0×10^{-6}	
Nitrous	HNO_2	4.5×10^{-4}		
Oxalic	$H_2C_2O_4$	5.9×10^{-2}	6.4×10^{-5}	
Paraperiodic	H_5IO_6	2.8×10^{-2}	5.3×10^{-9}	

Name	Formula	K_{a1}	K_{a2}	K_{a3}
Phenol	HC_6H_5O	1.3×10^{-10}		
Phosphoric	H_3PO_4	7.5×10^{-3}	6.2×10^{-8}	4.2×10^{-13}
Propionic	$HC_3H_5O_2$	1.3×10^{-5}		
Pyrophosphoric	$H_4P_2O_7$	3.0×10^{-2}	4.4×10^{-3}	2.1×10^{-7}
Selenous	H_2SeO_3	2.3×10^{-3}	5.3×10^{-9}	
Sulfuric	H_2SO_4	Strong acid	1.2×10^{-2}	
Sulfurous	H_2SO_3	1.7×10^{-2}	6.4×10^{-8}	
Tartaric	$H_2C_4H_4O_6$	1.0×10^{-3}	4.6×10^{-5}	

Appendix

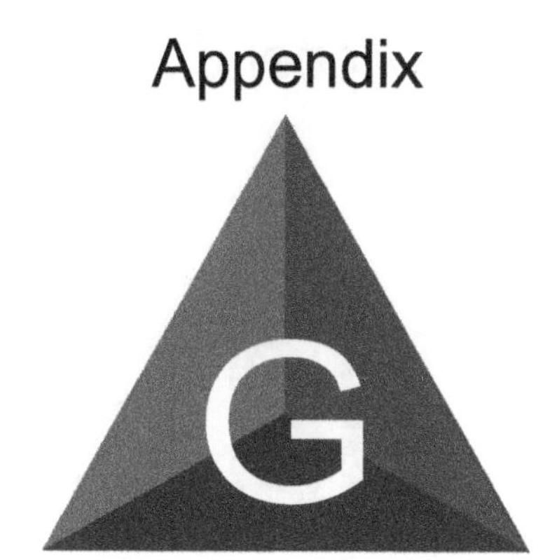

Dissociation Constants for Bases at 25 °C

Name	Formula	K_b
Ammonia	NH_3	1.8×10^{-5}
Aniline	$C_6H_5NH_2$	4.3×10^{-10}
Dimethylamine	$(CH_3)_2NH$	5.4×10^{-4}
Ethylamine	$C_2H_5NH_2$	6.4×10^{-4}
Hydrazine	H_2NNH_2	1.3×10^{-6}
Hydroxylamine	$HONH_2$	1.1×10^{-8}
Methylamine	CH_3NH_2	4.4×10^{-4}
Pyridine	C_5H_5N	1.7×10^{-9}
Trimethylamine	$(CH_3)_3N$	6.4×10^{-5}

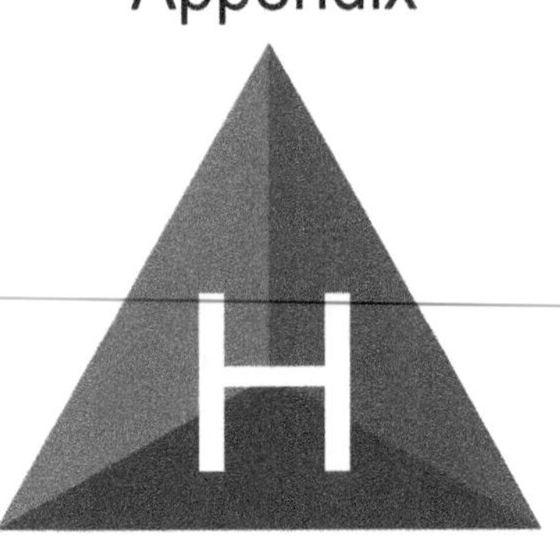

Appendix H

Selected Standard Reduction Potentials at 25 °C

Half-reaction	***E*° (V)**
$Ag^+(aq) + e^- \longrightarrow Ag(s)$	+0.799
$AgBr(s) + e^- \longrightarrow Ag(s) + Br^-(aq)$	+0.095
$AgCl(s) + e^- \longrightarrow Ag(s) + Cl^-(aq)$	+0.222
$Ag(CN)_2^-(aq) + e^- \longrightarrow Ag(s) + 2CN^-(aq)$	−0.31
$Ag_2CrO_4(s) + 2e^- \longrightarrow 2Ag(s) + CrO_4^{2-}(aq)$	+0.446
$AgI(s) + e^- \longrightarrow Ag(s) + I^-(aq)$	−0.151
$Ag(S_2O_3)_2^{3-} + e^- \longrightarrow Ag(s) + 2S_2O_3^{2-}(aq)$	+0.01
$Al^{3+}(aq) + 3e^- \longrightarrow Al(s)$	−1.66
$H_3AsO_4(aq) + 2H^+(aq) + 2e^- \longrightarrow H_3AsO_3(aq) + H_2O(l)$	+0.559
$Ba^{2+}(aq) + 2e^- \longrightarrow Ba(s)$	−2.90
$BiO^+(aq) + 2H^+(aq) + 3e^- \longrightarrow Bi(s) + H_2O(l)$	+0.32
$Br_2(l) + 2e^- \longrightarrow 2Br^-(aq)$	+1.065
$BrO_3^-(aq) + 6H^+(aq) + 5e^- \longrightarrow \frac{1}{2}Br_2(l) + 3H_2O(l)$	+1.52
$2CO_2(g) + 2H^+(aq) + 2e^- \longrightarrow H_2C_2O_4(aq)$	−0.49
$Ca^{2+}(aq) + 2e^- \longrightarrow Ca(s)$	−2.87
$Cd^{2+}(aq) + 2e^- \longrightarrow Cd(s)$	−0.403
$Ce^{4+}(aq) + e^- \longrightarrow Ce^{3+}(aq)$	+1.61
$Cl_2(g) + 2e^- \longrightarrow 2Cl^-(aq)$	+1.359
$HClO(aq) + H^+(aq) + e^- \longrightarrow \frac{1}{2}Cl_2(g) + H_2O(l)$	+1.63
$ClO^-(aq) + H_2O(l) + 2e^- \longrightarrow Cl^-(aq) + 2OH^-(aq)$	+0.89
$ClO_3^-(aq) + 6H^+(aq) + 5e^- \longrightarrow \frac{1}{2}Cl_2(g) + 3H_2O(l)$	+1.47
$Co^{2+}(aq) + 2e^- \longrightarrow Co(s)$	−0.277
$Co^{3+}(aq) + e^- \longrightarrow Co^{2+}(aq)$	+1.842
$Cr^{3+}(aq) + 3e^- \longrightarrow Cr(s)$	−0.74
$Cr^{3+}(aq) + e^- \longrightarrow Cr^{2+}(aq)$	−0.41

Half-reaction	$E°$ **(V)**
$Cr_2O_7^{2-}(aq) + 14H^+(aq) + 6e^- \longrightarrow 2Cr^{3+}(aq) + 7H_2O(l)$	+1.33
$CrO_4^{2-}(aq) + 4H_2O(l) + 3e^- \longrightarrow Cr(OH)_3(s) + 5OH^-(aq)$	−0.13
$Cu^{2+}(aq) + 2e^- \longrightarrow Cu(s)$	+0.337
$Cu^{2+}(aq) + e^- \longrightarrow Cu^+(aq)$	+0.153
$Cu^+(aq) + e^- \longrightarrow Cu(s)$	+0.521
$CuI(s) + e^- \longrightarrow Cu(s) + I^-(aq)$	−0.185
$F_2(g) + 2e^- \longrightarrow 2F^-(aq)$	+2.87
$Fe^{2+}(aq) + 2e^- \longrightarrow Fe(s)$	−0.440
$Fe^{3+}(aq) + e^- \longrightarrow Fe^{2+}(aq)$	+0.771
$Fe(CN)_6^{3-}(aq) + e^- \longrightarrow Fe(CN)_6^{4-}(aq)$	+0.36
$2H^+(aq) + 2e^- \longrightarrow H_2(g)$	0.000
$2H_2O(l) + 2e^- \longrightarrow H_2(g) + 2OH^-(aq)$	−0.83
$HO_2^-(aq) + H_2O(l) + 2e^- \longrightarrow 3OH^-(aq)$	+0.88
$H_2O_2(aq) + 2H^+(aq) + 2e^- \longrightarrow 2H_2O(l)$	+1.776
$Hg_2^{2+}(aq) + 2e^- \longrightarrow 2Hg(l)$	+0.789
$2Hg^{2+}(aq) + 2e^- \longrightarrow Hg_2^{2+}(aq)$	+0.920
$Hg^{2+}(aq) + 2e^- \longrightarrow Hg(l)$	+0.854
$I_2(s) + 2e^- \longrightarrow 2I^-(aq)$	+0.536
$IO_3^-(aq) + 6H^+(aq) + 5e^- \longrightarrow \frac{1}{2}I_2(s) + 3H_2O(l)$	+1.195
$K^+(aq) + e^- \longrightarrow K(s)$	−2.925
$Li^+(aq) + e^- \longrightarrow Li(s)$	−3.05
$Mg^{2+}(aq) + 2e^- \longrightarrow Mg(s)$	−2.37
$Mn^{2+}(aq) + 2e^- \longrightarrow Mn(s)$	−1.18
$MnO_2(s) + 4H^+(aq) + 2e^- \longrightarrow Mn^{2+}(aq) + 2H_2O(l)$	+1.23
$MnO_4^-(aq) + 8H^+(aq) + 5e^- \longrightarrow Mn^{2+} + 4H_2O(l)$	+1.51
$MnO_4^-(aq) + 2H_2O(l) + 3e^- \longrightarrow MnO_2 + 4OH^-(aq)$	+0.59
$HNO_2(aq) + H^+(aq) + e^- \longrightarrow NO(g) + H_2O(l)$	+1.00
$N_2(g) + 4H_2O(l) + 4e^- \longrightarrow 4OH^-(aq) + N_2H_4(aq)$	−1.16
$N_2(g) + 5H^+(aq) + 4e^- \longrightarrow N_2H_5^+(aq)$	−0.23
$NO_3^-(aq) + 4H^+(aq) + 3e^- \longrightarrow NO(g) + 2H_2O(l)$	+0.96
$Na^+(aq) + e^- \longrightarrow Na(s)$	−2.71
$Ni^{2+}(aq) + 2e^- \longrightarrow Ni(s)$	−0.28
$O_2(g) + 4H + (aq) + 4e^- \longrightarrow 2H_2O(l)$	+1.23

Half-reaction	***E*° (V)**
$O_2(g) + 2H_2O(l) + 4e^- \longrightarrow 4OH^-(aq)$	+0.40
$O_2(g) + 2H^+(aq) + 2e^- \longrightarrow H_2O_2(aq)$	+0.68
$O_3(g) + 2H^+(aq) + 2e^- \longrightarrow O_2(g) + H_2O(l)$	+2.07
$Pb^{2+}(aq) + 2e^- \longrightarrow Pb(s)$	−0.126
$PbO_2(s) + HSO_4^-(aq) + 3H^+(aq) + 2e^- \longrightarrow PbSO_4(s) + 2H_2O(l)$	+1.685
$PbSO_4(s) + H^+(aq) + 2e^- \longrightarrow Pb(s) + HSO_4^-(aq)$	−0.356
$PtCl_4^{2-}(aq) + 2e^- \longrightarrow Pt(s) + 4Cl^-(aq)$	+0.73
$S(s) + 2H^+(aq) + 2e^- \longrightarrow H_2S(g)$	+0.141
$H_2SO_3(aq) + 4H^+(aq) + 4e^- \longrightarrow S(s) + 3H_2O(l)$	+0.45
$HSO_4^-(aq) + 3H^+(aq) + 2e^- \longrightarrow H_2SO_3(aq) + H_2O(l)$	+0.17
$Sn^{2+}(aq) + 2e^- \longrightarrow Sn(s)$	−0.136
$Sn^{4+}(aq) + 2e^- \longrightarrow Sn^{2+}(aq)$	+0.154
$VO_2^+(aq) + 2H^+(aq) + e^- \longrightarrow VO^{2+}(aq) + H_2O(l)$	+1.00
$Zn^{2+}(aq) + 2e^- \longrightarrow Zn(s)$	−0.763

Appendix I

Spreadsheets

A computer spreadsheet program is a powerful tool for manipulating quantitative information.* In addition to simplifying repetitive calculations, spreadsheets allow you to plot such results as calibration curves and acid–base or redox titrations. Such graphs are critical for understanding and interpreting quantitative relationships.

Although nearly any spreadsheet program can be used for these purposes, the instructions given here apply to one of the more widely available spreadsheets—Microsoft Excel®. Specific steps outlined in the following pages will vary slightly between the different versions. You will need specific directions for other spreadsheet software, but they will not be very different from what is presented here.

GETTING STARTED: AN EXAMPLE

As an example, we will prepare a spreadsheet to compute the density of water from the equation

$$\text{density (g/mL)} = a_0 + a_1T + a_2T^2 + a_3T^3 \qquad (1)$$

where $a_0 = 0.99989$, $a_1 = 5.3322\times10^{-5}$, $a_2 = -7.5899\times10^{-6}$, $a_3 = 3.6719\times10^{-8}$, and T = temperature in °C. The blank spreadsheet in Figure I.1a has columns labelled A, B, C, and rows numbered 1, 2, 3, etc. The box in column B, row 4 is called *cell* B4.

It is helpful to begin each spreadsheet with a title to make it more readable and to remind you of what you are doing with the spreadsheet. Figure I.1b shows cell A1 highlighted along with the text **Calculating the Density of H_2O with Equation 1**. The computer automatically spreads the text so that it appears to over top of adjoining cells if those are empty.

The constants are placed in column A by selecting cell A4 and typing **Constants:** as a column heading. Cell A5 was then selected and includes the text $\mathbf{a_0}$ = to indicate that the constant a_0 will be written in the next cell in the same column. This was done by selecting cell A6 and typing the number **0.99989** (without extra spaces). In cells A7 to A12, the remaining constants were entered in the same way. You may need to increase the number of decimal places by using the button on the toolbar (or by FORMAT ⟶ cells …⟶ Number), click on Number in the list, and increase decimal places to five. Note, however, that powers of ten are written in an unusual format, for example, E-05 means 10^{-5}. Your spreadsheet should look similar to Figure I.1b.

---

*Spreadsheets for analytical chemistry: R. de Levie, *Principles of Quantitative Analysis*, New York: McGraw-Hill, 1997; D. Diamond and V. Hanratty, *Spreadsheet Applications in Chemistry Using Microsoft Excel*, New York: John Wiley & Sons, 1997; H. Freiser, *Concepts & Calculations in Analytical Chemistry: A Spreadsheet Approach*, Boca Raton, FL: CRC Press, 1992; R. de Levie, *A Spreadsheet Workbook for Quantitative Chemical Analysis*, New York: McGraw-Hill, 1992; B.V. Liengme, *A Guide to Microsoft Excel for Scientists and Engineers*, New York: John Wiley & Sons, 1997.

Columns

Rows

	A	B	C
1			
2			
3			
4		cell B4	
5			
6			
7			
8			
9			
10			
11			
12			

(a)

	A	B	C
1	Calculating the Density of H_2O with Equation 1		
2			
3			
4	Constants:		
5	a_0 =		
6	0.99989		
7	a_1 =		
8	5.3322E-05		
9	a_2 =		
10	−7.5899E-06		
11	a_3 =		
12	3.6719E-08		

(b)

	A	B	C
1	Calculating the Density of H_2O with Equation 1		
2			
3			
4	Constants:	Temp (°C)	Density (g/mL)
5	a_0 =	5	0.99997
6	0.99989	10	
7	a_1 =	15	
8	5.3322E-05	20	
9	a_2 =	25	
10	−7.5899E-06	30	
11	a_3 =	35	
12	3.6719E-08	40	

(c)

	A	B	C
1	Calculating the Density of H_2O with Equation 1		
2			
3			
4	Constants:	Temp (°C)	Density (g/mL)
5	a_0 =	5	0.99997
6	0.99989	10	0.99970
7	a_1 =	15	0.99911
8	5.3322E-05	20	0.99821
9	a_2 =	25	0.99705
10	−7.5899E-06	30	0.99565
11	a_3 =	35	0.99403
12	3.6719E-08	40	0.99223
13			
14	Formula:		
15	C5 = \$A\$6 + \$A\$8*B5 + \$A\$10*B5^2 + \$A\$12*B5^3		

(d)

▲FIGURE I.1 Evolution of a spreadsheet for computing the density of water.

Cell B4 includes the heading **Temp (°C)** and temperatures from 5 through 40 in cells B5 through B12. This completes the *data input* to the spreadsheet. The *data output* will be the computed values of the density and will be entered in column C. You now need to input the computation and determine where to put the result. First, highlight cell C4 and enter the heading **Density (g/mL)**. Then in cell C5, type the following formula to carry out the computation:

= A6 + A8 * B5 + A10 * B5 ^ 2 + A12 * B5 ^ 3

(You may omit the space before and after the arithmetic operators.) When you tap the RETURN or ENTER key, the number **0.99997** appears in cell C5. The formula above is the spreadsheet translation of Equation 1. A6 refers to the constant value in cell A6. (The dollar signs will be explained shortly.) B5 refers to the temperature given in cell B5. The times sign is *, and the exponentiation sign is ^. Thus, for example, the term "A12*B5^3" means "(contents of cell A12) × (contents of cell B5)^3."

Now comes the most magical and useful property of the spreadsheet. Highlight cell C5 and the empty cells below it from C6 to C12. Then select the FILL DOWN command from the Edit menu. This procedure copies the formula from C5 into the cells below it and computes the results from the above formula operating on the corresponding entries in columns A and B. Then the computed density of water at each temperature appears in column C, as in Figure I.1d.

This example included three types of entries. *Labels* such as $\mathbf{a}_0$ = were typed in as text. An entry that does not begin with a digit or an equal sign is treated as text. *Numbers*, such as 5 in cell B5, were typed in column B. The spreadsheet treats a number differently from text. In cell C5, a *formula* was entered to carry out the computation. Formulas necessarily begin with an equal sign. The mathematical operations of addition, subtraction, multiplication, division, and exponentiation have the symbols +, −, *, /, and ^, respectively. You can type *functions* such as Exp (.), or you can select them from the Insert menu or function menu. Exp (.) raises *e* to the power in parentheses. Other functions such as ln (.), log (.), sin (.), and cos (.) are also available. In some spreadsheet programs, functions are written in all capital letters.

The order of arithmetic operations in formulas is ^ first, followed by * and / (evaluated in order from left to right as they appear) and then followed by + and − (also evaluated from left to right). Make liberal use of parentheses so that the computer does what you intend. The contents of parentheses are evaluated before operations outside the parentheses. Some examples follow.

$$9/5*100+32=(9/5)*100+32=(1.8)*100+32$$
$$=(1.8)*100+32=(180)+32=212$$
$$9/5*(100+32)=9/5*(132)=(1.8)*(132)=237.6$$
$$9+5*100/32=9+(5*100)/32=9+(500)/32$$
$$=9+(500)/32=9+(15.625)=24.625$$
$$9/5\hat{}2+32=9/(5\hat{}2)+32=(9/25)+32=(0.36)+32=32.36$$

When in doubt about how an expression will be evaluated by the computer, use parentheses to get the program to do what you intend and double check one calculation by hand using a calculator.

ABSOLUTE AND RELATIVE REFERENCES

The **formula = A8*B5** refers to cells A8 and B5 in different ways. The symbol A8 is an *absolute reference* to the contents of cell A8. Wherever cell A8 is called in the spreadsheet, the computer goes to cell A8 to look for a number. In contrast, B5 is a *relative reference* in the formula in cell C5. When called from cell C5, the computer goes to cell B5 to find a number, but when called from cell C6, the computer goes to cell B6 to look for a number. Likewise, if it were called from cell C19, the computer would look in cell B19. This is why the cell address written without dollar signs surrounding it is called a relative reference. If you want the computer to look only in cell B5, type **B5**.

A SECOND EXAMPLE: CALCULATING STANDARD DEVIATION

Suppose you wanted to calculate the standard deviation of the following numbers: 17.4, 18.1, 18.2, 17.9, and 17.6. There are two ways to use a spreadsheet to do this. The first way involves entering the numbers and the required equations and then directing the computer to do the requisite calculations. The second way involves entering the numbers and using "built-in" functions to do the calculations. The following illustrates both ways.

Writing the equations to do the calculations

Begin by reproducing the spreadsheet template shown in Figure I.2a. Cells B4 to B8 contain the data (x values) whose mean and standard deviation you will compute.

In cells B9 and B10, you type the formulas to compute the sum and mean value of the x values, respectively, as in Figure I.2b. In cell C4, type the formula to compute the deviations from the mean, where x is in cell B4 and the mean is in cell B10. This formula appears in cell B16 of Figure I.2b. Use the FILL DOWN command to compute values in cells C5 to C8. Type a formula in cell D4 to compute the square of the value in cell C4. This formula appears in cell B17 of Figure I.2b. Use the FILL DOWN command to compute values in cells D5 to D8. Type a formula in cell D8 to compute the sum of the numbers in cells D4 to D8. (See cell B18 in Figure I.2b.) Type a formula in cell B11 to compute the standard deviation (cell B15, Figure I.2b). Then use cells B13 through B18 to document your formulas. Remember that all of the formulas start with an equal sign, not a cell number. For example, the formula used to calculate the sum is = B4 + B5 + B6 + B7 + B8, but your documentation is B9 = B4 + B5 + B6 + B7 + B8. The former would be typed in cell B9; the latter is just a reminder of what was done to get the number that appears in cell B9.

	A	B	C	D
1	Computing standard deviation			
2				
3		Data = X	X-mean	(X-mean)^2
4		17.4		
5		18.1		
6		18.2		
7		17.9		
8		17.6		
9	sum =			
10	mean =			
11	std dev =			
12				
13	Formulas:	B9 =		
14		B10 =		
15		B11 =		
16		C4 =		
17		D4 =		
18		D9 =		
19				
20	Calculations using built-in functions:			
21	sum =			
22	mean =			
23	std dev =			

(a)

	A	B	C	D
1	Computing standard deviation			
2				
3		Data = X	X-mean	(X-mean)^2
4		17.4	−0.44	0.1936
5		18.1	0.26	0.0676
6		18.2	0.36	0.1296
7		17.9	0.06	0.0036
8		17.6	−0.24	0.0576
9	sum =	89.2		0.452
10	mean =	17.84		
11	std dev =	0.3362		
12				
13	Formulas:	B9 = B4 + B5 + B6 + B7 + B8		
14		B10 =B9/5		
15		B11 = SQRT(D9/(5−1))		
16		C4 = B4-B10		
17		D4 = C4^2		
18		D9 = D4 + D5 + D6 + D7 + D8		
19				
20	Calculations using built-in functions:			
21	sum =	89.2		
22	mean =	17.84		
23	std dev =	0.3362		
24				
25	Formulas:	B21 = SUM(B4:B8)		
26		B22 = AVERAGE(B4:B8)		
27		B23 = STDEV(B4:B8)		

(b)

▲**FIGURE I.2** A spreadsheet to calculate standard deviation.

Using built-in functions

The calculations described above can be simplified by using formulas that are built into the spreadsheet program. In cell B21, type = **SUM(B4:B8)**, which means to find the sum of the numbers in cells B4 to B8. Cell B21 should display the same number as cell B9. In general, you will not know what functions are available and how to write them. Find the function menu in your program and find SUM in this menu. Now select cell B22 and go to the function menu; find a function called AVERAGE or MEAN. When you type = **AVERAGE(B4:B8)** in cell B22, the value that appears in that cell should be the same as the number in cell B10. For cell B23, to find the standard deviation, type = **STDEV(B4:B8)** in cell B23 and then press ENTER. The value that appears in cell B23 is the same as in cell B11. In future spreadsheets, you may make liberal use of built-in functions to save time.

STEPS FOR PRODUCING A GRAPH FROM A SPREADSHEET

1. Enter the Excel program.
2. Type the x-axis label in cell A1.
3. Type the y-axis label in cell B1.
4. Input x-axis data in Column A, beginning with cell A2.
5. Input y-axis data in Column B, beginning with cell B2.
6. Click and drag to select all data, including column headings (for example, cells A1 through B12).
7. Click on the ChartWizard icon in the upper right part of the screen. A crossbar with a small chart icon will appear as the mouse cursor.
8. Select the space where you want the graph to be drawn by clicking and dragging from the upper left to the lower right; this will create a box in which the graph will be placed. (You can alter the size of this box at a later time.)
9. ChartWizard will show a dialog box **(Step 1 of 5)**, asking you to verify that the cells you selected in step 6 above contain the data you want to graph. Click Next>.
10. Select a chart type (generally, XY Scatter). Click Next>.
11. Select a format for the XY Scatter plot (generally, 1). Click Next>.
12. You will see a sample chart. Check that the graph looks correct and that the choices to the right of the graph are as follows: Data series are in columns; use first 1 column for x data and use first 1 row for legend text. (At this point, the graph may have the y-axis label as the title. Don't worry; this will be changed in the next step. Also, you can delete the legend later if you like.) Click Next>.
13. In Step 5 of 5, you can delete the legend and give the chart a title and x- and y-axis labels. Click Finish.

The following are formatting preferences:

14. You can change the graph size at any time by clicking once anywhere on the graph area to get small black squares at the corners and on the sides of the graph area; click on a square and drag in any direction until you get the size you want.

15. The format of either axis can be changed by double-clicking on the axis. The following can be changed:

    Patterns: for tick marks

    Values: for numerical values printed on the axis

    Font: for numbers

    Formatting: for format of the numbers (general, scientific, etc.)

    Alignment of the numbers on the axis

16. You can change any text on the graph by clicking once to highlight the text and typing the changes. You can change text formatting by double-clicking on the text. Choices for changes are as follows:

    Patterns: for borders and colors

    Font: for type and size

    Alignment: for placement on the graph

17. You can change the format of the points displayed on the graph by double-clicking on any point. Choices for change are as follows:

    Patterns: change marker style/color, add a connecting line, delete markers.

    X Values: change your original data source for x values (from the spreadsheet).

    Name & Values: change your original data source for y values (from the spreadsheet).

    X/Y Error bars: add error bars.

    Data Labels: show value will print y values next to each data point; show label will print x values next to each data point (If data labels are added, you can format them by double-clicking on one of them; choices are Patterns, Font, Number, and Alignment.)

To print data and graph:

18. Make sure the graph does not cover any data (move it according to step 14 if necessary).
19. Go to File, Print Preview; arrange the page as desired; and print.

To print a graph only:

20. Double-click in the graph space to get a highlighted border.
21. Go to File, Print Preview; arrange the page as desired; and print.

To add a linear least-squares line to an existing graph:

22. Calculate a new y data set using $y = mx + b$ and place this in column C, with an appropriate heading in cell C1.
23. Click once in the graph area.
24. Click on the ChartWizard icon or insert chart.
25. Change the range of data set to include the new data. Click Next>.
26. View the new graph. Click OK.
27. To remove markers and add a line, double-click on any point of the new data set, set line to Automatic, and set Marker to None. Click OK. (To see the new line without the markers, click anywhere off the line.)

To create a graph with the linear least-squares data already calculated:

28. Make sure the x-axis data is in column A.
29. Begin at step 6 on the previous page and select all three columns of data (A, B, and C).
30. Follow steps 7 to 16 as desired. At step 17, select the linear least-squares data set, remove its markers, and add a line.
31. Use the embedded functions to obtain the SLOPE and INTERCEPT of the least-squares fit line of the form $y = \text{slope}\,x + \text{intercept}$.

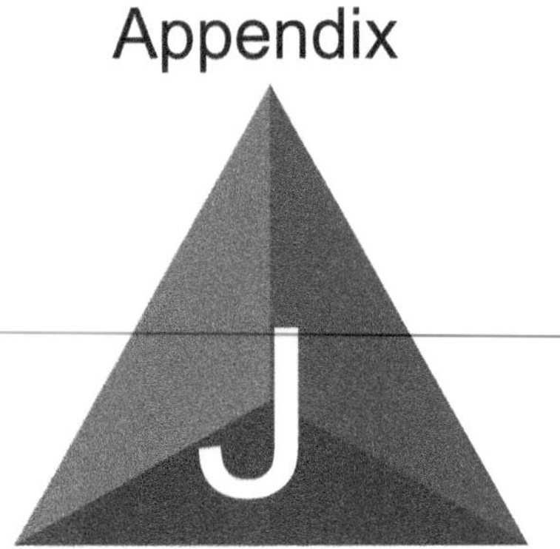

Appendix J

Qualitative Analysis Techniques

I. MIXING SOLUTIONS AND PRECIPITATION

When one solution is added to another in a small test tube, the two solutions must be thoroughly mixed. Mixing can be accomplished by using a *clean* stirring rod or by holding the test tube at the top in one hand and stroking or "tickling" it with the fingers of the other hand to generate a vortex, as shown in Figure J.1. When precipitation reagents are added to solutions in the test tube and you believe precipitation is complete, centrifuge the sample. Always balance the centrifuge by placing a test tube filled with water to about the same level as your sample test tube directly *across* the centrifuge head from your sample, as illustrated in Figure J.2. It usually requires only about 30 sec of centrifugation for the precipitate to settle to the bottom of the test tube. ***Caution: Do not slow down the centrifuge head with your hands. Instead, allow the centrifuge head to come to rest on its own.***

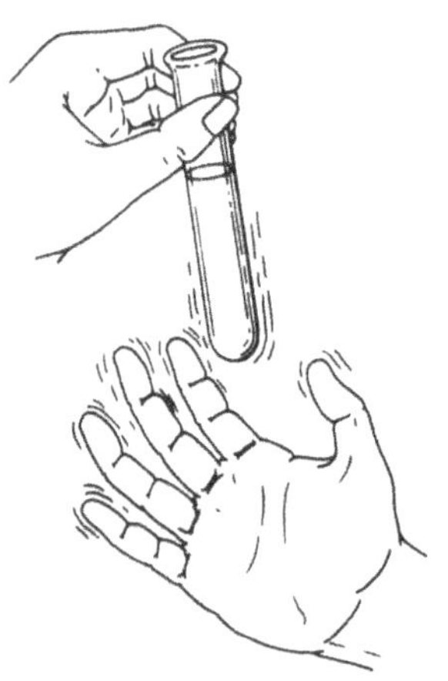

▲**FIGURE J.1** Mixing of a solution by hand.

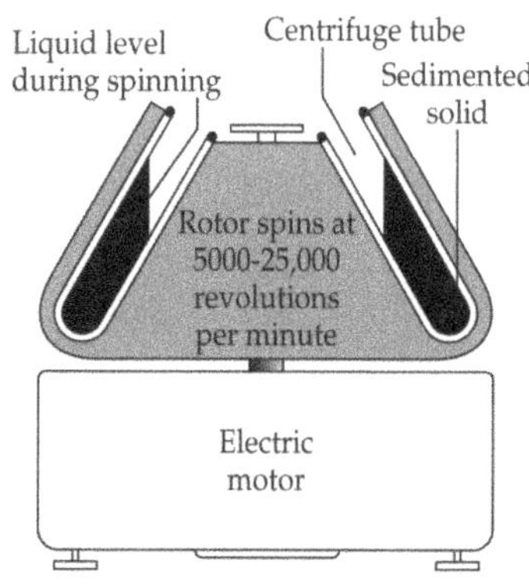

▲**FIGURE J.2** Centrifuge apparatus.

II. DECANTATION AND WASHING OF PRECIPITATES

The liquid above the precipitate is the *supernatant liquid*, or the *decantate*. The best way to remove this liquid without disturbing the precipitate is to withdraw it by means of a capillary pipet, as shown in Figure J.3. This operation is loosely referred to as *decantation*. Because the precipitate separated from the supernatant liquid using this technique will be wet with the decantate, it is necessary to *wash* the precipitate free of contaminating ions. Washing is usually accomplished by adding about 10 drops of distilled water to the precipitate, stirring with a stirring rod or mixing by rapping with a finger to generate a vortex, and repeating the centrifuging and decanting.

III. TESTING ACIDITY

Instructions sometimes require making a solution acidic or basic to litmus by adding acid or base. Always make sure the solution is thoroughly mixed after adding the acid or base; then by means of a clean stirring rod, remove a drop of the solution and apply it to litmus paper. Do not dip the litmus paper directly into the solution. Remember, just because you have added acid (or base) to a solution does not ensure that it is acidic (or alkaline).

IV. HEATING SOLUTIONS IN SMALL TEST TUBES

The safest way to heat solutions in small test tubes is by means of a water bath, as described in Experiment 1 and shown in Figure J.4. A pipe cleaner wrapped around the test tube serves as a convenient handle for placing the tube in or removing it from the bath. The test tubes can be stabilized in the beaker by using a piece of aluminum foil, placed over the top of the beaker, that has holes to accommodate each test tube. Stabilize the beaker with another iron ring placed around halfway up the beaker.

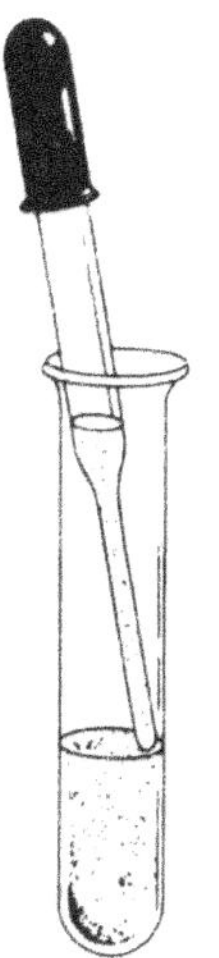

▲**FIGURE J.3** Use of a dropper (also called a capillary pipet) to withdraw liquid from above a solid.

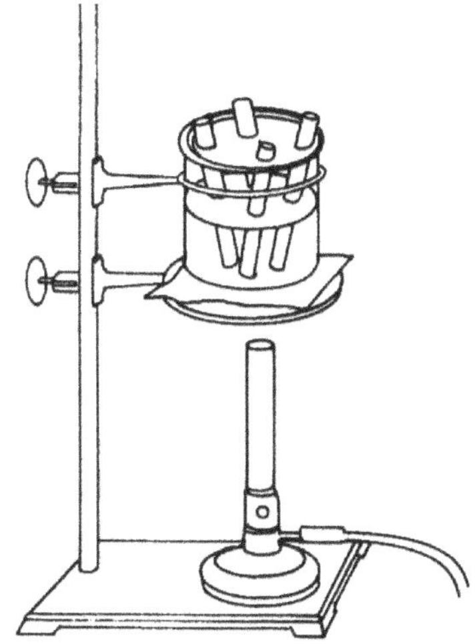

▲**FIGURE J.4** Water bath for heating solutions in test tubes.

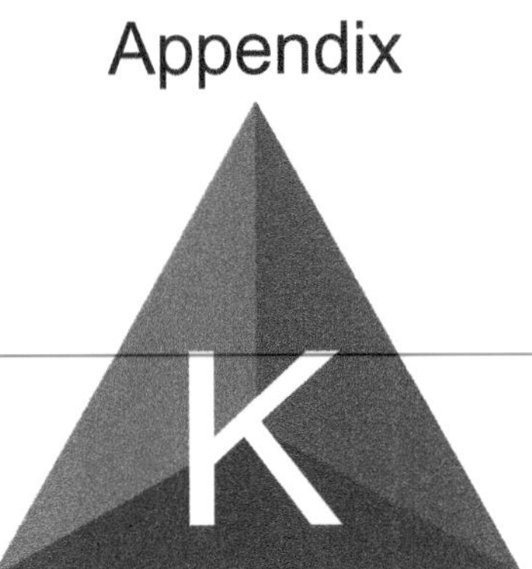

Appendix K: Answers to Selected Pre-lab Questions

EXPERIMENT 1

1. The basic units of measurement in the SI system are electric current, ampere; time, second; amount of substance, mole; and luminous intensity, candela.
2. (a) femto (b) nano (c) giga (d) centi (e) pico
4. (0.180 in.)(2.54 cm/in.)(10 mm/1 cm) = 4.57 mm
5. (a) 0.0723 mg (b) 6.0×10^{-7} mm (c) 3.25×10^{5} μm
7. 3.70×10^{3} mL or 3.70×10^{3} cm^3
8. Convection currents tend to buoy the object and thus lead to an inaccurately low mass. Moreover, hot objects can damage the balance.
9. Thermometers and volumetric glassware are designed to be accurate. But their manufacture is subject to human error, and they should be calibrated.
10. Precision is a measure of the internal consistency of a replicate set of data.
12. $$\text{mean} = \frac{8.2+8.1+8.0}{3} = 8.1$$
$$\text{average deviation from mean} = \frac{0.1+0.0+0.1}{3} = 0.07$$
14. $d = \frac{m}{v}$; $m = dv = (1.23\,\text{g/mL})(700\,\text{mL}) = 861\,\text{g} = 0.861\,\text{kg}$

EXPERIMENT 2

1. Melting points, boiling points, solubility properties, color, and densities are five physical properties.
3. Thymol, m.p. at 52 °C, and diphenyl, m.p. at 70 °C, are solids at room temperature because room temperature is normally about 20 °C and these melting points are above 20 °C.
4. The density of chloroform could be determined in water because it is insoluble in water.

6. Thermometers, pipets, and other pieces of laboratory equipment are mass-produced and subject to human error. Consequently, they should be calibrated.
7. Phenyl benzoate is not miscible with water but is miscible with ethyl alcohol.
11. The liquid is acetone assuming that it is one listed in Table 2.1.

EXPERIMENT 3

3. If a substance boils and melts at a fixed temperature, it is mostly considered a pure substance. If, on the other hand, it boils and melts over a range of temperature, it is considered a mixture. In reality, owing to various factors, m.p. and b.p. are more likely to be over a range rather than being fixed values for all substances. For impure substances, the ranges are however broader
4. Deposition or desublimation is the process by which a substance changes physical states directly from the gaseous state to the solid state without passing through the liquid state. The reverse of deposition is sublimation.
5. Filtration differs from decantation in that the liquid phase passes through a semipermeable substance such as a filter paper, whereas in decantation, the liquid phase is separated from the solid phase by carefully pouring off the liquid. Decantation does not involve the use of another substance to achieve the separation and is faster than filtration.
6. A hot object creates a buoyancy effect by radiating energy in the form of heat and will appear to have a reduced mass. In addition, the hot object may cause damage to the delicate balance and should be allowed to cool to room temperature before its mass is determined.
7. All of the original sample should be recovered; thus, the sum of the masses of the components, NH_4Cl, SiO_2, and NaCl, should precisely equal the total mass of the sample within experimental error, as no matter is being converted into energy in this experiment.
9. NaCl can be separated from acetone by simple filtration. Use a filter funnel fitted with a piece of filter paper, the residue collected is NaCl.

EXPERIMENT 4

1. Before a chemical equation can be written, you must know the reactants and products of the reaction, their formulas and their physical states.
2. A color change or formation of a solid or gas is indicative of a chemical reaction, as is evolution or absorption of heat.
3. NaCl is white, and $PbCl_2$ is bright yellow.
4. Metathesis (double replacement) reactions are atom (or group) exchange reactions. For example, $AgNO_3(aq) + NaCl(aq) \longrightarrow AgCl(s) + NaNO_3(aq)$ is a metathesis reaction in which chloride transfer occurs.
5. A precipitate is an insoluble substance that separates from a homogeneous chemical reaction mixture as a solid.
6. $2KBrO_3(s) \longrightarrow 2KBr(s) + 3O_2(g)$; $MnBr_2(aq) + 2AgNO_3(aq) \longrightarrow Mn(NO_3)_2(aq) + 2AgBr(s)$

EXPERIMENT 5

1. (a) Hg (b) Te (c) Pt (d) Br (e) Ne
3. A compound is composed of a definite number of whole atoms in a fixed proportion, which are united chemically.
4. CH_2O
5. $4Al + 3O_2 \longrightarrow 2Al_2O_3$
6. The percent composition of $PbCO_3$ is as follows:

$$\begin{array}{ccc} Pb & C & O \end{array}$$

$$207.02 + 12.01 + 3(16.00) = 267.21 \text{ u}$$

$$\% \text{ Pb} = \frac{207.2 \text{ u}}{267.21 \text{ u}} \times 100\% = 77.54\%$$

$$\% \text{ C} = \frac{12.01 \text{ u}}{267.21 \text{ u}} \times 100\% = 4.495\%$$

$$\% \text{ O} = \frac{48.00 \text{ u}}{267.21 \text{ u}} \times 100\% = 17.96\%$$

7. Assuming exactly 100 g of the compound, this would contain 47 g of oxygen and 53 g of aluminum. Thus,

$$\text{Moles of oxygen} = (47 \text{ g}/16) = 2.9375 \text{ mol}$$

$$\text{Moles of aluminum} = (53 \text{ g}/27) = 1.96 \text{ mol}$$

This implies that the smallest whole-number combining ratio of the constituent elements (that is, the empirical formula) is Al_3O_3.

8. The law of definite proportions states that compounds form from elements in definite proportions by mass such that all samples of the same compound will contain the same ratios of the constituent elements.
9. Molecular formulas are identical to or whole integer multiples of the empirical formulas because the empirical formulas represent the smallest whole-number combining ratios and the molecular formulas represent the actual numbers of the constituent elements in chemical compounds.

14. $$\text{Atoms sodium} = \left(6.022 \times 10^{23} \frac{\text{atoms}}{\text{mol}}\right)\left(\frac{0.1310 \text{ g}}{22.99 \text{ g/mol}}\right)$$

$$= 3.431 \times 10^{21} \text{ atoms}$$

EXPERIMENT 6

1. Numerous examples of redox and metathesis reactions could be cited. Two are $CuO + H_2 \longrightarrow Cu + H_2O$ (redox) and $AgNO_3 + NaI \longrightarrow AgI + NaNO_3$ (metathesis).

2. Reactions will proceed to completion whenever one of the products is physically removed from the reaction medium. This often occurs when a gas or solid is formed.
3. Percent yield = (actual yield/theoretical yield)(100%)
4. Materials can be separated from one another by for example; distillation, sublimation, filtration, decantation, sedimentation, chromatography, and extraction.
5. Exothermic reactions are usually accompanied by a temperature increase, while endothermic reactions are usually accompanied by a temperature decrease.
6. The balanced chemical equation is $Cu + 4HNO_3 \longrightarrow Cu(NO_3)_2 + 2H_2O + 2NO_2$. From this equation, moles $Cu(NO_3)_2$ = moles Cu and

$$\text{moles Cu} = \frac{2.25\ \text{g}}{63.55\ \text{g/mol}} = 0.0354\ \text{mol}$$

The theoretical yield is $Cu(NO_3)_2 = (0.0354\ \text{mol})(187.56\ \text{g/mol}) = 6.66\ \text{g}$

$$\%\ \text{yield} = \frac{3.35\ \text{g} \times 100\%}{6.66\ \text{g}} = 50.3\%$$

7. Methyl alcohol and acetone.
9. The maximum percent yield in any reaction is 100%. This occurs when the actual and theoretical yields are the same.

EXPERIMENT 7

1. Many chemicals are hazardous, and the haphazard mixing of chemicals can lead to the production of species with very dangerous properties.
2. Addition of a strong base to a solution containing NH_4^+ will release gaseous NH_3, which causes a piece of moist red litmus held above the reaction solution to turn blue.
3. Addition of an acid to a solution containing CO_3^{2-} will release gaseous CO_2, which reacts with a drop of $Ba(OH)_2$ held above the solution to precipitate white $BaCO_3$.
4. Solid chloride salts will release gaseous HCl when heated with concentrated H_2SO_4, and solutions of chloride salts will precipitate white AgCl when treated with $AgNO_3$ solution.
5. Solutions of sulfate salts will precipitate white $BaSO_4$ when treated with aqueous $BaCl_2$.
6. Solutions of iodide salts will react with Cl_2 to liberate I_2, which appears brown in H_2O and purple in mineral oil.
8. $2LiCl(s) + H_2SO_4(l) \longrightarrow 2HCl(g) + Li_2SO_4(aq)$

$NH_4^+(aq) + OH^-(aq) \rightleftharpoons NH_3(g) + H_2O(l)$

$AgNO_3(aq) + I^-(aq) \longrightarrow AgI(s) + NO_3^-(aq)$

$NaHCO_3(s) + H^+(aq) \longrightarrow CO_2(g) + H_2O(l) + Na^+(aq)$

EXPERIMENT 8

1. Gravimetric analyses involve a mass measurement as the determining measurement, whereas volumetric analyses involve a volume measurement as the determining measurement.
2. Stoichiometry is the mole or number ratio of atoms in a compound or compounds in a chemical reaction and can be related to the amounts (masses) of substances involved in reactions.
3. Silver chloride is photosensitive and reacts with light to produce silver metal and chlorine gas, which will lead to a low result if the silver chloride is not protected from light.
4. Indeterminate errors are just that, indeterminant. They cannot be ascertained or eliminated, but rather occur in a random fashions; so they are accounted for with statistics.
5. Standard deviation measures precision.
6. If the filter paper were opened this way, the solution and precipitate would simply pass through the opening between the sides of the filter paper and no filtration would result.
7. Because photodecomposition liberates Cl_2 from AgCl according to the equation $2AgCl \xrightarrow{h\nu} 2Ag + Cl_2$, the Ag will have a smaller mass than the AgCl and your results will be low.

EXPERIMENT 9

1. A solution that is in equilibrium with undissolved solute is saturated, a solution that contains less dissolved solute than the amount to be saturated is unsaturated, and a solution with a greater amount of solute than needed to form a saturated solution is supersaturated.
2. The solubility limit is at the point when crystals first begin to form. Typically, as the temperature is lowered, more crystals are formed.

EXPERIMENT 10

1. Chromatography is yet another technique that permits the separation of substances and sometimes allows for their identification.
2. An R_f value is the distance a substance moves on a chromatographic support relative to the distance the solvent moves on the same support under the same conditions. It aids in the qualitative identification of substances. Everything that affects the attractive forces between the mobile and stationary phases influences the R_f value.
3. The reactions are: for nickel, $Ni^{2+}(aq) + 2NH_3(aq) + 2H_2DMG(aq) \longrightarrow Ni(HDMG)_2(s) + 2NH_4^+(aq)$, where HDMG is the monoanion of dimethylglyoxime (H_2DMG); and for copper, $Cu^{2+} + 4NH_3 \longrightarrow [Cu(NH_3)_4]^{2+}$.
4. The solvent moves along the chromatographic support because of capillary action resulting from intermolecular attractions between the solvent and support material.

10. The petri dish (or beaker) should be covered to maintain equilibrium between the solvents in the liquid and gaseous phases. If the vessel were not covered, the more volatile liquid would escape and ultimately change the composition of a mixed solvent.

EXPERIMENT 11

1. Ionic bonding involves the transfer of one or more electrons from an atom with a low ionization potential to an atom with a high electron affinity and the resulting electrostatic attraction between the oppositely charged species. Covalent bonding results from the sharing of electrons between two atoms. Metallic bonding involves the sharing of electron density among several neighboring atoms in three dimensions.
2. HI, SO_2 and NH_3 possess polar covalent bonds.
3. HCl and HCN have molecular dipole moments.
4. AB_n molecules wherein the A atom is surrounded by 2, 3, 4, 5, and 6 electron pairs have linear, trigonal planar, tetrahedral, trigonal bipyramidal, and octahedral geometries, respectively.

EXPERIMENT 12

1. Violet, indigo, blue, green, yellow, orange, red.
3. $\lambda = c/v = 3.00\times10^8 \text{ ms}^{-1}/76 \text{ s}^{-1} = 3.9\times10^6 \text{ m}$;

$$3.9\times10^6 \text{ m}\left(\frac{1 \text{ km}}{1000 \text{ m}}\right)\left(\frac{1 \text{ mile}}{1.61 \text{ km}}\right) = 2.4\times10^3 \text{ mile}$$

4. $E = hv = 6.63\times10^{-34} \text{ J-s}\times76 \text{ s}^{-1} = 5.0\times10^{-32} \text{ J}$
5. Blue light has the higher frequency and greater energy because it has shorter wavelength.

EXPERIMENT 13

1. The volume of an ideal gas increases in direct proportion to the temperature increase at constant pressure.
2. The volume of an ideal gas decreases in inverse proportion to the pressure increase at constant temperature.
3. The pressure of an ideal gas increases in direct proportion to an increase in the number of molecules at constant temperature and volume.
4. $$n = \frac{PV}{RT} = \frac{6\times100\times10^3\times670\times10^{-5}}{8.31\times293} = 0.165 \text{ mol}$$
5. Combining the ideal gas law and the definition of moles, solving for M, you have

$$PV = nRT \text{ and } n = \frac{m}{\boldsymbol{m}}$$

$$n = \frac{PV}{RT} = \frac{m}{\boldsymbol{m}} \quad \text{or} \quad \boldsymbol{m} = \frac{mRT}{PV}$$

6. $PV = nRT$ where V and R are constant or

$\frac{n_1T_1}{P_1} = \frac{n_2T_2}{P_2}$ and $PCl_5(g) \longrightarrow PCl_3(g) + Cl_2(g)$ so that

$P_2 = \frac{n_2T_2P_1}{n_1T_1} = \frac{(2\times480)(25.5)}{(1\times338)} = 72.4$ kPa (ratio of n_2:n_1 is 2:1 since 1 mol of PCl_5 has dissociated into 2 moles of gases – 1 mol of PCl_3 and 1 mol of Cl_2)

7. STP is 273 K and 101.3 kPa so that

$$V = (200 \text{ mL})\left(\frac{273 \text{ K}}{333 \text{ K}}\right)\left(\frac{66.7 \text{ kPa}}{101.3 \text{ kPa}}\right)$$
$$= 108 \text{ mL}$$

EXPERIMENT 14

1. Methylamine, CH_3NH_2 will behave less ideally than mono-atomic argon.
2. From $PV = nRT$, $R = PV/nT$ and STP is 101.3 kPa and 0 °C, or 273 K, so that

$$R = \frac{(101.3 \text{ kPa})(22.4 \text{ L})}{(1.00 \text{ mol})(273 \text{ K})} = 8.31\frac{\text{L}-\text{kPa}}{\text{mol}-\text{K}}$$

3. Equalizing the water levels equalizes the pressures and ensures that the total pressure in the bottle is atmospheric and does not contain a contribution from the pressure due to the height of the water column. The pressure is then known and can be read from the barometer.
4. Because gaseous and liquid water are in dynamic equilibrium, there will always be some water vapor above a sample of liquid water or an aqueous solution. Because the vapor pressure of water is reasonably high at ambient temperature, it makes a significant contribution to the total pressure.
5. An error analysis allows you to judge the reliability of your data and gives an indication of the potential sources of error.
6. The ideal gas law assumes that there are no forces of attraction between the individual gaseous molecules. When this isn't the case, real molecules do not obey the ideal gas law. This would be expected to occur at very high pressures and at very low temperatures where molecules are so close to one another that they necessarily interact.
   The ideal gas law also assumes that the gas particles have no volume. At high pressures, their volume may become appreciable relative to the volume of the container.
7. $PV = nRT$

$$P = \frac{nRT}{V} = \frac{\left(\frac{0.05 \text{ g}}{2 \text{ g/mol}}\right)\left(\frac{8.31 \text{ L}-\text{kPa}}{\text{mol}-\text{K}}\right)(298 \text{ K})}{0.100 \text{ L}}$$
$$= 600 \text{ kPa}$$

assuming no gas in the void volume. Clearly, this presents a problem, as H_2 gas is explosive due to increased pressure as with any gas. Because lead storage batteries produce H_2, sealing them would be dangerous. These answers assume that no hydrogen can escape from the sealed battery, which is not the case.

EXPERIMENT 15

1. Most metals have high thermal and electrical conductivities, high luster, malleability, and ductility, whereas nonmetals usually have low thermal and electrical conductivities, low luster, low malleability, and low ductility. In addition, relative to nonmetals, metals have low ionization potentials and low electron affinities. Most metals are solids while the physical states of nonmetals is variable.
2. Ionization energy is the energy required to remove an electron from an atom in the gaseous state.
3. Electron affinity is the energy involved when an electron is added to a neutral species or strictly speaking an atom in the gas phase and is the inverse of the ionization energy in a physical sense. But the magnitudes are not just equal and of opposite sign because you are considering slightly different processes.
4. An oxidation must be accompanied by a reduction because the species being oxidized must transfer electron(s) to some other species that is reduced. The electron cannot just be given up to free space.
5. By systematically observing the displacement reactions among a series of metals and solutions of their cations, it is possible to determine the relative oxidation potentials of the metals. The metal with the lower reduction potential will reduce a cation of a metal with a higher reduction potential. Currently, all are listed as reduction potentials.
6. $Pb^{2+}(aq) + Mg(s) \longrightarrow Mg^{2+}(aq) + Pb(s)$; $2Al + Fe_2O_3 \longrightarrow Al_2O_3 + 2Fe$, and $Mg + 2HCl \longrightarrow MgCl_2 + H_2$

EXPERIMENT 16

1. Oxidation occurs at the anode to produce oxygen.
2. Faraday's constant is defined as the total charge carried by Avogadro's number of electrons.
3. In general, reduction occurs at the cathode and oxidation at the anode in an electrochemical cell regardless of whether the cell is producing energy (a galvanic cell) or consuming energy (an electrolytic cell).
4. The water column exerts a pressure that is directly proportional to the height of the water column, just as atmospheric pressure and altitude are related. The higher the altitude above sea level, the less the atmosphere and the lower the atmospheric pressure.
5. The sulfuric acid is present to increase the rate of the reaction by increasing the ability of the solution to conduct an electric current. According to Faraday's law, the more current that passes through this solution, the larger the number of species that will react. The more ions present in solution, the greater the conductivity of the solution and the greater the current flow. Sulfuric acid is a strong electrolyte and is almost completely dissociated in solution to produce the ions H_3O^+ and SO_4^{2-}. It does not directly enter into the electrochemical reaction.

EXPERIMENT 17

1. A faraday is the charge on 1 mol of electrons and equals 96,485 coulombs. A salt bridge is an ion-containing conducting medium used to physically and chemically connect the two half-cells in an electrochemical cell. An anode is an electrode at which oxidation occurs, and a cathode is an electrode at which reduction occurs in an electrochemical cell. A voltaic cell is an electrochemical cell that produces electrical energy by means of a spontaneous redox reaction. An electrolytic cell is a cell that requires energy to bring about a nonspontaneous redox reaction.
2. The cell $Cu|Cu^{2+}(aq)\|Ag^{+}(aq)|Ag$ represents the reaction $2Ag^{+}(aq) + Cu(s) \longrightarrow 2Ag(s) + Cu^{2+}(aq)$.
3. $E_{cell} = E_{cathode} - E_{anode} = 0.80 \text{ V} - 0.34 \text{ V} = 0.46 \text{ V}$
4. The cell $Zn|Zn^{2+}(aq)(0.10\ M)\|Cu^{2+}(0.20\ M)|Cu$ represents the reaction $Zn(s) + Cu^{2+}(aq) \longrightarrow Zn^{2+}(aq) + Cu(s)$, for which

   $$E = E^\circ - (0.0592/2)\log([Zn^{2+}]/[Cu^{2+}])$$

   so that

   $E = 1.100 \text{ V} - 0.0295 \log([0.10]/[0.20]) = 1.100 + 0.0089 = 1.109 \text{ V}.$
5. $E^\circ = -1.18 + 2.87 = 1.69$ V. To calculate K_{eq}, use the equation $E = E^\circ - (0.0592/n)\log K_{eq}$. Because at equilibrium $E = 0$, you have

   $$E^\circ = \left(\frac{0.0592}{n}\right)\log K_{eq}$$

   $$\log K_{eq} = \frac{nE^\circ}{0.0592} = \frac{(2)(1.69)}{0.0592} = 57.0946$$

   $$K_{eq} = \text{antilog}(57.0446) = 1.24 \times 10^{57}$$

   The free energy for the given cell is

   $$\Delta G^\circ = -nFE^\circ = -(2 \text{ mol } e^-)\left(\frac{96{,}485 \text{ J}}{\text{V-mol } e^-}\right)(1.69 \text{ V})$$
   $$= -326{,}000 \text{ J or } -326 \text{ kJ}$$
6. Many spontaneous reactions (ΔG negative) are exothermic (ΔH negative). Because voltaic cells have spontaneous reactions, you would expect ΔH to be negative for most voltaic cells.

EXPERIMENT 18

1. Ionic bonding results from transfer of electrons from one species to another to form a chemical bond, which results in an electrostatic attraction between two oppositely charged ions, whereas covalent bonding results from sharing electron density (equally or unequally) between the two partners forming the chemical bond. Ionic bonding will result from the union of atoms having low ionization energies with atoms of high electron

affinity. Covalent bonding will result from the union of atoms with similar or identical ionization energies and electron affinities, respectively.

2. An anhydride is literally a compound without water. Thus, the anhydride of sodium hydroxide, NaOH, is Na_2O.
3. According to the Brønsted-Lowry definition, an acid is a proton donor and a base is a proton acceptor.
4. The autodissociation of water is $2H_2O \rightleftharpoons H_3O^+ + OH^-$, for which $K_{eq} = [H_3O^+][OH^-]/[H_2O]^2$ and $K_w = [H_3O^+][OH^-]$. Because the molar concentration of water changes so little in this reaction, it is essentially a constant and is incorporated in K_w. $K_w = 55.6\ K_{eq}$ for dilute aqueous solutions.
5. No, the higher the pH the less acidic will be the solution.
6. The following table illustrates when aqueous solutions are acidic, basic, or neutral in terms of $[H_3O^+]$, $[OH^-]$, and pH, respectively.

	$[H_3O^+]$	$[OH^-]$	pH
Acidic	$>10^{-7}\ M$	$<10^{-7}\ M$	<7
Basic	$<10^{-7}\ M$	$>10^{-7}\ M$	>7
Neutral	$=10^{-7}\ M$	$=10^{-7}\ M$	$=7$

7. $[H^+] = 10^{-pH}$

   $[H^+]$ in $CH_3COOH = 10^{-2.4} = 3.98 \times 10^{-3}\ M$

   $[H^+]$ in $HCl = 10^{-1} = 0.100 M$
8. $MgO(s) + H_2O(l) \longrightarrow Mg(OH)_2(aq)$; $SO_3(aq) + H_2O(l) \longrightarrow H_2SO_4(aq)$

EXPERIMENT 19

1. Solutions are homogenous mixtures that contain two or more substances.
2. $\Delta T_f = -iK_f m$
3. freezing point, boiling point, osmotic pressure. Colligative properties depend on the amount of solute particles in a solution and not on the type of chemical species present.
4. A volatile substance has a measureable vapor pressure, whereas a novolatile substance has no measureable vapor pressure.
5. Lowers the freezing point.
6. Molarity is moles solute/liters of solution (mol/L) and molality is moles solute/kg of solvent (mol/kg).
8. No. of moles of urea = 0.1233

   Molality = 0.1233/0.300 kg benzene = 0.411 *m*

EXPERIMENT 20

1. Standardization is the process of determining the exact concentration of a solution. It is usually achieved by titrating the solution to be standardized against a known amount of a primary standard substance according to a known reaction and procedure.

2. Titration is the technique and procedure of accurately measuring the volume of a solution that is required to react with a known amount of another reagent.
3. Molarity is the number of moles of solute in a liter of solution.
4. You determine the mass by difference because this is generally more accurate than determining the mass directly.
5. An equivalence point is the point in a titration at which stoichiometrically equivalent amounts of the two reactants are brought together. An end point is the point in a titration where some indicator (such as a dye or an electrode) undergoes a discernible change and the titration is stopped. Ideally, the hope is that these two points coincide, but in practice, they differ slightly. Of primary importance is to minimize the difference if accuracy is desired. The equivalence point is a theoretical point, whereas the endpoint is a physical point in the lab procedure.
7. Parallax is the apparent displacement or the difference in apparent direction of an object as seen from two different points not on a straight line with the object. It should be avoided in titrations because it introduces an error in the volume measurements.
8. Carbon dioxide should be removed from the water because it is an acid anhydride and reacts with water to produce carbonic acid according to the reaction $CO_2 + H_2O \rightleftharpoons H_2CO_3$. Its presence would lead to an erroneous determination of the amount of a particular acid or base in solution. It is particularly detrimental to sodium hydroxide solutions as these absorb carbon dioxide to produce sodium carbonate.

EXPERIMENT 21

1. (a) Molecular: $CuSO_4(aq) + 2NaOH(aq) \longrightarrow Cu(OH)_2(s) + Na_2SO_4(aq)$;

   Ionic: $Cu^{2+}(aq) + 2OH^-(aq) \longrightarrow Cu(OH)_2(s)$;

   Net Ionic: $Cu^{2+}(aq) + 2OH^-(aq) \longrightarrow Cu(OH)_2(s)$

(b) Molecular: $AgNO_3(aq) + NaCl(aq) \longrightarrow AgCl(s) + NaNO_3(aq)$;

Ionic: $Ag^+(aq) + Cl^-(aq) \longrightarrow AgCl(s)$;

Net ionic: $Ag^+(aq) + Cl^-(aq) \longrightarrow AgCl(s)$

(c) Molecular: $K_2CO_3(aq) + 2HCl(aq) \longrightarrow 2\,KCl(aq) + CO_2(g) + H_2O(l)$;

Ionic: $CO_3^{2-}(aq) + 2H^+(aq) \longrightarrow CO_2(g) + H_2O(l)$;

Net ionic: $CO_3^{2-}(aq) + 2H^+(aq) \longrightarrow CO_2(g) + H_2(l)$

(d) Molecular: $NH_4NO_3(aq) + KOH(aq) \longrightarrow KNO_3(aq) + NH_3(g) + H_2O(l)$;

Ionic: $NH_4^+(aq) + OH^-(aq) \longrightarrow NH_3(g) + H_2O(l)$;

Net ionic: $NH_4^+(aq) + OH^-(aq) \longrightarrow NH_3(g) + H_2O(l)$

(e) Molecular: $Ba(NO_3)_2(aq) + H_2SO_4(aq) \longrightarrow BaSO_4(s) + 2\,HNO_3(aq)$;

Ionic: $Ba^{2+}(aq) + SO_4^{2-}(aq) \longrightarrow BaSO_4(s)$;

Net ionic: $Ba^{2+}(aq) + SO_4^{2-}(aq) \longrightarrow BaSO_4(s)$

2. The following are not water-soluble: MgO, $BaSO_4$, AgCl.
5. K_2CO_3, HCl, KCl, H_2SO_4 are strong electrolytes.
6. NH_3 and CH_3COOH are weak electrolytes.

EXPERIMENT 22

1. $Fe^{3+}(aq) + SCN^-(aq) \rightleftharpoons FeNCS^{2+}(aq)$

3. Solution 2: Assume that $[FeNCS^{2+}] = [SCN^-] = \dfrac{(1.00\text{ mL})(2.00\times10^{-3}\text{ m})}{50.00\text{ mL}}$

$$= 4.00\times10^{-5}\,M$$

Similarly, solutions 3, 4, 5, and 6 are $8.00\times10^{-5}\,M$, $1.20\times10^{-4}\,M$, $1.60\times10^{-4}\,M$, and $2.00\times10^{-4}\,M$, respectively. Solution 1 has no $FeNCS^{2+}$.

5. The Beer-Lambert law is $A = abc$, where A is absorbance, b is cell path length, C is molar concentration of the absorbing species, and a is molar absorptivity.

8. (a) A reversible reaction is a reaction that simultaneously proceeds in both directions.

(b) A state of dynamic equilibrium is reached when the reaction rates in the forward and reverse directions are equal.

(c) The equilibrium constant expression is the product of the molar concentrations of the reaction products raised to the power of their respective coefficients divided by the product of the reactants' molar concentrations raised to the powers of their respective coefficients in the balanced equation. In this experiment,

$$K_{eq} = \frac{[FeNCS^{2+}]}{[Fe^{3+}][SCN^-]}$$

(d) The equilibrium constant, K_{eq}, is the numerical value of the equilibrium constant expression.

EXPERIMENT 23

2. (a) Solution turns from yellow to orange to deeper, darker orange.
   (b) When more acid is added, the concentration of hydrogen ion increases. This favors the forward reaction.
   (c) When more alkali is added, the concentration of hydrogen ion decreases. This favors the backward reaction.
4. (a) $AgNO_3(aq) + HCl(aq) \rightleftharpoons AgCl(s) + HNO_3(aq)$;

   $$Ag^+(aq) + Cl^- \rightleftharpoons AgCl(s)$$

   (b) $NH_3(aq) + H^+(aq) \rightleftharpoons NH_4^+(aq)$

   $NH_4^+(aq) + Cl^-(aq) \rightleftharpoons NH_4Cl(aq)$

   (c) $Na_2CO_3(aq) + 2HNO_3(aq) \rightleftharpoons 2NaNO_3(aq) + CO_2(g) + H_2O(l)$;

   $$CO_3^{2-}(aq) + 2H^+(aq) \rightleftharpoons CO_2(g) + H_2O(l)$$

EXPERIMENT 24

1. According to the Brønsted-Lowry definitions, an acid is a proton donor and a base is a proton acceptor.
2. Cu^{2+}, Zn^{2+}, F^-, and SO_3^{2-} will undergo hydrolysis.
3. For example, $Cu^{2+}(aq) + 2H_2O(l) \rightleftharpoons Cu(OH)_2(s) + 2H^+(aq)$.
4. $K_b = 1.0\times10^{-14}/4.9\times10^{-10} = 2.0\times10^{-5}$
5. The conjugate base is CO_3^{2-}, and the conjugate acid is H_2CO_3.
6. For example, NaCl may be made from HCl and NaOH.
11. $pK_a = -\log(1.8\times10^{-5}) = 4.74$ and $pH = 4.74 + \log(0.40/0.20) = 4.74 + (0.30)$
    $= 5.04$

EXPERIMENT 25

1. According the Brønsted-Lowry definition, an acid is a proton donor and a base is a proton acceptor.
2. A weak acid dissociates in aqueous solution according to the equilibrium $HA + H_2O \rightleftharpoons H_3O^+ + A^-$, for which the equilibrium constant is $K_{eq} = [H_3O^+][A^-]/[HA][H_2O]$ and the dissociation constant is $K_a = [H_3O^+][A^-]/[HA]$ or $K_a = [H_2O]K_{eq}$.
3. The pH at the equivalence point in an acid–base titration depends upon the nature of the species present. For the titration of a strong acid and a

strong base, the pH will be 7 because a salt that does not hydrolyze will be formed (for example, $NaOH + HCl$). For the titration of a strong acid with a weak base (for example, $HCl + NH_4OH$), the pH will be less than 7. And for the titration of a weak acid with a strong base (for example $HOAc + NaOH$), the pH will be greater than 7. This is so because the salt formed in each case (NH_4Cl and NaOAc) will undergo hydrolysis reactions with water.

4. $pK_a = -\log K_a = -\log(4.2\times10^{-5}) = 5 - \log 4.2 = 5 - 0.62 = 4.38$
5. Two electrodes are necessary for an electrical measurement because some current flow must occur, and this requires both a donor and an acceptor for the electrons. Donation occurs at one electrode and acceptance at the other.
6. A buffer solution is a solution that is resistant to a pH change upon addition of an acid or a base. It always contains a weak electrolyte and normally is composed of two species (for example, a weak acid and one of its salts or a weak base and one of its salts). Two specific examples are acetic acid plus sodium acetate and ammonium hydroxide plus ammonium chloride. An example of a single-component buffer solution is disodium hydrogen phosphate, Na_2HPO_4.
7. At one-half equivalence point, $[HA] = [A^-]$. Because $HA \rightleftharpoons H^+ + A^-$ and $K_a = [H^+][A^-]/[HA]$, at one-half equivalence point, $K_a = [H^+]$. Therefore, $pK_a = pH$; so $pK_a = 5.67$, $K_a = \text{antilog}(-5.67) = 2.1\times10^{-6}$.

EXPERIMENT 26

1. A polyprotic acid is an acid that possesses more than one ionizable proton. For example, H_2SO_4 is a diprotic acid and H_3PO_4 is a triprotic acid because they possess two and three ionizable protons, respectively.
3. 75 mmol of NaOH is required.
4. 0.08 moles are present in 100 mL of a 0.40 *M* solution of H_2SO_4.
5. A pH meter must be standardized because it measures relative potentials and thus relative pH. A standard must be measured and the meter set to the known value for this standard. The pH values of other solutions are then measured relative to this standard.
6. See the answer to question 7, Experiment 25, for the method: $pK_1 = 4.20$; $K_1 = 6.3\times10^{-5}$; $pK_2 = 7.34$; $K_2 = 4.6\times10^{-8}$.

EXPERIMENT 27

1. $Ba(OH)_2(s) \longrightarrow Ba^{2+}(aq) + 2OH^-(aq)$; $K_{sp} = [Ba^{2+}][OH^-]^2$
2. $$\text{moles Ag}^+ = \left(\frac{5\text{ mL}}{1000\text{ mL/L}}\right)(0.0040\text{ mol/L})$$
$$= 2\times10^{-5}\text{ mol}$$

$$\text{moles } CrO_4^{2-} = \left(\frac{5 \text{ mL}}{1000 \text{ mL/L}}\right)(0.0024 \text{ mol/L})$$

$$= 1 \times 10^{-5} \text{ mol}$$

3. $2AgNO_3 + K_2CrO_4 \longrightarrow 2KNO_3 + Ag_2CrO_4$. The stoichiometry requires 2 mol of $AgNO_3$ for each mole of K_2CrO_4.

$$\text{millimoles } AgNO_3 = (5 \text{ mL})(0.0040 \text{ mmol/mL})$$
$$= 0.020 \text{ mmol}$$
$$\text{millimoles } K_2CrO_4 = (5 \text{ mL})(0.0024 \text{ mmol/mL})$$
$$= 0.012 \text{ mmol}$$

And 0.012 mmol K_2CrO_4 would require 0.024 mmol $AgNO_3$. Hence, K_2CrO_4 is in excess.

4. The hydroxide ion concentration, $[OH^-] = \dfrac{18.2/1000 \times 0.050}{20/1000} = 0.046$ mol/L.

From pOH = −lg 0.046 = 1.34, the pH of the original solution, pH = 14 − 1.34 = 12.7

9. A sparingly soluble salt will precipitate when the ion product Q exceeds K_{sp}.

EXPERIMENT 28

1. An exothermic reaction is a reaction that produces heat; its ΔH will be less than zero, or negative. An endothermic reaction is a reaction that absorbs heat; its ΔH will be greater than zero, or positive.
2. $Q = (50 - 10) \times 500 \times 4.184 = 83680$ J
3. The heat capacity of a substance is the amount of energy, usually in the form of heat, necessary to raise the temperature of a specified amount of the substance (usually 1 g) by 1 °C or 1K. If the mass is 1.00 g then this defines the specific heat.
4. $Q = (51.5 - 33.2) \times 20 \times 4.184 = 1531.344$ J

EXPERIMENT 29

1. Besides temperature and catalysts, the nature of the species undergoing the reaction, pressure, particle size, and concentration of reactants affect the rate of a chemical reaction.
2. The general form of the rate law for the reaction $A + B \longrightarrow$ is Rate $= k[A]^x[B]^y$.
3. A reaction that obeys the rate law of the form rate = $k[A]^3[B]^2$ is third order in A and second order in B.
4. The chemical reactions involved in this experiment are $S_2O_8^{2-} + 2I^- \longrightarrow I_2 + 2SO_4^{2-}$, $I_2 + 2S_2O_3^{2-} \longrightarrow 2I^- + S_4O_6^{2-}$, and $I_2 + \text{starch} \longrightarrow \text{starch} \cdot I_2(\text{blue})$, for which the rate of disappearance of $S_2O_3^{2-}$ is $k[S_2O_8^{2-}]^x[I^-]^y$ and the rate of appearance of blue color equals the rate of formation of the starch–iodine complex, which is proportional to the rate of appearance of iodine. Thus, the rate of appearance of iodine is $k[S_2O_8^{2-}]^x[I^-]^y$ and the rate of appearance of blue color is $k[I_2]^x$ $[\text{starch}]^y$, which is proportional to $k[S_2O_8^{2-}]^x[I^-]^y$.
5. $\text{Rate} = k[A]^2[B]^2$
6. $$\text{Rate} = \left(\frac{2\times10^{-4}\ \text{mol}}{0.050\ \text{L}}\right)\left(\frac{1}{188\ \text{s}}\right)$$
   $$= 2.1\times10^{-5}\ \frac{\text{mol}}{\text{L-s}} = 2\times10^{-5}\ \frac{\text{mol}}{\text{L-s}}$$
7. From a knowledge of the rates of chemical reactions, chemists can determine how long and under what conditions to run a reaction to obtain the desired products. Clearly, this is of practical value in the synthesis of new compounds.

EXPERIMENT 30

2. (a) A reaction with a rate law of the form k = [A][B] is second-order overall.

   (b) If the concentration of both A and B are tripled, the rate will increase by a factor of 9.

   (c) k is the known as the specific rate constant.
5. $\text{Rate} = k[A]^2[B]^2$

EXPERIMENT 31

CATIONS

I Cations

1. Ammonium, NH_4^+; silver, Ag^+; ferric or iron(III), Fe^{3+}; aluminum, Al^{3+}; chromic or chromium(III), Cr^{3+}; calcium, Ca^{2+}; magnesium, Mg^{2+}; nickel or nickel(II), Ni^{2+}; zinc, Zn^{2+}; and sodium, Na^+
2. Sometimes ions behave similarly. For example, Ba^{2+} and Pb^{2+} both form yellow precipitates with K_2CrO_4. Hence, other additional confirmatory tests would be required to distinguish between these ions. For example, Pb^{2+} forms a white precipitate with HCl, whereas Ba^{2+} does not.
3. Only three of the ions are colored: Fe^{3+}, rust to yellow; Cr^{3+}, blue-green; and Ni^{2+}, green.
4. Because only AgCl precipitates on the addition of HCl to a solution of the 10 cations, all chlorides of these ions except AgCl must be soluble.
5. Based on the group separation chart (Figure 31.1), Ag^+ can be separated from all of the other ions by the addition of HCl. Silver forms insoluble AgCl.
6. Examination of the group separation chart shows that in the presence of the buffer NH_3—NH_4Cl, Cr^{3+} precipitates as $Cr(OH)_3$ while Mg^{2+} remains in solution.
7. Addition of chloride ions to a solution containing Al^{+3} and Ag^+ would precipitate AgCl, while Al^{3+} would remain in solution. HCl would be a good source of chloride ions, and the H^+ would help retard hydrolysis of Al^{3+}.
8. $NH_4^+(aq) + OH^-(aq) \rightleftharpoons NH_3(aq) + H_2O(l)$; $AgCl(s) + 2\ NH_3$

   $(aq) \rightleftharpoons [Ag(NH_3)_2]^+(aq) + Cl^-(aq)$

ANIONS

II Anions

1. Sulfate, SO_4^{2-}; nitrate, NO_3^-; carbonate, CO_3^{2-}; chloride, Cl^-; bromide, Br^-; and iodide, I^-
2. See Table 31.1, Behavior of Anions with Concentrated Sulfuric Acid, H_2SO_4.
3. No. Because barium forms a precipitate with sulfate, but not with chloride.

EXPERIMENT 32

Part I

1. Ag^+, Hg_2^{2+}, and Pb^{2+}
2. $PbCl_2$
3. AgCl
4. (a) Add excess water to both; the one which dissolves readily is NaCl.
   (b) Add dilute silver nitrate to both solutions; the one which gives a white precipitate is HCl.
8. With the use of a small pipet to remove the supernatant without disturbing the solid.

Part II

1. Pb^{2+}, Cu^{2+}, Bi^{3+}, Sn^{4+}
2. CuS does not dissolve in aqueous $(NH_4)_2S$, whereas SnS_2 does.
3. Aqueous NH_3 precipitates $Bi(OH)_3$, and $Cu(NH_3)_4^{2+}$ remains in solution.

Part III

1. Fe^{3+}, Ni^{2+}, Mn^{2+}, Al^{3+}
2. Mn^{2+} (light pink), Ni^{2+} (green), MnO_4^- (purple) and Fe^{3+} (yellow)
3. $Fe(OH)_3$ (red-brown), MnS (salmon), $Al(OH)_3$ (colorless), $Ni(OH)_2$ (green)
4. Aqueous NaOH precipitates $Fe(OH)_3$, whereas $Al(OH)_4^-$ remains in solution.
7. (a) HCl or H_2SO_4
   (b) Excess NaOH

Part IV

1. Ba^{2+}, Ca^{2+}, NH_4^+, Na^+
2. $BaCrO_4$
3. $BaSO_4$ (white), $BaCrO_4$ (yellow), CaC_2O_4 (white)
4. (a) Ba^{2+} (green)
   (b) Ca^{2+} (orange-red)
   (c) Na^+ (yellow)

Part V

1. Sulfate, SO_4^{2-}; nitrate, NO_3^-; carbonate, CO_3^{2-}; chromate, CrO_4^{2-}; phosphate, PO_4^{3-}; chloride, Cl^-; bromide, Br^-; sulfide, S^{2-}; sulfite, SO_3^{2-}; and iodide, I^-.
2. See Table 32.1, Behavior of Anions with Concentrated Sulfuric Acid, H_2SO_4.
3. No. HCl reacts with carbonate to form carbon dioxide gas, in which case there will be effervescence. Gas is not produced when HCl reacts with chloride, so no effervescence will be seen.
4. (a) SO_4^{2-}

   (b) CO_3^{2-}, PO_4^{3-}, S^{2-}, CrO_4^{2-} or OH^-

   (c) Cl^-
5. Nitrate
6. CO_3^{2-}

EXPERIMENT 33

1. Sodium chloride is the solute. Water is the solvent.
2. Calculate the molar mass of $C_{12}H_{22}O_{11}$,

   $(12\times12.0107)+(22\times1.00794)+(11\times15.9994)=342.296$ g/mol

   Find the moles of sucrose,

   $(6.5532\text{ g } C_{12}H_{22}O_{11})\times(1\text{ mol } C_{12}H_{22}O_{11}/342.296\text{ g } C_{12}H_{22}O_{11})=$
   $0.019145\text{ mol } C_{12}H_{22}O_{11}$

   Convert volume of solution from mL to L.

   $(50.00\text{ mL})\times(1\times10-3\text{ L}/1\text{ mL})=0.050000\text{ L}$

   Calculate the molarity of the sucrose solution,

   $M=\text{mol}/V=0.019145\text{ mol}/0.05000\text{ L}=0.3829\text{ M}$
3. Calculate the molar mass of $FeSO_4$,

   $(55.845)+(32.065)+(4\times15.9994)=151.9076$ g/mol

   Find the moles of $FeSO_4$,

   $(2.85\text{ g})(1\text{ mol}/342.296\text{ g})=0.0188\text{ mol}$

   Convert volume of solution from mL to L.

   $(25.00\text{ mL})\times(1\times10-3\text{ L}/1\text{ mL})=0.02500\text{ L}$

   Calculate the molarity,

   $M=\text{mol}/V=0.0188\text{ mol}/0.05000\text{ L}=0.750\text{ M}$
4. $A=2-\log(\%T)=A=2-\log(68.6)=0.164$
5. Calculate the molarity of the solution "A". It is being prepared using the stock solution and diluting.

   $M_{conc}V_{conc}=M_{dil}V_{dil}$ solve for M_{dil}

$M_{dil} = M_{conc}V_{conc}/V_{dil} = (1.25\text{ M})(5.00\text{ mL})/(25.00\text{ mL}) = 0.2500\text{ M}$

Calculate the molarity of the solution "B". It is being prepared using the "B" solution and diluting.

$M_{conc}V_{conc} = M_{dil}V_{dil}$ solve for M_{dil}

$M_{dil} = M_{conc}V_{conc}/V_{dil} = (0.2500\text{ M})(2.00\text{ mL})/(50.00\text{ mL}) = 0.0100\text{ M}.$

EXPERIMENT 34

2. $PbI_2(s) \longrightarrow Pb^{2+}(aq) + 2I^-(aq)$

   $K_{sp} = [Pb^{2+}][I^-]^2$; only temperature can affect the magnitude of this equilibrium constant. Increasing the temperature will increase the K_{sp}.

4. $0.0010\text{ mol HCl/L} \times 0.00342\text{ L} = 3.42\times10^{-6}\text{ mol HCl} \times 1\text{ mol H}^+/1\text{ mol HCl} \times 1\text{ mol OH}^-/1\text{ mol H}^+ = 3.42\times10^{-6}\text{ mol OH}^-/0.05000\text{ L} = 6.84\times10^{-5}\text{ M}$

   $K_{sp} = [Mn^{2+}][OH^-]^2 = \frac{1}{2}[OH^-][OH^-]^2 = 1.60\times10^{-13}.$

6. $\Delta G°$ is positive when $K < 1$ (not spontaneous), and $\Delta G°$ is negative when $K > 1$ (spontaneous). This leads to the conclusion that AgCl(s) dissolving into $Ag^+(aq) + Cl^-(aq)$ [$K_{sp} = 1.8\times10^{-10}$] is not a spontaneous process, while $Co^{3+}(aq) + 6\,NH_3(aq)$ forming the $[Co(NH_3)_6]^{3+}$ complex ion [$K_f = 4.5\times10^{33}$] would be a spontaneous process. A larger positive K gives a larger negative $\Delta G°$.

EXPERIMENT 35

1. moles $Na_2S_2O_3 = (V)(M)$

   $$M = 0.0343\ M$$

2. Chloroform is an antibacterial agent and, as such, kills bacteria. Because the bacteria are killed, they do not consume oxygen by their growth; so the water sample is preserved by the addition of chloroform.

3. In the reaction, Fe^{2+} is oxidized and MnO_4^- is reduced.

4. The blue color returns because oxygen in the air oxidizes iodide to iodine according to the reaction $4I^- + O_2 + 4H^+ \longrightarrow 2I_2 + 2H_2O$.

5. $(4\times10^{-3}\text{ mol/L})(32\text{ g/mol})(1000\text{ mg/g}) = 128\text{ mg/L} = 128\text{ ppm} = 100\text{ ppm}.$

EXPERIMENT 36

3. A coordination compound is formed by the reaction of a Lewis acid with a Lewis base; it contains one or more coordinate covalent bonds. An example is the compound $[Co(NH_3)_3Cl_3]$.
4. Geometric isomers have the same empirical and molecular formulas, but they differ in their spatial arrangements of the constituent atoms. They are one type of stereoisomer.
5.

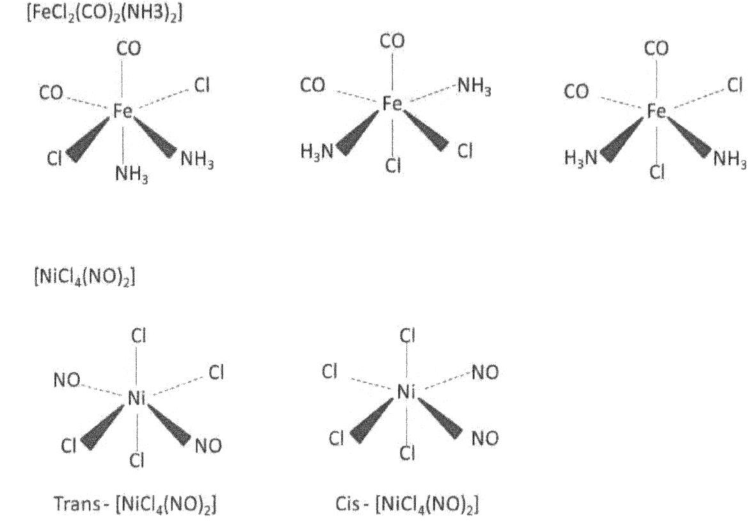

6. Yes. Because in both compounds, there are 2 isomers: cis- and trans-. In each case, the water molecules are in equivalent environments as the line and plane of symmetry reflect them onto each other.
7. Dichroism is a property of a crystal which results in the color of the crystal being different when viewed along two different axes.
8. Triturate means to cause a semisolid to become a solid through crushing or grinding.
9. You must find the answer to question 9 in the library.
10. You must find the answer to question 10 in the library.
11. It works by forming a soluble iron(II) oxalate complex, $[Fe(C_2O_4)_3]^{4-}$, by reacting with iron(II) oxide (rust), according to the reaction FeO + $3H_2C_2O_4 \longrightarrow H_4[Fe(C_2O_4)_3] + H_2O$. In the reactions of Fe_2O_3 or Fe_3O_4, iron is also reduced by oxalate and CO_2 is formed. Iron is reduced to iron (II), which then complexes with oxalate.

EXPERIMENT 37

1. 0.0464 *M*
2. The average is 15.55%, and the standard deviation is 0.02%.
3. The color change at the end point of permanganate titrations is colorless to pink.
4. The permanganate is filtered to remove insoluble MnO_2. Because rubber reduces MnO_4^- to MnO_2, the solution should not contact rubber or other organic materials.
5. Number of moles of potassium salt = 0.55/354 = 0.001554

   Number of moles of oxalate in 0.55 g = 0.003107

   Number of moles of $KMnO_4$ required = 0.001243

   Volume = 12.43 mL

EXPERIMENT 38

1. 5.6 mL
2. The markings on a graduated cylinder are designed for the shape of the meniscus of water in the standard vertical orientation. The correction is required for the inverted graduated cylinder.

EXPERIMENT 39

1. OCl^-

3. $S_2O_3^{2-}$ is a reducing agent
4. Iodide ions, I^-, are oxidized by air to I_2, and the oxidation is accelerated in acid solution. Hence, there would be iodine in addition to that formed from the KIO_3.

5. (a) Pale yellow

   (b) Blue-black

   (c) Colorless
7. 0.0993 *M*
8. A standard solution is a solution whose concentration is accurately known.

EXPERIMENT 40

1. Molecular formulas indicate the composition of molecules but do not indicate how the atoms are arranged; therefore, they do not distinguish between isomers. Structural formulas indicate how the atoms are arranged in molecules as well as the molecular composition. The molecular formula $C_2H_4Cl_2$ may refer to either isomer.

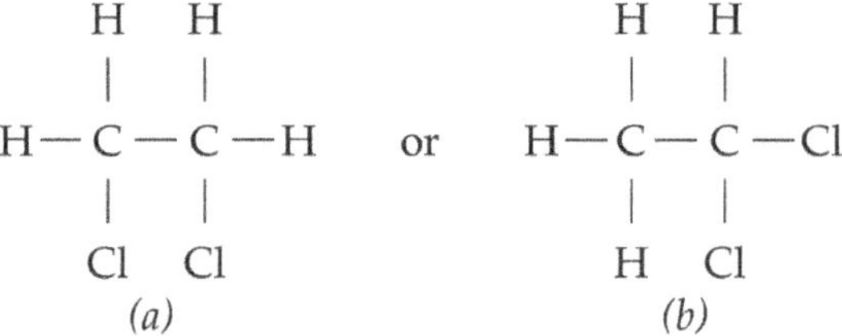

2. Condensed structural formulas are a convenient, simple way of writing structural formulas and are a “shorthand version” of structural formulas. For example, molecules (*a*) and (*b*) above may be written as (*a*) CH_2ClCH_2Cl and (*b*) CH_3CHCl_2.
3. Isomers are compounds with the same molecular formula but different structural formulas (different spatial arrangement or different connectivity of atoms); for example, (*a*) and (*b*) above are isomers.
4. Because atoms are arranged differently in isomers, the molecules have different properties. For example, ethyl alcohol, CH_3CH_2OH, has an —OH group and in some ways resembles H—OH. Dimethyl ether, CH_3OCH_3, is an isomer of ethyl alcohol and does not contain the —OH group. Hence, it differs from ethyl alcohol.

5.

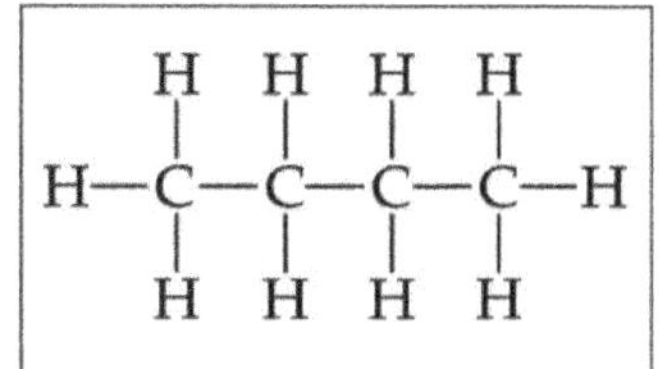

Butane

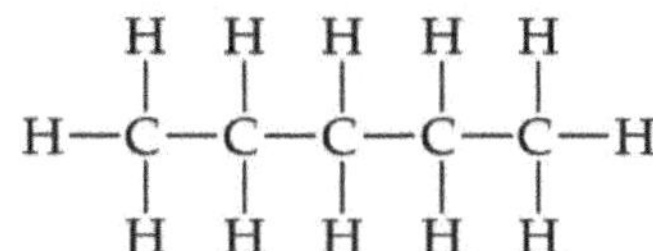

Pentane

EXPERIMENT 41

1. $2\ NaOH(aq) + SiO_2(s) \longrightarrow Na_2SiO_3(aq) + H_2O(l)$
2. $Na_2B_4O_7 \times 10H_2O\ (aq) \longrightarrow 2\ NaB(OH)_4(aq) + 2\ B(OH)_3(aq) + 3H_2O(l)$
3. $NaB(OH)_4(aq) \longrightarrow Na^+(aq) + B(OH)_4^-(aq)$
4. $B(OH)_3(aq) + 2\ H_2O(l) \rightleftharpoons B(OH)_4^-(aq) + H_3O^+(aq)$

EXPERIMENT 42

1. Calculate the mass of the hydrate of calcium chloride

   $26.4511\ g - 24.4272\ g = 2.0239\ g$ hydrate of calcium chloride

   After heating, the solid that remains is anhydrous $CaCl_2$, calculate the mass of $CaCl_2$.

   $26.4511\ g - 25.2280\ g = 1.2231\ g\ CaCl_2$

   Determine the mass of water lost after heating by subtracting the mass of $CaCl_2$ from the mass of the $CaCl_2$ hydrate.

   $2.0239\ g - 1.2231\ g = 0.8008\ g\ H_2O$

   Calculate moles of $CaCl_2$.

   $(1.2231\ g\ CaCl_2) \times (1\ mol\ CaCl_2 / 110.9834\ g\ CaCl_2) = 0.011021\ mol\ CaCl_2$

   Calculate moles of H_2O.

   $(0.8008\ g\ H_2O) \times (1\ mol\ H_2O / 18.01528\ g\ H_2O) = 0.04445\ mol\ H_2O$

   Compare the ratio of moles of water to $CaCl_2$, mol H_2O/mol $CaCl_2$

   $0.04445/0.011021 = 4.033$, which is approximately 4, thus $CaCl_2 \times 4H_2O$ is the formula

2. $MgSO_4(aq) + Na_2CO_3(aq) \longrightarrow Na_2SO_4(aq) + MgCO_3(s)$

   Use solubility rules to determine solubility of products

3. Mg: $1 + 3\times1 = 4$ Mg atoms

   O: $1\times2 + 3\times3 + 1\times3 = 14$ O atoms

   C: $1\times3 = 3$ C atoms

   H: $1\times2 + 3\times2 = 8$ H atoms

4. The balanced reaction involves equal moles of $Na_2CO_3$and $MgSO_4$, therefore whichever one has a smaller mole of reactants will be the limiting reagent.

$$\text{mol } Na_2CO_3 = (0.9278 \text{ g } Na_2CO_3)\times(1 \text{ mol } Na_2CO_3 / 105.98844 \text{ g } Na_2CO_3)$$
$$= 0.008754 \text{ mol } Na_2CO_3$$

$$\text{mol } MgSO_4 = (0.9278 \text{ g } MgSO_4)\times(1 \text{ mol } MgSO_4 / 120.3686 \text{ g } MgSO_4)$$
$$= 0.008351 \text{ mol } MgSO_4$$

$$5MgSO_4(aq) + 5Na_2CO_3(aq) + 5H_2O(l) \longrightarrow Mg(OH)_2\cdot3MgCO_3\cdot3H_2O(s) + Mg(HCO_3)_2(aq) + 5Na_2SO_4(aq)$$

	$MgSO_4$	Na_2CO_3	$Mg(OH)_2{\cdot}3MgCO_3{\cdot}3H_2O$
Before Reaction:	0.008351 mol	0.008754 mol	0 mol
Change (reaction):	−0.008351 mol	−0.008351 mol	+0.001670 mol
After reaction:	0 mol	0.000403 mol	+0.001670 mol

The limiting reactant is $MgSO_4$.

$(0.008351 \text{ mol } MgSO_4)\times(1 \text{ mol } Mg(OH)_2{\cdot}3MgCO_3{\cdot}3H_2O/5 \text{ mol } MgSO_4)$
$= 0.001670 \text{ mol } Mg(OH)_2{\cdot}3MgCO_3{\cdot}3H_2O$

$(0.001670 \text{ mol } Mg(OH)_2{\cdot}3MgCO_3{\cdot}3H_2O)\times(365.307 \text{ g}/1 \text{ mol}) = 0.6101$ g $Mg(OH)_2{\cdot}3MgCO_3{\cdot}3H_2O$

percent yield = actual yield/theoretical yield x 100 = 0.5422 g/0.6101 g $\times100\% = 88.87\%$

EXPERIMENT 43

2. Density = mass of atoms in the unit cell/volume of unit cell

   The number of Cu atoms (and consequently the mass of the Cu atoms in the unit cell) will vary from primitive to body-centered to face-centered.

   Primitive: 1 Cu atom per unit cell$\times$(1 mol Cu/6.022×10^{23} atoms)$\times$(63.546 g Cu/1 mol Cu) = 1.055×10^{-22} grams

   Body-Centered: 2 Cu atom per unit cell $\times$ (1 mol Cu/6.022×10^{23} atoms)$\times$ (63.546 g Cu/1 mol Cu) = 2.110×10^{-22} grams

   Face-Centered: 4 Cu atom per unit cell $\times$ (1 mol Cu/6.022×10^{23} atoms)$\times$ (63.546 g Cu/1 mol Cu) = 4.220×10^{-22} grams

   In each case, Volume $= a^3 = (3.61\text{Å})^3 = (3.61\times10^{-8} \text{ cm})^3 = 4.705 \times 10^{-23} \text{ cm}^3$.

Density Primitive: 1.055×10^{-22} grams/4.705×10^{-23} cm^3 = 2.24 g/cm^3

Density Body-Centered: 2.110×10^{-22} grams/4.705×10^{-23} cm^3 = 4.48 g/cm^3

Density Face-Centered: 4.220×10^{-22} grams/4.705×10^{-23} cm^3 = 8.96 g/cm^3

The experimentally determined density is 8.94 g/cm^3, which is consistent with the face-centered cubic unit cell.

5. In a close-packed metal, the nearest neighbors are the same atom. In an ionic solid, the coordination number includes only ions of opposite charge. The coordination number in a close-packed metal is generally larger than the coordination number in an ionic solid.

Vapor Pressure of Water at Various Temperatures

Temperature (°C)	Pressure (kPa)	Temperature (°C)	Pressure (kPa)
0	0.61	26	3.35
1	0.65	27	3.55
2	0.70	28	3.77
3	0.75	29	4.0
4	0.81	30	4.23
5	0.86	31	4.49
6	0.93	32	4.75
7	0.99	33	5.02
8	1.06	34	5.31
9	1.14	35	5.62
10	1.22	40	7.37
11	1.30	45	9.58
12	1.39	50	12.33
13	1.49	55	15.72
14	1.59	60	19.91
15	1.70	65	25.0
16	1.81	70	31.15
17	1.93	75	38.53
18	2.06	80	47.33
19	2.19	85	57.79
20	2.33	90	70.08
21	2.49	95	84.49
22	2.63	97	90.92
23	2.81	99	97.73
24	2.98	100	101.30
25	3.17	101	104.98

Names, Formulas, and Charges of Common Ions

Appendix

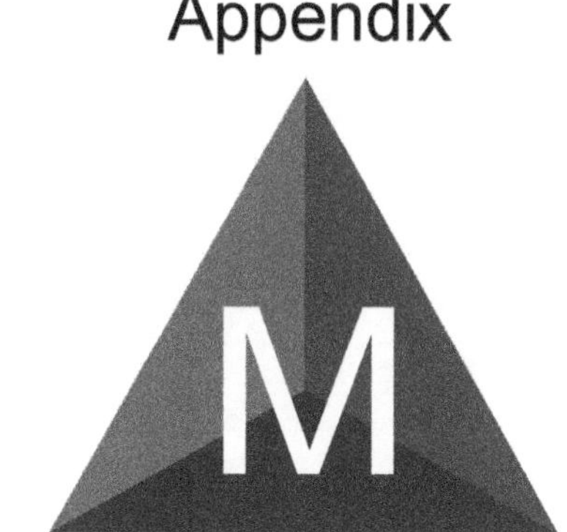

Positive ions (cations)	Negative ions (anions)
1+	1−
Ammonium (NH_4^+)	Acetate ($C_2H_3O_2^-$)
Cesium (Cs^+)	Bromide (Br^-)
Copper(I) or cuprous (Cu^+)	Chlorate (ClO_3^-)
Hydrogen (H^+)	Chloride (Cl^-)
Lithium (Li^+)	Cyanide (CN^-)
Potassium (K^+)	Dihydrogen phosphate ($H_2PO_4^-$)
Silver (Ag^+)	Fluoride (F^-)
Sodium (Na^+)	Hydride (H^-)
2+	Hydrogen carbonate or bicarbonate (HCO_3^-)
Barium (Ba^{2+})	Hydrogen sulfite or bisulfite (HSO_3^-)
Cadmium (Cd^{2+})	Hydroxide (OH^-)
Calcium (Ca^{2+})	Hypochlorite (ClO^-)
Chromium(II) or chromous (Cr^{2+})	Iodide (I^-)
Cobalt(II) or cobaltous (Co^{2+})	Nitrate (NO_3^-)
Copper(II) or cupric (Cu^{2+})	Nitrite (NO_2^-)
Iron(II) or ferrous (Fe^{2+})	Perchlorate (ClO_4^-)
Lead(II) or plumbous (Pb^{2+})	Permanganate (MnO_4^-)
Magnesium (Mg^{2+})	Thiocyanate (SCN^-)
Manganese(II) or manganous (Mn^{2+})	
Mercury(I) or mercurous (Hg_2^{2+})	2−
Mercury(II) or mercuric (Hg^{2+})	Carbonate (CO_3^{2-})
Strontium (Sr^{2+})	Chromate (CrO_4^{2-})

Positive ions (cations)	Negative ions (anions)
Nickel (Ni^{2+})	Dichromate ($Cr_2O_7^{2-}$)
Tin(II) or stannous (Sn^{2+})	Hydrogen phosphate (HPO_4^{2-})
Zinc (Zn^{2+})	Oxide (O^{2-})
	Peroxide (O_2^{2-})
3+	Sulfate (SO_4^{2-})
Aluminum (Al^{3+})	Sulfide (S^{2-})
Chromium(III) or chromic (Cr^{3+})	Sulfite (SO_3^{2-})
Iron(III) or ferric (Fe^{3+})	Thiosulfate ($S_2O_3^{2-}$)
	3−
	Arsenate (AsO_4^{3-})
	Phosphate (PO_4^{3-})

Some Molar Masses

Appendix

Formula	g/mol	Formula	g/mol	Formula	g/mol
AgBr	187.78	$HC_2H_3O_2$ (acetic)	60.05	$KHSO_4$	136.17
AgCl	143.32	$HC_7H_5O_2$ (benzoic)	122.12	KI	166.01
Ag_2CrO_4	331.73	HCl	36.46	KIO_3	214.00
AgI	234.77	$HClO_4$	100.46	KIO_4	230.00
$AgNO_3$	169.87	$H_2C_2O_4 \cdot 2H_2O$	126.07	$KMnO_4$	158.04
Al_2O_3	101.96	HNO_3	63.01	KNO_3	101.11
$Al_2(SO_4)_3$	342.14	H_2O	18.015	KOH	56.11
B_2O_3	69.62	H_2O_2	34.01	K_2SO_4	174.27
$BaCO_3$	197.35	H_3PO_4	98.00	$MgNH_4PO_4$	137.35
$BaCl_2$	208.25	H_2S	34.08	MgO	40.31
$BaCrO_4$	253.33	H_2SO_3	82.08	$MgSO_4$	120.37
$Ba(OH)_2$	171.36	H_2SO_4	98.08	MnO_2	86.94
$BaSO_4$	233.39	HgO	216.59	Mn_2O_3	157.88
CO_2	44.01	Hg_2Cl_2	472.09	NaBr	102.90
$CaCO_3$	100.09	$HgCl_2$	271.50	$NaC_2H_3O_2$	82.03
CaO	56.08	KBr	119.01	NaCl	58.44
$CaSO_4$	136.14	$K_2C_2O_4 \cdot H_2O$	184.24	NaCN	49.01
$[Co(C_5H_5N)_4(NCS)_2]$	491.05	$K_3[Cr(C_2O_4)_3] \cdot 3H_2O$	487.42	$Na_2C_2O_4$	134.0
CuO	79.54	$K_2[Cu(C_2O_4)_2] \cdot 2H_2O$	353.83	Na_2CO_3	105.99
Cu_2O	143.08	$K_3[Fe(C_2O_4)_3] \cdot 3H_2O$	491.27	$NaNO_3$	84.99
$CuSO_4$	159.60	$K_3[Al(C_2O_4)_3] \cdot 3H_2O$	462.40	Na_2O_2	77.98
$[Cu(C_5H_5N)_2(NCS)_2]$	337.54	KCl	74.56	NaOH	40.00
$Fe(NH_4)_2(SO_4)_2 \cdot 6H_2O$	392.14	$KClO_3$	122.55	NaSCN	81.07
FeO	71.85	K_2CrO_4	194.20	Na_2SO_4	142.04
Fe_2O_3	159.69	$K_2Cr_2O_7$	294.19	$Na_2S_2O_3 \cdot 5H_2O$	248.18
Fe_3O_4	231.54	$KHC_8H_4O_4$ (phthalate) KHP	204.23	NH_3	17.03
HBr	80.92	K_2HPO_4	174.18	NH_4Cl	53.49
$H_2C_2O_4$ (oxalic)	90.04	KH_2PO_4	136.09	NH_4NO_3	80.04

Formula	g/mol	Formula	g/mol	Formula	g/mol
$(NH_4)_2SO_4$	132.14	$PbSO_4$	303.25	SO_2	64.06
$[Ni(C_5H_5N)_4(NCS)_2]$	490.83	P_2O_5	141.94	SO_3	80.06
$PbCrO_4$	323.18	SiO_2	60.08	$ZnCl_2$	136.27
PbO	223.19	$SnCl_2$	189.60	$[Zn(C_5H_5N)_2(NCS)_2]$	339.37
PbO_2	239.19	SnO_2	150.69		

Appendix

Basic SI Units, Some Derived SI Units, and Conversion Factors

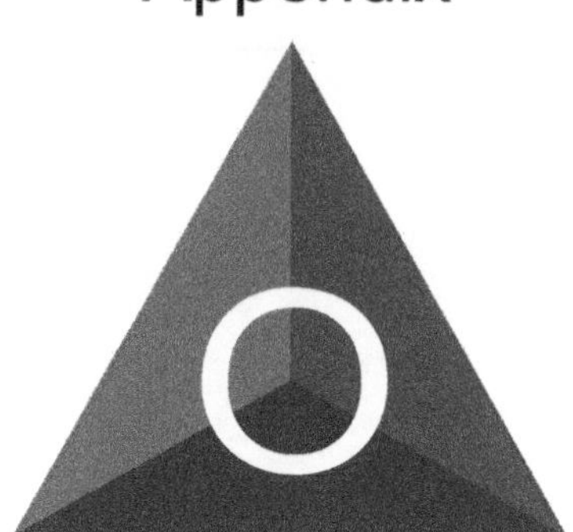

	SI unit	Conversion factors
Length	Meter (m)	1 m = 100 centimeters (cm)
		= 1.0936 yards (yd)
		1 cm = 0.3937 inch (in.)
		1 in. = 2.54 cm (exact)
		= 0.0254 m
		1 angstrom (Å) = 10^{-10} m
Mass	Kilogram (kg)	1 kg = 1000 grams (g)
		= 2.205 pounds (lb)
		1 lb = 453.6 grams (g)
		1 atomic mass unit (u) = 1.66053×10^{-24} g
Time	Second (s)	1 day (d) = 86,400 s
		1 hour (hr) = 3600 s
		1 minute (min) = 60 s
Electric current	Ampere (A)	
Temperature	Kelvin (K)	0 K = −273.15 ° Celsius (C)
		= −459.67 ° Fahrenheit (F)
		°F = (9/5) °C + 32°
		°C = (5/9)(°F − 32°)
		K = °C + 273.15°
Luminous intensity	Candela (cd)	
Amount of substance	Mole (mol)	
Volume (derived)	Cubic meter (m^3)	1 liter (L) = 10^{-3} m^3
		= 1.057 quarts (qt)
		1 $in.^3$ = 16.4 cm^3
Force (derived)	Newton ($N = m\text{-}kg/s^2$)	1 dyne (dyn) = 10^{-5} N
Pressure (derived)	Pascal ($Pa = N/m^2$)	1 atmosphere (atm) = 101, 325 Pa
		= 760 mm Hg
		= 14.70 $lb/in.^2$
		= 1.013×10^6 dyn/cm^2

	SI unit	Conversion factors
Energy (derived)	Joule	1 calorie (cal) = 4.184 J
	(J = N-m)	1 electron volt (eV) = 1.6022×10^{-19} J
		1 erg = 6.2420×10^{11} eV
		1 J = 10^{7} ergs
		1 kJ = 1000 J

Appendix

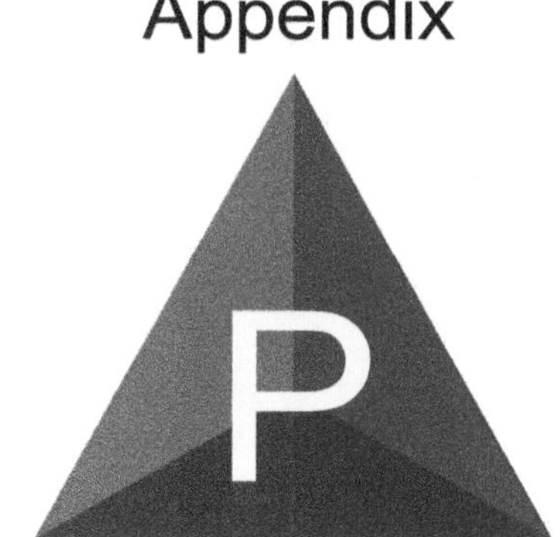

Composition of Commercial Reagent Acids and Bases

	Molar mass (g/mol)	Density (g/mL)	% Strength (w/w)	Molarity	Approx. mL conc. reagent/1 L 1 *M* solution
HCl	36.461	1.19	37.2	12.1	82.5
HNO_3	63.013	1.42	70.0	15.8	63.5
HF	20.006	1.19	48.8	29.0	34.5
$HClO_4$	100.458	1.67	70.5	11.7	85.5
$HC_2H_3O_2$ (glacial)	60.053	1.05	99.8	17.4	57.5
H_2SO_4	98.082	1.84	96.0	18.0	55.5
H_3PO_4	97.994	1.70	85.5	14.8	67.5
NH_4OH	35.046	0.90	57.6*	14.8	67.5
$NaOH$	39.997	1.53	50.5	19.3	52.0
KOH	56.106	1.46	45.0	11.7	85.5

*Equivalent to 28% NH_3.